Winds over the landscape and inside forests produce mechanical reactions in trees, and eventually failure in stems and roots when stressed by storms. The mechanics of these reactions and the physiological responses to wind in leaves, stems and root systems, and the important ecological consequences of windthrow, are described. Management techniques of forests in windy climates are detailed including the use of models predicting risk of wind damage. It is clear that the whole field of wind effects on trees has benefited from recent multi-disciplinary research, and significant advances in knowledge of most parts of the subject have been made in the last decade. This book brings the up-to-date theories, methodologies and results together, and gives the reader a sense of coherence in this complex but fascinating field.

WIND AND TREES

WIND AND TREES

Edited by

M. P. Coutts
Forestry Authority, Northern Research Station, Roslin, UK
and
J. Grace
Institute of Ecology and Resource Management, University of Edinburgh, Edinburgh, UK

CAMBRIDGE
UNIVERSITY PRESS

Published by the Press Syndicate of the University of Cambridge
The Pitt Building, Trumpington Street, Cambridge CB2 1RP
40 West 20th Street, New York, NY 10011-4211, USA
10 Stamford Road, Oakleigh, Melbourne 3166, Australia

© Cambridge University Press 1995

First published 1995

Printed in Great Britain at the University Press, Cambridge

A catalogue record for this book is available from the British Library

Library of Congress cataloguing in publication data
Wind and trees / edited by M.P. Coutts and J. Grace.
p. cm.
Selected papers from a conference held at Heriot-Watt University, Edinburgh, July 1993.
Includes index.
ISBN 0 521 46037 9 (hardback)
1. Trees – Effect of wind on – Congresses. 2. Trees – Physiology – Congresses. 3. Plants, Effect of wind on – Congresses. 4. Forest ecology – Congresses. I. Coutts, M. P. II. Grace, J. (John), 1945– .
SD390.7.W56W55 1995
634.9′616–dc20   94–48617 CIP

ISBN 0 521 46037 9   hardback

Contents

Preface		*page* ix
List of contributors		xi
Part I.	**Airflow over topography and in forests**	
1	Turbulent airflow in forests on flat and hilly terrain *J. J. Finnigan and Y. Brunet*	3
2	The interactions of wind and tree movement in forest canopies *B. A. Gardiner*	41
3	Edge effects on diffusivity in the roughness layer over a forest *B. Kruijt, W. Klaassen and R. W. A. Hutjes*	60
4	A wind tunnel study of turbulent airflow in forest clearcuts *J. M. Chen, T. A. Black, M. D. Novak and R. S. Adams*	71
5	Testing of a linear airflow model for flow over complex terrain and subject to stable, structured stratification *D. W. F. Inglis, T. W. Choularton, I. M. F. Stromberg, B. A. Gardiner and M. Hill*	88
6	Predicting windspeeds for forest areas in complex terrain *P. Hannah, J. P. Palutikof and C. P. Quine*	113
Part II.	**Mechanics of trees under wind loading**	
7	Understanding wind forces on trees *C. J. Wood*	133
8	Modelling mechanical stresses in living Sitka spruce stems *R. Milne*	165
9	Experimental analysis and mechanical modelling of wind-induced tree sways *D. G. E. Guitard and P. Castera*	182
10	Failure modes of trees and related failure criteria *C. Mattheck, K. Bethge and W. Albrecht*	195
11	An experimental investigation of the effects of dynamic loading on coniferous trees planted on wet mineral soils *M. Rodgers, A. Casey, C. McMenamin and E. Hendrick*	204

12	Measurement of wind-induced tree-root stresses in New Zealand *A. J. Watson*	220
13	New methods for the assessment of wood quality in standing trees *K. Bethge and C. Mattheck*	227

Part III. Tree physiological responses

14	Wind-induced physiological and developmental responses in trees *F. W. Telewski*	237
15	Responses of young trees to wind: effects on root growth *A. Stokes, A. H. Fitter and M. P. Coutts*	264
16	Wind stability factors in tree selection: distribution of biomass within root systems of Sitka spruce clones *B. C. Nicoll, E. P. Easton, A. D. Milner, C. Walker and M. P. Coutts*	276
17	Development of buttresses in rainforest trees: the influence of mechanical stress *A. R. Ennos*	293

Part IV. Impacts of wind on forests and ecology

18	Hurricane disturbance regimes in temperate and tropical forest ecosystems *D. R. Foster and E. R. Boose*	305
19	A comparison of methods for quantifying catastrophic wind damage to forests *E. M. Everham III*	340
20	Windthrow and airflow in a subalpine forest *G. Wooldridge, R. Musselman and W. Massman*	358

Part V. Risk assessment and management response

21	Assessing the risk of wind damage to forests: practice and pitfalls *C. P. Quine*	379
22	Forest wind damage risk assessment for environmental impact studies *G. C. Wollenweber and F. G. Wollenweber*	404
23	Recommendations for stabilisation of Norway spruce stands based on ecological surveys *C. C. N. Nielsen*	424
24	Thinning regime in stands of Norway spruce subjected to snow and wind damage *M. Slodičák*	436
25	A synopsis of windthrow in British Columbia: occurrence, implications, assessement and management *S. J. Mitchell*	448
26	Wind damage to New Zealand State plantation forests *A. Somerville*	460
27	The experience of and management strategy adopted by the Selwyn Plantation Board, New Zealand *W. P. Studholme*	468
Index		477

Preface

Wind not only causes extensive damage to trees in many parts of the world, it also has more subtle effects on their growth, form and ecology. This, the first symposium volume on the topic, contains a selection from the papers presented at a conference, *Wind and Wind-Related Damage to Trees,* held at Heriot-Watt University, Edinburgh, in July 1993. The conference, which was initiated by Chris Quine, was held under the auspices of the International Union of Forestry Research Organisations, and brought together about a hundred people from seventeen countries.

Wind damage to trees has historically been the province of the silviculturist, but increasing recognition of the importance and complexity of the subject has more recently involved people from many other discplines, and this has resulted in a greatly increased understanding of the main processes involved. The conference served to bring together the new approaches and methodologies; it enabled discussion between physicists, aerodynamicists, foresters, engineers, physiologists, ecologists, pathologists and modellers, and led to a remarkable cross-fertilisation of ideas between scientists who would not normally meet.

For convenience the papers have been gathered into five parts, each beginning with a review. The parts are: Airflow over topography and in forests; Mechanics of trees under wind loading; Tree physiological responses; Impacts of wind on forests and ecology; Risk assessment and management response. The coverage of different aspects of the subject is decidedly uneven. For example, tree root development and the way in which roots and soil anchor the tree have received less attention than what is going on above ground, and the contents of this book reflect such disparities. However, significant advances have been made in all aspects of the subject during the last decade and this book gives a clear view of the development of the whole area to date.

Preface

The members of the conference Organising Committee were C. P. Quine (chairman), M. P. Coutts, Monica Farm, M. Flannigan, B. A. Gardiner, J. Grace, D. Howat, C. Mattheck and R. Milne. We greatly appreciated the sponsorship provided by the European Community (Programme DGXII-E-2-AIR), Lothian and Edinburgh Enterprise Ltd and the Forestry Authority.

August 1994　　　　　　　　　　　　　　　　　　　　　　　　M. P. Coutts
　　　　　　　　　　　　　　　　　　　　　　　　　　　　　　　　J. Grace

Contributors

R. S. Adams
Kalkamilka Forestry Centre, BC Ministry of Forests, Vernon, BC, Canada V1B 2C7

W. Albrecht
Karlsruhe Nuclear Research Centre, Institute for Material Research II, PO Box 3640, D-76021 Karlsruhe 1, Germany

K. Bethge
Karlsruhe Nuclear Research Centre, Institute for Material Research II, PO Box 3640, D-76021 Karlsruhe 1, Germany

T. A. Black
Department of Soil Science, University of British Columbia, 139-2357 Main Mall, Vancouver, BC, Canada V6T 1Z4

E. R. Boose
Harvard Forest, Harvard University, PO Box 68, Petersham, MA 01366, USA

Y. Brunet
Laboratoire de Bioclimatologie INRA, BP81, F-33883 Villenave, D'Ornon, Cedex, France

A. Casey
Civil Engineering Department, University College, Galway, Republic of Ireland

P. Castera
Laboratoire de Rhéologie du bois de Bordeaux, Domaine de l'Hermitage-BP10, Pierroton, F-33610 Cestas Gazinet, France

J. M. Chen
558 Booth Street, 4th Floor, Global Monitoring Section, Canada Centre for Remote Sensing, Ottawa, Ontario, Canada K1A 0E4

T. W. Choularton
Department of Pure and Applied Physics, UMIST, PO Box 88, Manchester M60 1QD, UK

M. P. Coutts
Forestry Authority, Northern Research Station, Roslin, Midlothian EH25 9SY, UK

E. P. Easton
1 Saville Place, Edinburgh EH9 3EB, UK

A. R. Ennos
Department of Environmental Biology, University of Manchester, School of Biological Sciences, Williamson Building, Oxford Road, Manchester M13 9PL, UK

E. M. Everham III
301 Illick Hall, SUNY ESF, 1 Forestry Drive, Syracuse, New York, NY 13210-2778, USA

J. J. Finnigan
CSIRO Centre for Environmental Mechanics, PO Box 821, Act 2601, Canberra, Australia

A. H. Fitter
Department of Biology, University of York, York YO1 5DD, UK

D. R. Foster
Harvard Forest, Harvard University, PO Box 68, Petersham, MA 01366, USA

B. A. Gardiner
Forest Authority, Northern Research Station, Roslin, Midlothian EH25 9SY, UK

J. Grace
Institute of Ecology and Resource Management, University of Edinburgh, Darwin Building, Mayfield Road, Edinburgh EH9 3JU, UK

D. G. E. Guitard
Laboratoire de Rhéologie du bois de Bordeaux, Domaine de l'Hermitage-BP10, Pierroton, F-33610 Cestas Gazinet, France

P. Hannah
National Wind Power Ltd, Riverside House, Meadowbank, Furlong Road, Bourne End, Bucks SL8 5AJ, UK

E. Hendrick
Coillte Teoranta, 1–3 Sidmonton Place, Bray, Co Wicklow, Republic of Ireland

M. Hill
Holtech Associates, Roughrigg, Harwood-in-Teesdale, Barnard Castle, Co Durham DL12 0XY, UK

R. W. A. Hutjes
Department of Physical Geography, University of Groningen, Kerklan 30, NL-9751 NN Haren, The Netherlands

D. W. F. Inglis
Physics Department, UMIST, PO Box 88, Manchester M60 1QD, UK

W. Klaassen
Department of Physical Geography, University of Groningen, Kerklan 30, NL-9751 NN Haren, The Netherlands

B. Kruijt
Institute of Ecology and Resource Management, University of Edinburgh, Darwin Building, Mayfield Road, Edinburgh EH9 3JU, UK

W. Massman
Rocky Mountain Forest and Range Experimental Station, 240 West Prospect Road, Fort Collins, CO 80526, USA

C. Mattheck
Karlsruhe Nuclear Research Centre, Institute for Material Research II, PO Box 3640, D-76021 Karlsruhe 1, Germany

C. McMenamin
Civil Engineering Department, University College, Galway, Republic of Ireland

R. Milne
Institute of Terrestrial Ecology, Bush Estate, Penicuik, Midlothian EH26 0QB, UK

A. D. Milner
Forestry Authority, Northern Research Station, Roslin, Midlothian EH25 9SY, UK

S. J. Mitchell
2512 Melfa Lane, Vancouver, British Columbia, Canada V6T 2C6

R. Musselman
Rocky Mountain Forest and Range Experimental Station, 240 West Prospect Road, Fort Collins, CO 80526, USA

B. C. Nicoll
Forestry Authority, Northern Research Station, Roslin, Midlothian EH25 9SY, UK

C. C. N. Nielsen
Danish Forest and Landscape Research Institute, Skovbrynet 16, DK-2800 Lyngby, Denmark

M. D. Novak
Department of Soil Science, University of British Columbia, 139-2357 Main Mall, Vancouver, BC, Canada V6T 1Z4

J. P. Palutikof
Climatic Research Unit, University of East Anglia, Norwich NR4 7TJ, UK

C. P. Quine
Forestry Authority, Northern Research Station, Roslin, Midlothian EH25 9SY, UK

M. Rodgers
Civil Engineering Department, University College, Galway, Republic of Ireland

M. Slodičák
Vyzkumná Stanice Vulhm, CZ-517 73 Opocno, Czech Republic

A. Somerville
New Zealand Forest Research Institute, Private Bag 3020, Rotorua, New Zealand

A. Stokes
Department of Biology, University of York, York YO1 5DD, UK

I. M. F. Stromberg
Department of Pure and Applied Physics, UMIST, PO Box 88, Manchester M60 1QD, UK

W. P. Studholme
Selwyn Plantation Board Ltd, PO Box 48, Darfield, Canterbury, New Zealand

F. W. Telewski
Curator, W. J. Beal Botanic Garden, 412 Olds Hall, Michigan State University, East Lansing, MI 48824-1047, USA

C. Walker
Forestry Authority, Northern Research Station, Roslin, Midlothian EH25 9SY, UK

A. J. Watson
Manaaki Whenua, Landscape Research (NZ) Ltd, PO Box 31-011, Christchurch, New Zealand

F. G. Wollenweber
ECMWF, Shinfield Park, Reading, Berkshire RG2 9AX, UK

G. C. Wollenweber
University of Reading, Department of Meteorology, PO Box 239, 2 Earley Gate, Reading, Berkshire RG 2AU, UK

C. J. Wood
Department of Engineering Science, University of Oxford, Parks Road, Oxford OX1 3PJ, UK

G. Wooldridge
Rocky Mountain Forest and Range Experimental Station, 240 West Prospect Road, Fort Collins, CO 80526, USA

Part I
Airflow over topography and in forests

1

Turbulent airflow in forests on flat and hilly terrain

J. J. FINNIGAN and Y. BRUNET

Abstract

The turbulent velocity field in and above homogeneous forest canopies on flat ground has several universal features. These include a mean velocity profile with a point of inflexion; vertical inhomogeneity in second moments; positive u and negative w skewnesses; and integral length scales of order h, the canopy height. These features scale with u_* and h across a wide range of wind tunnel models and real canopies. Turbulence spectra also share common features such as vertical invariance of the position of the spectral peak and departures from classical inertial subrange behaviour. Many of these characteristics are more similar to what is observed in a plane mixing layer than in a boundary layer. The eddy structure of a plane mixing layer owes its origin to the instability of its inflexion point velocity profile, one of the features it shares with canopy flows. Linear perturbation models of this instability yield results that fit well with the eddy structure observed in canopies, strongly suggesting that it is instability of the canopy mean velocity profile, a profile set by the momentum absorption capacity of the canopy as a whole, that controls the universal turbulent structure of plant canopy flow. Turbulence produced directly by eddies shed from plant parts plays a minor role. Turbulent airflow over isolated hills also displays some characteristic features, including a large relative velocity speed-up at low levels above the hill crest. It can be demonstrated that this is caused by the pressure field set up by the whole of the flow field around the hill. In this sense it can be regarded as 'imposed' on any canopy growing on the hill. A simplified analysis shows that canopy airflow and airflow in the boundary layer above respond quite differently to such an applied pressure gradient. Reference to measurements made in a wind tunnel study of a tall canopy on a ridge reveals that this difference can explain observed modulations in the mean velocity profile as the hill is traversed. Specifically, the characteristic inflexion point profile of canopies on flat ground may be exaggerated or destroyed according to position on the hill. The reactions of turbulence moments to these changes in mean profile are profound and can be explained by supposing that the analogy between canopies and plane mixing layers continues to hold.

1.1 Introduction

The mid 1970s to the present day has seen a major reappraisal of the way we describe and model turbulent flow in plant canopies. Early models and descriptions of canopy flow assumed, quite reasonably, that the extremely high levels of turbulence observed were due to eddy shedding from leaves, stems, stalks or branches. Since turbulence produced in this way would be of relatively fine scale compared with the canopy height, which is the scale upon which mean profiles of scalar concentration or velocity vary, then a description of turbulent transport based on eddy diffusivities was an obvious step. Through the 1970s and 1980s, however, the conflict between measurements and the predictions of such models, particularly the observations of counter-gradient diffusion of mass, heat and momentum, sparked new approaches to measurement, analysis and modelling. The key innovations in these areas were multi-point measurements by accurate turbulence sensors; conditional sampling of the data; rigorous derivation of the flow equations using the volume averaging operator; the application of Lagrangian techniques to the modelling of scalar transport; and higher-order closure modelling of the wind field. Surveys of these developments have been published by Finnigan & Raupach (1987), Raupach (1988, 1989) and Kaimal & Finnigan (1994).

The picture of canopy turbulence that emerged is of large-scale, intermittent turbulent eddies or large structures dominating the transport of scalars and momentum in the canopy air space. Understanding of the precise nature of these large eddies, their origins and development and their connection to canopy morphology is still at an early stage, however. Furthermore, most detailed information has been gathered under conditions of horizontal homogeneity, or at least as close to them as the experimenters could manage.

The theme of this volume is wind and wind-related damage to trees. It is well known that such damage is often associated with hilly topography or canopy heterogeneity such as edges and clearings. The purpose of this chapter is to extend knowledge gained in simple situations to these more complex conditions by employing some recent advances in our understanding of the physical mechanisms that underlie canopy turbulence, specifically the model of turbulence generation proposed by Raupach *et al.* (1989) and Brunet *et al.* (1994).

We begin with a brief review of the key features of the turbulent flow in uniform forests and then go on to draw parallels between these and the turbulent structure of a plane mixing layer. We shall remind readers that many of the characteristics of the energy-containing eddies in canopy turbulence can be explained by supposing they result from hydrodynamic instability of the mean velocity profile within the roughness sublayer, the region extending from the ground at $z = 0$ to $z \simeq 2h$, z being the vertical coordinate.

To illustrate the effects caused by heterogeneity we shall consider a canopy on a hill. This complication is introduced by briefly reviewing the features of turbulent flow over an isolated hill, particularly the link between the perturbation pressure field and the ensuing changes in windspeed. The connection between these perturbations and canopy flow is provided by the different ways that airflow within and above the canopy reacts to the pressure gradient produced by the hill. By referring to measurements in a wind tunnel model of a uniform canopy on a two-dimensional ridge, we see that the characteristic canopy mean velocity profile may be exaggerated or destroyed, depending on position on the hill. Finally, by revisiting the turbulent statistics introduced in the first section, we see how this modulation of the mean velocity profile is reflected in changes in the turbulence.

Throughout this chapter we shall assume neutrally stratified flow in the canopy. This is consistent with our interest in strong winds when shear-driven turbulence will dominate the near surface region. It is worth noting, however, that over large hills, near surface wind patterns may be strongly influenced by stratification at higher levels leading to phenomena such as intense downslope winds in the lee of the hill (Durran, 1990).

We use standard meteorological notation with a right-handed coordinate system x, y, z or x_i with x, x_1 being the horizontal coordinate in the streamwise direction and z, x_3 the vertical coordinate with origin at the ground. The velocity components are u, v, w or u_1, u_2, u_3 in the x, y, z or x_1, x_2, x_3 directions respectively. Denoting the averaging operator by an overbar we may split each variable into its mean component denoted by a capital letter and its fluctuation denoted by a prime. Hence, for example:

$$\bar{u} = U \qquad (1.1a)$$
$$u = U + u' \qquad (1.1b)$$

In horizontally homogeneous conditions the only non-zero component of mean velocity is U. In later sections we shall introduce streamline coordinates which preserve this property in complex terrain.

A rigorous treatment of flow in plant canopies requires that scalar and vector quantities be averaged in space to remove the random point-to-point variations caused by the foliage. We do not wish to introduce this complication here. However, the reader should suppose that variables are sufficiently averaged spatially as to reflect average trends in the foliage distribution, but not the influence of every leaf. A proper treatment and derivation of the relevant equations may be found in Raupach & Shaw (1982), Finnigan (1985) and Raupach et al. (1986). With this warning, the overbar can be assumed to denote the time average and mean quantities to be time means.

1.2 Turbulent flow in horizontally homogeneous forests

A large number of canopy studies with reliable turbulence measurements has been published over the last two decades, many of them conducted in forests where the physical scale allows three-component sonic anemometers to be employed. They have allowed us to draw a consistent picture of canopy flow which has been presented in several recent articles, for example Raupach (1988) and Kaimal & Finnigan (1994). We present these data here in Fig. 1.1a–h as compendium plots drawn from eight experiments ranging from wind tunnel models to tall forests. Details of the experiments that furnished these data may be found in Table 1.1.

1.2.1 Single point velocity statistics

Fig. 1.1a displays mean velocity U normalised by $U(h)$, h being canopy height. The vertical coordinate is z/h. All the profiles display the characteristic feature of a point of inflexion near the canopy top, where $\partial U/\partial z$ is a maximum which divides the 'exponential' canopy velocity profile from the boundary layer profile above. In Fig. 1.1b we plot shearing stress, $-\overline{u'w'}$ normalised by friction velocity u_*, where $u_* = [-\overline{u'w'}(h)]^{\frac{1}{2}}$. Each profile has a constant stress region down to $z = h$ and then displays a rapid extinction of shear stress in the upper canopy as momentum is absorbed by the foliage.

The standard deviations of u and w, σ_u and σ_w, are shown in Fig. 1.1c and 1.1d, respectively. Normalised with u_* these show slow increases above $z = h$ and rapid attenuation in the canopy. Information in Figs. 1.1b–d can be combined to yield the correlation coefficient $R_{uw} = -\overline{u'w'}/\sigma_u\sigma_w$, a measure of the 'efficiency' of the turbulence in transporting momentum. R_{uw} is equal to 0.3 in the surface layer, increases through the roughness sublayer to peak at a value of 0.45 near $z = h$, then drops off within the canopy but less rapidly than $-\overline{u'w'}$.

Figs. 1.1e and 1.1f show the u and w skewnesses, which are both significant. The positive values of Sk(u) and negative Sk(w) indicate the prevalence of infrequent strong horizontal and vertical gusts relative to the boundary

Table 1.1. *Physical and aerodynamic properties of eight canopies in Fig. 1.1a–h*

Canopy	Site	Reference	h	Leaf area index	$U(h)/u_*$	Sensors Mean	Turbulence
Strips	WT	Raupach *et al.* (1986)	60 mm	0.23	3.3	T	T
Wheat	WT	Brunet *et al.* (1993)	47 mm	0.50	4.1	T	T
Rods	WT	Seginer *et al.* (1976)	19 cm	1.0	5.0	X	X
Corn	Elora	Shaw *et al.* (1974)	260 cm	3.0	3.6	C	F
Corn	Elora	Wilson *et al.* (1982)	225 cm	2.9	3.2	C, F	F
Forest (eucalypt)	Moga	Raupach *et al.* (unpublished)	12 m	1.0	2.9	C, S3	S3
Forest (pine)	Uriarra	Denmead & Bradley (1987)	16, 20 m	4.0	2.5	C	S1
Forest (pine)	Bordeaux	Brunet (unpublished)	20 m	3.0		C, S3	S3

WT, wind tunnel. Sensors: C, cup anemometer; F, split-film servo-driven anemometer; X, x-configuration hot-wire anemometer; T, co-planar triple hot-wire anemometer; S1, single-dimension vertical sonic anemometer; S3, three-dimensional sonic anemometer.

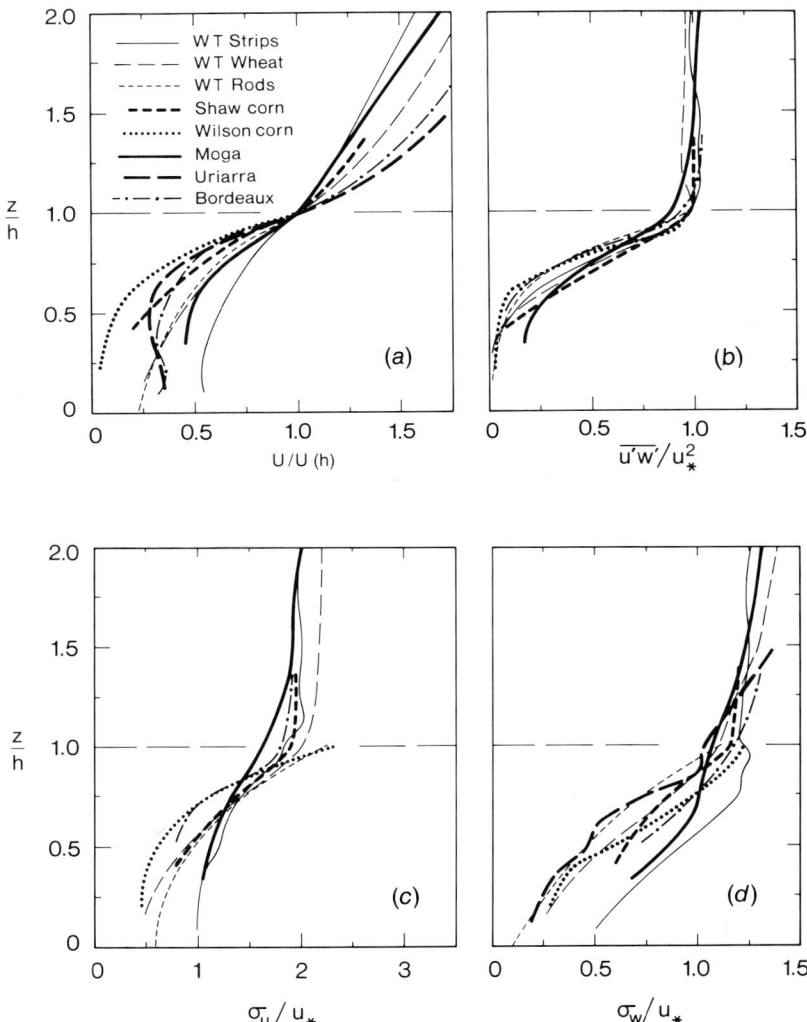

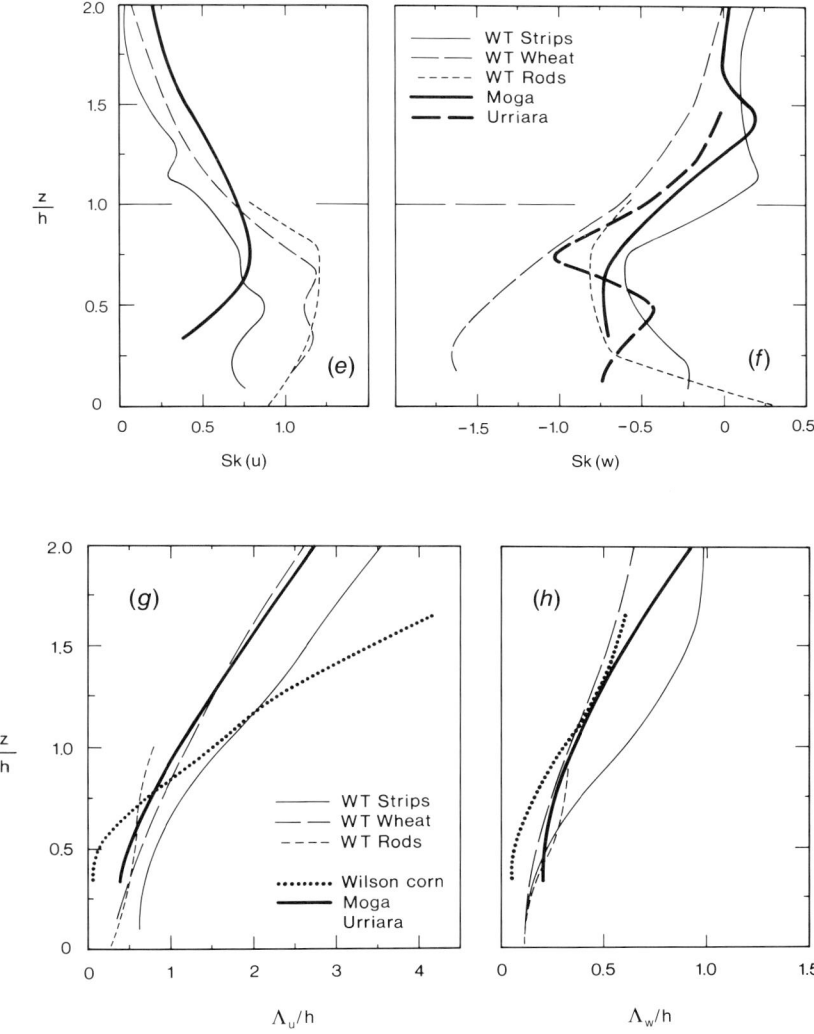

Fig. 1.1. Profiles of velocity moments drawn from eight canopies ranging from wind tunnel models to forests. Not all moments are available for all canopies. Details of the canopies may be found in Table 1.1. Moments are normalised with U, u_* or h and the vertical coordinate with h. (a) $U/U(h)$; (b) $\overline{u'w'}/u_*^2$; (c) σ_u/u_*; (d) σ_w/u_*; (e) u Skewness, $Sk(u)$; (f) w Skewness, $Sk(w)$; (g), (h) Eulerian integral length scales Λ_u, Λ_w.

layer above, where Sk(u) and Sk(w) are close to zero. Quadrant analysis (Shaw et al., 1983; Finnigan & Raupach, 1987) reveals that the vertical and horizontal gusts are combined in 'sweeps' – fast-moving downward incursions of air into the canopy which, as well as being intermittent, are responsible for most of the momentum transport. Typically, over half the momentum transfer occurs in sweeps occupying less than one-twentieth of the total time.

Gardiner (1992, 1994) and Stacey et al. (1994), amongst others, have emphasised the importance of short-lived intense gusts in causing mechanical damage to trees. Sk(u) values of order one indicate regions of intense gustiness (the Gaussian value is zero). For equal variance σ_u, the canopy with the higher Sk(u) will be more prone to wind damage. This is an important point. We shall return to it when we look at the modulation of canopy turbulence by a hill.

1.2.2 Length scales

The Eulerian integral length scales, Λ_u and Λ_w, are defined as:

$$\Lambda_u = U \int_0^\infty \frac{\overline{u'(t)u'(t+\tau)}}{\sigma_u^2} \, d\tau \quad (1.2a)$$

$$\Lambda_w = U \int_0^\infty \frac{\overline{w'(t)w'(t+\tau)}}{\sigma_w^2} \, d\tau \quad (1.2b)$$

where τ denotes time delay.

Despite the limitations of Taylor's hypothesis* in the high-intensity turbulence within a plant canopy, Λ_u and Λ_w are good indicators of the spatial scale of the dominant turbulent motion. As we see in Fig. 1.1g and 1.1h, these parameters indicate that this motion is typically of canopy scale.

Another important measure of eddy scale is the position of the peak of the u, v and w frequency spectra. These spectra, S_u, S_v and S_w respectively, are the Fourier transforms of the integrands in Eq. (1.2a) and (1.2b) (and in the equivalent expression for v), that is, the transforms of the temporal autocorrelation coefficients. By definition, the spectral peaks indicate the characteristic frequency of the energy-containing eddies. To observe a peak in these one-dimensional spectra it is necessary to plot $fS_u(f)$ versus $\ln(f)$,

---

* Taylor's 'frozen turbulence' hypothesis is that the transformation $x = Ut$ may be used to obtain a streamwise dimension for turbulent motions. See Kaimal & Finnigan (1994) for comments on its use in plant canopies.

logarithm of the frequency. Plotted in this 'area preserving' way, equal areas under the spectral curve make equal contributions to the total variance. (See Kaimal & Finnigan (1994) for a discussion of the relationship between one- and three-dimensional spectra and of the correspondence between the peak frequencies of these spectra and the integral length scales defined in Eq. (1.2).)

In the atmospheric surface layer, a height-dependent scaling is often used to non-dimensionalise the frequency axis so that different experiments may be compared. A dimensionless frequency, n, is defined as

$$n = fz/U(z) \qquad (1.3a)$$

or

$$n = f(z-d)/U(z) \qquad (1.3b)$$

where the second form which incorporates d, the zero plane displacement of the logarithmic velocity profile, is usual over tall vegetation.

In the canopy an analogous, height-independent non-dimensionalisation is

$$n = fh/U(h) \qquad (1.4)$$

With the choice of scaling defined by Eq. (1.4), the positions of the peaks in $fS_u(f)$, $fS_v(f)$ and $fS_w(f)$ do not vary as we descend through the roughness sublayer to mid-canopy height. This is confirmed by almost all of the more complete data sets available; for example, see Shaw et al. (1974) in corn, Wilson et al. (1982) in corn, Seginer et al. (1976) and Raupach et al. (1986) in the wind tunnel, Bergström & Högström (1989) in a pine forest, and Amiro (1990) in three forest canopies. This lack of variation also fits in well with the data of Fig. 1.1g which (with the exception of Wilson et al.'s data in corn) would display at most a weak variation of Λ_u or Λ_w with height, if the mean velocity $U(z)$ that appears in Eq. (1.2a) and (1.2b) were replaced by $U(h)$.

The position of the spectral peak is different for the three components u, v and w: $fS_u(f)$ peaks at a scaled frequency of $fh/U(h) \simeq 0.15$ (±0.05), whereas measured values of the $fS_w(f)$ peak cluster around 0.45 (±0.05). The peaks in $fS_v(f)$ are more variable; values range from $fh/U(h) \simeq 0.1$ in the Shaw et al. (1974) corn data to 0.35 in the Moga forest (see Table 1.1. for information on the Moga data set). The situation is illustrated in Fig. 1.2 where the data from the above mentioned canopies are plotted as dimensionless peak wavelengths, λ/h, where λ is defined as $\lambda = U(h)/f_{peak}$, versus z/h. d is the

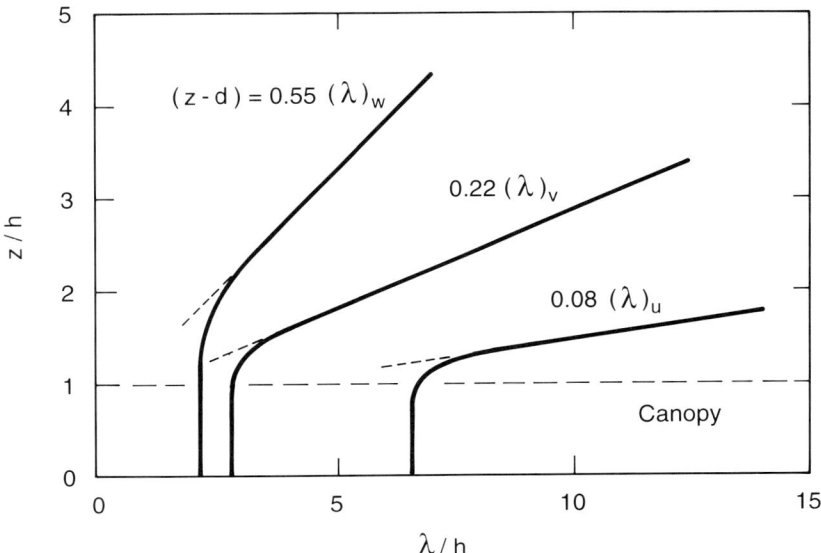

Fig. 1.2. Peak frequencies of the u, v, w power spectra expressed as a peak wavelength λ. The value of λ is constant in the canopy and then merges smoothly into standard height-dependent forms (Kaimal & Finnigan, 1994) in the surface layer.

displacement height of the logarithmic velocity profile that obtains above the roughness sublayer. In Fig 1.2 we see constant peak wavelengths in the canopy merging smoothly with standard expressions for $\lambda/U(h)$ applicable in the surface layer, viz. $(z - d) = 0.55\ (\lambda)_w$, $0.22\ (\lambda)_v$, $0.08\ (\lambda)_u$. It is evident that the component peak wavelengths are all much larger than h.

Explicitly or implicitly, the spatial information presented so far makes the transformation $x = Ut$ to provide a horizontal axis. Most presently available multi-point canopy data employ several sensors on a single mast and so must also employ this subterfuge to generate an x-axis. However, the wind tunnel experiment reported by Brunet *et al.* (1994) in an aeroelastic model wheat canopy contained multi-point measurements with separation in the x, y and z directions. In Fig. 1.3 (from Raupach *et al.* 1989) we present isocorrelation contours of the two-point, one-time u correlation r_{uu}. Here r_{uu} is defined as:

$$r_{uu}(x,y,z,0,z_R) = \frac{\overline{u'(0,0,z_R,t)\,u'(x,y,z,t)}}{\sigma_u(0,0,z_R)\,\sigma_u(x,y,z)} \qquad (1.5)$$

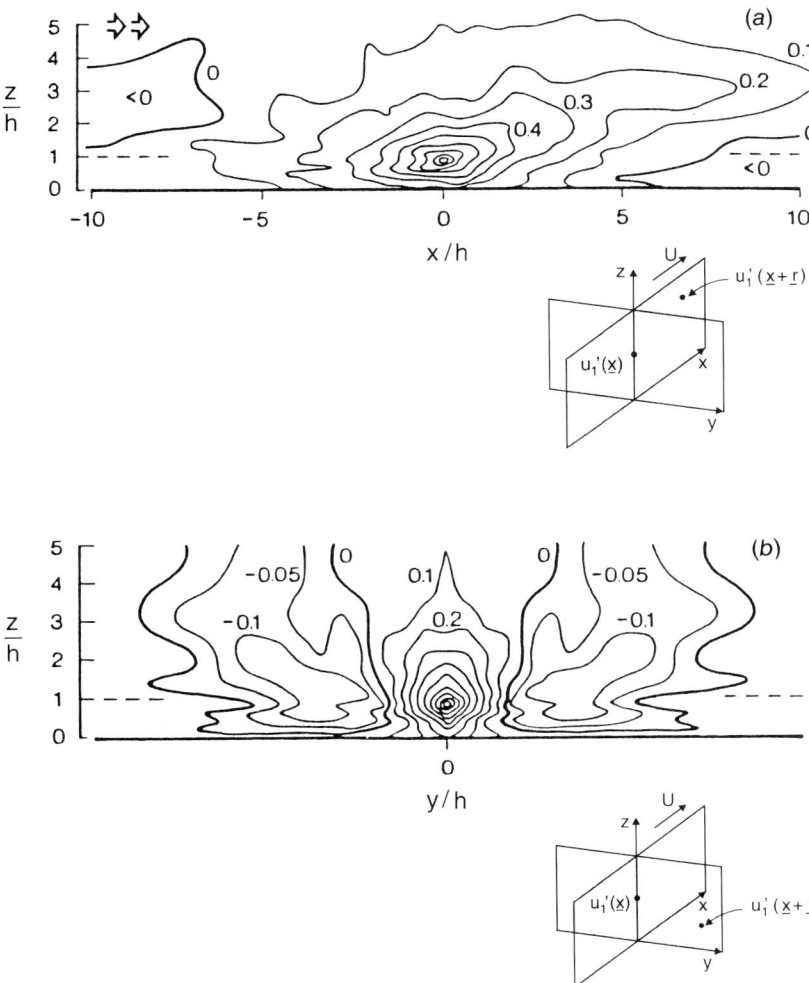

Fig. 1.3. Contours of constant values of the two-point, one-time correlation coefficient $r_{uu}(x, y, z, 0, z_R)$ with $z_R = h$. (a) In the x–z plane; (b) in the y–z plane.

To obtain r_{uu}, one sensor is held fixed at $x_i = (0, 0, z_R)$ while the other sensor is moved around. Their outputs are then cross-correlated with zero time delay. In the example presented in Fig. 1.3, the reference sensor was positioned at $z_R = h$. The contours are presented in the x–z and y–z planes and show good correlation ($r_{uu} > 0.3$) within a volume around the reference point defined approximately by $-2h \leq x \leq 2h$, $-h \leq y \leq h$, $0 < z \leq 2h$. Fig. 1.3 provides strong graphical evidence of the large scale of the main turbulent motion in a form uncontaminated by Taylor's hypothesis. It confirms the conclusion to be drawn from integral length scales and spectral peaks in both field and wind tunnel, that the main turbulent motion is of order h.

1.2.3 Spectra

Many of the contrasts between turbulence in forest canopies and that in the boundary layer above are conveniently illustrated by means of the energy spectrum, $E(k)$. Formally, $E(k)dk$ represents the contribution to turbulent kinetic energy (TKE) from all motions with wavenumber magnitude between k and $k + dk$. $E(k)$ is a useful theoretical construct but is impossible to measure except in homogeneous turbulence, where it can be calculated from one-dimensional spectra using symmetry arguments. The relationship between $E(k)$ and observable spectra is discussed by Kaimal & Finnigan (1994). In Fig. 1.4 we present a schematic diagram of the processes influencing $E(k)$ in a plant canopy.

First we note that the production of TKE through the interaction of existing turbulence and the strain in the mean velocity field peaks at the energy containing wavenumber $k \sim 1/\Lambda_u$. In horizontally homogeneous shear flows, the only component of mean strain is shear, so this process is often referred to as shear production. At higher wavenumbers, the direct effects of viscosity on the eddies start to become significant, and TKE is converted to heat. The characteristic length scale of such eddies is the Kolmogoroff microscale η, which is of order 1 mm in the roughness sublayer.

In the free air boundary layer above the canopy we can observe a region where wavenumbers are much larger than $1/\Lambda_u$ and much smaller than $1/\eta$ and where the production of energy from the mean velocity field is insignificant, as is the loss of TKE to viscous action. In this region of the spectrum, known as the inertial subrange (ISR), TKE 'cascades' from large- to small-scale motions through complex eddy–eddy interactions, but the absence of significant production or destruction of TKE enables the Kolmogoroff scaling laws to be derived (Kolmogoroff, 1941); in particular, in the ISR of the free air boundary layer, we can deduce and observe that $S(k) \propto k^{-5/3}$ and that the turbulent fluctuations are isotropic, whence (Batchelor, 1959):

$$S_v(k) = S_w(k) = \tfrac{4}{3}S_u(k)$$

In plant canopies several other processes complicate this picture. As well

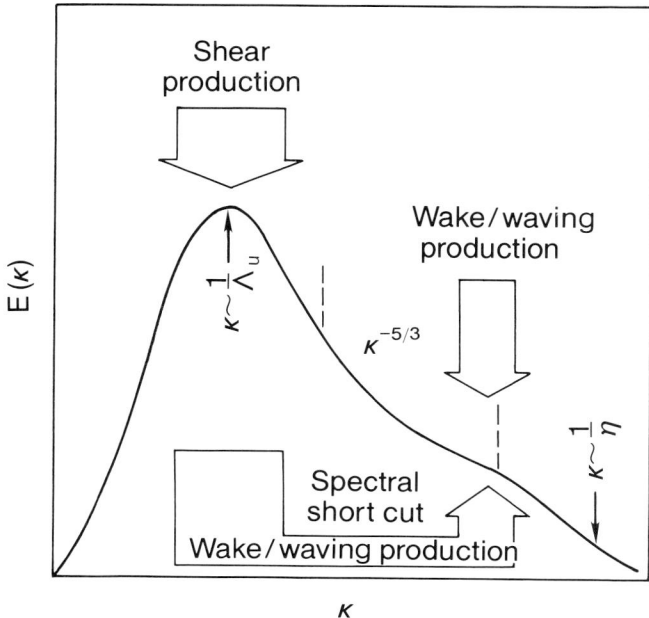

Fig. 1.4. Schematic diagram of the energy spectrum E as a function of wavenumber k in a plant canopy. Extra processes that are peculiar to canopies are marked, as is the inertial subrange ($k^{-5/3}$); the dissipation range occupies wavenumbers greater than η, η being the Kolmogoroff microscale.

as shear production, TKE may be produced in the wakes of plants or by the waving and fluttering of plants or plant parts. The production of TKE by this means is established by the rate of working of the mean flow against the aerodynamic drag of the canopy (Raupach *et al.*, 1986). Depending on the scale of eddy shedding or waving, TKE from this source may feed in anywhere along the spectrum, particularly in the inertial subrange. This is in contrast to the situation in free air where energy is, essentially, neither created nor destroyed in this range, but simply passed, on average, from lower to higher wavenumbers. A second process constitutes a 'spectral short cut' whereby eddies may bypass the usual cascade process in the inertial subrange (Baldocchi & Meyers, 1988). It occurs when eddies generated through shear production are themselves intercepted by leaves and branches and 'chopped up' into finer-scale eddies, through the wake and waving production mechanisms just described.

These extra processes have two consequences within the inertial subrange: they alter the rate of decrease of $S_u(k)$, $S_v(k)$ and $S_w(k)$ with increasing k and they disturb the isotropy observed in this region of the spectrum above the canopy.

Turning first to the spectral slopes, while some published data do not seem to depart much from the Kolmogoroff prediction of $S_u(k)$, $S_v(k)$, $S_w(k) \propto k^{-5/3}$ (e.g. Amiro, 1990), other data sets show individual component spectra rolling off more rapidly than $k^{-5/3}$ (e.g. $S_w(k)$ in Baldocchi & Meyers, 1988). The as yet unpublished data from Moga forest, a relatively open (leaf area index $\simeq 1$) and uniform canopy, show clear distinctions between the three components, with $S_u(k)$ decreasing more rapidly than $k^{-5/3}$ while $S_v(k)$ and $S_w(k)$, in contrast, roll off at less than $k^{-5/3}$. The wind tunnel study of Seginer et al. (1976) clearly points to wake production as the source of energy in the v and w spectra, whereas the more rapid roll off of $fS_u(k)$ might be explained in the same way. The aerodynamic drag in the canopy is predominantly pressure drag which is quadratic in the total velocity and therefore is much more efficient at extracting energy from u component fluctuations than from v and w. Unless this energy is dissipated almost immediately – and, of course, some of it is (Shaw & Seginer, 1985; Brunet et al., 1994) – the generally random orientation of eddies in plant wakes might ensure that v and w fluctuations are boosted at the expense of the u fluctuations.

Moving on to isotropy in the inertial subrange (ISR), we have noted this implies that $S_w(k) = S_v(k) = 4/3\, S_u(k)$. Within canopies, despite a good deal of scatter in the data, it is clear that isotropy is generally violated. Published ratios of $S_w(k)/S_u(k)$ in the ISR vary from 0.94 in a corn canopy to 1.7 in a eucalypt forest (Kaimal & Finnigan, 1994). The data of Amiro (1990) in three different forest canopies showed $S_u(f) : S_v(f) : S_w(f) \simeq 1 \pm 0.15$. We should not be too surprised at these large departures from classical isotropy, given that wake and waving production are so strongly coupled to each canopy's morphology and elastic properties.

TKE budgets obtained in a range of uniform canopies show that wake production rates in the canopy equal or exceed the peak in shear production which occurs at the canopy top (e.g. Raupach et al., 1986; Meyers & Baldocchi, 1990; Brunet et al., 1994). Despite this, measured spectra are not grossly dissimilar to those measured in the surface layer; the variations in slope and isotropy just described are generally minor effects. We ascribe this nonappearance of wake-generated turbulence to two causes: much of the turbulence generated in the near wakes and attached boundary layers of canopy elements is of very small scale and so is rapidly dissipated by viscosity; secondly, the resolution of most of the sensors employed in canopies is too coarse to resolve the greater part of the wake turbulence. These two arguments are discussed in detail by Brunet et al. (1994).

We are left with a picture of canopy airflow where eddies directly generated through vortex shedding from canopy elements are much less important than the characteristic large eddies that dominate the turbulence structure. The origin of these large eddies is the concern of the next section.

1.3 The mixing layer as a model of canopy turbulence

The mean flow and turbulence structure that characterises the canopy, while quite dissimilar to that in a wall boundary layer, is strikingly reminiscent of what is observed on the low-speed side of a plane mixing layer. A plane mixing layer forms when two co-flowing streams with different velocities, separated by a splitter plate when $x < 0$, are allowed to mix downstream of $x = 0$. The splitter plate is in the horizontal plane $z = 0$ (see Fig. 1.5a). The flow quickly becomes turbulent and then self-preserving as x increases, with a vertical length scale proportional to $x - x_0$, where x_0 is an appropriate virtual origin (Townsend, 1976, section 6.10). The behaviour of the velocity moments in the self-preserving region is indicated in Fig. 1.5b taken from Wygnanski & Fiedler (1970); the inflection point in $U(z)$ is clear. There is strong inhomogeneity in $\overline{u'w'}$ (and likewise in σ_u and σ_w, which are not shown). The skewnesses $Sk(u)$ and $Sk(w)$ are large and opposite in sign, both to each other and on opposite sides of the flow; their signs on the low-speed side of the mixing layer correspond to the canopy situation (Fig. 1.1e and f).

These similarities and the dissimilarities with attached boundary layers suggest that canopies and mixing layers might owe the essential features of their flows to a common eddy structure. This hypothesis has been explored by Raupach et al. (1989). They pointed out that the crucial difference between boundary layers and mixed layers was the point of inflexion in the mean velocity profile. It is well known that such a profile is inviscidly unstable to small perturbations, unlike boundary layer velocity profiles which are only unstable when the fluid is viscous (Drazin & Reid, 1981).

The predictions of a linear stability analysis of a plane mixing layer, which include both the spatial structure or modes, passage frequency, convection velocity and growth rate of emerging unstable structures, agree very well with observations and direct numerical simulations of mixing layers at the stage of incipient turbulence (Bayly et al., 1988). Not surprisingly, the agreement is not as good in fully developed turbulent mixing layers. For example, experiments reported by Ho & Huerre (1984) and Hussein (1983) indicate that in a fully developed mixing layer, the passage frequency of transverse vortices is about 1.3 times higher than the stability theory prediction for the fastest-growing mode. Even so, the departures of the fully developed turbu-

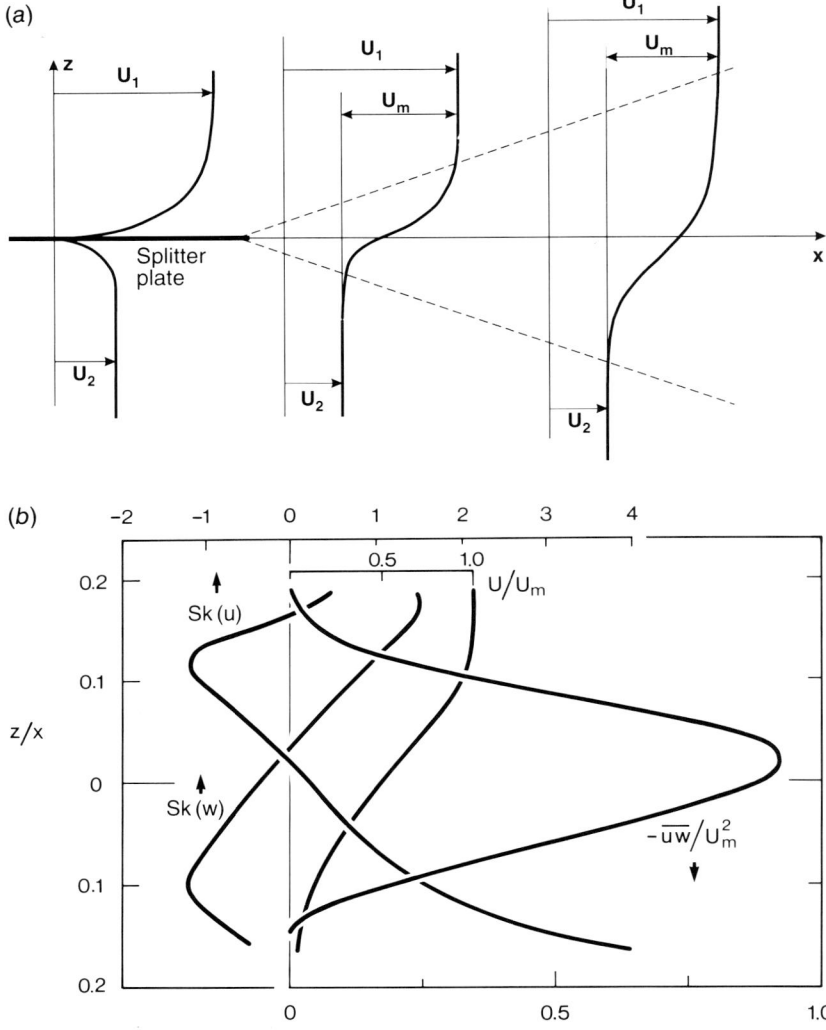

Fig. 1.5. (a) Idealised diagram of the mean velocity development in a plane mixing layer. (b) Velocity moments in a plane mixing layer after Wygnanski & Fielder (1970).

lent case from the linear inviscid prediction are surprisingly small and encourage us to believe that the instability model is relevant to the real situation.

The process of instability is described in detail by Raupach et al. (1989). The structure that initially emerges consists of transverse, Kelvin–Helmholtz waves. This part of the process is well described by linear stability theory, which tells us that the wavelength of the emerging waves is of the same order as the transverse scale of the shear layer, while their growth rate is set by the magnitude of the shear at the point of inflexion.

Following the primary instability process the subsequent development of the mixing layer towards a fully turbulent state includes several additional instability processes (Ho & Huerre, 1984), most of them non-linear and hence not describable by linear stability theory (though some can be described by non-linear theories). The transverse vorticity in the Kelvin–Helmholtz waves collects, under a non-linear self-interaction, into a string of concentrated blobs or 'cats-eyes' (Stuart vortices), linked by 'braids' of vorticity. The concentrated transverse vortices undergo a non-linear, stochastic pairing process which introduces irregularities into the spacing between vortices and provides the main mechanism for the vertical spread of the mixing layer. The main process leading to breakup of the transverse vortices is a three-dimensional secondary instability which produces longitudinal vorticity. Crucial features of this breakup have recently been elucidated by direct numerical simulations (Rogers & Moser, 1992; Knio & Ghoniem, 1992). The initial stages of the process can also be predicted by linear theory; the analysis of Pierrehumbert & Widnall (1982) shows that the lateral spacing of the longitudinal vortices is of the same order as the longitudinal spacing of the primary transverse vortices. Therefore, the scale of the primary, transverse vortices, which stability theory says is set by the scale of the shear, determines the scale of the emerging streamwise vortices also. Furthermore, the single distinct structure that emerges from the instability process spans vertically the entire shear layer resulting in constancy of measures of eddy scale across the layer, just as we have observed in the roughness sublayer.

This sequence of steps is illustrated in an idealised, schematic way in Fig. 1.6. Note especially that we have illustrated the dominant eddy structure emerging from this process as 'double roller' or hairpin vortices. These have been shown to be the characteristic eddy structure of a wide range of turbulent shear layers. Most powerfully, pattern recognition techniques (Mumford, 1982, 1983; Moin & Kim, 1985) confirmed earlier inferences of Townsend (1976) from rapid-distortion theory that double roller eddies give rise to characteristic two-point correlation maps of the form we have illustrated in Fig. 1.3.

Raupach *et al.* (1989) went on to compare the features of the instability derived from linear theory with a variety of measures of the characteristics of canopy eddies, including two-point space–time correlations, integral length scales Λ_u, peak wavelengths λ ($=U(h)/f_{peak}$) and eddy passage frequencies

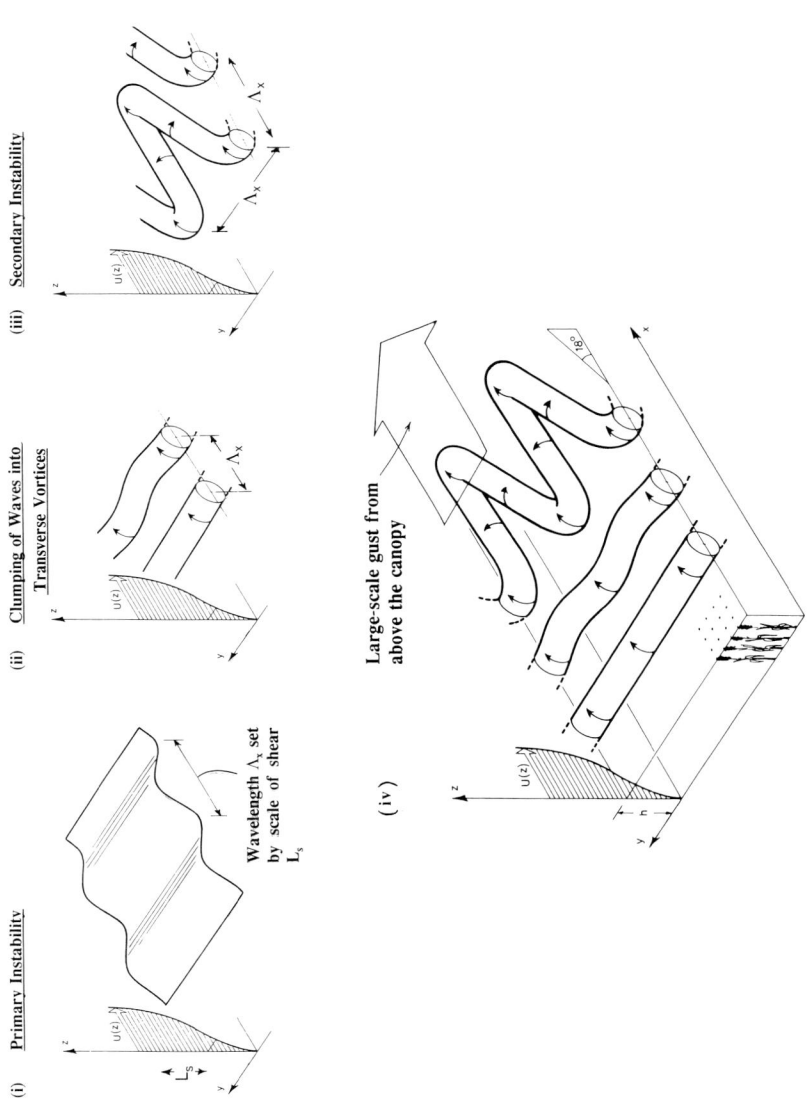

Fig. 1.6. Idealised diagram of the primary and secondary instability process hypothesised to lead to canopy large eddies.

directly measured in a wheat canopy. In all cases the agreement was well within what we should expect given the practical dissimilarities between idealised mixing layers and the canopy situation where the inflexion point shear layer is bounded by a boundary layer above and by the ground below.

We conclude, therefore, that the inflexion point instability process provides a plausible model for the generation of the canopy-scale turbulent large structures that dominate the roughness sublayer. The key features of the instability process – that the eddy size is set by shear layer depth, which in turn is of order h,* and that the rate of emergence of the eddy structure from the background turbulence is proportional to shear magnitude at the inflexion point – give rise to a further refinement of the conceptual model.

Intermittently, larger eddies from the boundary layer above produce surface gusts on scales much larger than the canopy height. Because of their size, these eddy motions are constrained to be almost parallel to the surface and hence, in themselves, transfer little momentum. Bradshaw (1967) has termed these large-scale horizontal velocity excursions 'inactive motion' since they increase σ_u and σ_v, but not σ_w or $\overline{u'w'}$, at the ground. However, they are able to increase the magnitude of $\partial U/\partial z$ at $z = h$, increasing the growth rate of the emerging instability to the point (presumably) where it can emerge from the turbulent background and be recognised as a canopy large eddy. This would be the primary mechanism coupling canopy airflow to that in the boundary layer above. It would also explain two other widely observed canopy phenomena: that the large structures have streamwise convection velocities higher than the local mean velocity, and that they have been observed to appear in sequences of three or four rather than singly (Finnigan, 1979; Raupach et al., 1989). This feature is also illustrated schematically in Fig. 1.6.

1.4 Airflow over isolated hills

The basic features of airflow over a low hill are illustrated in Fig. 1.7. Imagine we are recording the velocity along streamlines that approach an isolated two-dimensional ridge roughly at right angles. Close to the surface, we see the flow decelerate slightly at the upwind foot of the hill before accelerating to the hilltop. If the hill is steep enough, this deceleration may cause a small separation bubble to form at the foot of the hill. The wind

---

* Over tall canopies or those with dense crowns and open trunk spaces, $(h - d)$ is probably a better indication of shear layer depth and therefore a better measure than h to use in collapsing the totality of forest data. We have avoided this complication here by selecting illustrative data sets that do not require this refinement.

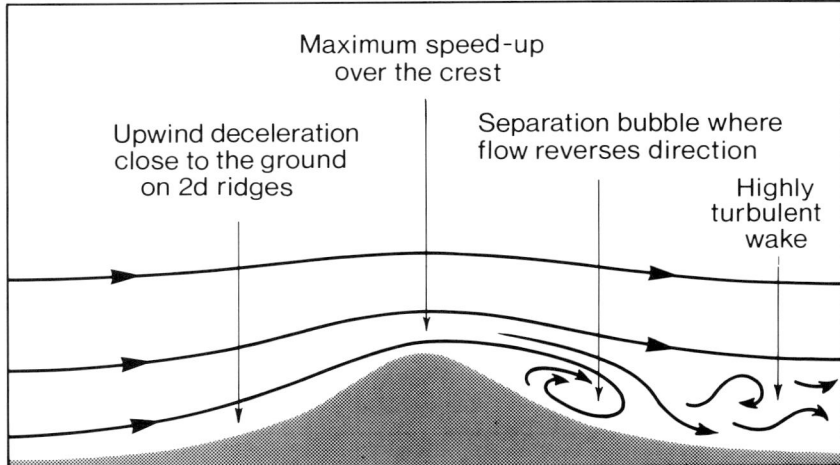

Fig. 1.7. Schematic diagram of the key elements of neutrally stratified windflow over a hill.

reaches its maximum velocity above the hilltop, then decelerates behind the hill. If the hill is steep enough downwind, a separation bubble forms in which the flow direction at the lowest level is opposite to that above. Whether or not a separation bubble forms, a wake region develops behind the hill with a marked velocity deficit extending for many hill heights downwind. If we follow the higher streamlines, we observe monotonic acceleration to the crest followed by deceleration behind the hill.

Following the streamlines over the crest of an axisymmetric hill, we observe the same features with one difference: the upwind deceleration is absent. It is replaced by a region of lateral flow divergence as the streamlines divide to pass around the hill. This lateral divergence decreases as we move up the hill centreline, disappearing completely at the crest.

This evolution of the mean wind is displayed in a more quantitative way in Fig. 1.8. Here we present typical vertical profiles upwind, on the crest and behind the hill. The heights, distances and velocities are normalised by parameters we will define below. The height scale is logarithmic to display the near surface structure in the profiles. Note the large increase in velocity at low levels over the hilltop and the velocity deficit in the hill wake.

To make the information displayed in Figs. 1.7 and 1.8 more concrete, we can imagine the profiles developing over a ridge or axisymmetric hill of height 100 m and proportions similar to those of the hill in Fig. 1.7. The vertical axis is normalised by the 'inner layer depth' l, the height at which mean flow acceleration, pressure gradient and turbulent stress divergence all

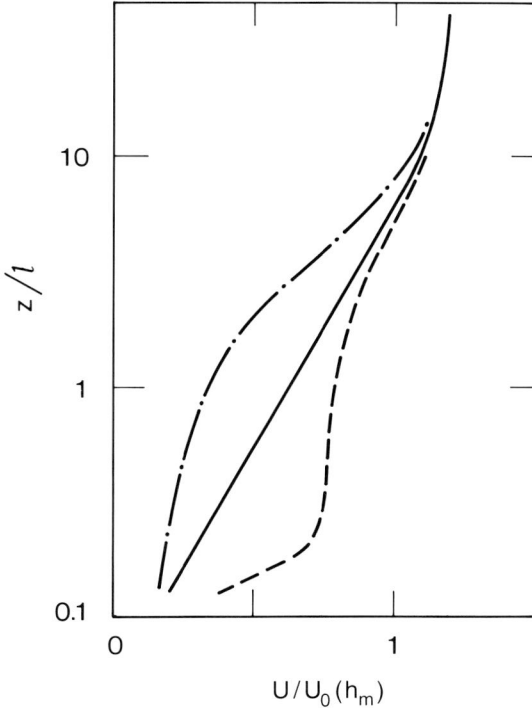

Fig. 1.8. Development of velocity profiles in a flow over an isolated hill. —, upwind, logarithmic profile; – – –, hilltop; – · – · –, behind the hill. Normalising scales are l, the inner layer depth; h_m, the middle layer height; $U_0(z)$, the upwind profile.

make comparable contributions to the budget of mean streamwise momentum. A formula for l is given by Hunt et al. (1988) for the case of low hills with a logarithmic upwind velocity profile. For our 100 m hill, l will be in the range 10–20 m. The mean streamwise velocity U is normalised by the value of the undisturbed upwind velocity at the 'middle layer height', h_m. This is also defined by Hunt et al. (1988), as the height above which mean flow vorticity which is predominantly vertical shear, $\partial U/\partial z$, ceases to play a significant role in the dynamics of mean flow over the hill. h_m is much larger than l and would be in the range 50–100 m over our exemplary hill. Both l and h_m are dependent on hill profile shape and surface roughness, which is why we have not specified them too exactly. A more complete discussion of flow regimes and scaling over hills may be found in Kaimal & Finnigan (1994).

We should also remember that if the hill flow is strongly influenced by buoyancy forces, which can occur with nocturnal inversion close to the crest of low hills or at any time of day over hills that occupy a significant fraction of the boundary layer, then the flow patterns we have presented may be strongly modified. In particular, the maximum velocity speed-up may move from the crest to the downwind slope and other phenomena such as lee waves may appear (Durran, 1990; Kaimal & Finnigan, 1994).

The point we wish to focus on is the emergence of a peak in streamwise velocity speed-up $\Delta U(z) = U(z) - U_0(z)$ which appears at $z \simeq l/3$ (Hunt et al., 1988) over the hill crest. (In this expression for speed-up the origin of z may be taken as the local ground surface or, alternatively, the expression may be interpreted in the streamline coordinate frame defined in the next section.) Below this level local shear, $\partial U/\partial z$, is increased relative to upwind values; above it, shear is reduced. Understanding the reasons for the emergence of this peak is crucial to understanding how canopy flow on a hill differs from that on flat ground.

As the mean flow is diverted around the hill, a perturbation pressure field is set up. Alternatively, we can view the pressure field as 'driving' the perturbations in mean flow – a point of view often adopted in calculations. The local perturbation in pressure $\Delta P(x_i')$ at some point x_i' over the hill can be shown to be a solution of a Poisson equation (Townsend, 1976). The solution takes the form:

$$\Delta P(x_i') = \frac{1}{4\pi} \int \left[\frac{\partial^2}{\partial x_i \partial x_j} \overline{(u_i u_j)} \right] \frac{dV(x_i)}{|x_i' - x_i|} \qquad (1.6)$$

where dV is an element of volume. (Note that if the integration in Eq. (1.6) is performed over all space, an image velocity field must be added below $z = 0$. Alternatively, if the integration is restricted to $z > 0$, boundary terms involving the surface pressure field must be included.) It is clear from Eq. (1.6) that the pressure perturbation at x_i' is not determined by the velocity field at x_i' only, but by the integral over the entire flow domain of the derivatives of velocity products, which contribute to the integral inversely as their distance from x_i'. The global nature of Eq. (1.6) ensures that in a shear flow, contributions to the pressure perturbation at x_i' will be weighted towards higher velocities than the mean velocity at x_i'.

The consequences of this have been worked out rigorously for low hills in the linear analytical model of Hunt et al. (1988). They observed that the flow field around a hill could be divided into three regions with essentially different dynamics: an outer region ($z > h_m$) where the flow was essentially inviscid and irrotational; a middle region ($h_m \geq z > l$) where the flow still behaved inviscidly but was rotational, i.e. affected by the shear in the approaching velocity profile; and an inner region ($l \geq z > 0$) where turbulence (parameterised in the theory by an eddy viscosity) was important. In the compass of this linear theory it is permissible to regard the velocity perturbations as being 'driven' by the perturbation pressure field, which is of order $\rho U_0^2(h_m)$. In the outer layer, where the background velocity is also of order $U_0(h_m)$, the resulting velocity perturbations, $\Delta U(z)$, are relatively small; close to the surface, where background velocity is of order $U_0(l)$, a value substantially smaller than $U_0(h_m)$, the relative perturbation in velocity is much larger.

If the flow response were entirely inviscid, then the velocity perturbation would peak at the surface. In practice, turbulent shear stresses in the inner layer prevent this and the theory predicts that the velocity perturbation will peak at $z \simeq l/3$; this is in good accord with most observations. The expression that Hunt et al. (1988) find for this peak in $\Delta U(z)$ is:

$$\Delta U(\ell/3) \simeq \frac{U_0^2(h_m)}{U_0(\ell)} \cdot \frac{h}{L} \cdot \sigma \tag{1.7}$$

where h/L is a measure of the hill steepness and σ is a constant of order 1 that factors in the precise hill shape and the surface roughness. Eq. (1.7) confirms that the pressure perturbation that drives velocity perturbations in the inner layer ($z < l$) over the hill is of order $\rho U_0^2(h_m)$. A further deduction from the theory is that, close to the surface, the streamwise pressure gradient, $\partial P/\partial x$, which is formally responsible for driving the velocity perturbation ΔU, varies only weakly with z.

Our point of departure for understanding the behaviour of a canopy on a hill is that changes in velocity are produced by a pressure gradient field; this in turn is produced by the flow field as a whole and hence can be regarded, in some sense, as imposed at any given point. This viewpoint imputes a false causality to events that really are interdependent, but is a useful way of making conceptual progress.

1.5 Response of mean canopy flow to a pressure gradient

The balance between forces accelerating or decelerating a particle of air along a streamline is expressed in the streamwise momentum equation. Written in streamline coordinates for the case of two-dimensional flow in a plant canopy this is:

$$U\frac{dU}{dx} = -\frac{1}{\rho}\frac{dP}{dx} - \frac{d}{dz}\overline{u'w'} + 2\frac{\overline{u'w'}}{R} - \frac{d}{dx}\overline{u'^2} + \frac{\overline{u'^2} - \overline{w'^2}}{La} - \tfrac{1}{2}C_d A U^2 \quad (1.8)$$

$$\phantom{U\frac{dU}{dx} = -}1 1 1 0.01 0.1 0.01 1$$

The x axes are streamlines and the z axes the orthogonal trajectories to the streamlines. R and La are the local radii of curvature of the streamlines and z lines, respectively. In Eq. (1.8) we have ignored viscous stresses and parameterised canopy aerodynamic drag as form drag through a dimensionless drag coefficient, C_d, and a function A which is the local area of canopy per unit volume of space.

The velocity components tangent to the x direction are U and u'; w' is the fluctuating component in the z direction. By definition, there is no mean component of velocity in the cross-streamline direction. In this sense, streamline coordinates extend the simplicity of rectangular Cartesian axes to complex shear flows.

Note that all derivatives are total, not partial. The reason for this, together with a more detailed description of this coordinate system, may be found in Kaimal & Finnigan (1994), while derivations of the flow equations in two and three dimensions can be found in Finnigan (1983) and Finnigan (1990), respectively. Formally, the aerodynamic drag term is a consequence of volume averaging the flow field in the canopy (see, for example, Raupach & Shaw (1982) and Brunet et al. (1994)). More information on the intricacies of parameterising canopy drag may be found in Monteith (1975, 1976), Finnigan & Raupach (1987) and Kaimal & Finnigan (1994). For the rest of this chapter all data will be presented in streamline coordinates.

The term on the left-hand side of Eq. (1.8) represents mean flow acceleration along the streamline, while the first term on the right-hand side is the mean pressure gradient. The next four terms are components of the divergence of the turbulent stress tensor, while the last term denotes aerodynamic drag on the foliage. The number beneath each term is a relative order of magnitude, typical of the upper half of a canopy on a hill. These values were derived from the experiment to be described in the next section; they are also typical of canopy flow close to a windbreak. Clearly, to first order, we can simplify the equation, writing:

$$\frac{d}{dx}(\tfrac{1}{2}U^2) = -\frac{1}{\rho}\frac{dP}{dx} - \frac{d}{dz}\overline{u'w'} - C_d A (\tfrac{1}{2}U^2) \quad (1.9)$$

In horizontally homogeneous flow through a canopy on flat ground, Eq. (1.9) reduces to:

$$0 = -\frac{d}{dz}\overline{u'w'} - C_d A \left(\tfrac{1}{2}U^2\right) \tag{1.10}$$

This is the familiar situation described in Section 1.2, where the aerodynamic drag on the vegetation is sustained by a downward flux of momentum through turbulent mixing. In the present case we wish to adopt the viewpoint that the pressure gradient, dP/dx, is 'imposed' upon the canopy. Can we then say anything of universal generality about how this imposed momentum flux is divided up amongst the three remaining terms in Eq. (1.9) representing, respectively, flow inertia (acceleration), turbulent flux divergence and aerodynamic drag? We can make some conceptual progress by dividing the question into two parts, asking first what the essential differences are between flow with a pressure gradient in a canopy and above a canopy, and then asking how any differences might combine to deliver what we observe in the roughness sublayer.

Let us address the first part with a thought experiment. Imagine a one-dimensional canopy – we could think of a duct full of foliage similar to that constructed by Seginer *et al.* (1976) – and suppose we subject it to an instantaneous change in pressure gradient. In the one-dimensional situation, all cross-stream derivatives vanish and Eq. (1.9) reduces to:

$$\frac{d}{dx}\left(\tfrac{1}{2}U^2\right) = -\frac{1}{\rho}\frac{dP}{dx} - C_d A \left(\tfrac{1}{2}U^2\right) \tag{1.11}$$

which is a linear, inhomogeneous equation in $\left(\tfrac{1}{2}U^2\right)$. We allow the pressure gradient to change instantaneously at $x = 0$ from:

$$\left(\frac{1}{\rho}\frac{dP}{dx}\right)_1 \text{ to } \left(\frac{1}{\rho}\frac{dP}{dx}\right)_2$$

(While this would be impossible to engineer in practice, we could manufacture a converging–diverging duct full of foliage which, since Eq. (1.11) is linear, could be subjected to an equivalent analysis and which would yield the same results as our thought experiment, albeit less transparently.)

To solve Eq. (1.11) subject to the forcing Eq. (1.12) we need some boundary conditions. These are that the flow is constant in x for $x < 0$ and $x \to +\infty$ and that U is bounded as $x \to \infty$. The upwind and far downwind boundary conditions correspond to a balance between pressure gradient and aerodynamic drag; this is the one-dimensional equivalent of Eq. (1.10).

With the given boundary conditions, the solution to Eq. (1.11) is easily seen to be:

$$(\tfrac{1}{2}U^2) = \frac{1}{A\,C_d}\left\{\left(-\frac{1}{\rho}\frac{dP}{dx}\right)_2 + \left[\left(-\frac{1}{\rho}\frac{dP}{dx}\right)_1 - \left(-\frac{1}{\rho}\frac{dP}{dx}\right)_2\right]\exp(-A\,C_d x)\right\} \quad (1.12)$$

The solution is illustrated in Fig. 1.9. Immediately after the change in pressure gradient at $x = 0$, the flow has not had time to change its velocity so that the aerodynamic drag term has not changed either; all of the change in pressure gradient is balanced by the flow acceleration. Subsequently, this falls as the velocity changes and the drag adjusts to balance the new pressure gradient.

The crucial point is that this relaxation is exponential, with the approach to the new steady state occurring with a distance constant $(A\,C_d)^{-1}$ as we see from Eq. (1.12). If we applied this same, quasi-inviscid dynamics to a duct without foliage, then no steady state could exist; application of a pressure gradient would result in a continuously accelerating flow.

If we could extend this one-dimensional thought experiment to a canopy on a hill, the pressure gradient imposed by the hill's presence would have quite different consequences above and within the canopy. Within the canopy we would see a constant readjustment to a new equilibrium with a distance constant given by $(A\,C_d)^{-1}$, which is a measure of the 'momentum absorption capacity' of the canopy, while above $z = h$ continual acceleration would occur in response to changing dP/dx. We hypothesis that this difference is the primary cause of modulation in the magnitude of the shear, dU/dz, at $z = h$.

In practice, the third term in Eq. (1.9), $(d/dz)\overline{u'w'}$, ensures that momentum is transferred between the air layers above and within the canopy, and that a

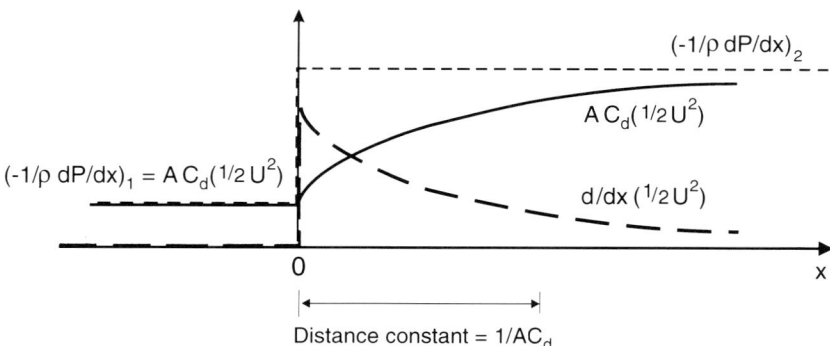

Fig. 1.9. Evolution of the terms in Eq. (1.11) after a step change in pressure gradient at $x = 0$.

smoothly varying mean velocity profile is established. This profile, however, we may expect to be strongly modulated as dP/dx changes over a hill. In the next section we shall see what changes actually occur.

1.6 A wind tunnel model of a canopy on a two-dimensional ridge: experimental arrangements

The model plant canopy used in the experiment is described in detail by Brunet *et al.* (1994). Fabricated from cylindrical nylon stalks, 50 mm long and 0.25 mm in diameter, the model was originally designed by Finnigan & Mulhearn (1978) to provide an aeroelastic simulation of a wheat field. An essential parameter in achieving aeroelastic similarity was the ratio of aerodynamic to elastic force on any stalk. The model canopy behaved like wheat when the tunnel windspeed was 30 m s^{-1}. The experiment described here was performed at a windspeed of 12 m s^{-1} when the stalks behave as though they were somewhat stiffer than wheat – more like a stylised forest canopy. The stalks are arranged on a square grid of side 5 mm yielding a canopy area density A of 10 m^{-1}. The canopy drag coefficient C_d has a marked height dependence (Brunet *et al.*, 1994), but its average value over the depth of the canopy is $C_d = 1.35$.

The experiment was carried out in the CSIRO Pye Laboratory wind tunnel (see Wooding (1968) for a full description), an open-return blower tunnel designed to simulate the flow in the neutral surface layer. It has a 5.5 : 1 two-dimensional contraction and a working section 11 m long, 1.8 m wide and 0.65 m high. A flexible roof permits adjustment of longitudinal pressure gradients. The tunnel arrangement is shown in Fig. 1.10. A 50 mm fence tripped the flow entering the working section, generating a deep turbulent boundary layer which developed initially over a 3 m long rough surface

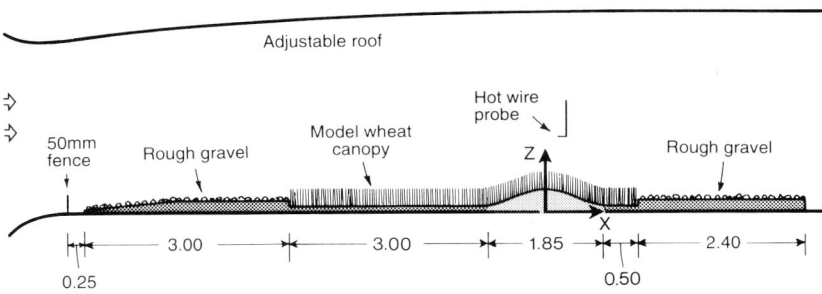

Fig. 1.10. Experimental arrangement in the wind tunnel.

formed by pieces of road gravel, of average diameter 14 mm, glued in a semi-random pattern to wooden baseboards. The upstream section of the stony surface was raised progressively, up to a point where its zero-plane displacement matched that estimated *a priori* for the model canopy (34 mm). The flow then encountered the canopy, which extended over 5.15 m in the streamwise direction and occupied the full width of the tunnel. Downstream of the canopy, a raised rough surface similar to that used for initial flow development covered the rest of the working section.

Rectangular Cartesian coordinates X and Z are used to locate measurement positions in the tunnel and to aid in the presentation of data. These coordinates are distinct from the curvilinear streamline coordinates x and z that appear in Eq. (1.8) onwards.

Three metres from the upwind edge of the canopy a two-dimensional ridge with its axis perpendicular to the wind direction was inserted beneath the canopy. The ridge profile, $Z_h(X)$, was defined by:

$$Z_h(X) = h \left[1 + \left(\frac{X}{L} \right)^2 \right]^{-1} \qquad (1.13)$$

with $X = 0$ at the crest of the ridge. The ridge height h was 150 mm and the length scale L was 420 mm. The profile shape of Eq. (1.13) was truncated at $X = \pm 925$ mm and faired smoothly into the upwind and downwind horizontal surfaces (see Fig. 1.10). The adjustable wind tunnel roof was set to give zero pressure gradient before the ridge was inserted beneath the canopy. After the hill was in place, the roof immediately above it was adjusted to match approximately the pressure field computed for a hill of this shape using potential flow theory.

Measurements of velocity components were made with co-planar triple hot wires (Legg *et al.*, 1984) mounted on a traverse gear. This triple-wire anemometer was designed to give better measurements of two velocity components – u and w in the present case – than are available from conventional x-wire probes. The addition of a third wire allows reliable results to be obtained in turbulent intensities up to about 60% instead of the 30% available from an x-wire.

Static pressure was measured at the surface below the canopy by 33 pressure taps and on the tunnel roof by 20 taps. Static pressure in mid-air was obtained both from a conventional static probe and by subtracting dynamic pressure $\frac{1}{2}\rho U^2$, obtained from the hot wires, from total pressure measured by a Pitot tube. Both conventional Pitot and static tubes are subject to errors in high-intensity turbulence; however, the two methods agreed within 10%

except in the separation region, where we did not attempt to use the Pitot tube method. The results obtained merged smoothly with pressure measured at the wall, even in the separation region, giving us some confidence that they are meaningful. Nevertheless, obtaining reliable measurements of static pressure continues to be one of the weak points of the kind of experimental analysis we are attempting here.

1.7 Flow in a canopy on a two-dimensional ridge

The data obtained from this experiment are the only ones we know of where systematic turbulence measurements have been made within a canopy on a hill. A model study of a forested hill has been described by Ruck & Adams (1991), but measurements were not made within the canopy. Similarly, Bradley (1980) and Bradley et al. (1986) measured turbulence above forested hills in the field, but not within the canopy. Inglis et al. (1991) have made measurements within and above a dense spruce forest in rugged terrain, but present only mean windspeeds. None of these data are in conflict with the results to be presented below and, where direct comparisons can be made, the present model results are useful in interpreting these earlier data.

We begin in Fig. 1.11 with vertical profiles of mean velocity U at 15 positions over the hill. Note first the existence of an extensive separation bubble extending to $X \simeq 2.2$ m. The maximum height of this region of reversed mean flow was approximately 200 mm. It is important to remember that x-wire anemometers will rectify reversing flows in an unpredictable way so that not only do we fail to record the reversed flow in the separation region, but the turbulence moments computed there cannot be regarded as more than a general indication of the level of turbulent activity. They certainly do not represent true velocity moments. The boundary of the separation region shown in Fig. 1.11 was determined by flow visualisation and we have also suggested with dashed lines the probable true mean velocity profiles in this region. It is also obvious, especially if we refer to these suggested velocity profiles, that a marked inflexion in the mean velocity profile, reminiscent of a mixing layer, bounds the separation bubble.

Turning to the attached flow upwind of the crest we should note three things. Well upwind we observe the standard canopy velocity profile recorded in Fig. 1.1. On the crest of the hill, the characteristic inflexion point profile is squeezed up into the upper third of the canopy and the magnitude of dU/dz at $z = h$ is greatly increased, while through almost all of the lower canopy there is no velocity shear at all. One-third of the way up the hill, the

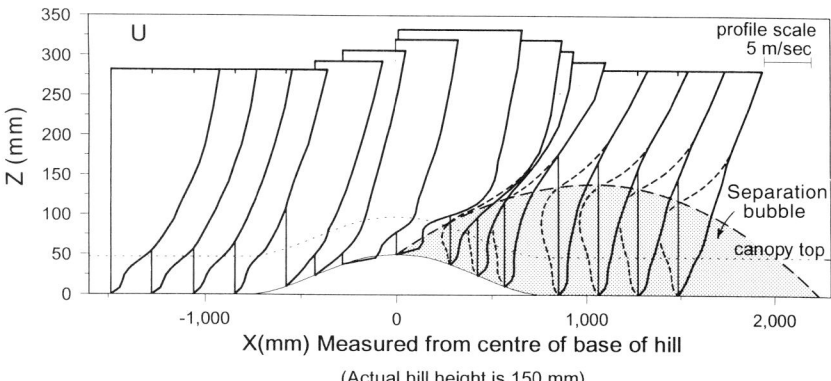

Fig. 1.11. Consecutive profiles of $U(z)$ at a series of X positions over the hill. The X position of each profile is given by the abscissa and the velocity scale indicated separately. The hill is not drawn to scale and serves merely to indicate profile locations. The boundary of the separation bubble is to scale, however. In the separation region, reversing flow renders the hot wire outputs unreliable; probable true velocity profiles in the region are indicated by dashed lines. The top of the canopy is indicated by a dotted line. Note that it is the intersection of this dotted line with the vertical axis that locates profile features relative to the top of the canopy.

canopy inflexion point profile has completely disappeared, leaving a boundary layer type velocity profile through the entire roughness sublayer.

To interpret these profile variations in the light of the conceptual model discussed earlier, we present in Fig. 1.12 the streamwise pressure gradient $1/\rho(dP/dx)$ averaged from $z = 0$ to $z = 1.5h$. The typical vertical variation of this quantity about the mean curve shown was ±15%. Also shown in Fig. 1.12 is the static pressure P along a streamline through the canopy (streamline 2 of Fig. 1.13 below). The length scale of streamwise adjustment in the canopy, $(AC_d)^{-1}$ was 74 mm (based on the canopy average value of $C_d = 1.35$), so the analysis of Section 1.5 suggests rapid response of the within-canopy flow to the pressure gradient; the smaller the value of $(AC_d)^{-1}$, the more closely is the canopy velocity coupled to the local pressure gradient. As a result, we should expect the mean velocity in the canopy to follow the curve of dP/dx in Fig. 1.12 with only a short lag. In other words, canopy velocities should be increasing almost from the foot of the hill and might peak before the hill crest. Indeed, this is the behaviour we observe.

In Fig. 1.13 we have plotted the height of six streamlines (numbered 1–6) as they pass over the hill. The heights are given relative to the local surface

Turbulent airflow in forests

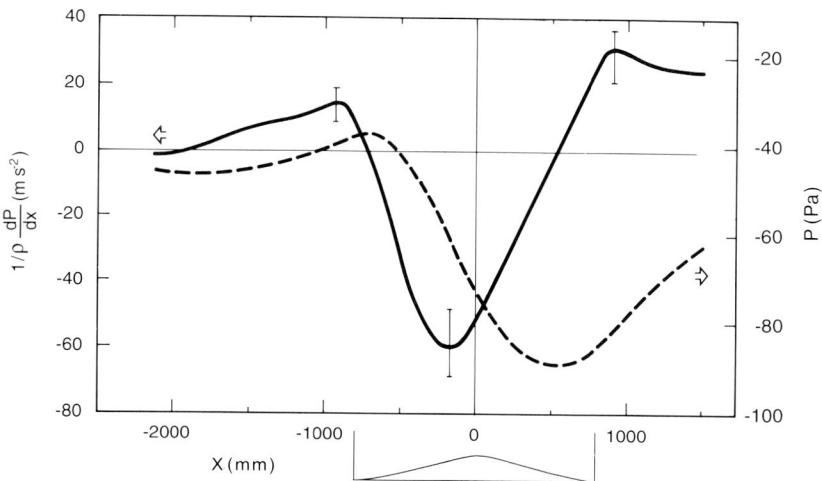

Fig. 1.12. Static pressure along streamline 2 of Fig. 1.13 and streamwise kinematic pressure gradient averaged from $z = 0$ to $z = 1.5h$. Typical profile scatter around the mean curve is shown by the error bars.

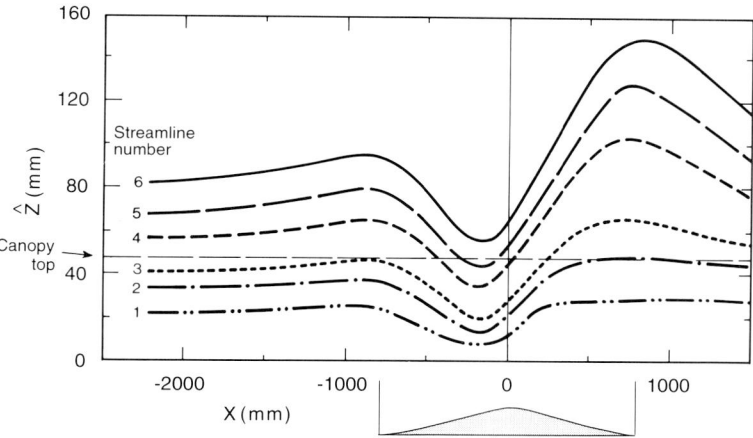

Fig. 1.13. Streamline heights above the local surface $\hat{Z}(X)$ for six streamlines.

so that the vertical coordinate of Fig. 1.13 is $\hat{Z} = Z - Z_h(X)$. Three of the streamlines are within the canopy far upwind and three above, but all save the highest one dip into the canopy at some point. In Fig. 1.14 we present the streamwise velocity U along these six streamlines. The brackets on streamlines 3, 4 and 5 indicate where they enter and leave the canopy.

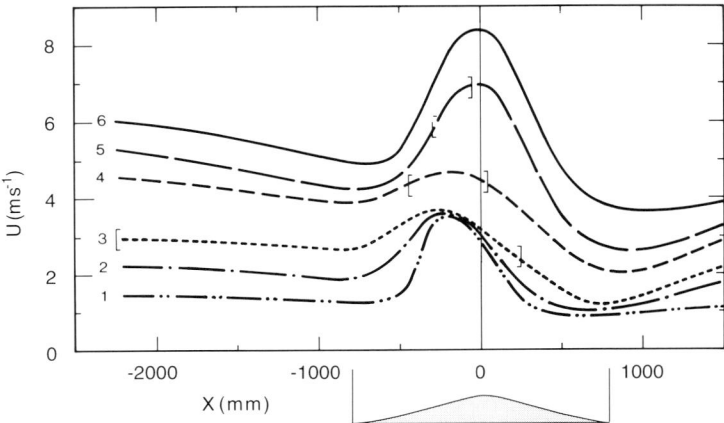

Fig. 1.14. Streamwise mean velocity U along the six streamlines of Fig. 1.13. Brackets on streamlines 3–5 indicate where they enter and leave the canopy.

It is clear that on streamlines 2 and 3, which are entirely within the canopy, acceleration has commenced almost immediately behind the point where the pressure gradient (Fig. 1.12) has become negative (favourable). Similarly, the peak in $U(x)$ on the lowest three streamlines occurs before the hill crest, as does the peak in pressure gradient. In contrast, above the canopy, deceleration continues for some distance after the pressure gradient has become negative and the velocity peak occurs over the hill crest, just as is normally observed in near surface flow over hills without tall canopies (Kaimal & Finnigan, 1994). (The delay in acceleration on the lowest streamline (1) at the foot of the hill is probably connected with the presence of shear at the solid surface below the canopy.)

These differing responses combine in an obvious way to produce the U profile modulations observed in Fig. 1.11. For example, the loss of the inflexion point profile one-third of the way up the hill occurs because the within-canopy flow has accelerated in response to the favourable pressure gradient more rapidly than has the wind above the canopy, while on the hilltop the deceleration within the canopy, coupled with continuing acceleration above, has exaggerated the inflexion point shear.

To deduce more than these broad responses from the simple reasoning of Section 1.5 is not possible. A more sophisticated model of the turbulence is required with proper consideration of the response of turbulent stress to all these forcings. Nevertheless, simple reasoning has clearly given us some important clues as to what is producing changes in the canopy velocity profile

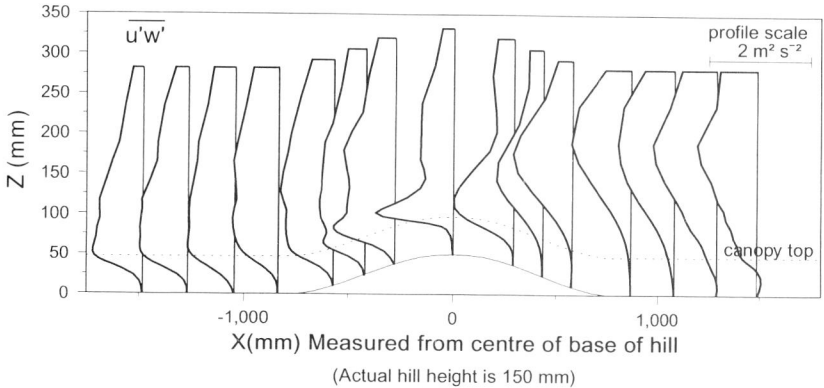

Fig. 1.15. As in Fig. 1.11, but for $\overline{u'w'}$.

over the hill. We can now ask whether turbulence moments react to these changes in the way the instability/shear layer analogy presented in Section 1.3 would suggest.

In Fig. 1.15 we present vertical profiles of shearing stress $\overline{u'w'}$. On the hilltop we observe a large peak in $-\overline{u'w'}$ accompanying the marked increase in canopy top shear there. At the same time, significant levels of $-\overline{u'w'}$ are confined to the upper third of the canopy, which is the region occupied by the inflexion point profile; in the lower canopy, where shear is absent, turbulent flux of momentum in the cross-streamline direction is negligible.

In contrast, one-third of the way up the hill, where the inflexion point profile has disappeared, significant $\overline{u'w'}$ levels can be seen much deeper in the canopy, the near-constant stress region actually penetrating below the canopy top. Clearly, turbulent coupling of the lower canopy with the flow above is much more active here than on the hill crest.

Finally, we should note the virtual absence of measured $\overline{u'w'}$ in the canopy below the separation bubble. Although we cannot interpret this as true shear stress because of sensor problems in the reversing flow, it is some measure of correlation between orthogonal turbulence components. Contrast this with the elevated peaks in $-\overline{u'w'}$ at the outer edge of the separation region. Finnigan et al. (1990) in a separate experiment suggested that these were also the result of an inflexion point instability, this time at the edge of the separation bubble.

In Fig. 1.16 we display vertical profiles of $\overline{u'^2}$ which mimic most of the features of the shearing stress with the exception of measurable levels of $\overline{u'^2}$

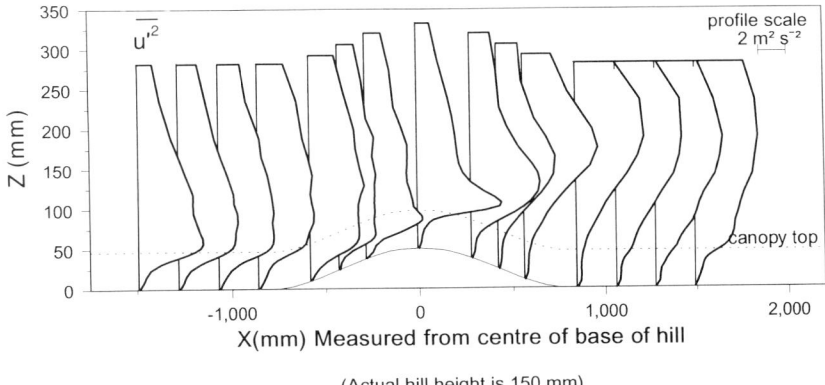

Fig. 1.16. As in Fig. 1.11, but for $\overline{u'^2}$.

in the canopy beneath the separation bubble. Interpreting this with some caution, we might deduce that some horizontal 'sloshing' motion is occurring there, possible driven by pressure pulsations in the separation bubble and being relatively ineffective in transforming momentum.

In the present context of damage to trees, Fig. 1.16 becomes more interesting when combined with Fig. 1.17 vertical profiles of u skewness. Care must be taken in interpreting the very high skewness at the bottom of the canopy, one-third of the way up the hill, as this is more a result of low σ_u values than high $\overline{u'^3}$. Nevertheless, two things are obvious: the variations in Sk(u) are qualitatively as we would expect if we were to assume that positive u skewness is associated with the inflexion point profile. For example, at the hill crest, where the inflexion point profile is squeezed into the upper third of the

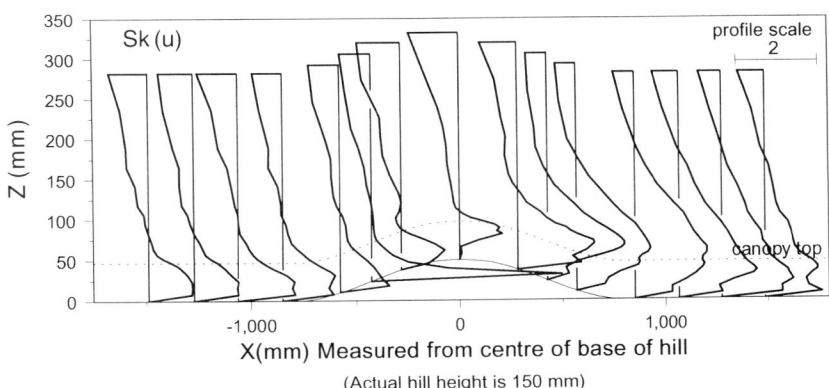

Fig. 1.17. As in Fig. 1.11, but for Sk(u).

canopy, Sk(u) also is reduced to an elevated peak, whereas one-third of the way up the hill, Sk(u) only attains appreciable values near the ground, and this for the reason given above. Secondly, the combination of this elevated peak in Sk(u) with the high values of $\overline{u'^2}$ we observed at the same location in Fig. 1.16, denotes extremely strong, highly intermittent gusts at crown height, which maximise the peak stem bending moment in this location, a parameter of crucial importance in windthrow as is emphasised elsewhere in this volume.

1.8 Conclusions

We have seen that universal features of canopy turbulence such as inhomogeneity of turbulence moments, positive u and negative w skewnesses, scaling with u_* and h, large size and height independence of integral scales of turbulence can all be explained by assuming that a distinct eddy structure separates canopy turbulence from that in the boundary layer above. These features of turbulent structure are reminiscent of a plane mixing layer rather than a boundary layer, leading naturally to the hypothesis that the eddy structure derives from instability of the inflexion point velocity profile, a characteristic of mixing layers and canopies but not of boundary layers. Comparison of the predictions of linear theory with measurements of eddy parameters in canopies has lent strong support to this hypothesis. This mechanism also explains the success of u_* and h scaling since the characteristic scale of the instability is set by the scale of mean shear which is h (or $h - d$) and its intensity by the shear magnitude at h.

We went on to see whether we might deduce simple guidelines to the way the canopy velocity profile would evolve in more complex situations, choosing as an example a canopy on a hill. Here, the driving pressure gradient was seen to be a key feature and a simplified analysis of the response of canopy flow to a changing pressure gradient showed that it reacted with a distance constant $(AC_d)^{-1}$, that is, the inverse of the canopy's capacity to absorb momentum.

Applying this reasoning to wind tunnel measurements of flow in a canopy on a two-dimensional ridge, we saw that the changes in the mean velocity profile reflected the differences in response of above- and within-canopy flow to the hill's pressure field in the expected way. Finally, we saw that the changes in turbulence moments over the hill reflected this modulation in profile shape just as the analogy with mixing layer flow would predict.

We would like to propose that this chain of reasoning might be extended to other inhomogeneous situations, particularly near forest edges and around windbreaks. In both those cases, one of the primary perturbations is to the pressure field and wind tunnel measurements in clearings (unpublished data of M. R. Raupach) and behind windbreaks (unpublished data of M. Judd, J. J. Finnigan, J. McAneny and M. R. Raupach) suggest that modulations in the turbulence field can be explained with the same level of confidence as the present hill results by the conceptual model we have laid out here.

Acknowledgements

The authors would like to acknowledge the central involvement of Dr M. R. Raupach in the analysis and measurements presented in this chapter.

References

Amiro, B. D. (1990). Drag coefficients and turbulence spectra within three boreal forest canopies. *Boundary-Layer Meteorology,* **52**, 227–46.

Baldocchi, D. D. & Meyers, T. P. (1988). A spectral and lag-correlation analysis of turbulence in a deciduous forest canopy. *Boundary-Layer Meteorology,* **45**, 31–58.

Batchelor, G. K. (1959). *The Theory of Homogeneous Turbulence.* Cambridge University Press, Cambridge.

Bayly, B. J., Orszag, S. A. & Herbert, T. (1988). Instability mechanisms in shear-flow transition. *Annual Review of Fluid Mechanics,* **30**, 359–91.

Bergström, H. & Högström, U. (1989). Turbulent exchange above a pine forest. II. Organized structures. *Boundary-Layer Meteorology,* **49**, 231–63.

Bradley, E. F. (1980). An experimental study of the profiles of wind speed, shearing stress turbulence at the crest of a large hill. *Quarterly Journal of the Royal Meteorological Society,* **106**, 101–23.

Bradley, E. F., Coppin, P. A. & Katen, P. C. (1986). Turbulent wind structure above very rugged terrain. In *Proceedings of the Ninth Australasian Fluid Mechanics Conference, Auckland, NZ,* 1986, pp. 428–31.

Bradshaw, P. (1967). 'Inactive' motion and pressure fluctuations in turbulent boundary layers. *Journal of Fluid Mechanics,* **30**, 241–58.

Brunet, Y., Finnigan, J. J. & Raupach, M. R. (1994). A wind tunnel study of air flow in waving wheat: single-point velocity statistics. *Boundary-Layer Meteorology,* **70**, 95–132.

Drazin, P. G. & Reid, W. H. (1981). *Hydrodynamic Stability.* Cambridge University Press, Cambridge.

Durran, D. R. (1990). Mountain waves and downslope winds. In *Atmospheric Processes over Complex Terrain,* ed. W. Blumen, American Meteorological Society Meteorological Monographs vol. 23, no. 45, pp. 59–82. American Meteorological Society, Boston, MA.

Finnigan, J. J. (1979). Turbulence in waving wheat. II. Structure of momentum transfer. *Boundary-Layer Meteorology,* **16**, 213–36.

Finnigan, J. J. (1983). A streamline coordinate system for distorted turbulent shear flows. *Journal of Fluid Mechanics,* **130**, 241–58.

Finnigan, J. J. (1985). Turbulent transport in flexible plant canopies. In *The Forest–Atmosphere Interaction,* eds. B. A. Hutchison & B. B. Hicks, pp. 443–80. D. Reidel, The Netherlands.

Finnigan, J. J. (1990). Streamline coordinates, moving frames, chaos and intergrability in fluid flow. In *Topological Fluid Mechanics*, ed. H. K. Moffatt & A. Tsinober, Proceedings of the IUTAM symposium, 13–18 Aug. 1989, Cambridge, UK, pp. 64–74.

Finnigan, J. J. & Mulhearn, P. J. (1978). Modelling waving crops in a wind tunnel. *Boundary-Layer Meteorology*, **14**, 253–77.

Finnigan, J. J. & Raupach, M. R. (1987). Transfer processes in plant canopies in relation to stomatal characteristics. In *Stomatal Function*, ed. E. Zeiger, G. D. Farquar & I. R. Cowan, pp. 385–429. Stanford University Press, Stanford, CA.

Finnigan, J. J., Raupach, M. R., Bradley, E. F. & Aldiss, G. K. (1990). A wind tunnel study of turbulent flow over a two-dimensional ridge. *Boundary-Layer Meteorology*, **50**, 277–317.

Gardiner, B. A. (1992). Mathematical modelling of the static and dynamic characteristics of plantation trees. In *Mathematical Modelling of Forest Ecosystems*, ed. J. Franke & A. Roeder, pp. 40–61. Sauerlander's Verlag, Frankfurt am Main.

Gardiner, B. A. (1994). Wind and wind forces in a plantation spruce forest. *Boundary-Layer Meteorology*, **67**, 161–86.

Ho, C.-M. & Huerre, P. (1984). Perturbed free shear layers. *Annual Review of Fluid Mechanics*, **16**, 365–424.

Hunt, J. C. R., Leibovich, S. & Richards, K. J. (1988). Turbulent shear flow over low hills. *Quarterly Journal of the Royal Meteorological Society*, **114**, 1435–70.

Hussein, A. K. M. F. (1983). Coherent structures: reality and myth. *Physics of Fluids*, **26**, 303–56.

Inglis, D.W.F., Gardiner, B.A. & Choularton, T.W. (1991) A model of airflow above and within a forest canopy in a region of complex terrain. In *Proceedings of the 20th Conference on Agricultrual and Forest Meteorology*, September 1991, Salt Lake City, Utah.

Kaimal, J. C. & Finnigan, J. J. (1994). *Atmospheric Boundary Layer Flows: Their Structure and Measurement*. Oxford University Press, New York.

Knio, O. M. & Ghoniem, A. F. (1992). The three-dimensional structure of periodic vorticity layers under non-symmetric conditions. *Journal of Fluid Mechanics*, **243**, 353–92.

Kolmogoroff, A. (1941). The local structure of turbulence in incompressible viscous fluid for very large Reynolds' numbers. In *Turbulence: Classical Papers on Statistical Theory*, ed. S. K. Friedlander & L. Topper, pp. 151–5. Interscience, New York.

Legg, B. J., Coppin, P. A. & Raupach, M. R. (1984). A three-hot-wire anemometer for measuring two velocity components in high intensity turbulent boundary layers. *Journal of Physics E: Scientific Instruments*, **17**, 970–6.

Meyers, T. P. & Baldocchi, D. D. (1990). The budgets of turbulent kinetic energy and Reynolds stress within and above a deciduous forest. *Agricultural and Forest Meteorology*, **50**, 1–16.

Moin, P. & Kim, J. (1985). The structure of the vorticity field in turbulent channel flow. I. Analysis of instantaneous fields and statistical correlations. *Journal of Fluid Mechanics*, **155**, 441–64.

Monteith, J. L. (1975). *Vegetation and the Atmosphere*, vol. 1, *Principles*. Academic Press, London & New York.

Monteith, J. L. (1976). *Vegetation and the Atmosphere*, vol. 2, *Case Studies*. Academic Press, London & New York.

Mumford, J. C. (1982). The structure of large eddies in fully developed turbulent shear flows. I. The plane jet. *Journal of Fluid Mechanics*, **118**, 241–68.

Mumford, J. C. (1983). The structure of large eddies in fully developed turbulent shear flows. II. The plane wake. *Journal of Fluid Mechanics*, **137**, 447–56.

Pierrehumbert, R. T. & Widnall, S. E. (1982). The two- and three-dimensional instabilities of a spatially periodic shear layer. *Journal of Fluid Mechanics*, **112**, 467–74.

Raupach, M. R. (1988). Canopy transport processes. In *Flow and Transport in the Natural Environment: Advances and Applications*, ed. W. L. Steffen & O. T. Denmead, pp. 95–127. Springer-Verlag, Berlin.

Raupach, M. R. (1989). Stand overstorey processes. *Philosophical Transactions of the Royal Society of London, Series B*, **324**, 175–90.

Raupach, M. R. & Shaw, R. H. (1982). Averaging procedures for flow within vegetation canopies. *Boundary-Layer Meteorology*, **22**, 79–90.

Raupach, M. R., Coppin, P. A. & Legg, B. J. (1986). Experiments on scalar dispersion within a model plant canopy. I. The turbulence structure. *Boundary-Layer Meteorology*, **35**, 21–52.

Raupach, M. R., Finnigan, J. J. & Brunet, Y. (1989). Coherent eddies in vegetation canopies. In *Fourth Australasian Conference on Heat and Mass Transfer*, Christchurch, New Zealand, 9–12 May 1989, pp. 75–90.

Rogers, M. M. & Moser, R. D. (1992). The three-dimensional evolution of a plane mixing layer: the Kelvin-Helmholtz rollup. *Journal of Fluid Mechanics*, **243**, 183–226.

Ruck, B. & Adams, E. (1991). Fluid mechanical aspects of the pollutant transport to coniferous trees. *Boundary-Layer Meteorology*, **56**, 163–95.

Seginer, I., Mulhearn, P. J., Bradley, E. F. & Finnigan, J. J. (1976). Turbulent flow in a model plant canopy. *Boundary-Layer Meteorology*, **10**, 423–53.

Shaw, R. H. & Seginer, I. (1985). The dissipation of turbulence in plant canopies. In *Seventh Symposium of the American Meteorological Society on Turbulence and Diffusion*, Boulder, Colorado, November 1985, pp. 200–3.

Shaw, R. H., Silversides, R. H. & Thurtell, G. W. (1974). Some observations of turbulence and turbulent transport within and above plant canopies. *Boundary-Layer Meteorology*, **5**, 429–49.

Shaw, R. H., Tavanger, J. & Ward, D. P. (1983). Structure of the Reynolds stress in a canopy layer. *Journal of Climate and Applied Meteorology*, **22**, 1922–31.

Stacey, G. R., Belcher, R. E., Wood, C. J. & Gardiner, B. A. (1994). Wind and wind forces in a model spruce forest. *Boundary-Layer Meteorology*, **69**, 311–34.

Townsend, A. A. (1976). *The Structure of Turbulent Shear Flow*, 2nd edn. Cambridge University Press, Cambridge.

Wilson, J. D., Ward, D. P., Thurtell, G. W. & Kidd, G. E. (1982). Statistics of atmospheric turbulence within and above a corn canopy. *Boundary-Layer Meteorology*, **24**, 495–519.

Wooding, R. A. (1968). *A Low-speed Wind Tunnel for Model Studies in Micrometeorology: II*. Technical Paper no. 25. CSIRO Centre for Environmental Mechanics, Canberra, Australia.

Wygnanski, I. & Fiedler, H. E. (1970). The two dimensional mixing region. *Journal of Fluid Mechanics*, **41**, 327–61.

2

The interactions of wind and tree movement in forest canopies

B. A. GARDINER

Abstract

The response of four trees to wind loading in a dense spruce plantation has been investigated. Tree movement is well correlated with the passage of coherent gusts over the forest which can be identified by applying the variable-interval time-averaging (VITA) method to the momentum flux signal within the canopy. The gusts are relatively long lived but with a small spatial scale (~1 tree height). The trees do not resonate with turbulent wind components close to their natural frequency but behave like damped harmonic oscillators responding to the intermittent impulsive loading during gust passage. The presence of trees severely modifies the turbulence spectra within the canopy by short-circuiting the normal energy cascade process and this may be an adaptive strategy developed by trees to efficiently lose energy absorbed from the wind.

2.1 Introduction

Strong winds associated with Atlantic depressions in Northern Europe, hurricanes in North America and the Caribbean or cyclones in Japan, New Zealand and the Pacific Islands cause extensive damage to forests world-wide every year (Savill, 1983). Considerable research has been carried out to understand the physical processes involved in order to improve silvicultural practices and to predict better the likelihood of damage.

It was soon realised that trees blow down at windspeeds considerably lower than those predicted from static pulling tests under calm conditions (Fraser & Gardiner, 1967; Oliver & Mayhead, 1974; Blackburn & Petty, 1988). One explanation is that trees are dynamic structures capable of resonating at their natural sway frequency with the turbulent wind field. This idea has been explored theoretically and in field measurements of tree movement and turbulence characteristics by Papesch (1974), Holbo et al. (1980), Mayer (1987) and Blackburn et al. (1988).

This chapter continues the investigation of the dynamic behaviour of trees. In particular it seeks to understand the importance of resonance and the role of the large coherent gust structures found close to rough surfaces (Grass et al., 1991). These coherent structures are intermittent and are known to dominate momentum transfer into canopies (Finnigan & Brunet, this volume). They are clearly visible as the 'Honami' waves in cereal crops on windy days (Inoue, 1955a).

2.2 Experimental arrangement

An experiment was carried out to study the movement of mature plantation trees in response to the wind in Rivox Forest, south-west Scotland, between March 1988 and January 1989. Gardiner (1994) describes the characteristics of the airflow within and above the canopy of this forest and provides a more complete description of the experimental arrangement.

2.2.1 Site description

The site, which is at an altitude of 360 m, is gently sloping and close to the bottom of a valley. The forest consisted of pure Sitka spruce (*Picea sitchensis* (Bong.) Carr.) planted at a density of 3584 stems ha^{-1} in 1962. The median tree height of a sample of 470 trees was 12.0 m with the largest trees close to 15 m. (A value of 15 m is taken as tree height, h, in this chapter.) The mean trunk diameter at breast height (dbh) was 12.9 cm and the leaf area index was 10.2.

2.2.2 Instrumentation

Two towers were erected along the mean wind direction, separated by a distance of 15 m (h). The upwind tower was 30 m tall, the downwind tower 16 m. The upwind tower was instrumented with three-axis anemometers at 12 heights within and above the canopy whilst the downwind tower had anemometers at three heights within the canopy (Table 2.1). Only windspeeds measured on the upwind tower are used here.

Close to the upwind tower four trees were instrumented with orthogonal pairs of displacement transducers (PST-900A, LCM Systems, Isle of Wight, UK) to monitor their movement. These transducers were mounted on wooden stakes driven into the ground and attached to the trees by very fine twisted

Table 2.1. *Anemometer heights on the two towers at Rivox Forest*

Upwind/downwind tower (U/D)	Anemometer height (m)	z/h
U	3.6	0.24
U	7.1	0.47
U	9.1	0.61
U	11.2	0.75
U	13.2	0.88
U	15.0	1.00
U	16.8	1.12
U	18.7	1.25
U	20.8	1.39
U	23.6	1.57
U	26.4	1.76
U	28.5	1.90
D	7.5	0.50
D	11.4	0.76
D	15.1	1.01

steel wire at an angle of 45°. Trees 1, 3 and 4 had transducers attached at close to half their heights whereas on tree 2 they were attached at 8, 6, 4 and 2 m to measure stem bending shape as well as tree movement. Displacement transducers are better than accelerometers for measuring movement because of the poor response of accelerometers to low frequencies. The physical and dynamic characteristics of the four instrumented trees are given in Table 2.2. Trees 2 and 3 were closest to the line between the two towers (2.5 and 4 m respectively) whilst trees 3 and 4 were further away (8.5 and 10 m respectively).

Instrument signals were recorded on ½-inch magnetic tape using a PC-based logging system. All signals were low-pass filtered at 5 Hz and recorded at 10 Hz except for those from the sonic anemometers which were recorded

Table 2.2. *Trees measured for displacement*

Tree	Height (m)	dbh (cm)	Height of canopy (m)	Displacement measurement heights (m)	Resonant frequency (Hz)	Damping ratio (ζ) Neighbour contact	Damping ratio (ζ) No contact
1	13.9	17.7	7.0	5.7	0.47	0.102	0.049
2	13.2	17.6	8.0	8.0, 6.0, 4.0, 2.0	0.48	0.054	0.042
3	13.9	14.8	5.8	6.65	0.36	0.059	0.050
4	10.2	9.7	5.6	5.0	0.39	0.074	0.077

dbh, diameter at breast height.

unfiltered at 40 Hz, twice their update frequency. Initial data analysis consisted of mathematically rotating the anemometers to point into the mean wind and tipping those above the canopy to account for any non-horizontal flow arising from the slope of the ground. Anemometers below canopy top were not tipped.

2.3 Results and discussion

Data from four windy days with neutral stability are presented. The general meteorological conditions for each day are given in Table 2.3.

2.3.1 Tree movement and wind loading

A plan view of the movement of tree 1 and 2 at very similar heights is given in Fig. 2.1. Tree 2 can be seen to describe an essentially elliptical orbit with the major axis along the prevailing wind direction. The general pattern is similar to that observed by Mayer (1987) for spruce trees and Maitani (1979) for a rush plant although the general downwind displacement from the rest position is more obvious than in Mayer's data. Interestingly tree 1, which is only 5 m from tree 2 and has almost identical physical characteristics, moves substantially less. The dynamic behaviour of these and similar spruce trees in the Rivox Forest has been studied by Milne (1991) and Gardiner (1992) who showed that they behave like forced damped harmonic oscillators. The damping characteristics of all four displacement trees (Table 2.2) were obtained by the method described by Gardiner (1989) and show that the damping ratio (ζ) of tree 1 is twice that of tree 2 when they are in contact with their neighbours. The resonant frequencies of the two trees are virtually identical and, assuming that the tree mass is similar (height and dbh are almost identical), we can deduce that at resonance a doubling in the damping ratio reduces the displacement amplitude by half for an identical driving force (Thompson, 1988, equation 3.1-7).

A knowledge of the mean displacement of the trees from their rest position should allow an estimate of the mean force acting on the trees since Gardiner (1992) has derived a general relationship for tree displacement in terms of the applied load, the height at which it acts, and tree height and dbh. The mean force can then be used to calculate the average drag on the canopy and compared with the Reynolds stress ($-\overline{\rho u'w'}$) obtained from the anemometers above the canopy. Unfortunately, the mean displacement of the trees is extremely small (typically <2 cm) and large errors are likely to occur due to the resolution of the transducers (~2 mm) and the difficulty of exactly

Table 2.3. *Meterological conditions and tree/gust correlations*

			Correlation (r_{ab})														
			Tree 1				Tree 2				Tree 3				Tree 4		
			Top of canopy		In canopy		Top of canopy		In canopy		Top of canopy		In canopy		Top of canopy		In canopy	
Date (1988)	Mean windspeed at h (m s^{-1})	Mean wind direction at h (degs mag)	$-u'w'$	u'_r	$-u'w'$	u'_r	$-u'w'$	u'_r	$-u'w'$	u'_r	$-u'w'$	u'_r	$-u'w'$	u'_r	$-u'w'$	u'_r	$-u'w'$	u'_r
24 Mar.	8.26	253	0.20	0.27			0.43	0.43							0.23	0.29		
10 Nov.	7.70	205	0.42	0.49	0.47	0.49	0.53	0.53	0.63	0.53	0.20	0.45	0.24	0.41	0.16	0.39	0.21	0.36
18 Dec.	3.54	283	0.07	0.03	0.17	−0.07	0.19	0.15	0.28	0.16	0.09	0.33	0.22	0.13	0.05	0.14	0.08	−0.03
21 Dec.	5.42	233	0.31	0.43	0.43	0.40	0.46	0.59	0.66	0.61	0.22	0.43	0.33	0.31	0.12	0.31	0.23	0.20

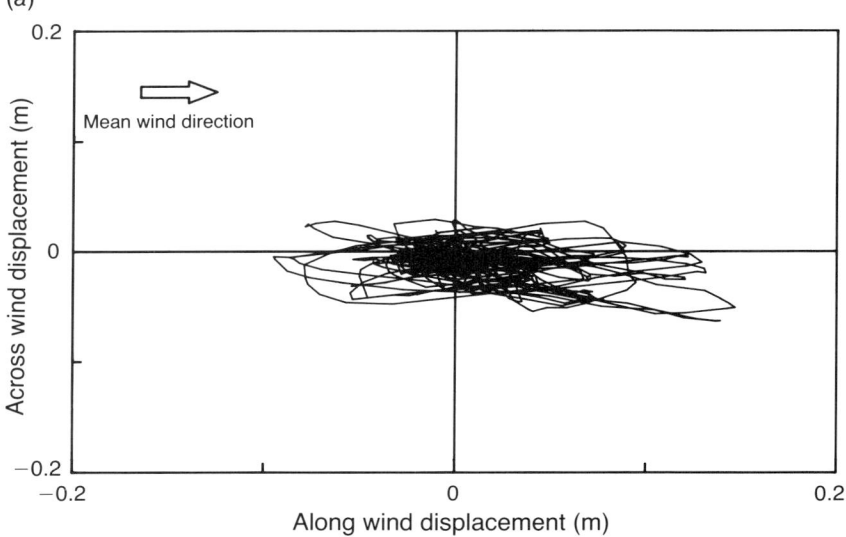

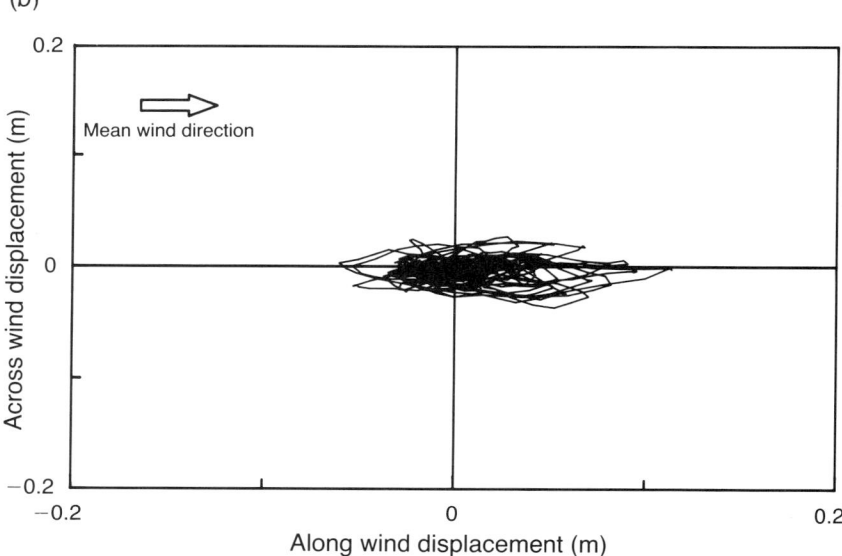

Fig. 2.1. (a) Bird's eye view of the movement of tree 2 over a 5 min period on 10 November 1988 measured at a higher of $0.4h$. (b) As for (a), but for tree 1 measured at a height of $0.38h$.

identifying the rest position of the tree. Complementary wind tunnel studies by Stacey *et al.* (1994) using dynamically correct model spruce trees mounted in a moment balance and field measurements using sensitive strain gauges (capable of resolving a strain of 2×10^{-6}) at another site (Gardiner *et al.*, unpublished) have found a consistent ratio of 10–12 between extreme and mean forces on plantation trees. Extreme displacement values, which can be measured more accurately (error ≈ 1%) than mean displacements, can be used to calculate extreme forces and a good estimate obtained of the mean forces by dividing the extreme forces by 11. On 21 December 1988 the stress measured above the canopy was constant with height with a value of 2.1 N m^{-2}. Thom (1971) suggested that the zero-plane displacement height (d) represents the level of action of the canopy drag; in other words the mean height at which the wind acts. The best estimate of d for this forest is 9.3 m (Gardiner, 1994), and since this is identical to the mean canopy height of the 470 sample trees we can assume that on average the wind force acts on trees at their mean canopy height. From the extreme displacements of trees 1, 2 and 3 (tree 4 was not used as it was much smaller than average), calculated over a 60 min period using Gumbel statistics, and using the canopy mid-points of each tree as the point of action of the wind, the average extreme force on the trees was estimated to be 94 N. Each tree occupies $10\,000/3584 = 2.8$ m^2 and, therefore, the average mean force per unit area = $94/(2.8 \times 11) = 3.0$ N m^{-2}. Although this is a crude calculation based on only three trees, it does suggest that we understand reasonably well the relationship between the stress on the canopy (mean momentum flux) and tree movement.

2.3.2 Intermittency

A description of the basic wind statistics for the Rivox Forest was presented in Gardiner (1994) and showed that the wind profile is highly sheared at canopy top and that within the canopy the streamwise velocity is positively skewed whilst the vertical velocity is negatively skewed. Raupach (1988) argued that this pointed to the dominance of intermittent large-scale coherent sweep/ejection events in canopy momentum and scalar transfer. These coherent events are a general feature of atmospheric boundary layer flow and their behaviour has been investigated over a number of different plant canopies [e.g. Finnigan, 1979 (wheat); Maitani, 1979 (wheat and rushes); Bergström & Högström, 1989 (pine forest); Cropley, 1990 (grass); Paw U *et al.*, 1992 (maize, almond orchard, walnut orchard and deciduous forest)]. The structure of these events was discussed in Gardiner (1994), where it was shown that their passage is indicated by a short-term (<5 s) increase in downward

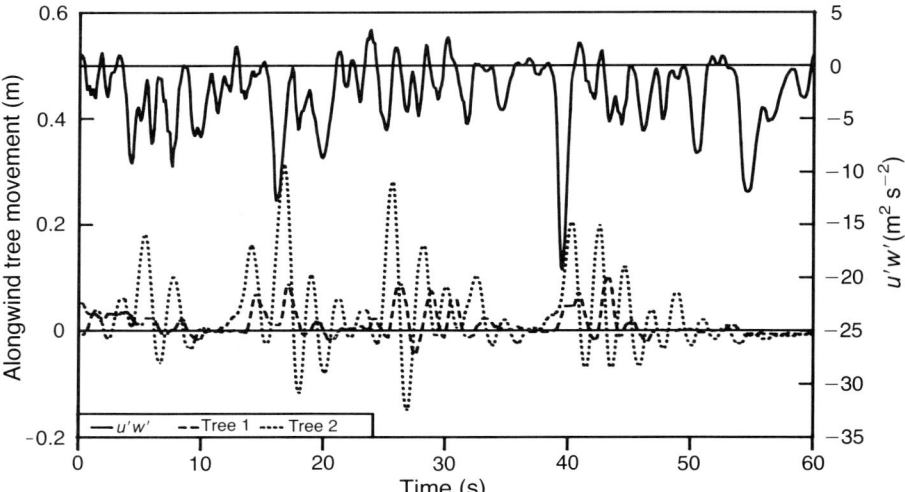

Fig. 2.2. A 60 s time series plot of the instantaneous moment flux ($u'w'$) at $z/h = 0.88$ and the alongwind movement of trees 1 and 2. (10 November 1988.)

momentum flux. This increase in momentum flux is observed throughout the depth of the canopy although the signal is attenuated and low-pass filtered by the presence of the canopy (see Gardiner, 1994, figure 5).

In Fig. 2.2 the instantaneous momentum flux in the top of the canopy over a 60 s period has been plotted with the alongwind movement of trees 1 and 2. The intermittent nature of the momentum transfer is clearly displayed as is the strong correlation between tree movement and momentum flux. The data look very similar to those obtained by Maitani (1979) with rush plants (see his figure 9). The mean wind direction on this day was in a direct line from the anemometer tower to trees 2 and 1, so that the movement of tree 2 lags behind increases in momentum flux with the movement of tree 1 delayed slightly longer.

2.3.3 Tree movement and gust passage

From a subjective analysis of Fig. 2.2, tree movement appears to be well correlated to the increase in momentum flux within the canopy associated with the passage of gusts. An objective method for detecting gust passage is required to measure how strong the correlation is. Shaw et al. (1989) explored a number of techniques used for the detection of coherent structures in laboratory flows to identify gusts over a deciduous forest. They relied particularly on the detection of temperature ramps, which was not possible in this study

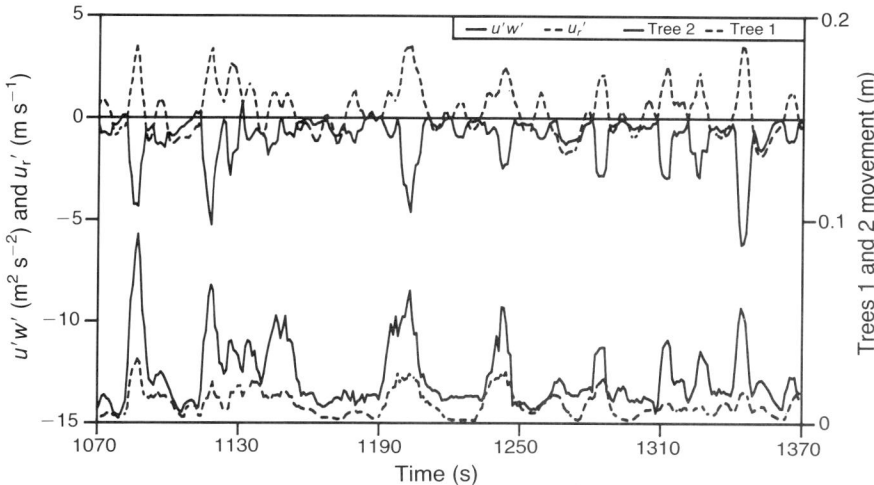

Fig. 2.3. A 300 s time series plot of the momentum flux ($u'w'$) and the variation in rotated velocity (u_r') at $z/h = 0.75$ (top lines) with the movement of trees 1 and 2 from their rest position (bottom lines). All signals have been smoothed by applying a running mean over 6 s. (21 December 1988.)

because the atmospheric stability was close to neutral and there was little temperature structure. Instead, their alternative approach of using the velocity signals for gust detection was used. An obvious velocity signal is the instantaneous momentum flux ($u'w'$), but use was also made of the variation in the rotated velocity vector u_r as defined by Shaw *et al.* (1989). This is formed by rotating the x, z axes in the vertical plane in line with the major axis of the elliptical joint probability contours of u and w at that height (see figure 4b in Gardiner, 1994). This has the effect of minimising the covariance between u and w. The advantage of this velocity over $u'w'$ is that it provides a single variable which makes a transition from negative to positive during gust passage.

Fig. 2.3 is a 300 s time series plot of $u'w'$ and u_r' within the canopy and the movement of trees 1 and 2 from their rest position. All signals have been smoothed by taking running averages over 6 s which acts like a low-pass filter and removes some of the background noise due to small-scale turbulence. The movement of both trees, particularly tree 2, corresponds extremely well to changes in $u'w'$ and u_r'. These signals can identify the passage of gusts by using the variable-interval time-averaging (VITA) technique (Blackwelder & Kaplan, 1976) adopted by Shaw *et al.* (1989). This identifies an event if the ratio of the short-term variance to the long-term (30 min) variance exceeds some threshold. Fig. 2.4 shows $u'w'$ at $z/h = 0.75$ and the

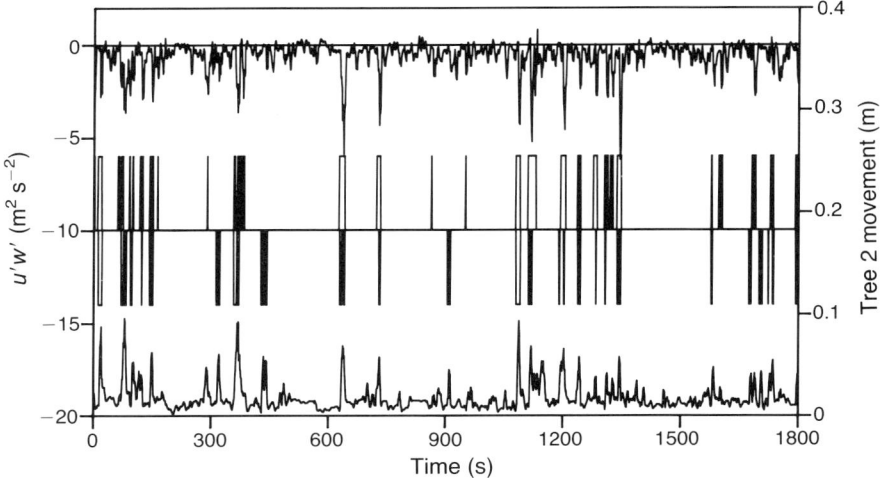

Fig. 2.4. A 30 min time series plot of momentum flux (top line), movement of tree 2 (bottom line) and the results of VITA detection on each signal (middle lines). The VITA averaging time was 11 s and the detection threshold was set at 0.75. (21 December 1988.)

movement of tree 2 over a 30 min period on 21 December 1988. In the centre of the figure the VITA detection function for both signals is displayed with the threshold set at 0.75. The short-term variance was calculated over 11 s, which we found, as did Shaw *et al.* (1989), to be the most successful interval for identifying gusts. Increases in the $u'w'$ signal and large movements of tree 2 are well identified and the two detection functions agree with each other most of the time.

To discover the strength of correlation between gust passage and tree movement linear correlation coefficients between the movement of all four trees and $-u'w'$ and u_r' at canopy top and within the top half of the canopy were calculated for each day. (Direct correlation between tree movement and variations in the streamwise windspeed, u', was found to be significantly weaker.) The values are given in Table 2.3. A number of general points emerge:

1. The closer to the mast the tree is, the higher is the correlation.
2. Within the canopy the highest correlation is with $-u'w'$; above the canopy the highest correlation is with u_r'.
3. The two trees closest to the line the wind makes through the tower (1 and 2) are best correlated to the within-canopy measurements, whilst the other two trees (3 and 4) are best correlated to the above-canopy measurements.

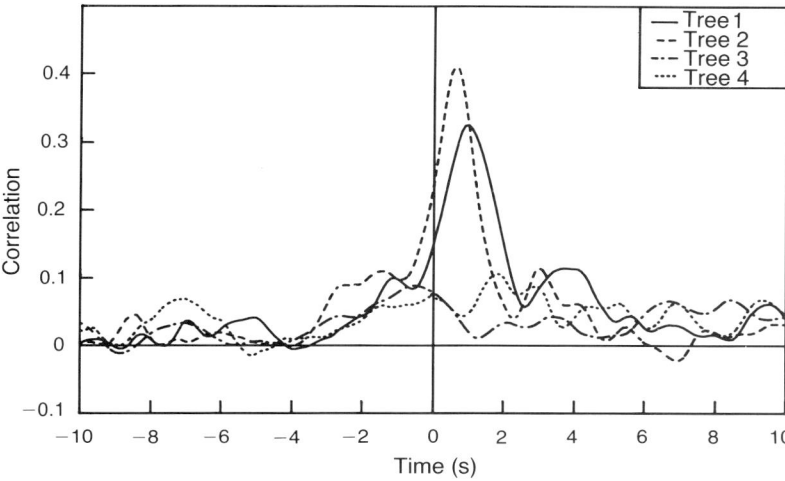

Fig. 2.5. Space–time correlation between momentum flux at $z/h = 0.88h$ and the movement of all four trees. (10 November 1988.)

4. On the day with the weakest winds (18 December 1988) the correlation is significantly reduced for all combinations.

The reduction in correlation as one moves away from the line made by the wind through the tower ($x = 0$ line) illustrates the small lateral extent of the gust events. This is similar to the observations of Finnigan & Brunet (this volume) in the wind tunnel which showed that the two-point correlation of the u wind component across the mean wind direction is reduced to zero in a distance of less than $2h$. Along the wind direction this correlation does not reduce to zero for $5h$. Analysis of the space–time correlation of the three wind components (u, v and w) and the temperature signal (T) measured on the two towers in this study (separation $= h$), and on the two towers separated by $10h$ in another similar spruce forest, found that the correlation was extremely high if the wind was blowing directly between the towers but that it rapidly diminished if the wind veered away from that line. This confirms the observation that gusts have a small lateral extent and indicates how long-lived and coherent they are.

Fig. 2.5 illustrates the space–time correlation between $-u'w'$ measured within the canopy and the movement of all four trees, calculated over a half-hour period on 10 November 1988. The correlation is defined as $r_{ab}(\tau) = a(t)b(t+\tau)/\sigma_a\sigma_b$ where $a(t)$ and $b(t)$ are the time series of $-u'w'$ and tree movement respectively and τ is the time lag. The delay in the movement of tree 2 relative to gust passage at the tower that we inferred in Fig. 2.2 is visible at +0.6 s, as is the slightly later delay of tree 1 (+1.0 s). The spacing between the mast and the trees indicates that the gusts are moving approximately 50% faster than the mean wind. The correlation of trees 3 and 4 is almost flat along the entire time axis and suggests that virtually none of the gusts passing over the measurement tower affected the movement of these two trees.

2.3.4 Spectral analysis

Spectral analysis allows us to infer which frequency components (or length scales) in the wind contribute most directly to tree movement. Spectra were calculated over half-hour periods by averaging five transforms calculated from 4096 points of 10 Hz data with cosine-bell tapering applied to the ends. The spectral densities are multiplied by the frequency and normalised by dividing by the variance.

Fig. 2.6a and b show the normalised power spectral densities (PSDs) of the streamwise windspeed and the instantaneous momentum flux ($u'w'$) at $z/h = 1.0$, 0.75 and 0.5. At canopy top ($z/h = 1.0$) the peak in the streamwise windspeed PSD occurs at a length scale equivalent to approximately 100 m whilst the peak in the momentum flux PSD occurs at a somewhat smaller length scale of around 18 m. The shape of the streamwise windspeed spectra at canopy top is as expected with a slope of $-2/3$ within the inertial subrange. Inside the canopy the spectral shape is markedly different and the slope of the inertial subrange is much steeper (-1.5 to -2). Similar observations have been made by other workers in a number of different canopies (see Gardiner, 1994, for a more complete discussion) and Kaimal & Finnigan (1994) have drawn attention to the phenomena. The presence of the trees appears to short-circuit the normal energy cascade (Baldocchi & Meyers, 1988). Either the trees are absorbing energy at low frequencies and subsequently emitting it at higher frequencies or energy is lost by work against plant-element form drag. In Gardiner (1994) it was shown that the natural resonant frequencies of trees in the forest coincide with the region occupied by the inertial subrange. Scannell (1984) discovered that the resonant frequency of the branches of spruce trees corresponds to a harmonic of the whole-tree frequency, the sub-branch resonance is a harmonic of the branch frequency, and so on down to the smallest scales on the tree. In this way energy absorbed by the whole tree can efficiently be transferred through the branches and sub-branches to the needles and re-emitted as wake turbulence at much higher frequencies. This would be an extremely effective method for rapidly dissipating energy absorbed from the wind to scales at which viscosity can act and could explain the steepness of the inertial subrange within the canopy.

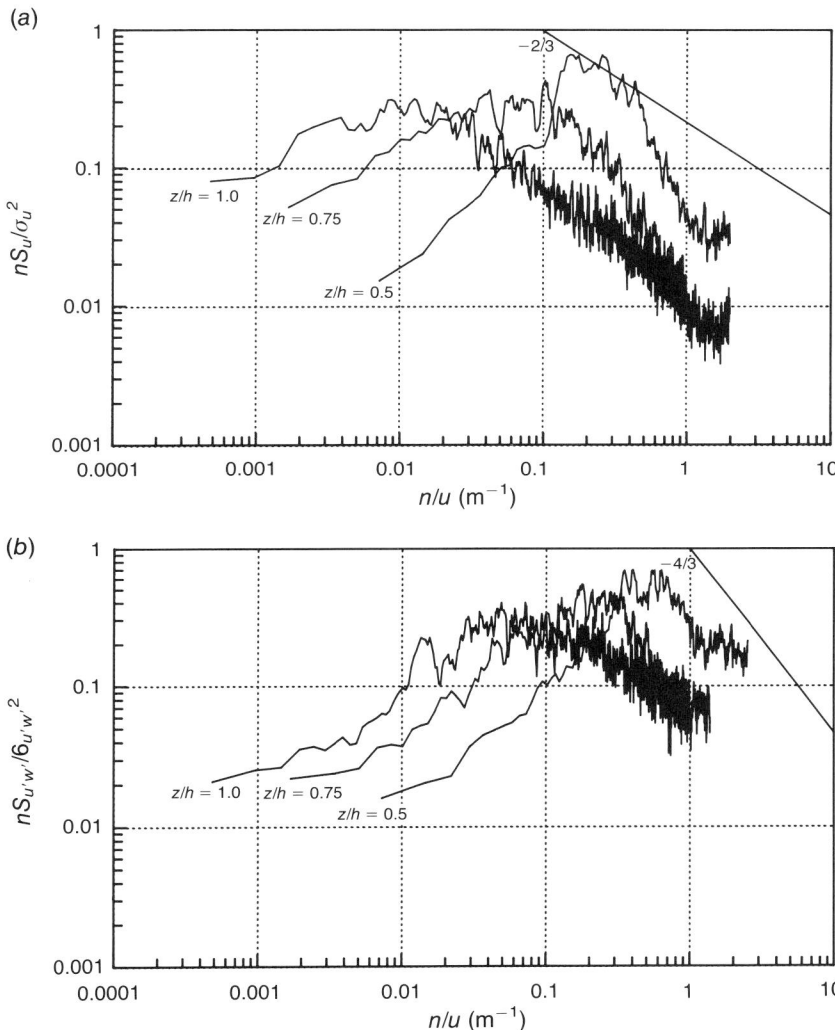

Fig. 2.6. (a) Normalised power spectra of the streamwise windspeed (u) at $z/h = 1.0$, 0.75 and 0.5 calculated over a 30 min period on 21 Dec 1988. The x-axis represents wavenumber, calculated by dividing the natural frequency by the local mean windspeed at $z/h = 1.0$ and 0.75 and by σ_u at $z/h = 0.5$ (5.05, 1.45 and 0.34 m s^{-1} respectively). (b) As for (a) but calculated for the momentum flux ($u'w'$).

The PSD for the displacement of tree 2 is shown in Fig. 2.7a with the peak corresponding to the tree resonant frequency ($f_n = 0.48$ Hz). At frequencies above the peak the rapid drop-off in power is much steeper than the −4/3 slope predicted by Inoue (1955b) and observed by Maitani (1979) in rush plants. The tree acts as a filter which transfers energy available in the

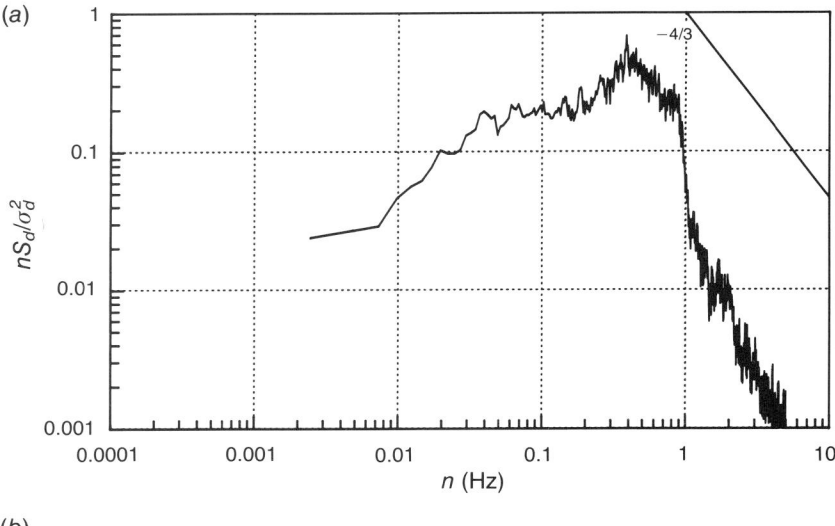

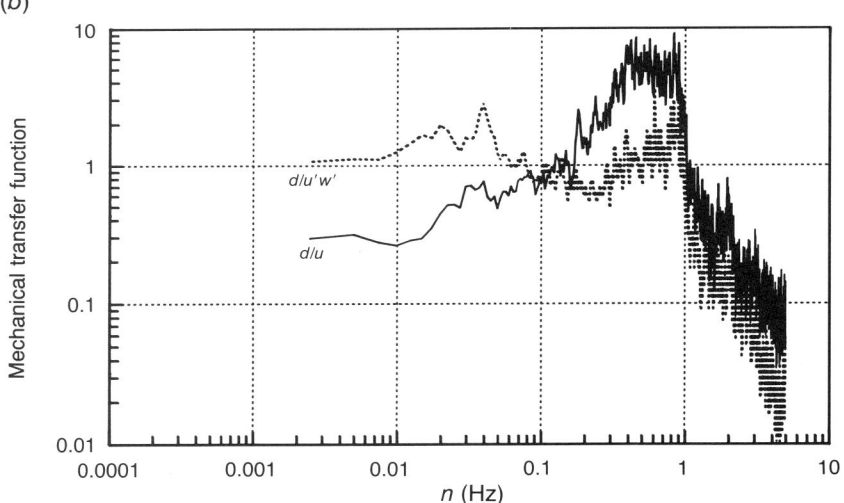

Fig. 2.7. (a) Normalised power spectra of the displacement of tree 2 obtained over the same period as Fig. 2.6. (b) Mechanical transfer functions for tree 2 obtained by dividing the displacement spectra (a) by the momentum flux spectra at $z/h = 0.75$ (Fig. 2.6b) and the streamwise windspeed spectra at $z/h = 1.0$ (Fig. 2.6a).

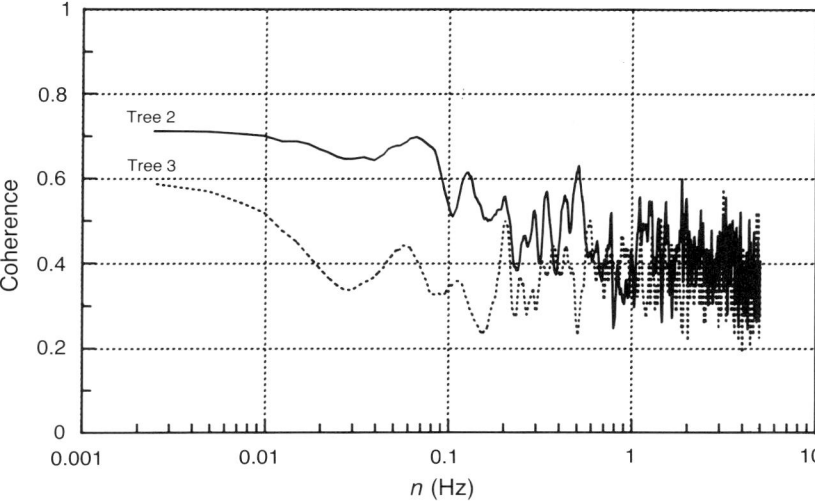

Fig. 2.8. Coherence between the momentum flux at $z/h = 0.75$ and the movement of trees 2 and 3.

wind spectrum into motion. The tree transfer function can be calculated by dividing the tree movement PSD by the wind component PSD. Transfer functions calculated using the windspeed at $z/h = 1.0$ and the momentum flux at $z/h = 0.75$ are presented in Fig. 2.7b. The transfer function calculated from the above-canopy windspeed rises to a peak at f_n and then rapidly drops off at higher frequencies. This behaviour has already been shown in Gardiner (1994) and by Stacey *et al.* (1994) to be almost identical to the transfer function of a forced damped harmonic oscillator (see, for example, Thompson, 1988, figure 3.1-3). The transfer function calculated from the momentum flux, which is more highly correlated with tree movement, is different. There is still a peak at f_n but there are equivalently high values between 0.01 and 0.06 Hz possibly corresponding to the arrival frequency of gusts in the canopy. This contrasts with the transfer functions calculated by Holbo *et al.* (1980) and Mayer (1987) which showed peaks only at tree resonance. A possible explanation is that the trees in this study were more heavily damped which reduced their ability to sway freely.

We now know that forest trees behave like lightly damped harmonic oscillators and a crucial question is whether damage results from trees resonating with the wind components at their natural frequency. To test for the presence of resonance we calculate the coherence between momentum flux and tree movement where:

$$\text{Coherence} = \sqrt{\{[Co_{xy}^2(n) + Q_{xy}^2(n)] / S_x(n) S_y(n)\}}$$

$Co_{xy}^2(n)$ is the cospectrum between x and y, $Q_{xy}^2(n)$ is the quadrature spectrum between x and y, and $S_x(n)$ is the power spectrum of x. [The cospectrum is the real part of the cross spectrum between x and y and the sum of its amplitude over all frequencies equals the covariance between x and y. The quadrature spectrum is the imaginary part of the cross spectrum (Stull, 1988).]

Coherence varies between 0 and 1 with a value close to 1 indicating a strong correlation between the two variables at that frequency regardless of any phase shift. The coherence between $-u'w'$ ($z/h = 0.75$) and the movement of trees 2 and 3 is shown in Fig. 2.8. It is clear in the plot for tree 2, which is very similar to one obtained by Maitani (1979) with rushes, that the highest coherence occurs at the lowest frequencies. Tree movement is best correlated with changes in momentum flux occurring at intervals of 10 s or longer and although there is a small peak at the resonant frequency of tree 2 (0.48 Hz) there is no indication of strong resonance. Tree movement is impulse driven, with trees responding to the intermittent arrival of gusts sweeping into the canopy. Such a deduction can also be made by re-examining Fig. 2.2. The trees react to the sudden increase in $-u'w'$, reach their greatest displacement on the first cycle and then decay back to their rest position under the influence of damping. There is no evidence for an increase in amplitude over successive cycles. Resonance of the trees with turbulent components at their natural frequency is not as important as the frequency of arrival of coherent gusts. Since the gust arrival frequency increases as the windspeed rises (Paw U et al., 1992) the gust loading can get into sympathy with tree motion leading to large and destructive displacements. However, as the windspeed increases, damping also increases (Wood, this volume) and this will help to reduce the impact of the stronger and more frequent gust loading.

The coherence of $-u'w'$ and tree 3, which was further from the mast than tree 2, is small at all but the lowest frequencies. This suggests that correlation over the distance separating the mast and tree 3 (8.5 m) exists only at long time scales and provides further evidence for the small lateral extent of the gusts moving over the forest.

2.4 Summary and conclusions

In an earlier paper (Gardiner, 1994) the intermittency of the momentum transfer ($u'w'$) into a spruce canopy was investigated and was shown to be correlated to tree movement. In this study, a more detailed investigation of the response of four spruce trees to this intermittent wind loading has been carried out.

Tree movement has been correlated to two turbulence parameters ($u'w'$ and u_r') which can be used as indicators of the passage of coherent gusts. The best correlation was obtained between tree movement and momentum flux ($u'w'$) measured within the canopy and between the turbulence parameters and those trees closest to the line made by the mean wind through the anemometer tower. The correlation drops off very rapidly with lateral distance away from this line, illustrating the small spatial scale of gusts. Gust passage was identified by applying the VITA technique (Blackwelder & Kaplan, 1976) to the momentum flux and tree movement measurements. Strong correspondence between gusts identified using momentum flux and those identified from tree movement provides further evidence for the crucial role of coherent gusts in causing forest damage.

Spectral analysis was used to identify the frequency components in the turbulence mainly responsible for tree movement. Although the trees respond well to frequencies close to their resonant frequency, their behaviour is closer to that of damped harmonic oscillators subjected to intermittent impulsive loading. The periodicity of gust arrival is more important than the energy available at tree resonance. As the wind speed increases, gusts become more frequent, enhancing the probability of gusts becoming in phase with the motion of trees and reinforcing the increasing wind loading. The trees themselves severely modify the turbulence spectra within the canopy by short-circuiting the normal energy cascade process. This is probably a design feature which allows energy absorbed by the tree to be efficiently re-emitted as high-frequency wake turbulence from needle and branch tips and which provides a method for reducing the growth of destructive resonance.

The evidence from many sources points to the crucial role of coherent gust structures in canopy momentum and scalar transport and forest wind damage. Finnigan & Brunet (this volume) have identified the importance of the inflexion in the windspeed profile at canopy top in initiating the development of coherent gusts while their size and strength is determined by the magnitude of the wind shear. Although coherent gusts occur over all plant canopies, the strength and size of the gusts will be reduced if the windspeed transition within the canopy is made more gradual. Forests with a mixture of age classes will have less severe windspeed profiles within the canopy than even-aged, uniform stands. So long as the trees are not suddenly exposed to wind conditions to which they have not mechanically adapted (Wood, this volume), such as by late thinning or large-scale felling, an irregular forest structure may have stability benefits over more uniform stands. An enhanced understanding of the relationship between forest structure and coherent gust formation is a research priority.

Acknowledgements

The analysis presented in this paper was prompted by a suggestion made by Roger Shaw; I would like to thank him and Chris Baker, who reviewed this article, and John Grace and Mike Coutts who made many useful editorial comments. I would also like to thank the British Antarctic Survey and the Macaulay Land Use Research Institute for the loan of a number of sonic anemometers; Bryce Reynard and the Mabie Outstation staff of the Forestry Commission for helping establish the experiment and the Forestry Commission Research Workshop for constructing the Leda anemometers; and Martin Hill of Holtech Associates for designing and constructing the data acquisition system and for helping to run the experiment.

References

Baldocchi, D. D. & Meyers, T. P. (1988). A spectral and lag-correlation analysis of turbulence in a deciduous forest canopy. *Boundary-Layer Meteorology*, **45**, 31–58.

Bergström, H. & Högström, U. (1989). Turbulence exchange above a pine forest. II. Organized structures. *Boundary-Layer Meteorology*, **49**, 231–63.

Blackburn, P. & Petty, J. A. (1988). Theoretical calculations of the influence of spacing on stand stability. *Forestry*, **61**, 235–44.

Blackburn, P., Petty, J. A. & Miller, K. F. (1988). An assessment of the static and dynamic factors involved in windthrow. *Forestry*, **61**, 29–43.

Blackwelder, R. F. & Kaplan, R. E. (1976). On the bursting phenomena near the wall in bounded turbulent shear flows. *Journal of Fluid Mechanics*, **162**, 89–112.

Cropley, F. (1990). Coherent vortical structures in the atmospheric boundary layer near ground. PhD Thesis, Cranfield Institute of Technology, Bedford, UK.

Finnigan, J. J. (1979). Turbulence in waving wheat. I. Mean statistics and honami. *Boundary-Layer Meteorology*, **16**, 181–211.

Fraser, A. I. & Gardiner, J. B. H. (1967). *Rooting and Stability of Sitka Spruce*. Forestry Commission Bulletin 40. HMSO, London.

Gardiner, B. A. (1989). *Mechanical Characteristics of Sitka Spruce*. Forestry Commission Occasional Paper 24. Forestry Commission, Edinburgh.

Gardiner, B. A. (1992). Mathematical modelling of the static and dynamic characteristics of plantation trees. In *Mathematical Modelling of Forest Ecosystems*, ed. J. Franke & A. Roeder, pp. 40–61. Sauerländer's Verlag, Frankfurt am Main.

Gardiner, B. A. (1994). Wind and wind forces in a plantation spruce forest. *Boundary-Layer Meteorology*, **67**, 161–86.

Grass, A. J., Stuart, R. J. & Mansour-Tehrani, M. (1991). Vortical structures and coherent motion in turbulent flow over smooth and rough boundaries. *Philosophical Transactions of the Royal Society of London, Series A*, **336**, 35–65.

Holbo, H. R., Corbett, T. C. & Horton, P. J. (1980). Aeromechanical behaviour of selected Douglas-fir. *Agricultural Meteorology*, **21**, 81–91.

Inoue, E. (1955a). Studies of phenomena of waving plants ('Honami') caused by wind. I. Mechanism and characteristics of waving plants phenomena (in Japanese). *Journal of Agricultural Meteorology (Tokyo)*, **11**, 18–22.

Inoue, E. (1955b). Studies of phenomena of waving plants ('Honami') caused by wind. II. Spectra of waving plants and plants 'vibration' (in Japanese). *Journal of Agricultural Meteorology (Tokyo)*, **11**, 87–9.

Kaimal, J. C. & Finnigan, J. J. (1994). *Atmospheric Boundary Layer Flows: Their Structure and Measurement*. Oxford University Press, New York.

Maitani, T. (1979). An observational study of wind-induced waving plants. *Boundary-Layer Meteorology*, **16**, 49–65.

Mayer, H. (1987). Wind-induced tree sways. *Trees*, **1**, 195–206.

Milne, R. (1991). Dynamics of swaying of *Picea sitchensis. Tree Physiology*, **9**, 383–99.

Oliver, H. R. & Mayhead, G. J. (1974). Wind measurements in a pine forest during a destructive gale. *Forestry*, **47**, 185–95.

Papesch, A. J. G. (1974). A simplified theoretical analysis of the factors that influence the windthrow of trees. In *Fifth Australasian Conference on Hydraulics and Fluid Mechanics*, Christchurch, New Zealand.

Paw U, K. T., Brunet, Y., Collineau, S., Shaw, R. H., Maitani, T., Qui, J. & Hipps, L. (1992). On coherent structures in turbulence above and within agricultural plant canopies. *Agricultural and Forest Meteorology*, **61**, 55–68.

Raupach, M. R. (1988). Canopy transport process. In *Flow and Transport in the Natural Environment: Advances and Application*, ed. W. L. Steffan & O. T. Denmead, pp. 95–127. Springer-Verlag, Berlin.

Savill, P. S. (1983). Silviculture in windy climates. *Forestry Abstracts*, **44**, 473–88.

Scannell, B. (1984). Quantification of the interactive motions of the atmospheric surface layer and a conifer canopy. PhD Thesis, Cranfield Institute of Technology, Bedford, UK.

Shaw, R. H., Paw U, K. T. & Gao, W. (1989). Detection of temperature ramps and flow structures at a deciduous forest site. *Agricultural and Forest Meteorology*, **47**, 123–38.

Stacey, G. R., Belcher, R. E., Wood, C. J. & Gardiner, B. A. (1994). Wind and wind forces in a model spruce forest. *Boundary-Layer Meteorology*, **69**, 311–34.

Stull, R. B. (1988). *An Introduction to Boundary Layer Meteorology*. Kluwer, Dordrecht.

Thom, A. S. (1971). Momentum absorption by vegetation. *Quarterly Journal of the Royal Meteorological Society*, **97**, 414–28.

Thompson, W. T. (1988). *Theory of Vibration with Applications*. Allen and Unwin, London.

3

Edge effects on diffusivity in the roughness layer over a forest

B. KRUIJT, W. KLAASSEN and R. W. A. HUTJES

Abstract

Above a small forest stand, close to an upwind edge, windspeeds, momentum fluxes and turbulence spectra and their dependence on distance to the edge were investigated. Measurements were made at a single tower, but with a range of fetches associated with various wind directions. The adjustment rate of the air downwind of an edge strongly depends on the variation of local turbulent diffusivity. To investigate this variation, the deviations of diffusivity from the expected equilibrium value were determined in the same way as in previous studies of flux-profile relationships in forest roughness layers. These studies generally showed diffusivity to be enhanced close to a canopy. In the present case, diffusivity was found to be suppressed down to 60%. The spectral analysis showed that the low diffusivities were associated with poorly developed horizontal low-frequency turbulence. Spectral densities were slightly enhanced at high frequencies. Vertical turbulent motions were not affected in this way. The study suggests that the adjustment of turbulence to a new surface is more rapid in vertical motions and high frequencies. The slower development of low-frequency structures, such as shear-induced local gusts, is probably the reason for the suppression of diffusivity close to the edge.

3.1 Introduction

Exchange of momentum, heat and moisture between the atmosphere and a horizontally homogeneous earth surface can be described with one-dimensional Monin–Obhukov similarity theory (Monin, 1959; Businger *et al.*, 1971). If, however, a landscape is composed of many elements differing in exchange characteristics, a situation which is quite common in North-Western Europe, the surface layer is constantly adjusting to changing con-

Edge effects on diffusivity in the roughness layer

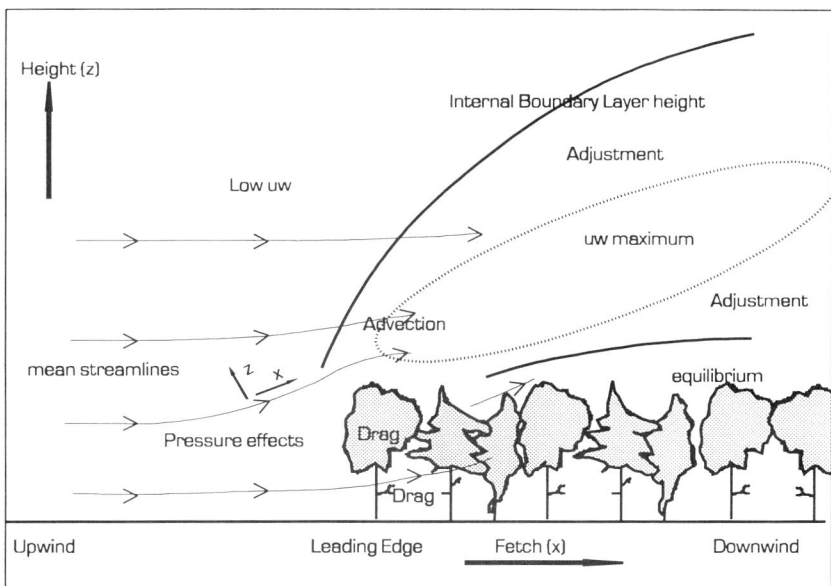

Fig. 3.1. Conceptual drawing of a transition from low crops to forest. The vertical scale is exaggerated.

ditions in its flow over the earth's surface. Adjustment at the transitions is not instantaneous but rather propagates upward and downwind gradually, forming internal boundary layers (IBL; Fig. 3.1). In these conditions standard similarity theory is not applicable. Other chapters in this volume show that the risk of windthrow is enhanced just behind a forest edge. It is likely that this is associated with the properties of turbulence close to edges. This chapter investigates the adjustment of turbulence downwind of a forest edge and its effects on momentum fluxes.

The rate of adjustment of the flow depends on the turbulent mixing rate. In local exchange, according to 'K-theory', a flux F_c of a quantity c at any point in the surface layer can be considered to depend on the local vertical (z) gradient of the transported quantity c and on the mixing rate or *turbulent diffusivity* K_c:

$$F_c = -K_c \frac{\partial c}{\partial z} = -v_* l_m \frac{\partial c}{\partial z} \qquad (3.1)$$

where v_* is a velocity scale and l_m is a length scale or *mixing length*. Over homogeneous surfaces, v_* is a constant and l_m, and hence also K_c, a function of height only (e.g. Thom, 1975). Several studies of diffusivity close to a rough (forest) surface have shown that K_c can be higher than expected from this relationship with height (Raupach, 1979; Högström et al., 1989). It is the objective of this study to compare these results with the variation and adjustment of K_c, in particular K_m, the diffusivity for momentum, downwind of a forest edge, also close to a forest canopy.

It was also investigated how eddies of different sizes advect and adjust downwind of an edge. It can be expected that large eddies adjust to a new surface at a slower rate than smaller eddies, because they persist over longer time scales. The turbulence downwind of an edge may therefore be partly in equilibrium with the surface (the small eddies) and partly still representative of the upwind surface (the large eddies). An efficient way to analyse the sizes and velocities of the eddies is to derive turbulence frequency spectra from measured time series. Højstrup (1981) studied the adjustment of spectra downwind of a roughness change, and indeed observed that spectra tend to adjust more rapidly at high than at low frequencies.

3.2 Site and instrumentation

Measurements were made from a 33 m tower in the experimental forest stand of the SHEAR project (Kruijt, 1994). The stand is of mixed deciduous composition, with an average tree height of 22 m. This small stand (about 200×300 m) is surrounded by various, mostly shorter stands in three directions, and borders agricultural land to the West. A map of the site and its surroundings is presented in Fig. 3.2.

The average roughness of the upwind surface varied with wind direction. The distance from the stand edge (fetch) also varied with wind direction, because of the eccentric location of the tower within the stand. This situation allowed the definition of two wind direction sectors within which the dependence on fetch of the flow properties could be analysed. One sector was associated with a large transition in roughness and canopy height, when the wind blew from the agricultural land (grass, Gr) onto the forest, while the other sector represented a smaller transition with winds from a low forest stand (low forest, Lf).

Turbulence data were measured with three-dimensional Gill propeller arrays and thermocouples at three heights over the canopy. The Gills were tilted to avoid stalling on the 'vertical' axes. Here only the upper-level ($z/h = 1.5$, h being canopy height) and the lowest-level ($z/h = 1.2$) arrays were used in the analysis of flux variability. Turbulence spectra were found from 10 Hz time series measured with the Gills at all three levels ($z/h = 1.2$, 1.3

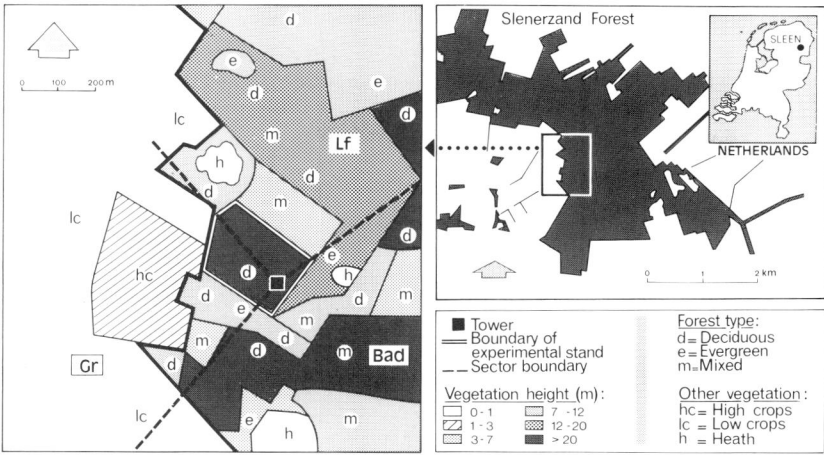

Fig. 3.2. Map of the site and its surroundings near Sleen, The Netherlands. The boundaries between the sectors Gr (grass and agricultural crops), Lf (low forest) and Bad (complex fetch) are indicated by dashed lines.

and 1.5), combined with a simultaneously measured series at 14.3 m ($z/h =$ 0.7) which was measured with a three-dimensional sonic anemometer (Applied Technologies, 'K-probe').

An intercomparison study (Kruijt, 1994) showed that fluxes measured by the different arrays of Gill propellers were comparable within 9% for individual half-hour statistics. Although average statistics of Gills and the sonic differed more strongly, turbulence spectral shapes found with the two instruments did not differ (apart from their inertial subrange).

3.3 Data analysis

At all instrument levels, half-hour averages have been determined of mean windspeed U (m s^{-1}), temperature T (K), kinematic momentum flux $\overline{uw}$ (m^2 s^{-2}) and the kinematic sensible heat flux $\overline{wt}$ (K m s^{-1}) as well as of variances of all windspeed components. The variables u, w and t denote the fluctuating components of U, W (vertical windspeed) and T. These statistics have been corrected for trends and rotated to a coordinate system aligned with the mean wind.

In equilibrium conditions, K_m, which we will denote as $K_{m,eq}$ here, can be parameterised with u_*, the square root of the kinematic momentum flux magnitude, and $k(z-d)$, with k being Von Kármán's constant, z being height, and d the zero-plane displacement. Effects of stability ζ ($=(z-d)/L$, L being the Monin–Obhukov length) are then taken into account by an empirical factor $1/\Phi_m(\zeta)$ (Dyer & Hicks, 1970; Businger et al., 1971):

$$K_{m,eq} = \frac{k(z-d)u_*}{\Phi_m(\zeta)} \tag{3.2}$$

Analogous to studies of flux-profile relationships close to rough surfaces (e.g. Raupach, 1979), the ratio γ of the 'theoretical' dimensionless momentum gradient $\Phi_m(\zeta)$ and its measured value ϕ_m has been determined to investigate the variation of diffusivity:

$$\gamma = \frac{\Phi_m(\zeta)}{\phi_m} = \frac{-\overline{uw}\,\Phi_m(\zeta)}{u_*k(z-d)\frac{\partial U}{\partial z}} = \frac{u_*\Phi_m(\zeta)}{k(z-d)\frac{\partial U}{\partial z}} \tag{3.3}$$

This ratio may also be interpreted as the ratio of the measured diffusivity to the equilibrium first-order $K_{m,eq}$ for homogeneous forest, well above the roughness layer.

The value of $\Phi_m(\zeta)$ was calculated as $(1-20\zeta)^{-\frac{1}{4}}$ for unstable conditions, and as $(1+5\zeta)$ for stable conditions. In calculating γ, momentum fluxes and profiles were interpolated between the upper and the lower instrument level. The height of d has been arbitrarily fixed at $0.7h = 15.5$ m for the foliated season, and at 10 m for the unfoliated season. Over a heterogeneous surface, d is poorly defined. Eq. (3.3) shows that γ is very sensitive to the value of d. A wrong value of the zero plane will give a systematic error in γ. As a consequence, in the analysis most attention should be given to *variation* of diffusivity, and not to a systematic deviation from equilibrium.

Spectra and cospectra have been determined from measured 10 Hz time series that had been smoothed to 3 Hz. These ranged in length from 40 to 90 min, while the spectra have been computed over 20 min windows. The spectra have been normalised with the associated total variance or flux at the lowest measurement level above the canopy (at $1.2h$), and were smoothed over logarithmically equal frequency intervals of about half a decade.

All measured spectra have been compared with the neutral Kansas spectra as specified by Kaimal et al. (1972, their equations 21a–g). With the variance spectra, the compatibility of spectral magnitudes may be imperfect, because the ratio of fluxes over variances in the measured and Kansas spectra may not have been the same. The spectral shape, however, is not affected by this difference in normalisation.

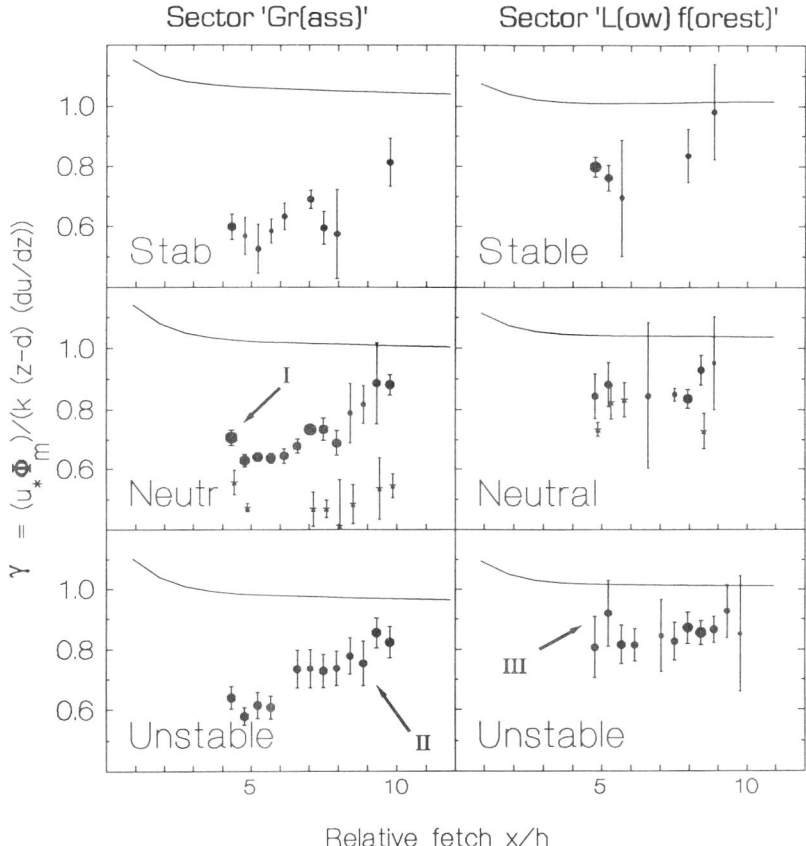

Fig. 3.3. The ratio γ of measured diffusivity and the equilibrium value as a function of fetch, stability and wind direction class. Filled circles refer to the foliated season, while stars refer to the unfoliated season. Arrows and roman numbers refer to the corresponding spectra. Error bars are 95% confidence intervals, and symbol sizes relate to the number of samples in each class (varying from 3 to 70).

3.4 Results

About 1000 half-hour statistics, measured in three consecutive years, were analysed, after having been subjected to a quality filter. The filter typically removed low windspeed and extreme stability conditions from the data set (Kruijt, 1994). Three stability classes were defined: unstable $(-0.2 < \zeta < -0.03)$, neutral $(-0.03 < \zeta < 0.03)$ and stable $(0.03 < \zeta < 0.2)$.

Fig. 3.3 shows the dependence of the turbulent diffusivity on fetch, relative to its equilibrium value: the ratio γ. The data for the large roughness transition

(Gr) show that at a relative fetch of $x/h = 5$, diffusivity is about 60% of its equilibrium value. Towards a fetch of 10 tree heights diffusivity increases, but the ratio does not reach unity within the measurement domain. Also, for the smaller transition diffusivity does not reach equilibrium, and it increases from $x/h = 5$ downwind, but here the deviations are smaller. The few data for the unfoliated season show even lower values, which also stay relatively low for the sector Lf. This may be partly because the value of d was set to 10 m for this season, which may be too low. These results are contrary to observations by Raupach (1979), Högström et al. (1989) and Bosveld (1991), who showed diffusivity to be enhanced close to homogeneous forest canopies.

Fig. 3.4 shows three sets of normalised turbulence spectra, which correspond to different conditions with respect to upwind surface, fetch and stability. The conditions have been ranked with increasing fetch and decreasing size of the roughness transition at the edge. These have been labelled (I) to (III) and are indicated in Fig. 3.3 with arrows. For each set the spectra of horizontal and vertical velocity variance (uu and ww) and for momentum and sensible heat fluxes (uw and wt) are shown. Fig. 3.4 shows, for each spectrum type, the vertical variation of the spectral shape at the measurement level within the canopy and at three levels above. For condition (I), it can be seen that the momentum flux changes with height mainly at middle frequencies, near the peak (see Kruijt, 1994). In the following, emphasis will be put on the shape of the spectra rather than on their magnitude. All spectra are compared with the neutral Kansas spectra, which have been normalised to the integral of the spectrum at $z/h = 1.2$.

Taking the relative spectral statistical confidence intervals (about 20%, near the peak) into account, we see that the spectral shape is nearly always significantly different from the Kansas spectra. In almost all the above-canopy spectra except the ww spectra the spectral peak is located at a higher frequency than in the Kansas spectra. The location of the peak of the flux cospectra does not differ much among the four conditions. At first sight this discrepancy would suggest that all values of $(z - d)$ are too large by a factor of 2 to 3, but this would imply that the zero plane is near to, or above, the canopy height. Furthermore, such a relative error in $(z - d)$ should decrease with height, but the peaks are nearly equally shifted at all heights over the canopy. In the uu spectra the peak is at higher frequencies for condition (I) than it is for other conditions. A more important trend can be observed for the uu and flux spectra: in condition (I) a large excess (co)variance relative to the Kansas spectra exists at high frequencies, while at lower frequencies a deficit exists. These differences with the equilibrium spectra are increasingly reduced in conditions (II) and (III), i.e. with decreasing assumed edge influence and increasing γ.

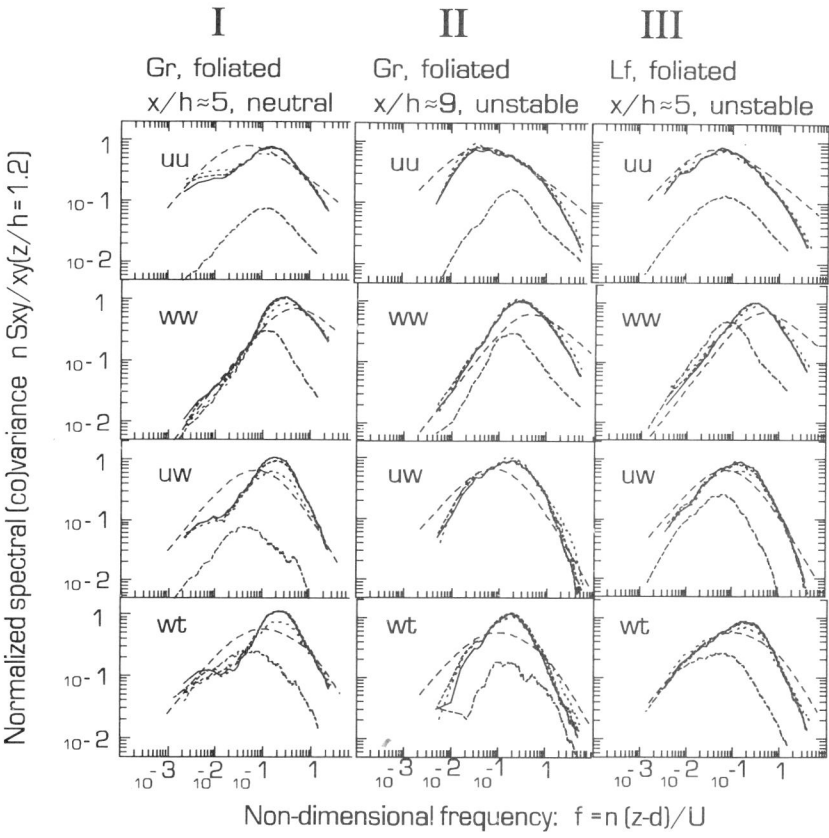

Fig. 3.4. Smoothed spectra and cospectra measured with different wind directions and stability, at several heights. The spectra are compared with the respective Neutral Kansas Spectra (Kaimal et al., 1972).

The *ww* spectra are an exception to the observed trends in spectral shape. These show little variation with fetch conditions (in condition (I) the low-frequency part is slightly depressed). Also, their peak is always located at *lower* frequencies than predicted by the Kansas spectra. The differences in peak locations compared with the Kansas spectra are significant. This shift to lower frequencies suggests that the *vertical* variance over the foliated forest is associated with larger length scales than it would be over a smooth surface.

The excesses and deficits at lower frequencies are most probably associated with the adjustment of the spectra to the new surface. It can be seen qualitatively from Fig. 3.4 that in condition (I) the (co)variances are on average located at higher frequencies, i.e. these are associated with smaller turbulent length scales than in equilibrium conditions. This average length scale increases towards condition (III), although even at the largest fetch and smallest transition the spectra are not yet fully adjusted. When the spectral shape is interpreted in this way to estimate average length scales, comparison of the average length scales with the variation of γ in Fig. 3.3 shows that these quantities are correlated. This suggests that the apparent variations in diffusivity with fetch scale with the variation in the turbulent length scale and are caused by inertia of the spectral adjustment. The similarity in shape between *uw* and *wt* spectra suggests that the diffusivity for heat is affected in the same way as for momentum.

3.5 Discussion and conclusions

The present study shows that in the flow over forest downwind of an edge with agricultural land, the diffusivity is depressed to about 60% of its equilibrium value. This low value occurs at a fetch of about five tree heights, and increases to approach the equilibrium value only beyond ten tree heights.

The drop, as well as the subsequent relaxation of turbulent diffusivity with increasing fetch, was correlated with deficits of low-frequency (large eddy) turbulence, as inferred from comparing (co)variance spectra with established smooth-terrain spectra (Kaimal *et al.*, 1972). Because the low-frequency deficits occurred in both the variance and in the covariance spectra, these large eddies carried momentum and probably contributed to the (depressed) diffusivity. The spectral analysis also showed spectral peaks to occur at smaller eddy sizes than expected, which indicated that turbulent length scales on average were smaller than in equilibrium conditions. This variation in length scales appeared to explain the variation in diffusivity quite well. These observations are remarkable since several studies of flux-profile relationships close to extensive forest surfaces have shown enhanced diffusivity instead (e.g. Raupach, 1979). The observations are, however, qualitatively consistent with predictions of 'Turbulent Kinetic Energy closure' models of the flow over roughness changes (Claussen, 1988).

Wang (1989) also studied the flow over and within a forest close to an edge. In his spectra of horizontal wind variance, with flow from the edge, he also observed peaks to be located at higher frequencies than with flow from extensive forest. He attributed the peak shift to near-canopy effects. Anderson *et al.* (1986), in a study of spectra close to an extensive deciduous forest canopy, found spectral peaks of all wind variance components to be at *low* frequencies relative to the Kansas spectra, instead of at higher frequencies. This agrees with the present results only for the *ww* spectra. If we extend the analogy of spectral shape and diffusivity found in the present study to the experiment by Anderson *et al.* (1986), their observations would predict diffusivity to be larger than the equilibrium value close to the canopy. The analogy suggests that the peak shifts observed by Anderson *et al.* result from the 'roughness effect', described by Raupach (1979) and Högström *et al.* (1989). The assumption seems justified that these 'flux-gradient anomalies' are mainly associated with average length scales that are larger than the smooth-terrain value, probably as a result of exchange with the forest stem space. The spectra presented in this study suggest that near the edge, vertical turbulent movement, measured by the *ww* spectrum, penetrates more rapidly into the canopy, while horizontal movement and fluxes have not yet established 'contact' with the full depth of the forest.

This study leads to questions about the nature of the suppressed low-frequency turbulence. It may be associated with shear-induced gusts and sweeps, developing to a new equilibrium with the rough canopy, where the dominant length scale is of order h (Finnigan & Brunet, this volume). If this hypothesis is true, the observed rate of development of these structures is not consistent with observations by Raupach (cited by Finnigan & Brunet, this volume), who found a rapid development of such gusts within a few tree heights, associated with enhanced windthrow at that fetch. An explanation might be that Raupach's observation resulted from a very local effect associated with the tall vertical edge itself. Further downwind of the edge the interaction with the horizontal canopy surface possibly develops at a slower rate, and this might represent the phenomena that were described here.

Acknowledgements

The stimulating support of Professor Arthur Veen, University of Groningen, is greatly acknowledged. Backing by Jan van den Burg, Henk de Groot and others was indispensable. The study has benefited from discussions with: Martin Claussen, Hamburg; Chadran Kaimal, Boulder, Colorado; David Miller, Storrs, Connecticut; and Barry Gardiner, Roslin, Scotland. Paul Jarvis, Edinburgh, generously provided the opportunity to complete this work. The study was partly sponsored by the Netherlands Organisation for Scientific Research.

References

Anderson, D. E., Verma, S. B., Clement, R. J., Baldocchi, D. D. & Matt, D. R. (1986). Turbulence spectra of CO_2, water vapour, temperature and velocity over a deciduous forest. *Agricultural and Forest Meteorology*, **38**, 81–99.

Bosveld, F. C. (1991). *Turbulence Exchange Coefficients over a Douglas Fir Forest*. Scientific Report KNMI WR-91-02. De Bilt, The Netherlands.

Businger, J. A., Wyngaard, J. C., Izumi, Y. & Bradley, E. F. (1971). Flux-profile relationships in the atmospheric surface layer. *Journal of Atmospheric Sciences*, **28**, 181–9.

Claussen, M. (1988). Models of eddy viscosity for numerical simulation of horizontally inhomogeneous, neutral surface-layer flow. *Boundary-Layer Meteorology*, **42**, 337–69.

Dyer, A. J. & Hicks, B. B. (1970). Flux-gradient relationships in the constant flux layer. *Quarterly Journal of the Royal Meteorological Society*, **96**, 715–21.

Högström, U., Bergström, H., Smedman, A., Halldin, S. & Lindroth, A. (1989). Turbulent exchange above a pine forest. I. Fluxes and gradients. *Boundary-Layer Meteorology*, **49**, 197–217.

Højstrup, J. (1981). A simple model for the adjustment of velocity spectra in unstable conditions downstream of an abrupt change in roughness and heat flux. *Boundary-Layer Meteorology*, **21**, 341–56.

Kaimal, J. C., Wyngaard, J. C., Izumi, Y. & Coté, O. R. (1972). Spectral characteristics of surface-layer turbulence. *Quarterly Journal of the Royal Meteorological Society*, **98**, 563–89.

Kruijt, B. (1994). Turbulence over forest downwind of an edge. PhD Thesis, University of Groningen, The Netherlands.

Monin, A. S. (1959). Smoke propagation in the surface layer of the atmosphere. *Advances in Geophysics*, **6**, 331–43.

Raupach, M. R. (1979). Anomalies in flux-gradient relationships over forest. *Boundary-Layer Meteorology*, **16**, 467–86.

Thom, A. S. (1975). Momentum, mass and heat exchange of plant communities. In *Vegetation and the Atmosphere*, ed. J. L. Monteith, vol. 2. Academic Press, London.

Wang, Y. (1989). *Turbulence Structure, Momentum and Heat Transport in the Edge of Broad Leaf Tree Stands*. PhD Thesis, University of Connecticut, Storrs, CT. Research in Microclimate at the Forest–Urban Interface, Progress Report 15.

4

A wind tunnel study of turbulent airflow in forest clearcuts

J. M. CHEN, T. A. BLACK, M. D. NOVAK
and R. S. ADAMS

Abstract

A wind tunnel was used to investigate the wind regime in forest clearcuts. The model forest had a uniform height of 15 cm and a density of 500 stems m^{-2} to simulate an Engelmann spruce stand of height 15 m and density 500 stems ha^{-1}. It was found that the mean windspeed near the surface at 22 tree heights downwind of the forest edge was about 65% of the potential value obtained when no forest was upwind. From the measured mean windspeed, turbulence intensity and integral scale in the vertical direction, the vertical turbulent diffusivity was calculated. According to the spatial distribution of the vertical diffusivity at one-fifth of the tree height, a quiet zone, a wake zone and a readjustment zone were identified downwind of the forest edge.

4.1 Introduction

Despite many environmental concerns, clearcutting still remains the major timber harvesting method in British Columbia, Canada, and in many regions of the world. Forest regeneration in clearcuts is an issue of great public and scientific concern. Tree seedlings in clearcuts often experience temperature extremes, increased water loss due to exposure to the wind, and winter desiccation as a result of snow removal by wind. Blowdown along the edges of clearcuts is also a significant problem in British Columbia and little information is available to assess the effects of clearcut size and shape on wind regime. The purpose of this chapter is to provide information on the wind regime in forest openings. This information is essential to evaluate the effect of clearcut size on seedling microclimate and blowdown potential of clearcut edges.

There have been a limited number of studies of wind regimes associated with forest edges or openings (Geiger, 1966; Raynor, 1971; Bergen, 1975;

Gash, 1986). In a comprehensive review on wind regimes near forest edges, McNaughton (1989) concluded that our understanding of the complicated wind field near forest edges remained anecdotal. Recently, a first-order closure model was developed for calculating the mean flow field in a forest clearcut (Li et al., 1990; Miller et al., 1991). Little information is available on turbulence, which plays an important role in altering the microclimate after a forested area is cleared. There is a lack of systematic measurements to address issues such as how windspeed in a clearcut recovers with distance from the forest edge.

This shortage of quantitative information is probably due to the difficulty of conducting field experiments. We have chosen to use the wind tunnel technique for this study because it would otherwise be impossible to carry out a systematic study in the field with presently available instrumentation. Measurements obtained indoors can be reliably scaled up to assess practical problems outdoors. The objectives of this chapter are (1) to provide detailed information on the spatial distribution of the mean windspeed associated with forest edges and (2) to quantify the effects of turbulence on the vertical diffusion of scalars from the clearcut surface.

4.2 Principles in using a scale model

There are two basic requirements in using a scale model of a forest stand. First, the model must be placed in a fully turbulent atmospheric environment like the outdoor conditions. The setup in the wind tunnel shown in Fig. 4.1 serves this purpose. The spires and roughness elements were used to shorten the distance at which a fully turbulent boundary layer is obtained and to generate the required turbulence scales. Several criteria

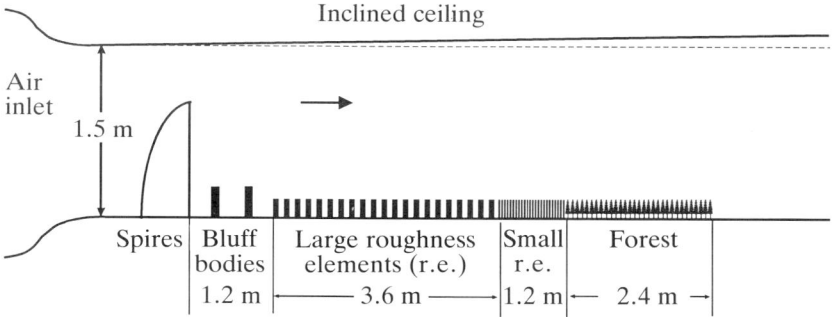

Fig. 4.1. Experimental setup in an open-circuit blowing-type wind tunnel.

are used to determine whether a boundary layer is fully turbulent. The main criteria are (Macdonald *et al.*, 1988): (1) the mean windspeed profile, (2) the turbulence intensity profile, (3) the Reynolds stress profile and (4) the turbulence length scale and spectra. The minimum requirement is that there is a logarithmic velocity profile close to the forest canopy, from which the friction velocity (u_*) and the canopy roughness length (z_0) can be determined. The roughness Reynolds number, $u_* z_0/\nu$, where ν is the kinematic viscosity, must be larger than a certain value to ensure that the surface is aerodynamically rough and the boundary layer becomes fully turbulent within a short distance from the leading edge. Generally the value should be larger than 5.0 (Finnigan *et al.*, 1990) but sometimes 2.5 is considered adequate (Isyumov & Tanaka 1979; Kim *et al.*, 1990; see also Wood, this volume).

The second requirement is dynamic similarity, which includes both geometrical and kinematic similarity (Fox & McDonald, 1985). Geometrical similarity requires that, for example, the dimensions of a scale model are reduced by a constant factor. Kinematic similarity requires forces on parts of a scale model be related to those at full scale by the same ratio. When geometrical similarity is satisfied, kinematic similarity is obtained under the condition that the Reynolds number based on the characteristic dimension of the model and the velocity in its vicinity is larger than a certain value so that effects of the viscous sublayer on the model surface can be ignored. This value was found to be of the order of 10 000 (Britter, 1982). Snyder (1981) and Kim *et al.*, (1990) found values of 11 000 and 5500, respectively, for flow around buildings. To satisfy the condition that the Reynolds number be greater than 10 000, the airflow speed near the tops of model trees should be greater than 1.1 m s^{-1} and the speed of the free stream more than 3.5 m s^{-1} when the tree height is 0.15 m. All measurements shown here were made with a free stream speed of about 8.7 m s^{-1}.

4.3 Experimental methods

4.3.1 Wind tunnel

The wind tunnel, which belongs to the Department of Mechanical Engineering, University of British Columbia, has a working section 2.4 m wide, 1.5 m high and 25 m long (Fig. 4.1). Counihan spires (Counihan, 1969) were placed upstream to generate large-scale turbulence and to produce a boundary layer with a thickness approximately equal to the height of the spires. Boards 250 mm in height and 50 mm in thickness, serving as bluff bodies, were placed across the tunnel at 0.5 and 1 m behind the spires to reduce the windspeed at the lower heights and to generate turbulence of intermediate scales. The large roughness elements extending from 1.2 m to 6.0 m downstream of the spires were used to create small-scale turbulence appropriate for the downstream model forest. These elements were 150 mm in height, 50 mm in width (facing the wind) and 25 mm in thickness (in the wind direction). They were arranged in a diamond grid pattern having a density of 21 m^{-2}. Following these large roughness elements was a section occupied by small roughness elements which were used to simulate the model forest so that the requirement for the length of the model forest could be reduced. These elements were 150 mm in height, 25 mm in width and 13 mm in thickness. They were fixed in a regular pattern with a density of 120 m^{-2}. The model forest was placed immediately behind the small roughness elements. It occupied a streamwise length of 3.2 m. The forest almost extended from wall to wall except for a 25 mm open space on each side to allow air passage close to the tunnel walls so that any edge effects were removed.

4.3.2 Model forest

The model forest, which was built to simulate an Engelmann spruce stand, consisted of individual flexible model trees inserted into drilled holes in two 1.2×2.4 m sheets of plywood with a density of 500 stems m^{-2} – which is equivalent to 500 stems ha^{-1} when scaled up by 100 times. The trees had a uniform height of 15 cm and were 'planted' in an interlock regular grid. We simplified the structure of the forest stand so as to study the basic flow features behind the forest edge.

The foliage of the model trees consisted of plastic strips (1 mm in width and 30 mm in length) distributed radially and supported at the centre by two 0.9 mm interwound steel wires. These wires formed the stem of each tree. The foliage strips had an upward angle of 30°. The upper third of the foliage was trimmed into a conical shape and the lower fifth of the tree trunk was cut bare (no foliage). The leaf area index of the model stand was 6.3. The model trees were made using materials from a Christmas tree manufacturer (Barcana, Granby, Quebec, Canada). Each tree had a natural oscillation frequency of 2.2 Hz. The frequency was obtained by recording the signal from a quantum sensor (Model LI-190SB, LI-COR, Lincoln, Nebraska, USA) with a response time of 10 μs placed in the shadow of a model tree in the sun after the model was displaced and allowed to oscillate freely. The frequency was close to the desired frequency of 4 Hz, which was calculated using the

Strouhal number with a natural oscillation frequency of 0.5 Hz for trees at full scale (Stacey *et al.*, 1991). The models fluttered slightly at a wind speed of 9 m s^{-1}, and it is felt that the aeroelasticity of the models had only a very small effect on the flow in the opening behind the forest stand.

4.3.3 Instrumentation

The three components of the instantaneous wind vector were measured using a hot-film constant temperature anemometer (CTA; model 56C01, Dantec Electronik, Tonsbakken 16-18, DK-2740 Skovlunde, Denmark). The probe (model 55R91) has three orthogonal sensing quartz fibres (coated with a nickel film) which are 1.25 mm in length and 75 µm in diameter. The sensing fibres are all at the same angle of 54.6° to the axis of the probe, which was aligned with the mean wind direction. The probe has a frequency response up to 17 kHz. Voltage signals from the CTA were acquired using a 486-based PC with an A/D board (multi-functional carrier PCI-200981C-1, Intelligent Instrumentation, Tucson, Arizona, USA). Longitudinal, lateral and vertical velocities were obtained after digitisation through two coordinate transforms. It was found using our software that the velocities were not sensitive to the 1–2° alignment or rotation errors which may occur with our probe mounting device. The system was calibrated against a system consisting of a Pitot tube and manometer in the wind tunnel over a windspeed range from 0.7 to 16 m s^{-1}. The probe was mounted on a traversing system which had electrical control in the lateral and vertical directions and manual control in the longitudinal direction. The electrical control had a resolution of 0.25 mm in probe displacement. During the experiments, signals from the probe were sampled at a location for 16.4 s at 500 Hz, creating 3×8196 data points. Mean windspeeds, turbulence intensities and variance spectra were calculated using these 16.4 s time series.

There are several limitations of the hot-film probe. First, the acceptance angle for wind direction relative to the probe axis is smaller than 90°, i.e. the probe can detect wind direction only within a hemisphere. In locations where either the mean flow reverses from the free stream direction or the probability exists of backward fluctuations in wind direction, the measurements of both the mean velocity and the turbulence statistics would be in error. Second, in highly turbulent flows, significant errors may result from the assumption of small perturbation in calculating the cooling power of turbulent flow (Hinze, 1959). However, errors are very small for flows with turbulence intensity less than 30% (Legg *et al.*, 1984; Chen *et al.*, 1988).

4.4 Results and discussions

4.4.1 Flow above the model forest

The flow above the model forest indicates the suitability of the wind tunnel setup for simulating airflow in the field. The velocity profile measured in the middle of the model forest is similar to that measured immediately upwind of the forest (Fig. 4.2). This suggests that the roughness elements in front of the forest stand created a boundary layer similar to that over the forest. The logarithmic velocity distribution with height, or the inertial sublayer, is largely confined to the middle portion of the profiles. At the heights close to the forest, the effect of the wake from the tree tops is significant, i.e. a roughness sublayer distorts the otherwise logarithmic profile. At the upper levels, the flow may not have fully adjusted to the underlying surface because of the short distance of the artificial environment. The curvature displaced in Fig. 4.2 is typical of artificial canopies (Raupach *et al.*, 1986). To increase the height at which the flow is fully adjusted to the underlying surface, the section with the roughness elements would have to be extended, but increas-

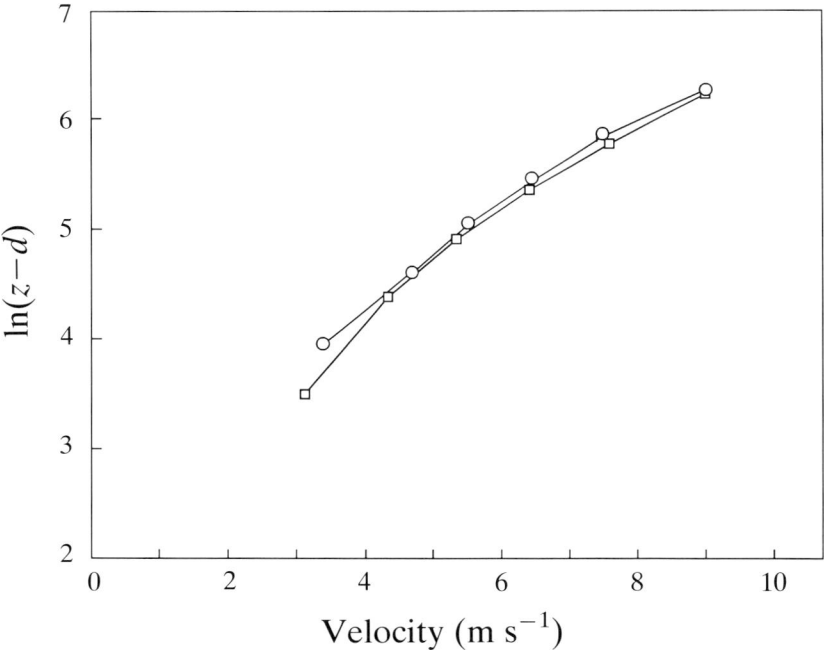

Fig. 4.2. Longitudinal velocity profiles measured immediately upwind (circles; $d = 100$ mm) and at the middle (squares; $d = 115$ mm) of the model forest.

ing the distance from the spires would result in a substantial loss of turbulent kinetic energy of large eddies created by the spires. We felt that the fully adjusted boundary-layer thickness is appropriate for the purpose of this study, i.e. the flow at more than 2 tree heights above the canopy would not affect significantly the flow in the opening because the shear stress near the top of the canopy is the main source of turbulent kinetic energy responsible for the mixing near the ground surface in the opening.

Displacement heights d were calculated to be 115 mm and 100 mm for the profiles above the forest and roughness elements, respectively, shown in Fig. 4.2. These values were obtained using the following equation based on the theorem of centre of pressure (Thom, 1971; Jackson, 1981):

$$d = \frac{\int_0^h zF(z)\mathrm{d}z}{\int_0^h F(z)\mathrm{d}z} \qquad (4.1)$$

where h is the height of the canopy and $F(z)$ is the drag force exerted on the airflow by the foliage at height z calculated from

$$F(z) = c_\mathrm{d} a(z)[u(z)]^2 \qquad (4.2)$$

where c_d is the drag coefficient, taken to be a constant, $a(z)$ is the distribution of foliage area density determined by the structure and density of the trees, and $u(z)$ is the mean windspeed at height z.

The boundary-layer development can also be seen in profiles of the vertical momentum flux $\overline{(u'w')}$ with height (Fig. 4.3). Below $z/h = 3$, the values for both the profiles corresponding to Fig. 4.2 were reasonably constant, beginning to decrease at $z/h = 3$. This agrees with Fig. 4.2, which shows that the flow at the highest measurement position was not fully adjusted to the underlying surface. The scatter of the data points in Fig. 4.3 is due to the finite length of the time series used in the calculation. In a separate experiment, the scatter of a similar distribution was very much reduced when the length of record was doubled.

Fig. 4.4 shows samples of the turbulence kinetic energy spectra for the flow at $z = 275$ mm over the middle of the forest. Each spectrum was obtained from a time series of 8196 data points at 500 Hz using a fast Fourier transform routine (Brigham, 1988) and a computer routine to reduce the output data to 16 numbers per order of magnitude increment in the frequency. No signal and data conditioning was imposed except for detrending, which was hardly necessary because the time trend was almost non-existent in the controlled environment. There is a small inertial sub-range in each of the spectra from which the slope of $-5/3$ can be

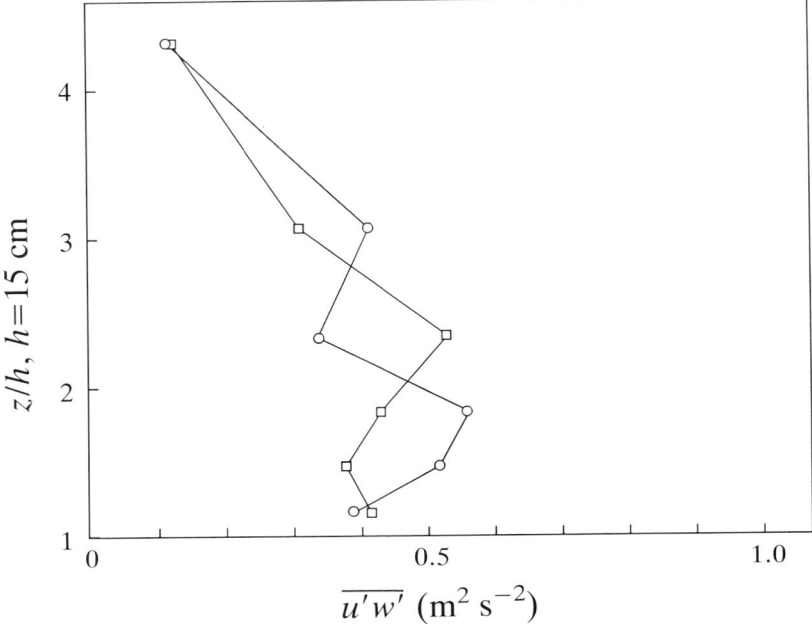

Fig. 4.3. Distribution of the vertical momentum flux ($\overline{u'w'}$) with height upwind (circles) and at the middle (squares) of the model forest. An approximate constant flux layer extends to about 3 tree heights from the forest floor.

identified. At the high-frequency end of the spectra there are a few inconsistent points showing the effect of aliasing. An important feature of the spectra plotted in this way is that the distribution becomes asymptotically flat as the frequency decreases and the values of $E_u(n)$, $E_v(n)$ and $E_w(n)$ at $n = 0$ can be reliably determined. These values are used to calculate the integral scales, L_u, L_v and L_w in the longitudinal, lateral and vertical directions, respectively (Hinze, 1959), as follows:

$$L_u = \frac{uE_u(0)}{4\overline{u'^2}} \tag{4.3}$$

$$L_v = \frac{uE_v(0)}{4\overline{v'^2}} \tag{4.4}$$

$$L_w = \frac{uE_w(0)}{4\overline{w'^2}} \tag{4.5}$$

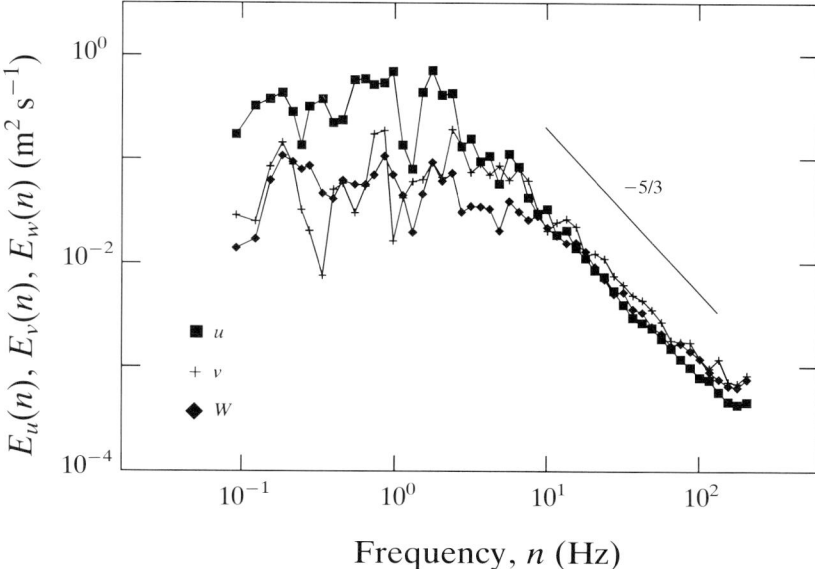

Fig. 4.4. Spectra of velocity variances measured at $z = 275$ mm at the middle of the model forest, where the subscripts u, v and w denote the longitudinal, lateral and vertical components of the total variance.

where $\overline{u'^2}$, $\overline{v'^2}$, and $\overline{w'^2}$ are the variances in the longitudinal, lateral and vertical directions, respectively.

Fig. 4.5 shows the profiles of longitudinal, lateral and vertical integral scales with height above the middle of the model forest. The relative error is smaller than ±30%. The linear increase of the scales with height is expected for flow in the surface layer over an extensive surface under neutral stratification conditions (Stull, 1988). Wilson et al. (1982) found the vertical integral scale over a corn field increased linearly with height. Seginer et al. (1976) and Raupach et al. (1986) obtained linear distributions of the longitudinal and vertical integral scales up to one canopy height above their model canopies in a wind tunnel. The linear distribution extended to higher levels in the present study because of the use of the spires and roughness elements. Evidence for the inconsistency at the top level is also found in the distributions of the scales, especially in the case of the w component. The longitudinal scale was about twice the lateral and vertical scales, indicating that the turbulent flow is nearly locally isotropic (Hinze, 1959). It is interesting to note that the ratio of the vertical integral scale to the height above the tree tops (i.e. $z - h$) is about 0.4, the value of the von Karman constant.

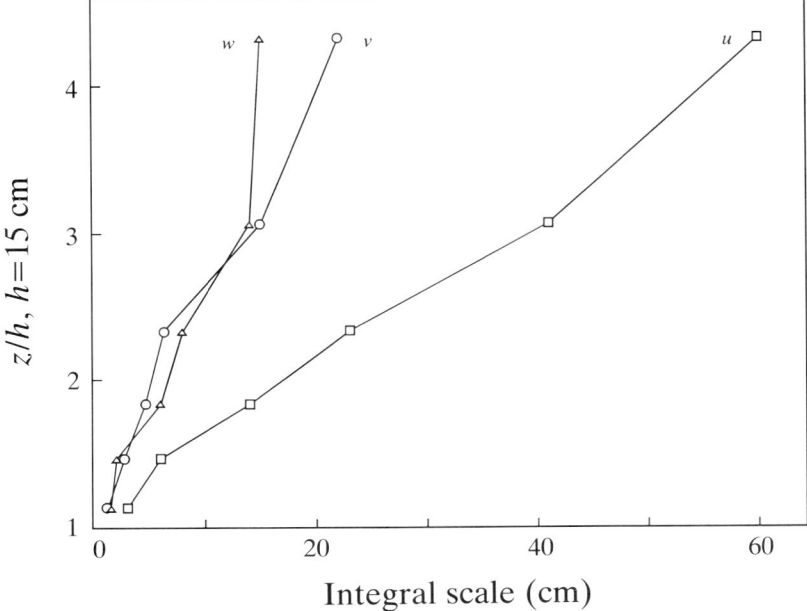

Fig. 4.5. Height distributions of the longitudinal (u), lateral (v) and vertical (w) integral scales above the middle of the model forest.

4.4.2 Flow in the opening

4.4.2.1 Mean flow field

Figure 4.6 shows profiles of the mean velocity vector measured at several distances downwind of the forest edge. The lateral component of the velocity was always zero because the flow was perpendicular to the edge (i.e. a two-dimensional case). In the lower portion of the profiles near the forest edge, the flow was essentially horizontal and the speed was only about 5% of the free stream value. There was a bulge near the bottom of the profile where foliage was not present. This slight velocity increase indicates the effect of the efficient vertical momentum transport through the forest rather than the effect of the finite distance from the leading edge. (When we blocked the wind at the lower heights with a horizontal board 4 cm in height placed on the floor, we found that the secondary wind maxima recovered within 1.5 tree heights from the board.) As the distance from the edge increased, the mean velocity below the tree height increased rapidly. The increase in the velocity was largely due to the vertical transport of momentum by the mean

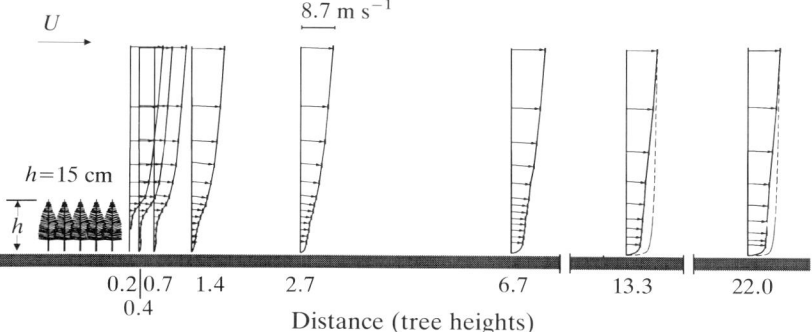

Fig. 4.6. Velocity vectors measured in an opening with an infinite downwind distance. The free stream velocity was 8.7 m s^{-1}.

flow, which can be seen from the downward vertical component in the velocity vectors. The mean vertical velocity was largest at distances of 1–4 tree heights from the forest edge and gradually became small as the distance increased. At a distance of about 22 tree heights, the effect of the upwind forest can still be identified from the following features: (1) the mean velocity was considerably smaller than that (potential value) measured with no upstream obstacles, which is indicated by the broken lines; (2) the largest velocity deficit occurred below 2 tree heights, indicating that there was still significant momentum transport to the lower portion of the profiles. The roughness length of the tunnel floor was found to be 0.005 mm, equivalent to 0.5 mm at the full scale, which is within the range of 0.21–1.5 mm for bare soil surfaces.

4.4.2.2 Vertical diffusivity

The following formula derived by Taylor (1921) is frequently used to calculate vertical diffusivities in homogeneous turbulent flows (Thomson, 1987; Degrazia & Moraes 1992):

$$D_w = \sigma_w^2 \, T_\mathrm{L} \qquad (4.6)$$

where σ_w is the standard deviation of the vertical velocity, equal to $\sqrt{\overline{w'^2}}$ and T_L is the Lagrangian time scale. This formula is adopted in this study to calculate the diffusivity at various locations in the opening. We realise that the turbulence immediately behind the forest edge is considerably non-homogeneous, and significant errors may result from using this formula. For non-homogeneous turbulence, it is suggested that T_L be replaced with a turbulence time scale τ involving an energy dissipation rate ϵ (Thomson, 1987;

Weil, 1990). However, ϵ can not be measured directly and the error involved in the parameterisation of ϵ may be larger than that due to non-homogeneity. We define the vertical turbulence intensity I_w as follows:

$$I_w = \frac{\sigma_w}{u} \quad (4.7)$$

The Lagrangian time scale is related to the Eulerian time scale, T_E, by

$$T_L = \beta T_E \quad (4.8)$$

where β varies from 2 to 6 depending on the atmospheric stabilty, and under neutral conditions, $\beta = 4$ (Panofsky & Dutton 1984). T_E can be calculated from:

$$T_E = L_w/u \quad (4.9)$$

From Eqs. (4.6)–(4.9), we have

$$D_w = 4I_w^2 \, u \, L_w \quad (4.10)$$

where all the variables are obtained from turbulence measurements.

Figure 4.7 shows the distribution of the mean windspeed, vertical turbulence intensity and integral scale for three heights at various distances from the forest edge. The velocity at $z/h = 1/5$ increased almost linearly with the distance (x) within 4–6 tree heights from the edge (Fig. 4.7a). The increase became smaller as distance increased. At $x/h = 13$, velocities below tree height were about 50% of their respective potential values obtained when the upwind forest was removed, and at $x/h = 22$, they were 65% of these values. At $z/h = 1.0$, the turbulence intensity decreased almost exponentially with the distance from the forest edge (Fig. 4.7b). The decrease in the intensity may be caused by (1) the increase in the mean windspeed, (2) the reduction in the production of turbulent kinetic enrgy (TKE) by the shear stress near the tree tops as the distance from the tree edge increased, and (3) the decay of TKE with the distance, especially the TKE of small eddies. At the lower heights, there were initial increases in the intensity up to about 3 tree heights downwind of the edge, followed by gradual decreases for the same reason given above. The initial increases suggest that at these distances the vertical transport of TKE was more efficient than that of momentum by the mean flow. The vertical integral scale increased rapidly at all heights with distance downwind for $x/h < 6$ (Fig. 4.7c). This rapid increase indicates that the turbulent wake from the tree tops spread out gradually to lower levels and became the dominant source of TKE at the lower levels after only 2–3 tree heights from the edge. The integral scale was small at the average tree height immediately behind the edge, suggesting that the shear layer there was

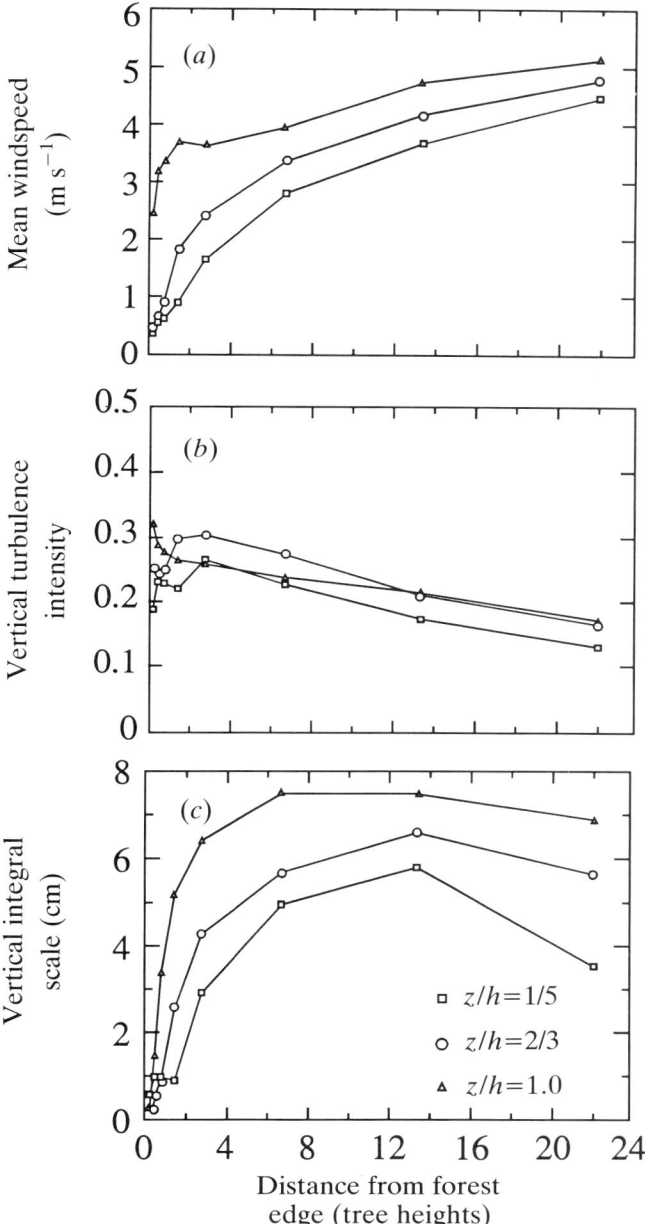

Fig. 4.7. Variations of the mean windspeed (*a*), vertical turbulence intensity (*b*) and vertical integral scale (*c*) at three heights in a forest opening with an infinite downwind distance.

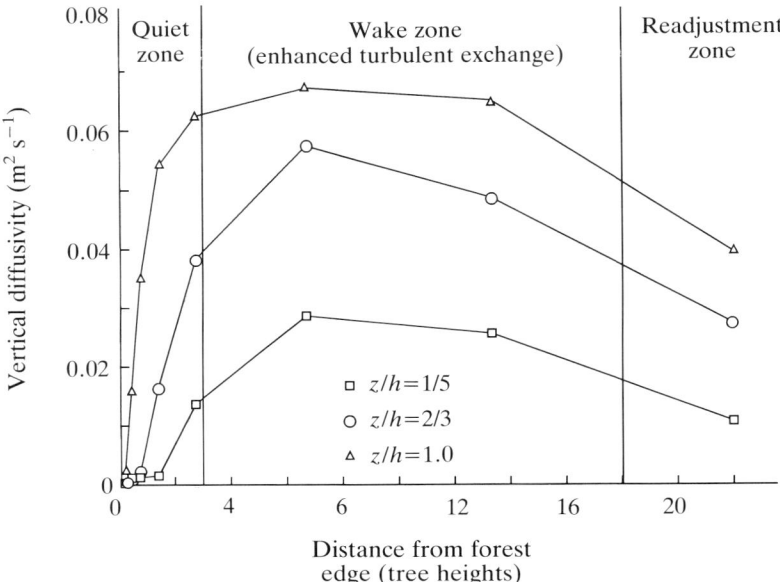

Fig. 4.8. Distributions of the vertical diffusivity at three heights in a forest opening with an infinite downwind distance and three zones divided according to the curve at $z/h = 1/5$.

quite thin. The energetic eddies in the wake gradually increased in size as the shear layer spread out vertically with increasing distance from the edge. The vertical integral scale would have reached an asymptotic value of about half the tree height at the higher levels had no significant decay in TKE occurred. In reality, the decay resulted in a decrease of the scale, especially at the lower levels. Flow at the lowest height was the first to adjust to the opening surface and hence exhibited a more pronounced decrease in scale than flow at the higher levels.

The values in Figs. 4.7a–c were used in Eq. (4.10) to calculate the vertical diffusivities shown in Fig. 4.8. The overall distributions of the diffusivity at these three heights were similar but the magnitudes differed. At distances close to the edge, the values at all heights were very small mainly because the integral scale was small. The increase in the diffusivity was fastest near the tree tops and slowest near the floor. In order to assess the rate of heat and mass exchange between the surface and the atmosphere, the diffusivity at the lowest height was used to identify the different zones indicated in the figure. The *quiet zone* is the region close to the upwind edge where the vertical diffusivity is less than half the maximum value found at about $x/h =$

6 in the wake zone. The quiet zone is found to be within 3 tree heights from the edge. In the *wake zone*, which extends from 3 to 18 tree heights downwind, the wake from the upwind tree tops greatly enhances the vertical exchange. The *readjustment zone* is a region far downstream from the edge where the diffusivity decreases to a value smaller than half the maximum value and the flow has considerably adjusted to the conditions of the underlying surface. The readjustment zone is found at distances larger than 18 tree heights. It is expected that for surfaces rougher than the wind tunnel floor, the readjustment zone may start nearer to the forest edge.

These results have many implications in silvicultural practice. For example, seedlings in the wake and readjustment zones will be more vulnerable to wind damage. Since the vertical diffusivity is large in the wake zone, daytime seedling temperatures will probably be lower than in the quiet zone and evaporation rates in wet conditions will probably be higher. However, until the interactions between the vertical diffusivity as determined by the wind field and the interception of solar irradiance by the clearcut surface (Chen *et al.*, 1993) are better understood, the effect on seedling microclimate of reducing clearcut size is unclear. Further research may find an optimum size in which windspeeds are reduced but penetration of solar radiation is adequate to permit soil warming and seedling photosynthesis.

4.5 Conclusions

Flow regimes over a forested surface were adequately simulated using individual flexible model trees and additional artificial roughness elements in a wind tunnel. Quiet, wake and readjustment zones in a clearcut downwind from the forest edge were delineated according to the spatial distribution of the vertical diffusivity calculated from turbulence measurements. The size of the quiet zone decreased with height above the ground surface, and at one-fifth of the tree height it extended to 3 tree heights from the forest edge. Further research is required to determine how these zones would affect overall seedling microclimate.

Acknowledgements

This research was funded by the BC Science Council and the BC Ministry of Forests. Dr Ian Gartshore in the Department of Mechanical Engineering, UBC, provided valuable advice and support during the experiment. We appreciate the constructive discussions with Dr Alberto Orchansky and Rick

Ketler. Jun Luo and Robert Sagar helped in constructing the model forest, and Jingxian Liu and Jun Luo assisted in the data analysis.

References

Bergen, J. D. (1975). Air movement in a forest clearing indicated by smoke drift. *Agricultural Meteorology*, **15**, 165–79.

Brigham, E. O. (1988). *The Fast Fourier Transform and its Applications*. Prentice-Hall, Englewood Cliffs, NJ.

Britter, R. (1982). Modeling flow over complex terrain and implications for determining the extent of adjacent terrain to be modeled. In *Wind Tunnel Modeling for Civil Engineering Applications*, ed. T. A. Reinhold. Cambridge University Press, Cambridge.

Chen, J. M., Ibbetson, A. & Milford, J. R. (1988). Boundary layer resistances of artificial leaves in turbulent air. I. Leaves parallel to the mean flow. *Boundary-Layer Meteorology*, **45**, 371–86.

Chen, J. M., Black, T. A., Price, D. T. & Carter, R. E. (1993). Model for calculating photosynthetic photon flux densities in forest openings on slopes. *Journal of Applied Meteorology*, **32**, 1656–65.

Counihan, J. (1969). An improved method of simulating an atmospheric boundary layer in a wind tunnel. *Atmospheric Environment*, **3**, 197–214.

Degrazia, G. A. & Moraes, O. L. L. (1992). A model for eddy diffusivity in a stable boundary layer. *Boundary-Layer Meteorology*, **58**, 205–14.

Finnigan, J. J., Raupach, M. R., Bradley, E. F. & Aldis, G. K. (1990). A wind tunnel study of turbulent flow over a two-dimensional ridge. *Boundary-Layer Meteorology*, **50**, 277–317.

Fox, R. W. & McDonald, A. T. (1985). *Introduction to Fluid Mechanics*, 3rd edn. Wiley, New York.

Gash, J. H. C. (1986). Observations of turbulence downwind of a forest–heath interface. *Boundary-Layer Meteorology*, **36**, 227–37.

Geiger, R. (1966). *The Climate near the Ground*. Harvard University Press, Cambridge, MA.

Hinze, J. O. (1959). *Turbulence*. McGraw-Hill, New York.

Isyumov, N. & Tanaka, H. (1979). Wind tunnel modeling of stack gas dispersion: difficulties and approximations. *Wind Engineering*, **5**, 989–1001.

Jackson, P. S. (1981). On the displacement height in the logarithmic velocity profile. *Journal of Fluid Mechanics*, **111**, 15–25.

Kim, S., Brandt, H. & White, B. R. (1990). An experimental study of two-dimensional atmospheric gas dispersion near two objects. *Boundary-Layer Meteorology*, **52**, 1–16.

Legg, B. J., Coppin, P. A. & Raupach, M. R. (1984). A three-hot-wire anemometer for measuring two velocity components in high intensity turbulent boundary layers. *Journal of Physics E*, **17**, 970–6.

Li, Z. J., Lin, J. D. & Miller, D. R. (1990). Air flow over and through a forest edge: steady-state numerical simulation. *Boundary-Layer Meteorology*, **51**, 179–97.

MacDonald, P. A., Kwok, K. C. S. & Holmes, J. D. (1988). Wind loads on circular storage bins, silos and tanks. *Journal of Wind Engineering and Industrial Aerodynamics*, **31**, 165–88.

McNaughton, K. G. (1989). Micrometeorology of shelter belts and forest edges. *Philosophical Transactions of the Royal Society of London, Series B*, **324**, 351–68.

Miller, D. R., Lin, J. D. & Lu, Z. N. (1991). Air flow across an alpine forest clearing: a model and field measurements. *Agricultural and Forest Meteorology*, **56**, 209–25.

Panofsky, H. A. & Dutton, J. A. (1984). *Atmospheric Turbulence: Models and Methods for Engineering Applications*. Wiley, New York.

Raupach, M. R., Coppin, P. A. & Legg, B. J. (1986). Experiment on scalar dispersion within a model plant canopy. I. The turbulence structure. *Boundary-Layer Meteorology*, **35**, 21–52.

Raynor, G. S. (1971). Wind and temperature structure in a coniferous forest and a contiguous field. *Forest Science*, **17**, 351–63.

Seginer, I., Mulhearn, P. J., Bradley, E. F. & Finnigan, J. J. (1976). Turbulent flow in a model plant canopy. *Boundary-Layer Meteorology*, **10**, 423–53.

Snyder, W. H. (1981). *Guidelines for Fluid Modelling of Atmospheric Diffusion*. Environmental Protection Agency Report EPA-600/8-009.

Stacey, G. R., Wood, C. J. & Belcher, R. E. (1991). Wind flow and forces in model conifer plantations. In *Proceedings of the Twentieth Conference on Agricultural and Forest Meteorology*, September 1991, Salt Lake City, USA.

Stull, R. B. (1988). *An Introduction to Boundary Layer Meteorology*. Kluwer, Boston.

Taylor, G. I. (1921). Diffusion by continuous movements. *Proceedings of the London Mathematical Society*, **20**, 196.

Thom, A. S. (1971). Momentum absorption by vegetation. *Quarterly Journal of the Royal Meteorological Society*, **97**, 414–28.

Thomson, D. J. (1987). Criteria for the selection of stochastic models of particle trajectories in turbulent flows. *Journal of Fluid Mechanics*, **180**, 529–56.

Weil, J. C. (1990). A diagnosis of the asymmetry in top-down and bottom-up diffusion using a Lagrangian stochastic model. *Journal of Atmospheric Science*, **47**, 501.

Wilson, J. D., Ward, D. P., Thurtell, G. W. & Kidd, G. E. (1982). Statistics of atmospheric turbulence within and above a corn canopy. *Boundary-Layer Meteorology*, **24**, 495–519.

5

Testing of a linear airflow model for flow over complex terrain and subject to stable, structured stratification

D. W. F. INGLIS, T. W. CHOULARTON,
I. M. F. STROMBERG, B. A. GARDINER and M. HILL

Abstract

The model (Flowstar) is tested against new airflow data from a region of complex terrain and roughness cover (Kintyre, south-west Scotland). The dataset consists of mean flow data at 10 m above the surface at up to 14 sites supported by airborne measurements of the upstream temperature profile and the flow aloft. The structured, stable stratification is shown to exert a strong influence on the mean flow close to the surface. Far from the surface Flowstar reproduces the main features of the stratification-dependent flow. However, additional Flowstar stratification options may improve the model in cases where the upstream profile is complicated. At 10 m above the surface and in near-neutral conditions the reliability of Flowstar predictions of mean flow are comparable with those from similar models applied to simpler terrain shapes. Errors are observed in wake regions where the model overpredicts windspeeds. This is due to a previously known limitation of linear models. Flowstar modelling of the effects of stable stratification on the mean flow at 10 m is encouraging, although some dependence of accuracy on stratification type is noted. Recommendations for the use of Flowstar in practical applications are made.

5.1 Introduction

In many areas of the United Kingdom the wind climate is the limiting factor to forest growth (Cannell & Coutts, 1988). A model of wind flow over complex terrain (i.e. terrain that does not approximate to any simple shape) would help foresters to plan planting patterns to minimise damage. Such a model must be adaptable enough to represent a wide range of realistic conditions and accurate enough to allow confident forest planning. It is the aim of this chapter to use a new data set (from Kintyre, 1989 and 1990) to test an airflow model (Flowstar) in these two regards.

Since the 1970s theoretical modelling of airflow over hills has followed two main avenues: numerical solutions to the full non-linear equations of motion, and analytical solutions to the linearised equations. Non-linear models avoid many of the restrictions of linear models but the numerical solutions are far more expensive in terms of computer time and so resolution is lost. Linear models are more efficient in terms of computing time owing to their analytical solution. This is achieved through greater approximation and so they are less generally applicable. However, experimental studies have shown that linear models can provide useful results for a wide range of simple terrain shapes (e.g. Mason & Sykes, 1979; Bradley, 1980; Salmon et al., 1988).

Flowstar is one of a family of linear airflow models derived from the two-layer model of Jackson & Hunt (1975). Far from the surface Jackson & Hunt define the outer layer where the velocity perturbations are small, and non-turbulent flow can be assumed for the perturbations. Close to the surface is the inner region where it is assumed that the turbulence is in local equilibrium so a mixing length approximation can be used to close the set of equations. There followed a series of developments to the Jackson & Hunt model which aimed to extend its applicability and refine its mathematics, many of which are reviewed by Finnigan (1988). In the inner region Flowstar makes use of one of these solutions developed by Hunt et al. (1988a).

Far from the ground buoyancy forces dominate and the stratification can have a large effect on the flow pattern over hills and windspeeds in the inner region (Carruthers & Choularton, 1982). Despite this, most models which aim to predict the wind field close to the surface have considered neutral stratification only. Flowstar makes use of an upper layer solution from Hunt et al. (1988b) which takes account of different stratification types such as constant stable stratification and elevated inversions.

All the major full-scale experiments designed to test airflow models have been sited on simple terrain shapes such as two-dimensional ridges or axisymmetric, isolated hills. These are reviewed by Taylor et al. (1987) and Finnigan (1988). Linear theory was found to provide a reasonably accurate estimate of the mean wind field on upwind slopes and hilltops. Errors were confined to overpredictions of windspeed in the wake of hills in the inner region. This is due to non-linear components of the flow becoming important in these regions. Non-linear models perform more accurately in these areas although this is at the cost of poorer spatial resolution.

Several models, including Flowstar, now have the capability to consider grids of arbitrary terrain and so are not limited to simple terrain shapes. This feature and the ability to consider the effects of stratified layers on the flow makes Flowstar a potentially useful tool in practical situations such as forest planning. The Kintyre dataset provides an opportunity to test the model in these two regards.

5.2 Kintyre airflow experiments 1989 and 1990

The Kintyre airflow experiments of February/March 1989 and February/March 1990 were carried out through collaboration between UMIST and the Forestry Authority. The ground-based observations were made using largely Forestry Authority land and resources. Additional airborne measurements were carried out by the UMIST instrumented aircraft.

The object was to observe the effects of complex terrain, roughness and stratification changes on mean and turbulent airflow characteristics. This work differs from many previous studies in that the choice of terrain is not dictated by a desire to approximate to a two-dimensional or axisymmetric hill shape.

5.2.1 Choice of experimental region

To collect data in such a way as to maximise their utility in terms of testing the performance of airflow models requires a careful choice of site. The Kintyre peninsula runs approximately north to south on the south-west coast of Argyll, Scotland. A contour plot of the chosen piece of terrain is shown in Fig. 5.1.

The lowest contour represents the coastline and the south-west corner of the square defines the origin (National grid point NR 630 208). The peninsula is about 12 km wide at the experimental site and the highest terrain lies in the centre of the peninsula forming a series of peaks of slightly less than 400 m in altitude. The nearest coastlines are northern Ireland to the south-west (35 km), Islay to the north-west (40 km) and Arran to the east (10 km). Directly to the west the fetch is unobstructed as far as the Atlantic Ocean. Thus, the airflow at Kintyre from any westerly direction can be expected to be unaffected by other terrain.

A major advantage is the proximity of the Machrihanish air base and Meteorological Office station (shown as site 14 in Fig. 5.1). By agreement with the RAF, Machrihanish was made available as a base for the UMIST aircraft, allowing it to react rapidly to changes in the weather. Detailed local weather forecasts and analysis were also available from the Meteorological Office station. The Building Research Establishment wind direction analysis (Cook, 1985) for this site shows the prevailing wind direction to be between west and south-west.

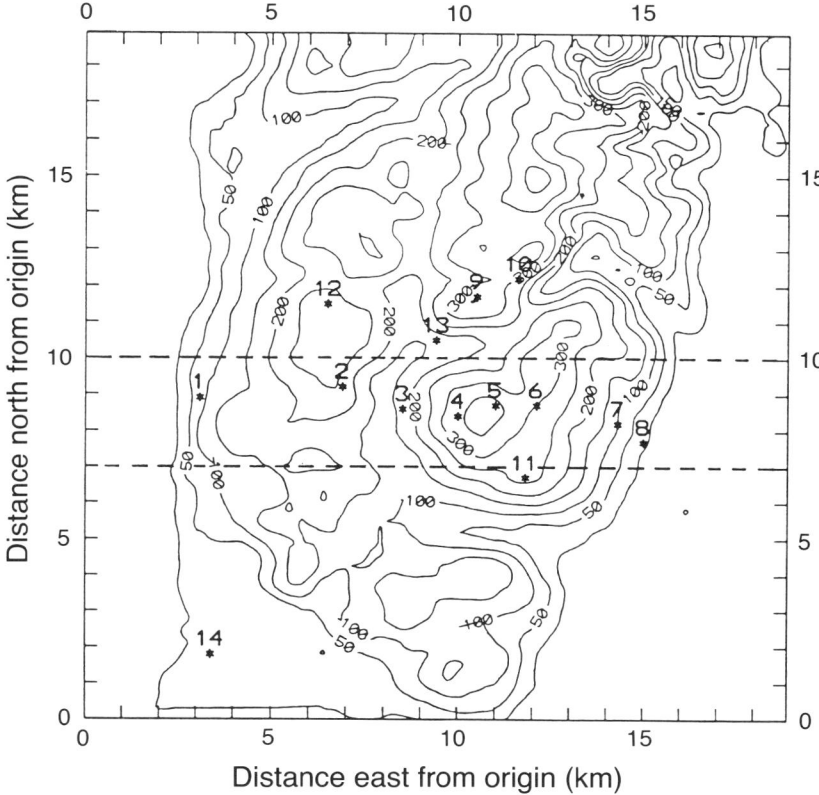

Fig. 5.1. Contour plot of the Kintyre terrain. The contour interval is 50 m and the lowest contour represents the coastline. The positions of the mean flow monitoring sites are indicated by an asterisk and numbered. The dashed lines show the band inside which aircraft data used to construct potential temperature isopleths were accepted.

5.2.2 The upstream site

It is important to monitor the conditions at an upstream site so that the data from any other site can be normalised against unperturbed flow. A knowledge of the unperturbed conditions is also necessary to initialise an airflow model.

Unfortunately, it was not possible to place a site over the ocean to the west of the peninsula. The best alternative is to use the data recorded at the Machrihanish air base where windspeed and direction were recorded as hourly means with an uncertainty of ± 0.5 m s^{-1}. There is no significant terrain between the coast and site 14, so that the unperturbed flow conditions can be recovered by using a simple adjustment for the surface roughness change.

This method assumes that the hills to the north and south have no effect on the airflow at Machrihanish (site 14). This is supported by Cook (1985), who successfully predicted direction-dependent windspeeds at Machrihanish using an assessment of surface roughness changes only. This increases confidence in the use of Machrihanish to estimate upwind velocity profiles. However, modelling results from Inglis (1992) indicate that under certain stratification conditions a simple roughness change method may underestimate the upstream windspeed for westerly flow over Kintyre. For this reason the following iteration method is used to calculate the upstream wind profile for all the cases presented here. The logarithmic upstream mean wind profile, $U_0(z)$, is defined by the upstream friction velocity, u_*, and roughness length, z_0.

Over the ocean the roughness length is dependent on the windspeed and is calculated following Large & Pond (1981). An initial estimate of $U_0(z = 10) = 1.1 \times U_{14}(z = 10)$ is made, where $U_{14}(10)$ is the observed mean windspeed at site 14, and the corresponding upstream profile calculated. The model is run and the values of u_* and z_0 are iterated until the observed value of $U_{14}(10)$ is reproduced by Flowstar.

5.2.3 Instrumentation

Fig. 5.1 shows the position of the mean flow monitoring sites. In 1989 sites 1, 2, 3, 5, 6, 8, 9, 11 and 14 were used and in 1990 all sites were used. The equipment at each site comprised a Vector A100R anemometer and a Vector W200G wind vane mounted vertically on a light-weight mast 10 m above local ground level. The masts were aligned to magnetic north (±2°). Data were logged continuously on battery-powered Holtech loggers.

For windspeed the number of pulses in a 2 s period is recorded and converted to a velocity. Two hundred and twenty-five such values (covering 7.5 min) are stored in a histogram of bin width 2 m s^{-1} ranging from 0 m s^{-1} to 50 m s^{-1} in 1989 and 0 m s^{-1} to 96 m s^{-1} in 1990. A similar histogram of bin width 22.5° is constructed for the wind direction. The mean, variance maximum and minimum values can be calculated from these histograms. The uncertainty in a mean windspeed or direction arising from the logging method is small due to the turbulent nature of the flow.

Cup anemometers suffer from overspeeding in turbulent flows. Estimates of this error were made using the expression given by Busch & Kristensen (1976). The errors range from a maximum of 13% for cases were the most turbulent flows are observed (lee side sites) to a minimum of 1% for hilltop sites. These values agree well with direct measurements of overspeeding error for four short periods in 1990 when turbulence measurements were made alongside cup anemometer measurements (Inglis, 1992). All cup anemometer data presented here have been corrected for these errors.

All data from the mean flow monitoring sites are presented here in terms of the fractional speed-up ratio, $\Delta S = [U_n(10) - U_0(10)]/U_0(10)$, where $U_n(10)$ refers to the mean windspeed for 1 h at site n. However, the hourly means from a mean flow site (constructed from eight histograms) do not necessarily coincide with the hourly means from site 14. This gives rise to offset errors in ΔS. The number of minutes offset is known for each hourly mean chosen at each site. Hence, the magnitude of these errors can be estimated if the mean windspeed is assumed to be changing linearly with time. This effect and the uncertainty in an hourly mean speed from site 14 are the two significant uncertainties in observed values of ΔS.

The UMIST aircraft is a Cessna 182, extensively modified for research purposes. It is used for airborne studies into a wide range of research topics; the equipment is configured specifically for each study. For the Kintyre 1989 and 1990 projects, mean wind field, temperature, dewpoint, horizontal position and altitude were measured.

Horizontal position was measured using a DME/VOR radial system; this normally has a resolution of 1° in angle and 0.1 km in distance. Horizontal wind vectors were calculated using a Litton system. This calculates windspeeds from the difference between true airspeed and speed over the ground. The vector typically has a resolution of ±1 m s^{-1} and ±5°. Temperature and dewpoint are measured to a resolution of 0.1 K and altitude is measured by means of a pressure transducer to a resolution of 10 m.

The vertical motion of the atmosphere as it flows over the terrain is important in determining the flow type. However, the most problematic aircraft measurement is vertical windspeed. At the time it could not be measured directly and had to be inferred from changes in airspeed. To do this it must be assumed that the aircraft flies with constant power. There are two main problems with this method. Firstly the airspeed of the aircraft responds slowly to changes in vertical windspeed. Secondly, the assumption of constant power is often not justified due to aircraft manoeuvres to avoid cloud, etc. For these reasons it was felt that a more reliable picture of the flow far from the surface could be gained by utilising a potential temperature contouring technique as used by Stromberg *et al.* (1989).

A parcel of cloud-free air displaced vertically as it crosses the terrain will heat and cool adiabatically, thus maintaining its potential temperature. So long as the time scale of the turbulence is much longer than the transit time over the terrain, streamlines and potential temperature isopleths are coincident. To ensure a valid comparison between data and model, data used to construct the isopleths were only accepted if the aircraft was in cloud-free air. Also, all measurements were made well within the outer layer where the turbulence time scale criterion is met.

Data from four flights were used, out of a total of 10 for both projects. Ideally we wish to compare the flow on days when the wind direction is exactly the same so that the effect on the airflow of different stratification types can be isolated from the effect of altered terrain shape (relative to wind direction). Also, in all the flights the majority of passes over the peninsula were in a west/east orientation. This means that only in an approximately westerly or easterly airstream can a potential temperature contour be tracked through a continuous band of data points across the peninsula. The four flights selected all have approximately westerly wind directions. The aircraft was also used to make soundings of the atmosphere upstream of the peninsula to measure the stratification profile.

Potential temperature contours were drawn from 5 s averages (roughly corresponding to a distance of 250 m over the ground) recorded during aircraft passes over the summit of Sgreadan Hill (site 5). Only data points which fell inside a band 3 km wide over the summit were accepted. This band is indicated by the dashed lines in Fig. 5.1.

5.3 The airflow model (Flowstar)

A very brief description of Flowstar is given here. The full mathematical solution can be found in Hunt *et al.* (1988*a*, *b*). Carruthers & Hunt (1990) summarise the main points of the model, as does Inglis (1992).

Flowstar is a solution of the linearised equations of turbulent fluid flow applied to arbitrary three-dimensional fields of terrain and roughness. The atmosphere is split into layers which are defined by their height above the surface. In each of these layers a different approximation is appropriate and allows an analytical solution for the wind, pressure and turbulence fields.

In the outer layer shear stresses are small and buoyancy forces caused by the upstream stratification profile are important. The equations collapse to the Scorer equation (Scorer, 1953) and can be solved for a number of upstream stratification profiles. Fig. 5.2*a–d* shows stratification options currently available in Flowstar. Fig. 5.2*c* shows a scenario observed frequently in the United Kingdom. The lowest (neutral) layer is capped by an inversion which is itself capped by a slightly stable layer. The pressure field calculated at the base of the outer layer can be substituted into the equations for the inner region, thereby transmitting the effect of the stratification.

The inversion and the stable layer above it can be described in terms of their Froude numbers: F_i for the inversion and F_L for the stable layer. These are defined as

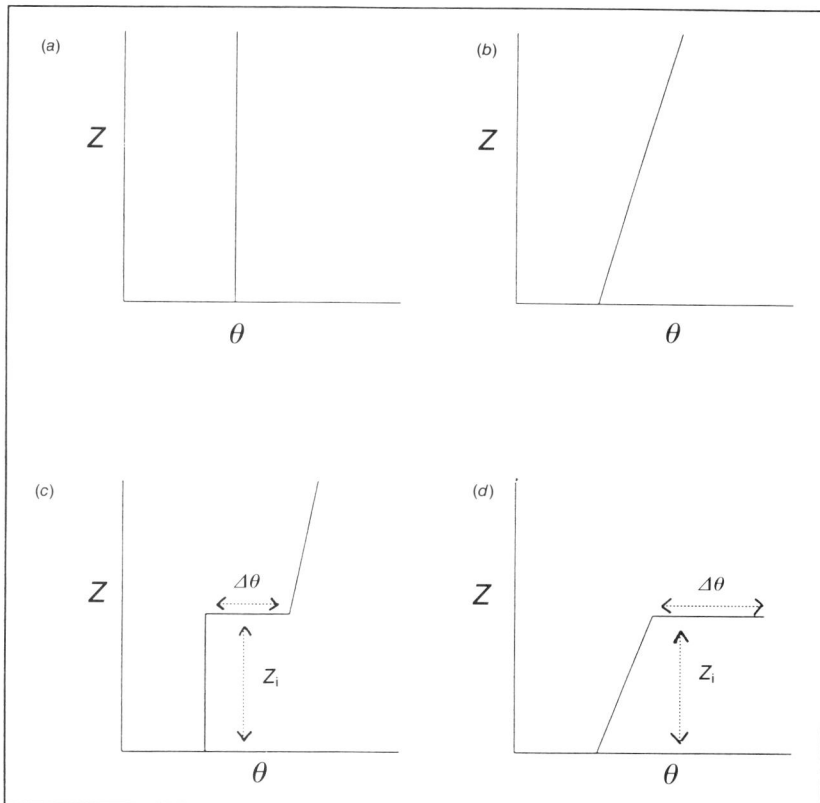

Fig. 5.2. Upstream stratification options available in Flowstar. (*a*) Neutral stratification. (*b*) Constant stable stratification. (*c*) Neutral layer capped by an inversion of potential temperature step $\Delta\theta$ with a slightly stable layer above that. (*d*) Stable layer beneath a strong inversion.

$$F_i^{-2} = \frac{z_i}{U_0^2(z_i)} \frac{g\Delta\theta}{\theta_i}, \quad F_L = \frac{U_0(z_i)}{NL} \quad (5.1)$$

where z_i is the height of the inversion, g is gravitational acceleration, θ_i is the potential temperature at the inversion and $\Delta\theta$ is the potential temperature step at the inversion. N is the Brunt–Väisälä frequency of the stable layer and L is the half-length of the terrain (taken as $L = 2500$ m throughout).

In the inner region a first-order 'mixing length' closure is used to solve the equations. This relates the shear stress to the mean velocity profile and assumes that the turbulence is always in local equilibrium with the surface.

The depth of the layers depends on the terrain and the roughness length. For Kintyre the outer layer exists above approximately 500 m. For the inner region depth Flowstar uses the original definition by Jackson & Hunt (1975) resulting in a depth of close to 100 m. Between these layers Flowstar includes a middle layer which allows the correct matching of the solution for the velocity perturbations.

Terrain and surface roughness information is input as square grids of 64×64 points. For the Kintyre terrain this results in a grid spacing of 300 m, placing an absolute limit on the resolution of the model.

The roughness length over the land surface is estimated according to land use type and crop height. The terrain is partly forested with conifers of varying age. A maximum roughness length of 1 m is used for the most mature plantation. Apart from site 12 in 1990, it was not possible to site equipment in forest plantations of significant height. This means that the roughness length of the surface close to all of the mean flow sites is that appropriate for rough heathland (0.3–0.05 m).

The perturbation to the wind profile due to the roughness changes is calculated in a similar way to the perturbations caused by the terrain (Belcher *et al.*, 1990). Since the theory is linear the two perturbations can be added to the unperturbed profile producing the final wind vector.

The wind vector is calculated at each grid point and at a range of user-defined heights. Flowstar calculates the position of streamlines passing over the terrain starting from a specified position and height on the upstream edge of the grid. The velocity vector is worked out at each intersection of the streamline with a grid line by linear interpolated and the next position of the leading edge of the streamline is inferred.

5.4 Results

The method of substituting the pressure field from the outer to the inner layer is valid only if the velocity and pressure fields do not interact. This may not be the case downstream of a hill. If the terrain is steep enough to cause a steady region of separated flow then the model will be invalid through the domain. However, if the terrain is only steep enough to produce intermittent separation or a region of enhanced turbulence downstream of the summit then the model will be invalid only in this region.

Although, when compared with previous experiments (detailed by Finnigan, 1988), we would not expect the gross shape of the Kintyre terrain to cause a steady separation region it is worth examining the data to confirm this. Fig. 5.3*a* shows a scatter plot of mean wind direction at site 8 versus mean wind direction at site 14. Site 8 is downstream of Sgreadan Hill for all westerly wind directions and if a large, steady separation bubble existed we would expect to observe reverse flow. This is not the case. However, a wake region with high turbulence and low speed-up could still be generated. Fig. 5.3b and c show scatter plots of speed-up and gustiness at site 8 versus wind direction at site 14. Gustiness here is defined as $(U_{max} - U_{min})/U_{mean}$ for a 1 h period and is a measure of turbulent intensity. These plots show that when site 8 is downstream of Sgreadan Hill the turbulence is enhanced and speed-up reduced, indicating the presence of a wake region. Thus, we expect the model assumption to be valid throughout most of the domain although errors may be encountered in the wake regions generated by the gross shape of the terrain.

5.4.1 The synoptic situation

The strong westerly flow at Kintyre is geostrophically driven. It is produced by the rapid passage of low-pressure systems to the north of Scotland from west to east. Fig. 5.4a shows a typical surface chart. Conditions were nearly always overcast due to the fronts associated with the depressions, and bands of rain and showers were frequent. Windspeeds were often increased by the presence of an anticyclone over the mid-Atlantic or south-western Europe. Any components of the flow generated by heating or cooling of the surface are insignificant in comparison with the geostrophically driven flow.

Finer weather was required for the aircraft flights. All the flights discussed here were carried out in periods when a ridge of high pressure interposed between two low-pressure systems. Fig. 5.4b shows a typical surface chart. There were brief spells of sunshine during some of the flights. However, the moderate to strong winds and the stable stratification profile observed during all the flights (discussed below) indicate that it is acceptable to assume that all of the turbulence is mechanically generated.

5.4.2 Outer layer flow

Fig. 5.5a–d show comparisons of model streamlines to potential temperature contours for four flights. There are no potential temperature contours below approximately $z = 1000$ m. In these regions little or no potential temperature gradient was observed, making the contouring technique locally inapplicable.

Fig. 5.6a–d show the stratification profiles recorded by the aircraft during soundings at the start and end of each flight. These must be approximated to one of the Flowstar stratification options. In each case the existence of a

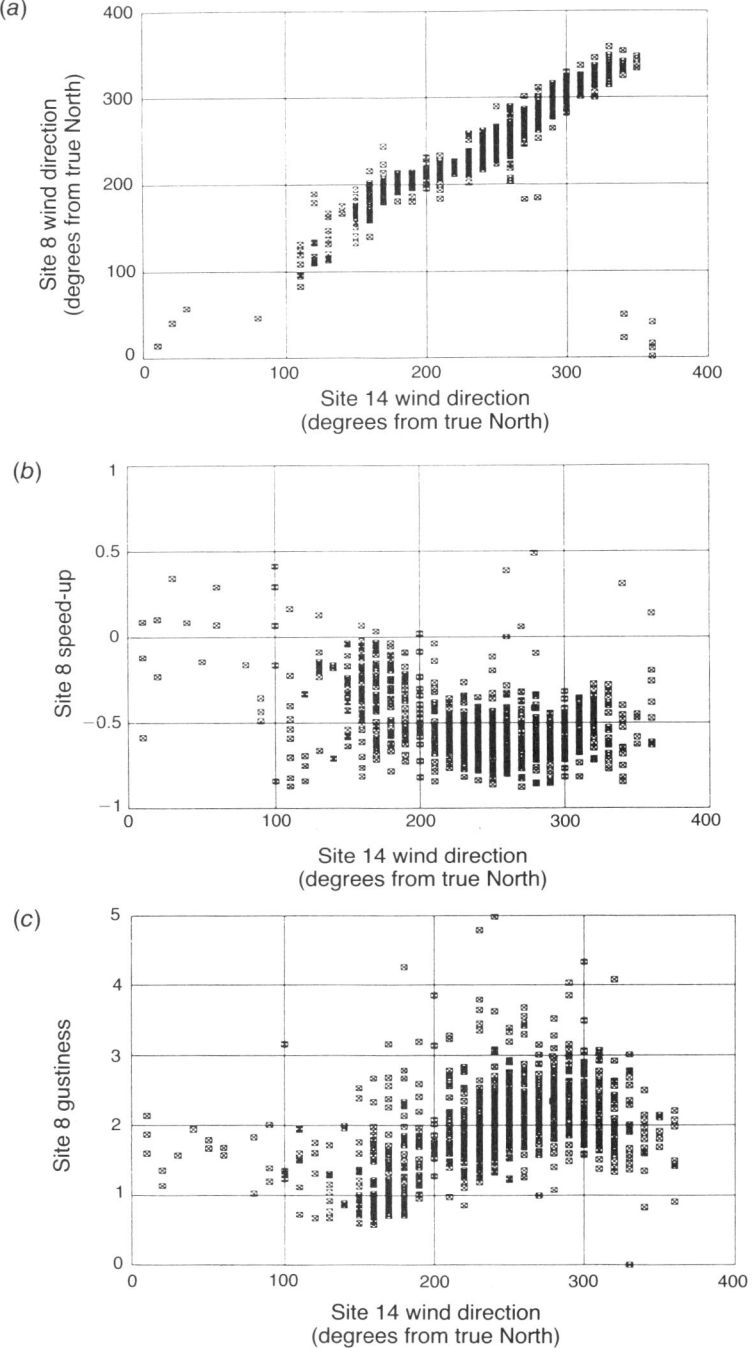

Fig. 5.3. Scatter plots of site 8 mean wind data versus site 14 wind direction. (*a*) Site 8 wind direction; (*b*) site 8 speed-up; (*c*) site 8 gustiness.

Model for flow over complex terrain

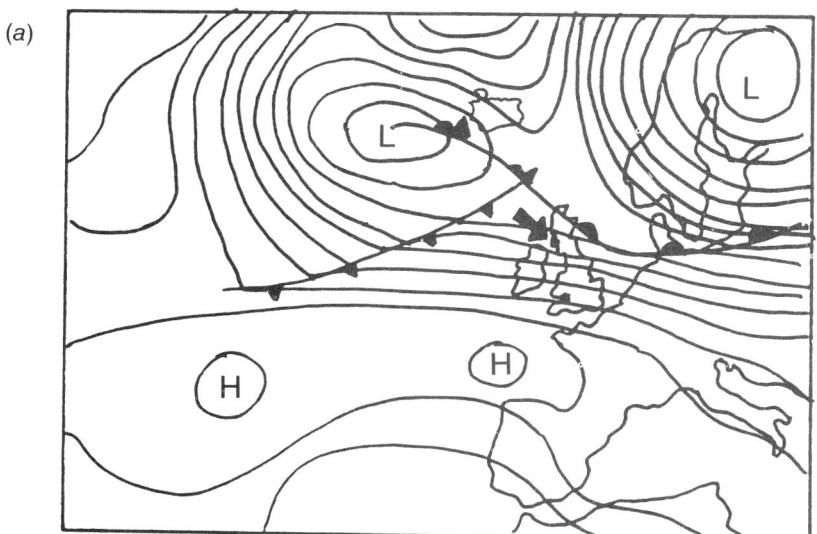

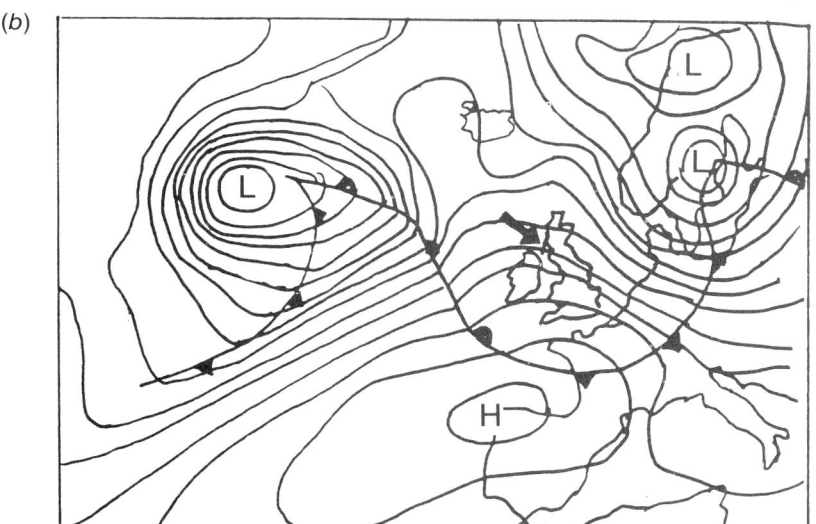

Fig. 5.4. Surface charts showing typical synoptic situations: (*a*) for strong westerly flow, (*b*) for aircraft flights. The approximate position of Kintyre is indicated by an arrow.

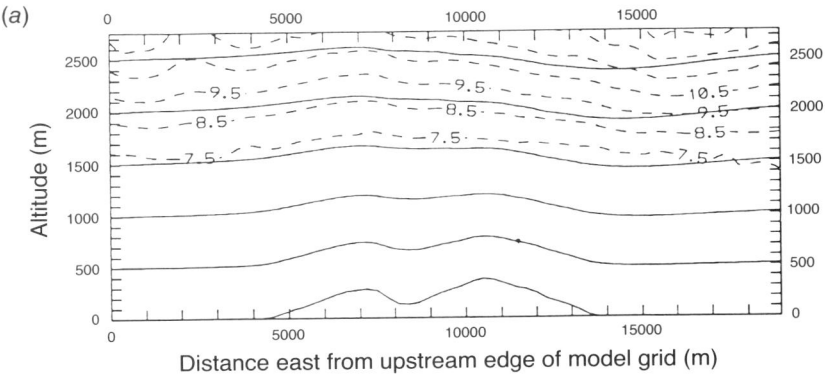

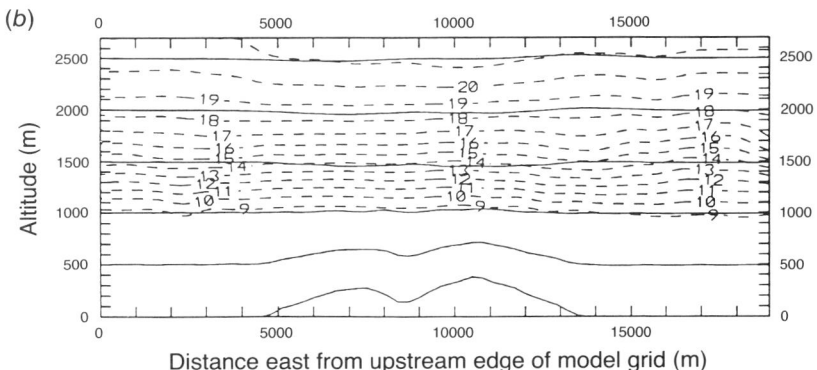

neutral layer next to the surface makes the Flowstar option shown in Fig. 5.2c the most appropriate. Input parameters such as the size of the potential temperature step, $\Delta\theta$, and the Brunt–Väisälä frequency, N, of a stable layer are calculated as a mean value from the two soundings. These input parameters can be found in Table 5.1. Wind directions and Froude numbers for the flights (calculated from Eq. 5.1) are shown in Table 5.2.

For some cases the observed stratification profile is more complex than the Flowstar options allow for. On 9 March 1990, 22 March 1990 a.m. and 22 March 1990 p.m. this is the case. A four-layer stratification profile is observed with a stable layer existing between the inversion and the lowest (neutral) layer. There are two possible options in representing this situation: either ignore the stable layer and model the inversion above a deep neutral layer, or include the inversion in a deep stable layer which begins at the top of the

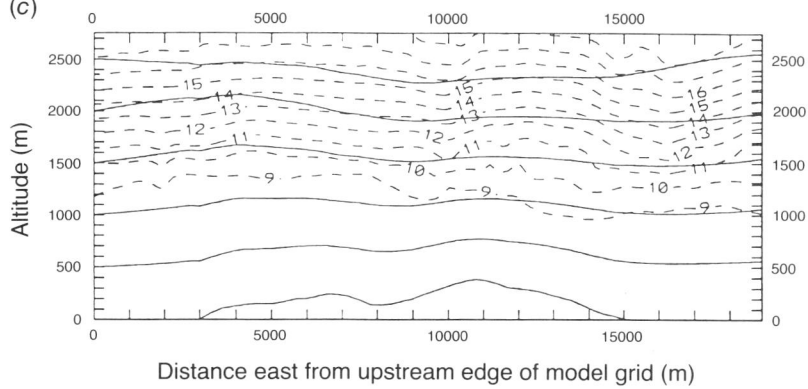

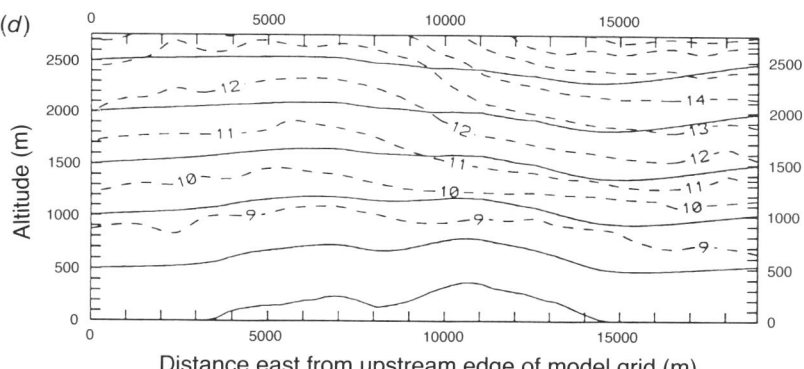

Fig. 5.5. Comparisons of observed potential temperature isopleths (dashed lines) with model streamlines in the outer layer (continuous lines). The lowest continuous line represents a west/east cross-section of the Kintyre peninsula through site 5. (a) 9 March 1990; (b) 12 March 1990; (c) 22 March 1990, a.m.; (d) 22 March 1990 p.m.

neutral layer. It was found that the observed flow patterns could be more accurately reproduced if the first of these options was used when $F_i \leq 1$ and the second option used when $F_i > 1$. All the model results presented were produced using this guideline.

Attempts to model the flow pattern from the four flights are discussed briefly below. Table 5.3 compares quantitatively some of the major features of the observed and modelled flows.

On 9 March 1990 the observed potential temperature contours reached a maximum at $x = 8000$ m and descended to a minimum at $x = 16\,000$ m. Flow-star reproduces this pattern well although it places the minimum approximately 2000 m further upstream than observed and the amplitude of the

(a)

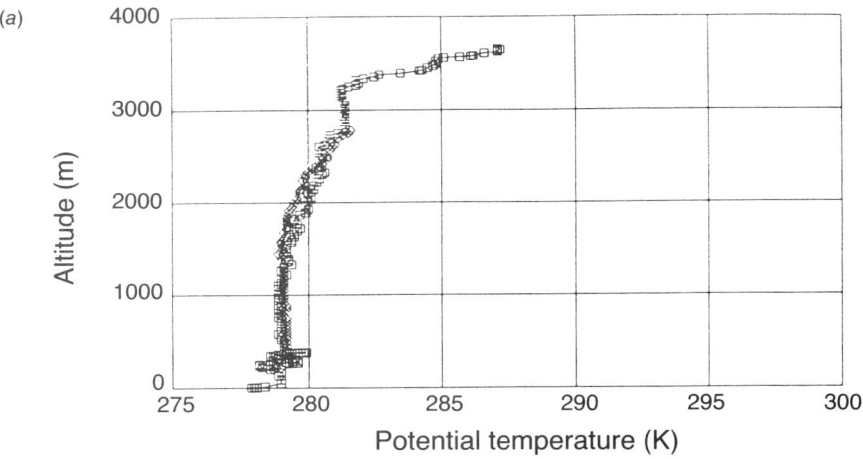

(b)

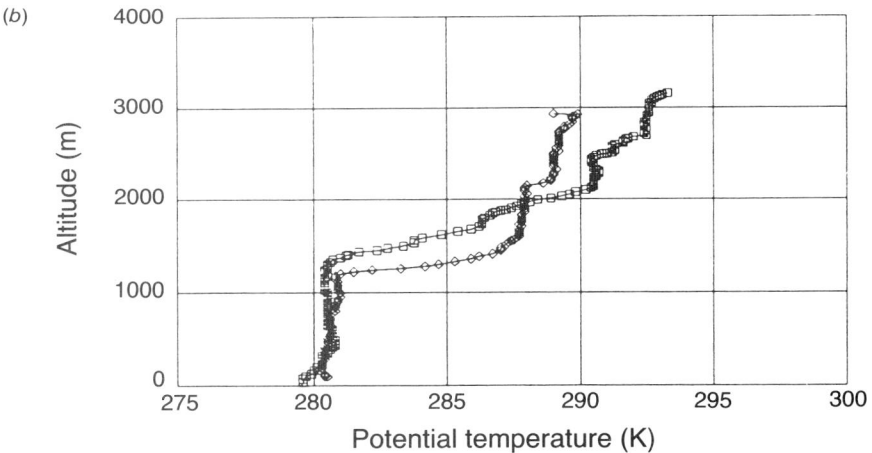

perturbation is somewhat underestimated. There is also some detail upstream of the peninsula which is not reproduced. The sounding shows the inversion layer to be at $z = 3200$ m. It is not visible in Fig. 5.5a since the aircraft made passes over the peninsula only up to an altitude of $z = 2500$ m.

On 12 March 1990 the most striking feature of the observed flow was the lack of deflection of the inversion layer. The strong inversion at $z = 1200$ m acts as a 'lid' on the boundary layer. Fig. 5.5b shows that the model reproduces this very well. The two-dimensional stratified flow model of Carru-

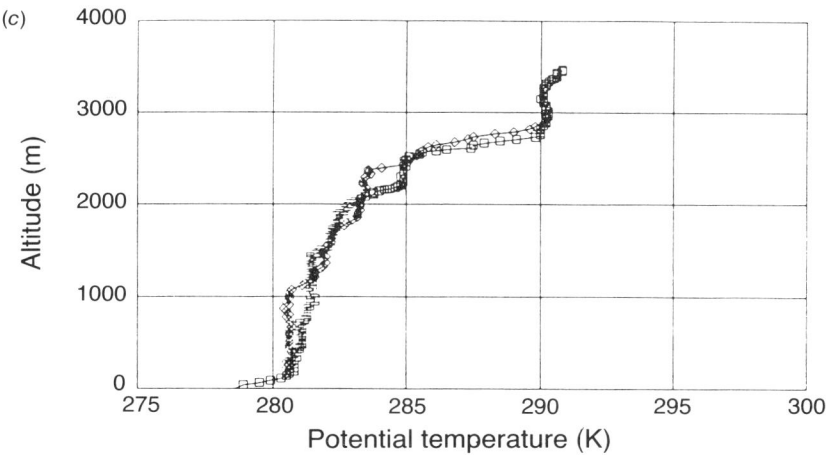

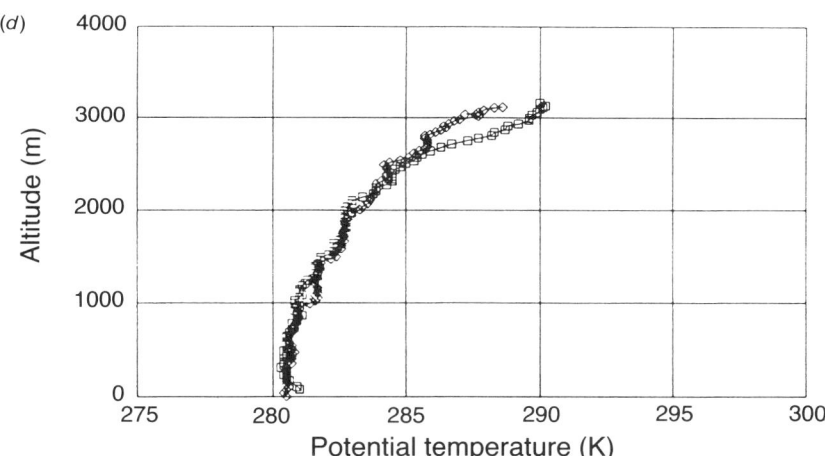

Fig. 5.6. Profiles of potential temperature recorded upstream of the terrain by the aircraft at the beginning (squares) and end (diamonds) of each flight. (a) 9 March 1990; (b) 12 March 1990; (c) 22 March 1990, a.m.; (d) 22 March 1990, p.m.

thers & Choularton (1982) produced a similar pattern for this type of stratification profile.

On 22 March 1990 a.m. Fig. 5.5c shows that a more complex flow pattern was observed during this flight. A wave develops above $z = 1500$ m. It shows minima at approximately $x = 10\,000$ m and $x = 16\,000$ m. The model reproduces the position of these features well and their amplitude reasonably well,

Table 5.1. *Upstream atmospheric conditions during four aircraft flights*

Date (1990)	N (s^{-1})	z_i (m)	$\Delta\theta$ (K)	u_* (m s^{-1})	z_0 (m)
9 Mar.	—	3200	6	0.444	0.00017
12 Mar.	0.0082	1200	8	0.156	0.00007
22 Mar., a.m.	—	2500	7	0.489	0.00017
22 Mar., p.m.	0.0117	1500	0	0.493	0.00017

Table 5.2. *Wind directions and Froude numbers during four aircraft flights*

Date (1990)	Wind direction	F_i	F_L
9 Mar.	285°	0.719	—
12 Mar.	289°	0.355	0.315
22 Mar., a.m.	273°	0.765	—
22 Mar., p.m.	273°	—	0.657

Table 5.3. *Comparison of peaks and troughs in modelled and observed outer layer flow*

Date of flight (1990)	Maximum upward displacement of streamline				Maximum downward displacement of streamline			
	Horizontal position (m)		Magnitude (m)		Horizontal position (m)		Magnitude (m)	
	Obs.	Model	Obs.	Model	Obs.	Model	Obs.	Model
9 Mar.	7000	7000	200	150	16 000	14 000	300	140
22 Mar., a.m.[a]	4500, 12 500	4000, 12 500	100, 150	200, 70	10 000, 16 500	9500, 15 500	50, 200	100, 100
22 Mar., p.m.	6000	6000	330	150	16 000	14 500	480	200

[a]For 22 Mar., a.m., there are two significant maxima and minima. The figures for the first and second of these are given, separated by a comma.

although the slope of the observed contours close to the troughs is greater than for the model streamlines.

On 22 March 1990 p.m. Fig. 5.5d shows that the flow has developed since the morning flight. The wave pattern is replaced by strongly supercritical flow. In supercritical flow the inversion layer rises upstream of the summit then descends to flow down the leeward slope. Subcritical flow means that the inversion layer dips below its equilibrium level as it passes the summit then rises as it flows over the leeward slope. The model reproduces the general flow type but underestimates the amplitude of the perturbations.

5.4.3 Inner region results

Eleven separate hours of moderate to strong westerly flow were isolated from the dataset and the model run to simulate ΔS at the mean flow monitoring sites. These periods were selected on the grounds of wind direction and windspeed only. Sonde data recorded at the Meteorological Office station at Long Kesh in Northern Ireland were used to determine the upstream stratification profile. In each case a neutral layer exists next to the ground capped by varying stable layers, making the option shown in Fig. 5.2c the most appropriate. The upstream wind profile was estimated using mean wind data from site 14 as explained in Section 5.2.2. The input parameters for the 11 hours are shown in Table 5.4.

The observed speed-up data can be interpreted, to some extent, in terms of the Froude numbers and the stratification type (see Table 5.5). The 11 hours fall into two main groups: those with and those without an elevated inversion. At summit sites (e.g. site 5) the speed-up is increased in the presence of an elevated inversion. Hours 6 to 11 can be categorised further. According to Carruthers & Choularton (1982) subcritical flow is expected when $F_i < 1$ and supercritical flow when $F_i > 1$ or $F_i \sim 1$. Also, $F_L < 1$ causes increased asymmetry in the flow. Hours 6 and 11 have $F_i < 1$ and $F_L < 1$, hours 8 and 9 have $F_i > 1$ and $F_L < 1$ and hours 7 and 10 have $F_i \sim 1$ and $F_L < 1$. The data show a dependence on these categories at lee side sites. At

Table 5.4. *Upstream atmospheric conditions for 11 hours of westerly flow*

Hour	Date	Time (GMT)	N (s^{-1})	z_i (m)	$\Delta\theta$ (K)	u_* (m s^{-1})	z_0 (m)
1	4 Mar. 1989	18 : 00	0.014	1752	0	0.617	0.00042
2	5 Feb. 1989	04 : 00	0.015	1752	0	0.425	0.00015
3	11 Feb. 1989	17 : 00	0.013	500	0	0.387	0.0001
4	13 Feb. 1989	12 : 00	0.012	988	0	0.963	0.00140
5	20 Feb. 1989	13 : 00	0.011	597	0	0.535	0.00027
6	4 Mar. 1990	09 : 00	0.011	1457	6.5	0.295	0.00007
7	4 Mar. 1990	21 : 00	0.011	1363	6.5	0.482	0.00014
8	5 Mar. 1990	18 : 00	0.013	998	6	0.460	0.00010
9	6 Mar. 1990	12 : 00	0.013	2500	5	0.551	0.00027
10	6 Mar. 1990	18 : 00	0.026	700	13	0.427	0.00008
11	7 Mar. 1990	18 : 00	0.04	580	15	0.253	0.00007

Table 5.5. *Froude numbers and ΔS at selected sites for 11 hours of westerly flow*

Hour	F_i	F_L	ΔS_5	ΔS_7	ΔS_8
1	—	0.79	0.14	—	−0.67
2	—	0.58	0.02	—	−0.69
3	—	0.60	0.01	—	−0.79
4	—	1.37	0.20	—	−0.63
5	—	0.92	0.06	—	−0.6
6	0.68	0.54	0.18	−0.42	−0.67
7	1.10	0.84	0.20	−0.2	−0.59
8	1.28	0.70	0.22	−0.27	−0.58
9	1.05	0.80	0.25	−0.24	−0.56
10	0.95	0.33	0.30	−0.22	−0.46
11	0.58	0.13	0.27	−0.52	−0.76

site 7 hours 6 and 11 have the lowest speed-up, which corresponds to the subcritical flow pattern most likely to result in separation in this region. At site 8 higher speed-ups are observed for hours 7, 8, 9 and 10 when $F_i > 1$ or $F_i \sim 1$ causes supercritical flow, inhibiting separation. These results demonstrate the need to take the outer region stratification into account when attempting to model the inner region flow over complex terrain.

Observed speed-ups are compared with model speed-up predictions for the stratified flow in Fig. 5.7. The error bars in the figure show the estimated uncertainty in each observation, as discussed in Section 2.1. The offset errors are calculated by making the somewhat pessimistic assumption that the true mean windspeed is changing by 50% per hour. Speed-up predictions for neutral flow are also shown for comparison.

The model output is for a line connecting sites 1 to 8 running approximately west to east. Most of the main points can be discussed with reference to these eight sites. However, the results from site 10 are interesting in terms of flow separation and are discussed later.

For hours 1 to 5 (Fig. 5.7a–e) the predicted speed-up is close to that for neutral flow. For these cases the performance of the model can be compared with previous attempts to model neutral flow. At summit and upstream sites (1, 2, 4, 5 and 6) Flowstar predicts the majority of the speed-ups to within or close to the confidence interval. Salmon *et al.* (1988) and Walmsley & Salmon (1984) modelled the neutral flow over Askervein using a linear model. They correctly predicted a similar proportion of the speed-ups on upwind and hilltop sites.

For valley and lee side sites (sites 3, 7 and 8) Flowstar overpredicts windspeeds consistently. The most severely affected is site 8, where the prediction is typically too high by a factor of 2. This confirms that the wind field in the wake region is not well modelled by Flowstar and agrees with previous comparisons of linear models with data (Salmon et al., 1988).

The performance of the model when an inversion is present (hours 6 to 11) seems to be dependent on the altitude of the inversion layer. When the inversion is lower than twice the hilltop height (hours 10 and 11) the model reproduces the large speed-up at the hilltop sites 4, 5 and 6 quite convincingly. When the inversion is higher (hours 6, 7 and 8) Flowstar produces hilltop speed-ups slightly greater than the near-neutral cases but between 0.1 and 0.2 lower than the observed values.

The only case which contradicts the arguments above, which are based on stratification type, is hour 9. The speed-up data for hour 9 are similar to those for hour 10. This strongly suggests that the real stratification profile for hour 9 was also similar to that for hours 10 and 11 when a strong, low inversion was present and the speed-up was well modelled by Flowstar. We believe that the stratification profile taken for hour 9 (the 12:00 sonde on 6 March 1990 from Long Kesh) was probably not appropriate to Kintyre at that time. If the next sonde were used (00:00 on 7 March 1990) the model performance for hour 9 would be similar to that for hour 10.

The gross shape of the Kintyre terrain does not appear to cause a steady region of separated flow. However, since the terrain is complex, this may occur in localised areas associated with steeper slopes. Directly to the east of site 10 the terrain descends steeply (Fig. 5.1). The maximum angle of the slope is approximately 40°, which is steep enough to create a localised region of separated flow. Under these circumstances the linear model is not applicable and is expected seriously to overpredict the windspeeds upstream of the separated region, i.e. at site 10 (Finnigan, 1988). Fig. 5.8 shows that this is the case for each of the 11 hours. This demonstrates the importance of carefully assessing the slopes in the immediate vicinity of a site before accepting the model results. Linear theory has previously been shown to be useful for slopes of up to 25%.

5.5 Discussion and conclusions

In the outer region Flowstar can reproduce the flow type quite convincingly when the stratification profile approximates readily to one of the existing Flowstar stratification options. When the stratification profile is more complicated than allowed for by the Flowstar options qualitative agreement is achieved. A Flowstar option allowing a four-layer stratification structure such as observed on 22 March 1990 may improve quantitative agreement for these cases.

108 *D. W. F. Inglis et al.*

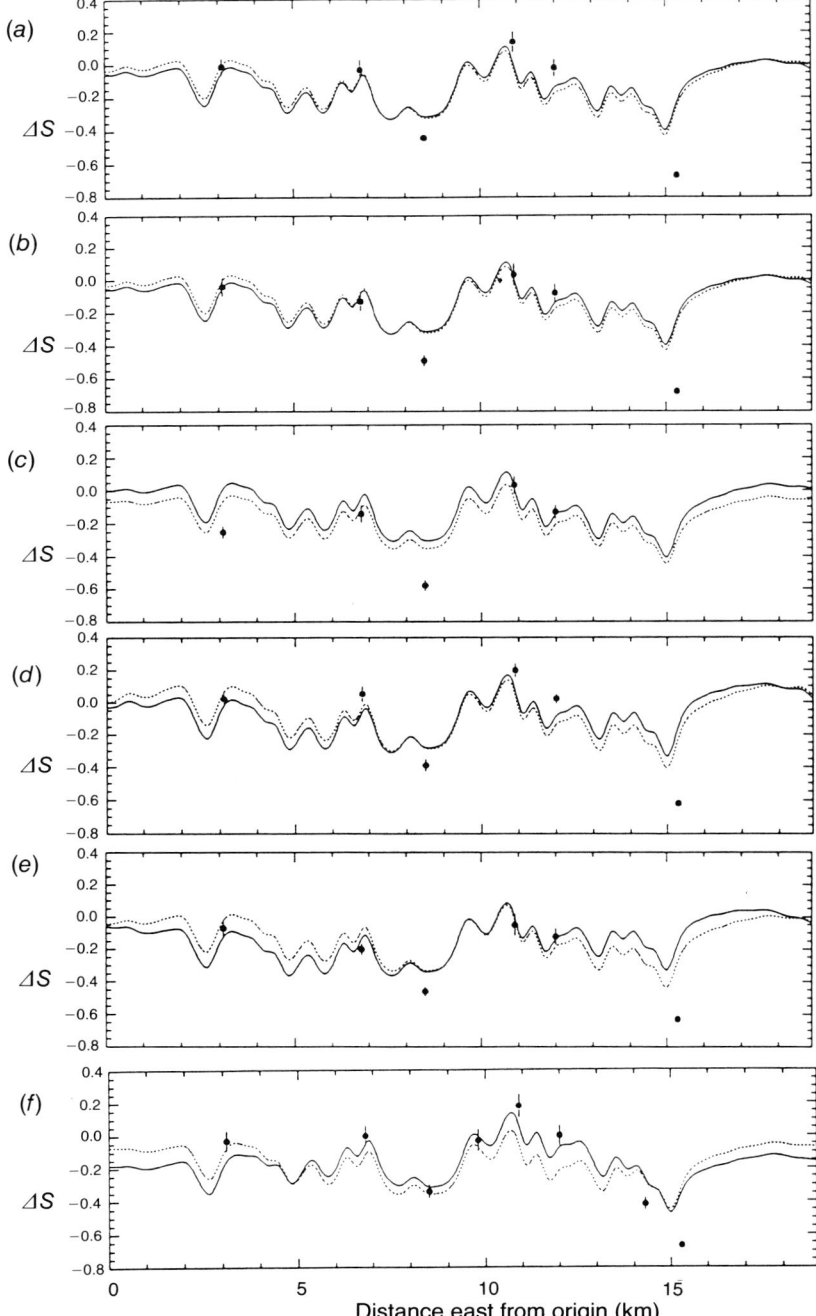

Model for flow over complex terrain 109

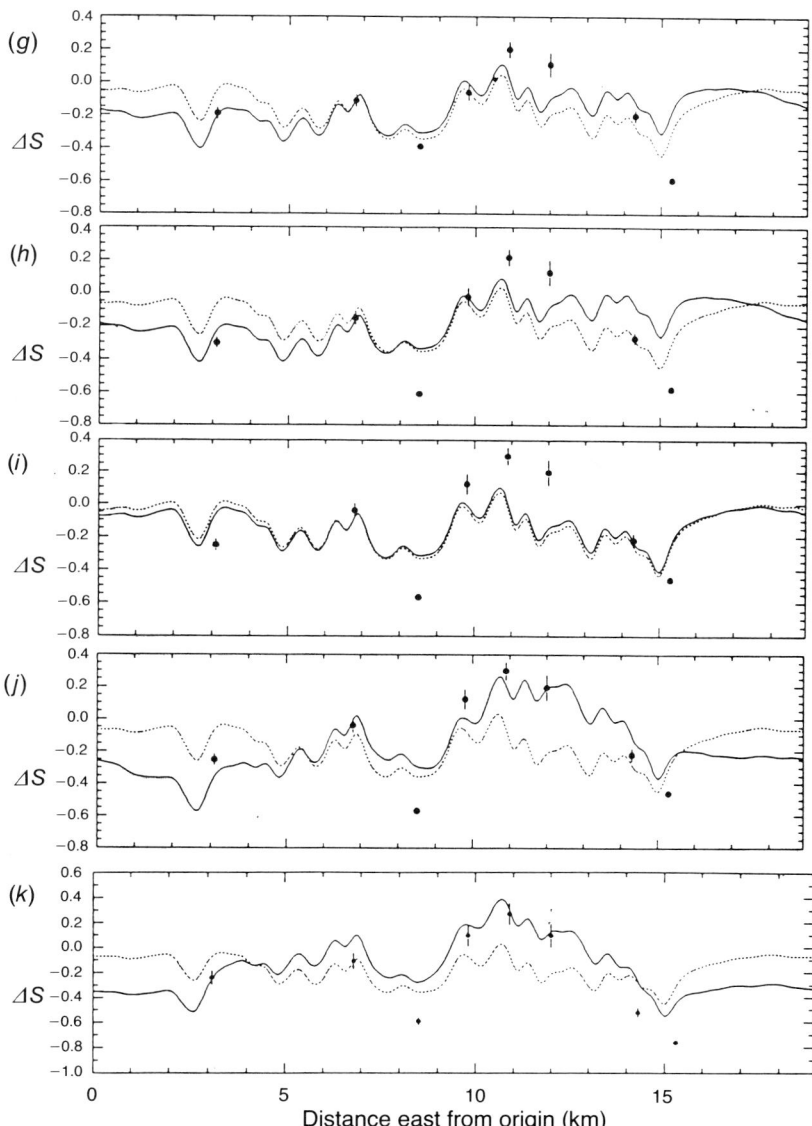

Fig. 5.7. Comparisons of observed (filled circles) and modelled (continuous line) speed-up (ΔS) along a line connecting sites 1 to 8 for 11 hours of westerly flow. Error bars indicate the estimated uncertainty inbserved values. Predictions for neutral flow (dotted lines) are also shown for comparison. (*a*) hour 1; (*b*) hour 2; (*c*) hour 3; (*d*) hour 4; (*e*) hour 5; (*f*) hour 6; (*g*) hour 7; (*h*) hour 8; (*i*) hour 9; (*j*) hour 10; (*k*) hour 11.

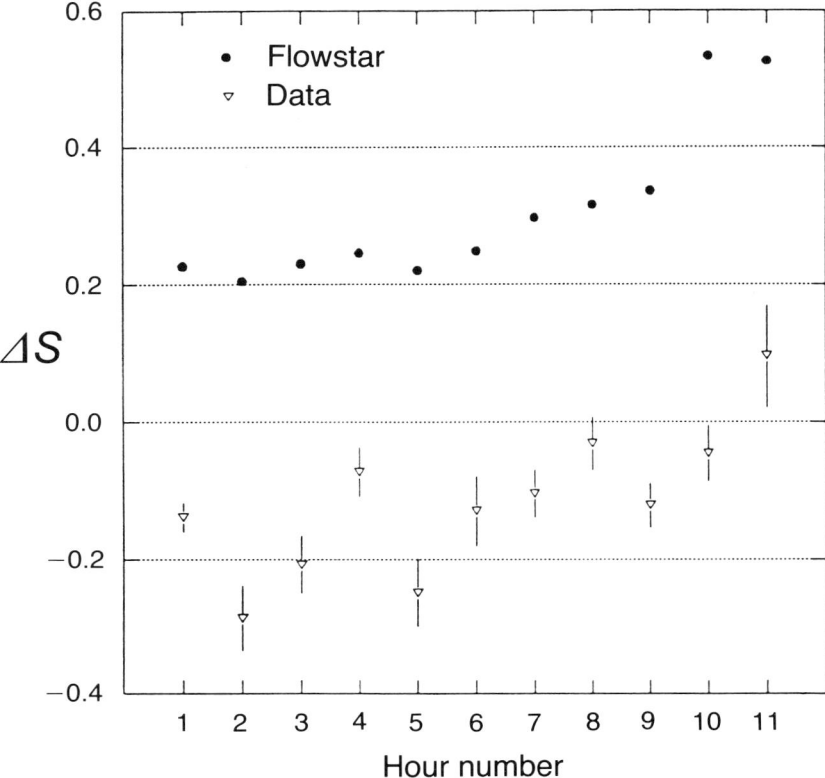

Fig. 5.8. Comparison of observed with modelled speed-up at site 10 for 11 hours of westerly flow. Errors bars indicate the estimate uncertainty in observed values.

In the inner region, for near-neutral conditions, the performance of Flowstar is comparable to the performance of other linear models for much simpler terrain. This confirms that the linear, mixing length theory, previously tested for simple terrain shapes, can be used successfully for complex terrain.

As expected, model overpredictions of windspeeds were encountered on lee side slopes and in valleys; however, the gross shape of the terrain is not steep enough to invalidate the model outside these areas. Steep localised slopes, however, are capable of causing errors at individual sites.

The data demonstrate the need to take the stratification profile into account when attempting to model inner region winds. The response of the inner region windspeeds to the stratification profile is most obvious at sites close to hilltops. Flowstar attempts to model this were quite encouraging, especially in the presence of an inversion lower than twice the hilltop height. When the inversion is higher than this the model underestimates the hilltop speed-up somewhat.

The potential usefulness of flowstar and models like it in terms of forest planning appears to be considerable. However, the flow in the wake remains a problem which must be addressed for the model to become fully robust. This consideration should not prevent the careful use of this model for real applications.

Acknowledgements

The authors wish to thank the many Forestry Authority staff, especially Angus Mackie and Alistair Strang, at NRS and in Kintyre who made the airflow experiments possible. We are also indebted to the staff at RAF Machrihanish and at the Meteorological Office station for their help and the use of their facilities. D.W.F.I. was supported by a NERC studentship during this work.

References

Belcher, S. E., Xu, D. P. & Hunt, J. C. R. (1990). The response of a turbulent boundary layer to arbitrarily distributed two dimensional roughness changes. *Quarterly Journal of the Royal Meteorological Society*, **116**, 611–37.

Bradley, E. F. (1980). An experimental study of the profile of wind speed, shearing stress and turbulence at the crest of a large hill. *Quarterly Journal of the Royal Meteorological Society*, **106**, 101–23.

Busch, N. E. & Kristensen, L. (1976). Cup anemometer overspeeding. *Journal of Applied Meteorology*, **15**, 1328–32.

Cannell, M. & Coutts, M. (1988). Growing in the wind. *New Scientist*, **117**, 42–6.

Carruthers, D. J. & Choularton, T. W. (1982). Airflow over hills of moderate slope. *Quarterly Journal of the Royal Meteorological Society*, **108**, 603–24.

Carruthers, D. J. & Hunt, J. C. R. (1990). Fluid mechanics of airflows over hills, turbulence, fluxes and waves in the boundary layer. In *Atmospheric Processes Over Complex Terrain*, chap. 5. AMS Meteorological Monographs vol. 23 no. 45. American Meteorological Society, Boston, USA.

Cook, N. J. (1985). *The Designer's Guide to Wind Loading of Building Structures*, part 1. Butterworth (for the Building Research Establishment), London.

Deaves, D. M. (1981). Computations of wind flow over changes in surface roughness. *Journal of Wind Engineering and Industrial Aerodynamics*, **7**, 65–94.

Finnigan, J. J. (1988). Airflow over complex terrain. In *Flow and the Natural Environment: Advances and Applications*, ed. W. K. Stiffan & O. T. Denmead. Springer-Verlag, Berlin.

Hunt, J. C. R., Leibovich, S. & Richards, K. J. (1988a). Turbulent shear flow over low hills. *Quarterly Journal of the Royal Meteorological Society*, **114**, 1435–70.

Hunt, J. C. R., Richards, K. J. & Brighton, P. W. M. (1988*b*). Stably stratified shear flow over low hills. *Quarterly Journal of the Royal Meteorological Society*, **114**, 859–86.

Inglis, D. W. F. (1992). A field and modelling study of airflow over complex terrain and in forest canopies. PhD Thesis, Department of Pure and Applied Physics, UMIST, Manchester, UK.

Jackson, P. S. & Hunt, J. C. R. (1975). Turbulent wind flow over a low hill. *Quarterly Journal of the Royal Meteorological Society*, **101**, 929–55.

Large, W. G. & Pond, S. (1981). Open ocean momentum flux measurements in moderate to strong winds. *Journal of Physical Oceanography*, **11**, 324–36.

Mason, P. J. & Sykes, R. I. (1979). Flow over an isolated hill of moderate slope (Brent Knoll). *Quarterly Journal of the Royal Meteorological Society*, **105**, 393–5.

Salmon, J. R., Bowen, A. J., Hoff, A. M., Johnson, R., Mickle, R. E., Taylor, P. A., Tetzlaff, G. & Walmsley, J. L. (1988). The Askervein Hill Project: mean wind variations at fixed height above ground. *Boundary-Layer Meteorology*, **43**, 247–71.

Scorer, R. S. (1953). Theory of airflow over mountains. II. The flow over a ridge. *Quarterly Journal of the Royal Meteorological Society*, **79**, 70–83.

Stromberg, I. M., Mill, C. S., Choularton, T. W. & Gallagher, M. W. (1989). A case study of stably stratified airflow over the Pennines using an instrumented glider. *Boundary-Layer Meteorology*, **46**, 153–68.

Taylor, P. A., Mason, P. J. & Bradley, E. F. (1987). Boundary-layer flow over low hills: a review. *Boundary-Layer Meteorology*, **39**, 107–33.

Walmsley, J. L. & Salmon, J. R. (1984). A boundary layer model for wind flow over hills: a comparison of model results with Askervein 1983 data. In *Proceedings of the European Wind Energy Conference*, Hamburg, October 1984.

6

Predicting windspeeds for forest areas in complex terrain

P. HANNAH, J. P. PALUTIKOF and C. P. QUINE

Abstract

The determination of the wind regime in a potential or existing forest area is an important aspect of any site assessment. Wind affects the growth rate of the trees and determines the occurrence of windthrow in the later years of the development of a forest. Consequently the prediction of tree growth, and the forecast of the financial viability of any forest project, are dependent upon an accurate assessment of the windspeed. Many forestry locations have sparse windspeed records due to difficulties of siting equipment in remote areas. In Britain, methods of windspeed assessment for such purposes have included the use of tatter flags and, more recently, numerical and statistical modelling techniques. An attempt is made here to relate windspeed to the topographic and geographic characteristics of the site in order to avoid the necessity for long-term on-site wind measurements. The variables related to windspeed are altitude, which has a recognised relationship with windspeed, topex (a measure of the exposure of a site), roughness length (related to the height of the surface elements), United Kingdom Ordnance Survey grid position, and distance to the coast. Initially, each geographic variable is related individually to annual mean windspeed at a set of 21 sites, taken from both the Forestry Commission and United Kingdom Meteorological Office networks. Subsequently, the variables are related in multiple regression equations to the annual mean windspeeds of 1989 and these equations are then used to predict the windspeeds in 1990. A test of the method has been made on an independent site not used in the previous stages. The results are encouraging.

6.1 Introduction

The growth of plants and trees is influenced by soil conditions and a number of climatic variables, most notably temperature, rainfall and wind (Jarvis &

Mullins, 1987; Worrell & Malcolm, 1990a). These climatic variables vary across sites and this variation may be explained by factors such as site location, altitude, aspect of the slope (e.g. south-facing) and distance to the sea (Bendelow & Hartnup, 1980). Because of a lack of windspeed measurements in upland Britain, tatter flags (Miller et al., 1987) were used by Worrell & Malcolm (1990b) as a surrogate for windspeed, who then related tatter by multiple regression equations to geographic variables.

Work of a similar nature to that described here has been carried out by White (1979), who applied principal components analysis (PCA) to 39 widely distributed United Kingdom Meteorological Office (UKMO) stations, with data covering 1960–9. The analysis attempted to predict a range of climate parameters using topographic and geographic values. Altitude and grid position were found to be influential on some climate variables, including windspeed and direction.

White & Smith (1982) extended the work of White (1979) from an annual analysis to predict the quarterly variability of individual climate parameters such as temperature, precipitation and windspeed, using multiple regression equations. The use of quarterly data, instead of a seasonal analysis, is unhelpful in climatic terms since the variability of parameters such as temperature, rainfall and windspeed is more closely associated with the seasons than with quarters. The same dataset of 39 sites from 1960–9, used in the earlier study, was employed.

To represent the wind climate the predicted variable was the windspeed at 0900 hours. The performance of the method was poor, with the greatest variance explained for the 0900 hours windspeed being 44% for the first quarter. The choice of variables is not limited to those which could be expected to have some influence, but includes many with dubious physical significance. For example, the authors used the highest elevation within a 10 km radius, which fails to take account of the actual distance of the peak from the site: higher ground will have more effect nearer to the site. The levels of variance explained are so low that the use of this method does not seem appropriate here for studies requiring accurate estimates of windspeed.

Murakami & Komine (1983) attempted to predict windspeed from topographic factors including altitude, highest and lowest points within a 10 km radius, slope at the site, and many of the other variables already used by White & Smith (1982). Results were not quoted, but they did suggest that topographic data within 300 m of a site were the most influential on windspeed characteristics.

Baker (1984) evaluated the topographic effects on windspeeds over railway embankments. He defined a value for the topographic exposure, K_L, as the ratio of the windspeed at a height z m above the embankment to the windspeed at a similar height above reference terrain some distance upwind. Subsequent analysis (Baker, 1985) produced values of K_L ranging from 0.85–0.95 for valley bottom sites to 1.15–1.25 for hilltop sites. Yamazawa & Kondo (1989) related windspeed to roughness length and a similar topographic parameter, Ω. This second variable was derived from the aspect ratio of the hills under study, but errors of ±20% in the prediction of the mean windspeed were recorded.

Klinka et al. (1991) used PCA to revise the bioclimatic classification for British Columbia, Canada. The method was similar to that of the studies by White (1979) and White & Smith (1982) quoted previously. Ecological data were used to define the regions best suited for growth of particular species within the province. The results produced by these methods gave much more detail in the climatic zones of the region than had been previously possible. Far more ecological data were available (1299 0.04 ha sample plots of conifers) to test their hypotheses than for White & Smith (1982) but there were fewer climatic data. Linear extrapolation of climatic data from low-altitude base stations was performed to extend the data coverage to alpine and sub-alpine regions. Although some variables are known to have a linear relationship with altitude (e.g. temperature), many do not (e.g. windspeed, rainfall), and this kind of simple approach has been called into question by Smithson (1987). Klinka et al. (1991) revised the classification for British Columbia by extending the range of defined bioclimatic zones from 6 to 14.

The research described above used statistical techniques to attempt to relate topographic and geographic variables to climatic indicators, including windspeed. Little success has been recorded in the ability to predict windspeed with any degree of certainty. The work presented here attempts to use topographic and geographic variables, easily measured at the site, or from maps and photographs, to predict the annual windspeeds at a range of sites in a variety of terrain types.

6.2 The sites

The sites used here (shown in Fig. 6.1) are taken from UKMO and Forestry Commission (FC) (Quine & Reynard, 1990) sources. All relevant data for the UKMO sites could be taken from publicly available sources (UKMO Monthly Weather Reports, Ordnance Survey maps, etc.). The locations were chosen for the variety of types of local terrain, and to increase the data coverage across Scotland.

116 P. Hannah et al.

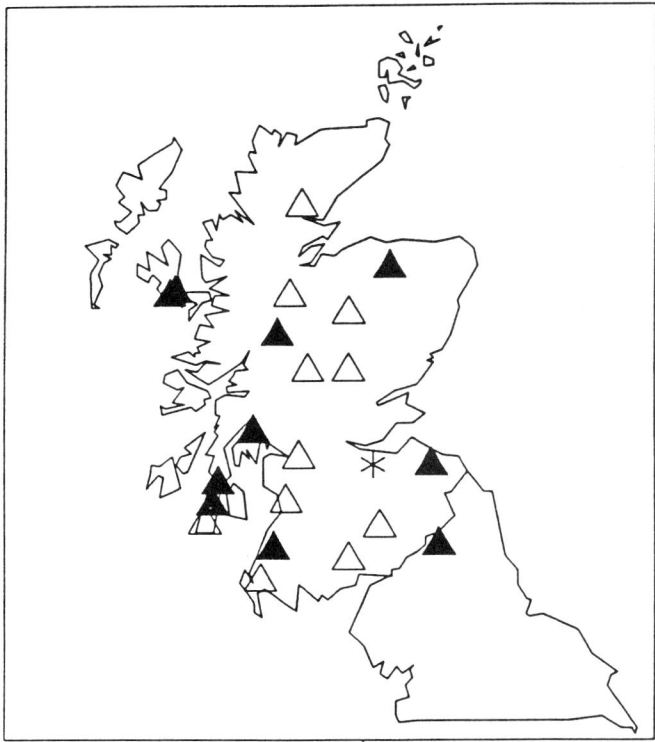

Fig. 6.1. Locations of sites used in the analysis. Filled triangles, Forestry Commission sites; open triangles, UK Meteorological Office sites; asterisk, Edinburgh Royal Observatory.

The sites are concentrated on the west coast of Scotland, both on the mainland and on the island of Skye; mainland coverage in the north and the south-east is poor, and there are no sites on the east coast. UKMO sites on the east coast are generally located on airfields in moderate terrain. It was considered that the dataset already contained enough of this category. The windspeed data used in the analysis described in Sections 6.4.1 and 6.4.2 are annual mean windspeeds, as recorded in the Monthly Weather Report (UK Meteorological Office, 1990, 1991) for the UKMO locations, and calculated from the raw data in the case of the FC sites.

6.2.1 Topex-to-distance

The concept of TOPographic EXposure, or topex, is the summation of the measured skyline angles at eight points of the compass (N, NE, E, etc.), with

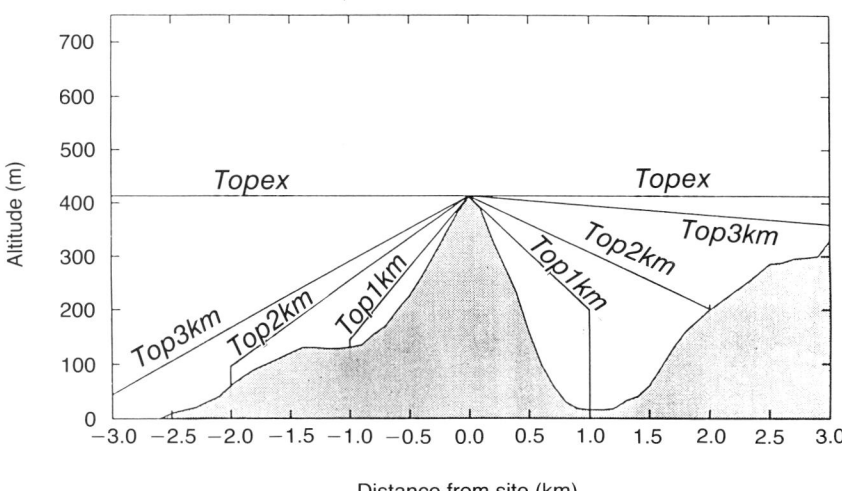

Fig. 6.2. Derivation of topex-to-distance.

no value less than zero (Pyatt et al., 1969). This standard topex has been used in British forestry for many years as a partial indication of the exposure of a site. Harrison (1988) used topex to assess the exposure of meteorological recording sites; negative values were permitted and measurements derived from non-landform skylines (vegetation, forestry, etc.) were also included. Miller et al. (1987) argue that the use of topex is limited by the lack of weighting for site aspect and wind distribution, but Quine & White (1993) and Harrison (1988) have developed the use of the eight individual sector values to define aspect effects.

For the purposes of this study, four topex variables are used. The variable *Topex* is a map-derived value (Wilson, 1984), with no declination (i.e. a minimum value of 0°). The boundary is arbitrarily set to the horizon, and the distance of this from the measurement point varies with the severity of the topography. The variables *Top1km*, *Top2km* and *Top3km* all allow for declination since there is a definite boundary at which the measurement stops. In allowing declination, it is now possible to differentiate between a hilltop site and a flat plain, both of which would have *Topex* values of 0° using standard forestry practice. Fig. 6.2 shows the differences between the four topex variables for a West–East transect at Beinn Staic on Skye.

Table 6.1 shows the variation of all topex variables for all 21 sites. Values quoted are the summation of the eight sector values. The values of the topex-to-distance variables increase with distance, as would be expected, and as

Table 6.1. Values of the characteristics at all sites

Site	Topex (deg)	Top1km (deg)	Top2km (deg)	Top3km (deg)	DTC (km)	Alt (m)	z_0 (m)	GridEast	GridNor
Abbotsinch	1	0	0	0	166	16	0.41	250	665
Aviemore	23	14	16	19	131	240	0.56	289	897
Beinn Staic	7	−69	−44	−29	71	411	0.27	145	820
Caplestone Fell	0	−22	−12	−9	185	479	0.60	365	590
Charter Hall	4	1	2	2	159	112	0.17	370	650
Deucharon	1	−29	−18	−10	17	329	0.28	170	640
Drungans	10	5	6	8	136	15	0.45	295	575
Eskdalemuir	20	7	10	10	138	259	0.65	285	605
Fort Augustus	30	26	27	28	106	42	0.65	236	809
Glentrool	9	−23	−18	−16	67	411	0.15	230	580
Leanachan	31	1	6	14	108	140	0.67	220	715
Machrihanish	10	0	0	2	11	10	0.25	165	620
Meall Bhuidhe	0	−23	−17	−8	17	374	0.66	170	620
Prestwick	2	0	0	0	109	21	0.26	235	620
Rannoch Moor	11	6	9	10	136	307	0.25	245	755
River Brittle	74	54	66	70	72	10	0.16	150	822
Rosarie	2	−6	−4	−2	100	330	0.34	333	850
Shin Falls	30	19	25	26	107	14	0.50	257	897
Strathlachlan	14	−31	−14	−7	152	317	0.44	203	690
Tummel Bridge	40	34	35	37	141	145	0.58	277	759
West Freugh	3	0	0	0	59	16	0.15	215	555

can be seen from the example in Fig. 6.2. For the site at River Brittle, with high positive values, the *Top3km* value is very close to the *Topex* value. The largest differences between the standard *Topex* and the three topex-to-distance values occur at the hilltop sites of Beinn Staic, Deucharen, Glentrool and Strathlachlan. These sites all have small positive values of *Topex*, but the range of values for the topex-to-distance is much greater, with large negative values now recorded. For example, Beinn Staic shows a value of −69° for *Top1km*.

6.3. Analysis of the topographic variables

There are two distinct groups of data used here. Firstly, there are those parameters such as altitude (*Alt*) and grid position (*GridEast* and *GridNor*) which have only a single value at each site. The values of altitude, grid easting and grid northing were taken from Ordnance Survey (OS) maps of scale 1 : 50 000 and less. The other variables (topex, roughness length and distance-to-coast, as explained in the following) were initially calculated as sector-specific parameters, the sectors consisting of eight 45° divisions.

Roughness length (z_0) is proportional to the height of the surface roughness elements, such as vegetation or forestry. Here, the value of roughness length is averaged over an area 3 km around each site to arrive at a single value, following the methods of Wieringa (1992). Distance-to-coast (*DTC*) is defined as the average distance to a major body of water (greater than 10 km across) on each of the eight compass bearings.

For *Topex*, z_0 and *DTC* the eight sectored values of each variable were combined, using simple arithmetic means (or a summation in the case of *Topex*), to produce single values which could be used in the analysis together with *Alt*, *GridEast* and *GridNor*. These variables were estimated from OS maps, 1 : 24 000 scale aerial photographs and site visits.

The values of all topographic and geographic variables are presented in Table 6.1. The inter-variable correlations described in the following are shown in Table 6.2. The variables are related to annual mean windspeeds at all sites for 1989 and 1990 (Table 6.3). The choice of 21 sites used here was made to include as much variety in the topographic characteristics as possible. The locations are typical of the region, although extension of the number of sites covered by the study would be needed to confirm the validity of the relationships.

As can be seen from Tables 6.2 and 6.3, the relationships between the topex-to-distance variables and windspeed have the highest correlations. Many of the other relationships, however, although appearing important, are

Table 6.2 *Inter-variable correlations*

	Topex	Top1km	Top2km	Top3km	DTC	Alt	z_0	GridEast
Top1km	0.71							
Top2km	<u>0.84</u>	<u>0.97</u>						
Top3km	<u>0.90</u>	<u>0.93</u>	<u>0.99</u>					
DTC	0.02	0.20	0.18	0.04				
Alt	−0.37	−<u>0.67</u>	−<u>0.61</u>	−<u>0.57</u>	−0.06			
z_0	0.14	0.12	0.12	0.14	0.31	<u>0.53</u>		
GridEast	−0.22	0.19	0.09	−0.01	<u>0.74</u>	0.07	0.23	
GridNor	<u>0.51</u>	0.29	0.39	<u>0.46</u>	0.09	−0.05	0.16	−0.03

Underlined values are 99% significant

Table 6.3 *Correlations with windspeed, 1989 and 1990*

	Topex	Top1km	Top2km	Top3km	DTC	Alt	z_0	GridEast	GridNor
Mean 89	−0.53	−<u>0.81</u>	−<u>0.84</u>	−<u>0.84</u>	−0.26	<u>0.68</u>	−0.43	−0.26	−0.29
Mean 90	−0.58	−<u>0.86</u>	−<u>0.83</u>	−<u>0.78</u>	−0.46	<u>0.73</u>	−0.30	−0.32	−0.35

Underlined values are 99% significant.

a direct result of the choice of, and location of, sites in this study (for example, the high correlation between *GridEast* and *DTC*).

Since the prevailing wind direction in the United Kingdom is south-west, it might be thought more appropriate either to take the south-westerly value of those topographic characteristics which vary with direction (topex in all its forms, *DTC* and z_0), or to weight the characteristic score to take account of the distribution of the wind at each of the sites. Harrison (1988) suggested that the exposure of a site in the British Isles (derived from topex measurements) should be weighted according to the wind direction distribution. No windspeed distribution data were available for the UKMO sites, so this extension of the method could not be followed. It was possible, however, to examine the relationship between site mean windspeed and the components from the south-west sector of all topex variables, *DTC* and z_0. All correlation coefficients were lower than the combined values used in Table 6.3, with an average magnitude of 0.70 for the topex-to-distance variables, and below 0.20 for *DTC* and z_0. This confirmed the use of the all-sector values for these parameters as being more suitable in relating each parameter to windspeed.

6.4 Regression analysis

The results presented here are divided into three parts. In the first, simple regression equations are formed between the mean windspeed for 1989 and each of the individual topographic characteristics. The equations are then used to predict the 1990 values at the sites, and the results are compared with the observed values. A linear approach is used because the data did not suggest any curvilinear relationships were present. In the second part, suitable combinations of variables are used in multiple regression equations. Again, the 1989 windspeed data are used to create the equations and the 1990 data are used to test the results. Finally, the best single and multiple regression techniques are tested on an independent site.

6.4.1 Simple regression techniques

The results in this section utilise the variables whose relationships with mean annual windspeed, as shown in Table 6.2, are statistically significant to the 99% level. These are the three topex-to-distance variables and altitude. Since the topographic variables do not change from 1989 to 1990, it is necessary to define one site which can be used as a reference site for the relationships between windspeed and topographic variables. The mean annual windspeed for the 21 sites is 5.14 m s^{-1} in 1989 and 5.78 m s^{-1} in 1990. The site which has the closest windspeeds to these averages in both years is West Freugh (UKMO), with a 1989 mean of 5.30 m s^{-1} and a 1990 mean of 5.76 m s^{-1}. This site was therefore chosen as the reference site for the prediction method. The difference in the mean windspeed at West Freugh between 1989 and 1990 was +0.46 m s^{-1}. This value was added to the constant term in each of the regression equations formed (one for each of the predictor variables).

Using the regression constants derived from the 1989 windspeed data, predictions were made for the windspeeds in 1990. In Fig. 6.3 the predicted and observed values of the annual mean windspeeds are plotted for each of the four predictor variables.

The results in Table 6.4 and Fig. 6.3 show that the predictions of the annual mean windspeeds are generally good. The three topex-to-distance variables produce predictions with correlations of 0.80 or higher, and only the *Top3km* prediction has an error in the prediction of the mean (calculated over all the sites) greater than 5%. Although the error in the prediction of the mean using altitude is the lowest of the four, the correlation is the poorest at 0.71 (less than 50% of the variance in the annual mean windspeed explained), as can be seen in Fig. 6.3.

Fig. 6.3 also shows that one site consistently under-performs for all three topex-to-distance variables. This site is River Brittle, the FC site in Skye, which has very large positive topex values (up to 74°) due to its sheltered position. The large differences between the values of topex-to-distance at River Brittle and the other sites (at least 40°) mean that this site is probably in a different statistical population.

6.4.2 Multiple regression techniques

The use of multiple regression techniques has been the basis of previous studies in this field (White, 1979; White & Smith; 1982). In the latter work, combinations of variables were used to predict the windspeeds at a set of 39 sites across Britain. For statistical correctness, Spiegel (1961) recommends that the variables used in a multiple regression equation should be highly correlated with the dependent variable (in this case, annual mean windspeed) but independent of each other. On examination of Tables 6.2 and 6.3, it can be seen that each of the four variables used, namely *Top1km*, *Top2km*, *Top3km* and *Alt*, is highly correlated with the others, so only one of these can be included in any multiple regression equation. Of the remaining variables, *DTC* comes nearest to fulfilling the requirements of independence with the main independent variables, with both *GridEast* and *GridNor* also close.

Windspeeds in forest in complex terrain

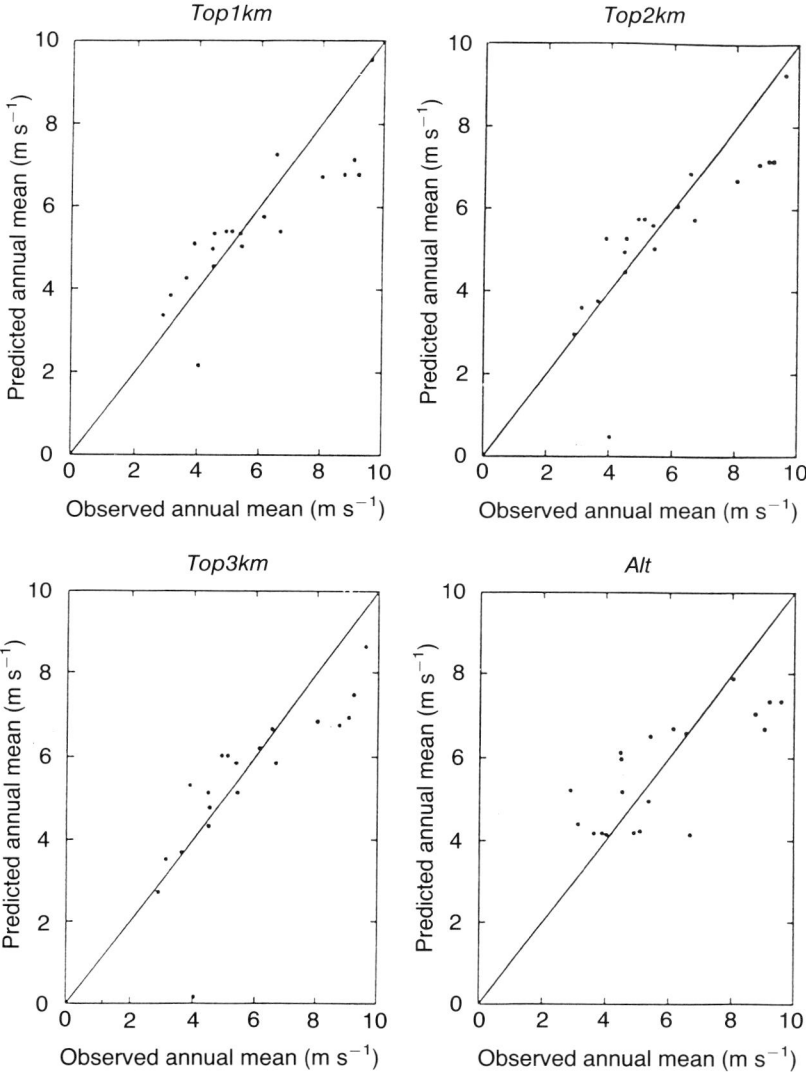

Fig. 6.3. Annual windspeed predictions using simple regression techniques plotted against observed values.

The high correlations between *DTC* and *GridEast*, and *Alt* and z_0, mean that neither *GridEast* nor z_0 is suitable. Therefore, the equations are formed with one of the four variables (*Top1km*, *Top2km*, *Top3km* and *Alt*) together with *DTC* and *GridNor*. The correlation of these two latter terms with the dependent variable, annual mean windspeed, is not really high enough for them to be included, but the attempt was made nevertheless.

Table 6.4. *Performance of single predictor variables*

Predictor variable	1990 Mean (m s^{-1})	SD (m s^{-1})	r	% error
Observed	5.78	2.12	NA	NA
Top1km	5.56	1.62	0.81	−3.9
Top2km	5.49	1.86	0.84	−5.0
Top3km	5.43	1.30	0.84	−6.1
Alt	5.70	1.30	0.71	−1.4

NA, not applicable.

The results are plotted in Fig. 6.4 and the statistics are shown in Table 6.5. The correlations are all above 0.80 and two of them, those for *Top1km* and *Alt*, are above 0.90. Estimation of the mean windspeed for the sites is also better than for the single predictor method, with improvements for all main variables apart from *Top3km*. The results, with the exception of those produced using *Top3km*, improve the accuracy of the prediction of the mean windspeed with the addition of *DTC* and *GridNor*, despite the loss of two degrees of freedom. All correlations are significant to the 99% level.

6.4.3 A test on independent data

To test both the simple and multiple regression methods for the prediction of annual mean windspeed, a location not previously used in the creation/validation exercises was chosen. Edinburgh Royal Observatory (RO) is 159 m above sea level and has near-average values of the topographic variables compared with the dataset previously used in the creation and validation of the regression equations. The site is close to the coast in the north and east sectors, but the long distances to the south and the west mean that the value of *DTC* (112 km) is again close to the average. Roughness lengths are high, since the site is located on a hill surrounded by urban areas. This was not thought to be a problem, however, since roughness length is not used in either the single or multiple regression equations. The grid position is central both in the North–South and East–West ranges of the dataset.

Using regression equations derived from both the 1989 and 1990 data from the other 21 sites, predictions, of the annual mean windspeed at Edinburgh RO were made, using the same combinations of topographic characteristics that were found to be important previously (i.e. *Top1km*, *Top2km*, *Top3km*, *Alt*, *DTC* and *GridNor*).

The results presented in Table 6.6 show that all variants of both the simple and multiple methods produce good results for the mean windspeeds for 1989

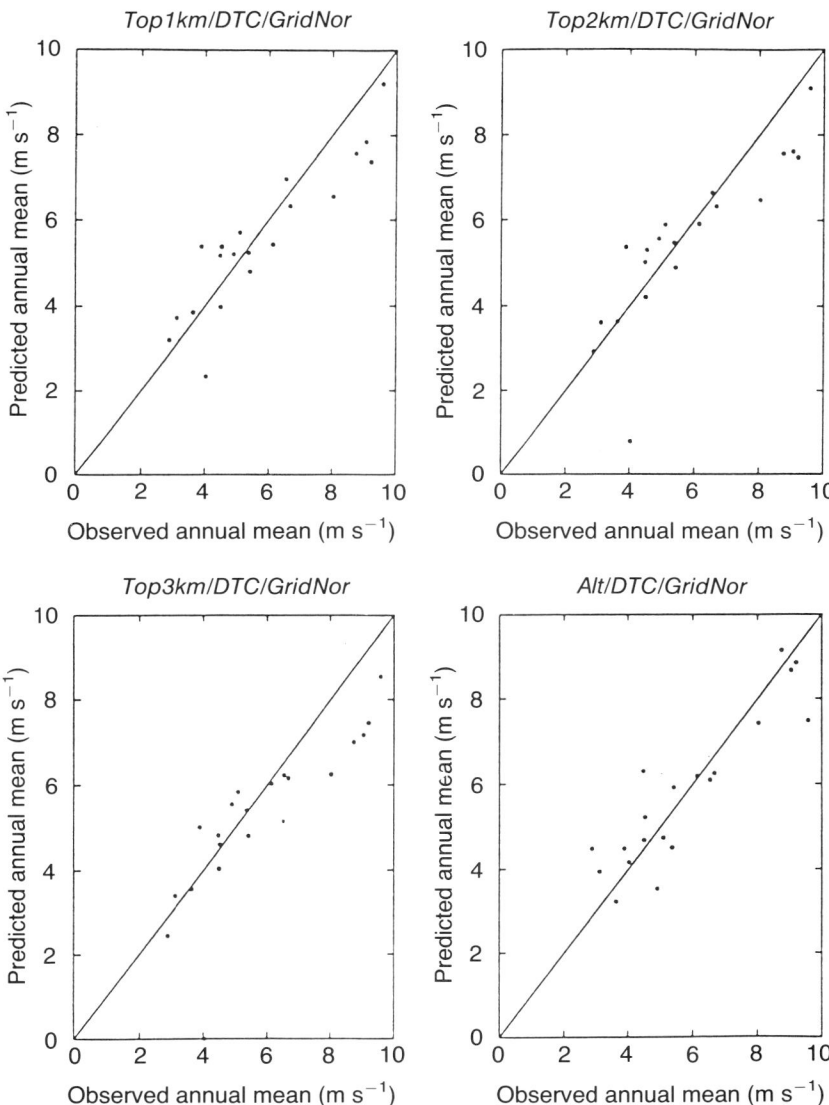

Fig. 6.4. Annual windspeed predictions using multiple regression techniques plotted against observed values.

and 1990. The exception is the use of *Alt* as the main predictor variable; performance is poorest of the four predictor variables, with errors greater than 15% in all cases. The predictions using the topex-to-distance variables, with the exception of *Top3km* in the multiple regression, all have errors of less than 7.5%.

Table 6.5. *Performance of multiple predictor variables*

Main predictor variable	1990 Mean (m s^{-1})	SD (m s^{-1})	r	% error
Observed	5.78	2.12	NA	NA
Top1km	5.62	1.71	0.91	−3.0
Top2km	5.54	1.88	0.86	−4.2
Top3km	5.26	1.94	0.83	−9.0
Alt	5.82	1.78	0.90	−1.0

NA, not applicable.

Table 6.6. *Observed and predicted windspeed data for Edinburgh Royal Observatory, 1989–90*

	1989		1990	
	Mean (m s^{-1})	Error %	Mean (m s^{-1})	Error %
Observed	6.13	NA	6.75	NA
Predictor				
Top1km	6.39	−4.2	7.23	−7.1
Top2km	6.05	−1.3	6.72	−0.4
Top3km	5.96	−2.8	5.49	−1.9
Alt	4.97	−18.9	5.49	−18.7
Top1km/DTC/GridNor	6.42	−4.7	7.05	4.4
Top2km/DTC/GridNor	6.06	−1.1	6.56	−2.8
Top3km/DTC/GridNor	5.48	−10.6	6.43	−4.7
Alt/DTC/GridNor	5.17	−15.7	5.51	−18.3

NA, not applicable.

The use of topographic variables therefore appeared to be successful in this case. The method is simple and requires no on-site monitoring. The results for Edinburgh RO show that the windspeed at a site which has been excluded from the equation creation stage, but whose characteristics all lie within the limits for the sites that are used to create the equations, can be predicted with reasonable accuracy.

6.5 Conclusions

The methods outlined here appear to work well, with the predictions by both single predictors and combinations of the variables producing good estimates of the annual mean windspeed that are highly correlated with the observed values and have low errors. The use of annual mean windspeeds has applications in forestry, particularly in the prediction of growth rates and the determination of planting limits. The prediction of tree damage caused by extreme windspeeds requires higher temporal resolution. The methods outlined here are of limited use in that they can be applied to the topographic variables at a site to estimate the annual mean windspeed. This may then be of value in determining whether or not to continue the study of the wind climatology in the area with more time-consuming and expensive data gathering techniques. The innovative use of topex-to-distance has been found to be more strongly related to windspeed than standard topex.

Previous studies using the same basic criteria and variables have found little success in the prediction of windspeeds. It is important to note that the results are region-specific. If a prediction were required for a site in Wales, for example, where no wind data were available, then it would be necessary to re-define the inter-variable relationships using data from other Welsh sites. However, within a region the method has been shown successfully to predict a windspeed at a site which was not used in the construction and validation of the equations. Smith (1982) proposed an index of windiness for the United Kingdom based on the standardised anomaly of the windspeed at a site to its long-term mean, but found that the results had to be regionalised.

In this study, the lack of high-altitude sites in the north of Scotland also affected the results. The use of topex-to-distance variables, on the other hand, was successful, and may yet be adopted as a replacement for the more limited topex that is in forestry use. The use of digital terrain models to calculate topex allows easy adoption of distance limits whereas field definition would be impractical. The most useful topex-to-distance variable of the three used may be dependent upon the severity of the topography. In the example of Edinburgh RO, *Top2km* was the best predictor, whilst for the analysis of the 21 sites in the dataset predictions using *Top1km* were more accurate.

Both the simple and multiple regression methods produce good estimates of the mean windspeeds for 1989 and 1990 at Edinburgh Royal Observatory, a site not included in the original analysis. The site has average characteristics in almost all cases when compared with the original 21 sites used in the regression equation and validation stages and the results are very encouraging. Further analysis will be required from a larger range of sites to assess the true strengths of the relationships and to see whether a predictive method can be derived.

Acknowledgements

The work reported here was carried out by P. H. during a CASE research studentship at the Climatic Research Unit, University of East Anglia, funded by the Science and Engineering Research Council and the Forestry Commission. The authors are very grateful to Beki Barthelmie of Risø National Laboratory, Denmark, for her comments and advice.

References

Baker, C. J. (1984). Determination of topographical exposure factors in complicated hilly terrain. *Journal of Wind Engineering and Industrial Aerodynamics*, **17**, 239–49.

Baker, C. J. (1985). The determination of topographical exposure factors for railway embankments. *Journal of Wind Engineering and Industrial Aerodynamics*, **21**, 89–99.

Bendelow, V. C. & Hartnup, R. (1980). *Climatic Classification of England and Wales*. Soil Survey Technical Monograph no. 13. Rothamsted Experimental Station.

Harrison, S. J. (1988). Numerical assessment of local shelter around weather stations. *Weather*, **43**, 325–30.

Jarvis, N. J. & Mullins, C. E. (1987). Modelling the effects of drought on the growth of Sitka Spruce in Scotland. *Forestry*, **60**, 13–30.

Klinka, K., Pojar, J. & Meidinger, D. V. (1991). Revision of biogeoclimatic units of coastal British Columbia. *Northwest Science*, **65**, 32–47.

Miller, K. F., Quine, C. P. & Hunt, J. (1987). The assessment of wind exposure for forestry in upland Britain. *Forestry*, **60**, 179–92.

Murakami, S. & Komine, H. (1983). Prediction method for surface wind velocity distribution by means of regression analysis of topographic effects on winds. *Journal of Wind Engineering and Industrial Aerodynamics*, **15**, 217–30.

Pyatt, D. G., Harrison, D. & Ford, A. S. (1969). *Guide to Site Types in Forests of North and mid-Wales*. Forestry Commission, Forest Record no. 69. HMSO, London.

Quine, C. P. & Reynard, B. R. (1990). *A New Series of Windthrow Monitoring Areas in Upland Britain*. Forestry Commission Occasional Paper 25. Forestry Commission, Edinburgh.

Quine, C. P. & White, I. M. S. (1993). *Revised Windiness Scores for the Windthrow Hazard Classification: The Revised Scoring Method*. Forestry Commission Research Information Note 230. Forestry Authority.

Quine, C. P. & Wright, J. A. (1993). *The Effects of Revised Windiness Scores on the Calculation and Distribution of Windthrow Hazard Classes*. Forestry Commission Research Information Note 231. Forestry Authority.

Smith, S. G. (1982). An index of windiness for the United Kingdom. *Meteorological Magazine*, **111**, 232–47.

Smithson, P. A. (1987). An analysis of windspeed and direction at a high-altitude site in the southern Pennines. *Meteorological Magazine*, **116**, 76–85.

Speigel, S. (1961) *Theory and Problems of Statistics*. McGraw-Hill, New York.

UK Meteorological Office (1990). *Monthly Weather Report, 1989: Annual Summary*. HMSO, London.

UK Meteorological Office (1991). *Meteorological Glossary*. HMSO, London.
White, E. J. (1979). The prediction and selection of climatological data for ecological purposes in Great Britain. *Journal of Applied Ecology*, **16**, 141–60.
White, E. J. & Smith, R. I. (1982). *Climatological Maps of Great Britain*. Institute of Terrestrial Ecology, Edinburgh.
Wieringa, J. (1992). Updating the Davenport roughness classification. *Journal of Wind Engineering and Industrial Aerodynamics*, **41**, 357–68.
Wilson, J. D. (1984). Determining a TOPEX score. *Scottish Forestry*, **38**, 251–6.
Worrell, R. & Malcolm, D. C. (1990*a*). Productivity of Sitka Spruce in Northern Britain. I. The effects of elevation and climate. *Forestry*, **63**, 105–18.
Worrell, R. & Malcolm, D. C. (1990*b*). Productivity of Sitka Spruce in Northern Britain. II. Prediction from site factors. *Forestry*, **63**, 119–28.
Yamazawa, H. & Kondo, J. (1989). Empirical-statistical method to estimate the surface wind speed over complex terrain. *Journal of Applied Meteorology*, **21**, 996–1001.

Part II
Mechanics of trees under wind loading

7
Understanding wind forces on trees

C. J. WOOD

Abstract

This overview chapter considers some key topics in the description, measurement and analysis of tree elasticity, dynamics and aerodynamics. Engineering mathematics and the hypothesis of adaptive growth are shown to be in agreement, and it is acknowledged that Nature achieves, by adaptive growth, an optimisation of strength that is invariably more elegant than that achieved by the human design process. The engineering contribution to tree biomechanics includes not only recent developments in the understanding of turbulent wind, but also some classical mechanics in the description of trees as vibrating systems exposed to random excitation. Some exact solutions for bending under load and for sway vibrations are examined in relation to the adaptive growth hypothesis, in order to draw some tentative conclusions about tree failure mechanisms. Finally, the importance of sway resonance is reconsidered in the light of the probable effect of windspeed on aerodynamic damping.

7.1 Introduction

The purpose of this chapter is to survey the status of current research on tree mechanics in relation to windthrow. This requires more than a bland acknowledgement of the many notable research achievements. It also demands interpretation that will stimulate future thought. To achieve this, the discussion is focused upon a few distinct topics. They are chosen particularly to highlight the interdisciplinary interaction between biology and engineering. This cross-fertilisation has yielded great benefits already and is likely to exert a powerful influence on future research.

The central thread in this discussion is adaptive growth. This is illustrated in Fig. 7.1, which compares the cross-section of an unusual tree

Fig. 7.1. Comparison of tree root and engineering beam.

root with the cross-section of an engineering 'rolled steel joist'. The root came from a Sitka spruce that had been poorly anchored, and had swayed from the base. The root in Fig. 7.1 was horizontal, just under the surface of the ground. It had been bent back and forth, and under this regime of repeated strain, growth was optimised and the wood fibres were laid down at the top and bottom where the tensile and compressive stresses were greatest. The steel beam also is of economical design, with a concentration of material in the regions of greatest stress. However, the engineering design cannot compare with the simple elegance of the root. In Nature, this stress equalisation develops adaptively. The living material grows in response to stresses that occur. Thus trees compensate for damage or for a changing environment.

The 'adaptive growth hypothesis' has been championed by Mattheck (1989), who has produced computer designs for many living structural elements. In each case, the shape is computed which fulfils the design function with a uniform distribution of stress along the surface. Invariably, the computed shapes are remarkably similar to those observed. Fig. 7.2, from Mattheck (1989), shows a tree branch junction. The weight F of the branch is supported, and the graph shows that the tensile stress σ is approximately constant between $S_0 = 0$ and $S_0 = 1$ in the curved upper surface.

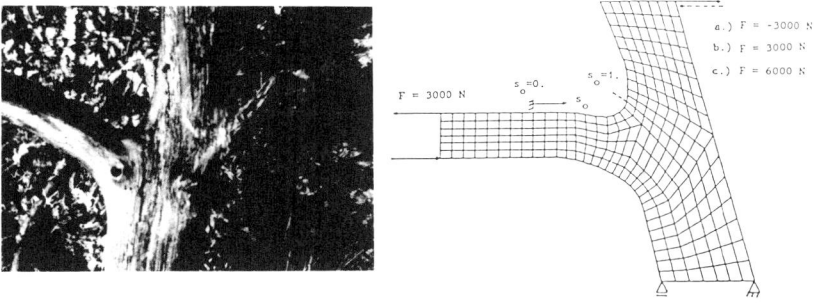

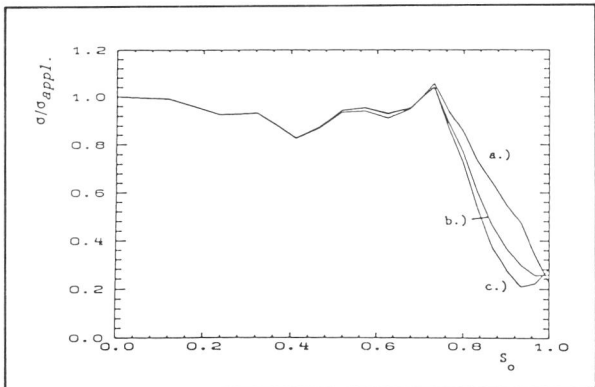

Fig. 7.2. Uniform stress in a tree branch. From Mattheck (1989).

7.2 Engineering theories of bending

Wind damage usually arises from excessive bending. To discuss this, it is necessary to examine the relevant equations.

7.2.1 Bending in a circular arc

The Euler Bernoulli (1705) theory of bending appears in many textbooks. It is valid for beams of uniform cross-section and uniform stiffness EI, which are bent by a constant moment M to a uniform radius of curvature R.

$$\frac{1}{R} = \frac{M}{EI} \tag{7.1}$$

The stiffness is the product of Young's modulus E and the second moment of area I of the cross-section. For a tree stem of diameter d, this is given by:

$$I = \frac{\pi d^4}{64} \tag{7.2}$$

The tensile stress σ in the wood fibres is proportional to the distance y from the centre of the stem. Thus the greatest stress (tensile or compressive) is in the surface fibres where $y = d/2$:

$$\sigma = \frac{Md}{2I} \tag{7.3}$$

7.2.2 Bending with small deflections

The circular arc condition may be relaxed if the curvature of the beam is so small that the deflected shape $x(z)$ remains close to and approximately parallel to the undeflected axis. The curvature is then approximately equal to the second derivative of the deflection x:

$$\frac{1}{R} = \frac{d^2 x}{dz^2}$$

so that Eq. (7.1) becomes

$$\frac{d^2 x}{dz^2} = \frac{M}{EI} \tag{7.4}$$

7.2.3 The Second Moment Area Theorem

These equations lead to the extremely convenient Second Moment Area Theorem. This gives the lateral deflection $x(z)$ at height z on a rigidly rooted tree, directly from the curvature $d^2x/d\zeta^2$ at all heights ζ below z (Fig. 7.3). The advantage is that it requires only one integration:

$$\int_0^z (z - \zeta) \frac{d^2 x}{d\zeta^2} d\zeta = x(z) \tag{7.5}$$

7.2.4 Bending with large deflections

In some cases, small deflection theory may be inadequate for the large movements of trees in the wind. For this reason Morgan (1989) has promoted a computer model which tolerates large deflections and slopes. It is based upon Eq. (7.4), applied to very short lengths of stem or branch, as illustrated in

Understanding wind forces on trees

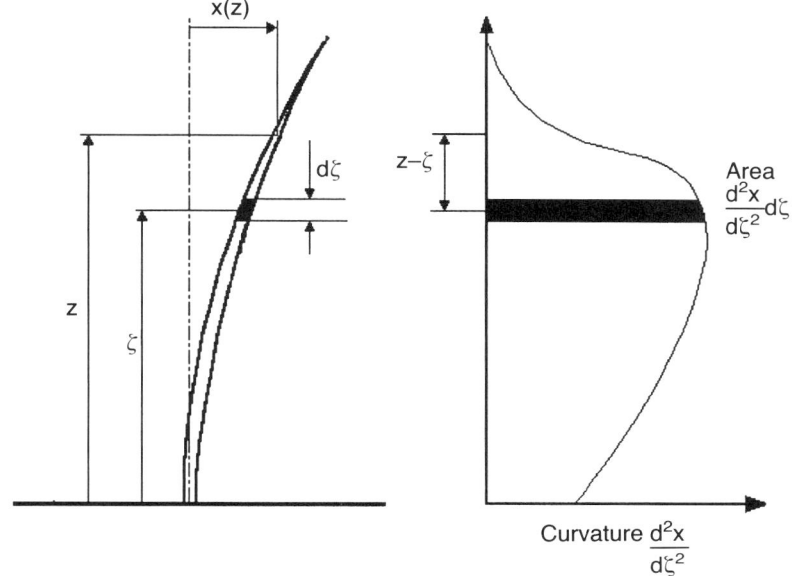

Fig. 7.3. Diagram for Moment Area Theorem.

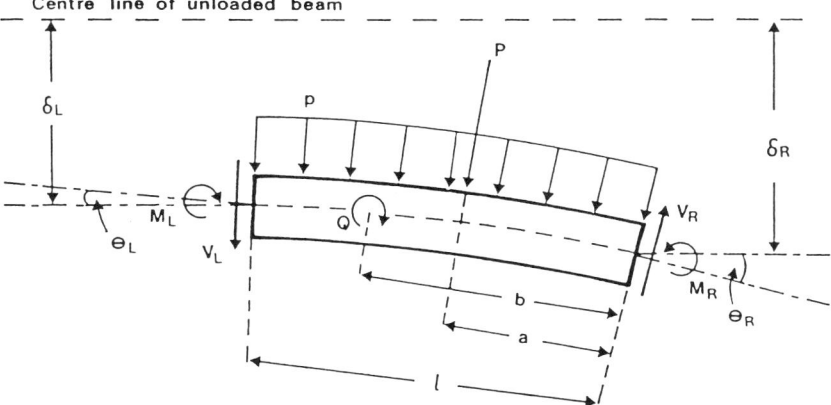

Fig. 7.4. Large deflection model. From Morgan (1989).

Fig. 7.4. Transmitted moments M and shear forces V are allowed at each end, together with the possibility of a transverse point load P or moment Q or a distributed load p applied to the element itself. The displacement $\delta_R - \delta_L$ and rotation $\theta_R - \theta_L$ of one end relative to the other are then always small, and the computation simply adds up the deflections and slopes for a series of elements, jointed end to end.

The method has been applied by Morgan, Cannell, Milne and others, to a number of problems in tree mechanics (e.g. Cannell & Morgan, 1988; Morgan & Cannell, 1990; Milne & Blackburn, 1989; Milne, 1991). In particular, Milne & Blackburn used the method for the analysis of pull tests, and an application to vibrations is expected shortly (J. Morgan, personal communication 1993).

7.3 Field measurements

Before applying any of these calculations, it is necessary to define the distributions of mass and stiffness of a tree. Inevitably, the measurements involved are far more difficult in a forest than in a laboratory.

7.3.1 Mass distribution

The mass per metre is usually measured by cutting up the stem and weighing the logs. The branches are also weighed. If the branches are horizontal, the mass is assigned at the attachment level. If they slope, then the appropriate level for the mass will be more difficult to estimate.

7.3.2 Distribution of bending stiffness

The bending stiffness of a section of tree stem is the product EI in Eq. (7.1). It is the moment required to cause unit radius of curvature (Eq. 7.4). The stem diameter d is needed in Eq. (7.2) to calculate the second moment of area I.

One way to measure the bending stiffness EI, is to support a cut length of stem on trestles and hang a weight in the middle. Unfortunately the deflections are difficult to measure accurately in the field. Also, the calibration requires a careful analysis of the taper.

It may be better to pull a live tree sideways. Fig. 7.5 shows the technique used by Milne (1991), using a number of measuring strings to map the deflected shape. Milne used Morgan's (1989) computation procedure and deduced Young's modulus by matching the theoretical and measured deflections at the load point. He reports, however, that his measured deflections at other points then differed apparently by up to 20 cm from the predicted curve. This raises questions about the assumption that E is constant for the whole of a tree, about the accuracy of the representation of diameter in the analysis,

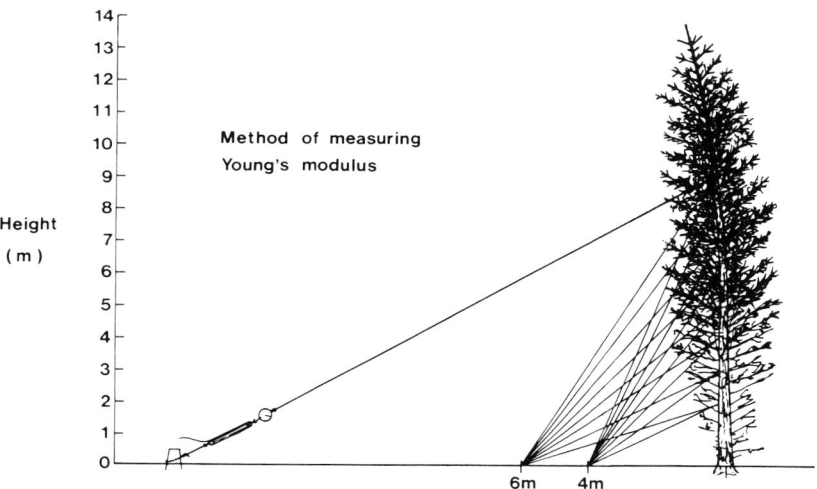

Fig. 7.5. Diagram of tree pull rig. From Milne (1991).

and also about the assumption of a rigid ground anchorage. Burden (1993) has represented the ground anchorage by a hinge with its own separate stiffness.

7.3.3 Sway frequency and damping

Measuring the natural frequency of a tree is easy. After a little trial and error, a resonant sway vibration can be established by periodic hand pressure, and the cycles counted against time.

A more challenging measurement is the damping which attenuates the motion. The 'damping ratio' δ is defined formally in the universal vibration equations for a mass m on a spring of stiffness k. The term $c\,dx/dt$ allows for a force *proportional to the velocity* and opposing it:

$$m\frac{d^2x}{dt^2} + c\frac{dx}{dt} + kx = 0 \qquad (7.6)$$

Eq. (7.6) is often written in the form

$$\frac{d^2x}{dt^2} + 2\delta\omega_n\frac{dx}{dt} + \omega_n^2 x = 0 \qquad (7.7)$$

where ω_n is the natural frequency

$$\omega_n^2 = k/m$$

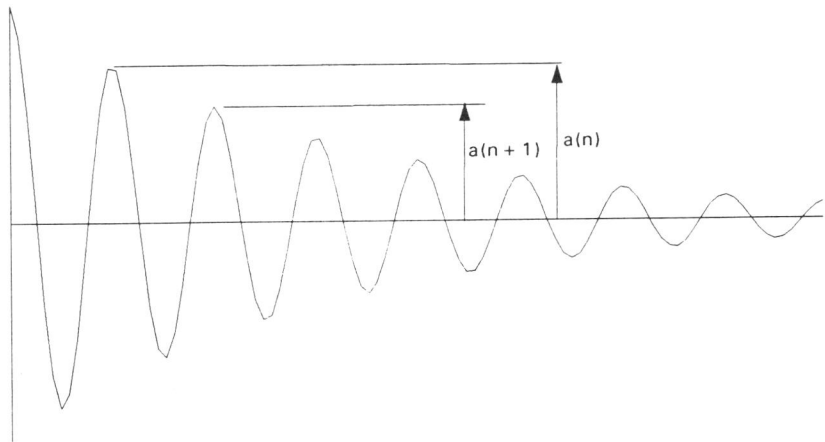

Fig. 7.6. Measurement of damping.

and δ is the damping ratio

$$\delta = \frac{c}{2\sqrt{km}}$$

This may be determined from the ratio of successive displacement peaks $a(n)$ and $a(n+1)$ as shown in Fig. 7.6:

$$a(n)/a(n+1) = \exp\left(\frac{2\pi\delta}{\sqrt{1-\delta^2}}\right) \qquad (7.8)$$

Damping arises from hysteresis in the timber and the soil and the movement of foliage through the air. Also, energy may be dissipated by contact between the crowns of adjacent trees. Actually, none of these are linear functions of velocity, so the equations are not strictly applicable. Nevertheless, Eqn (7.8) remains a practical definition of damping, and measuring the decay of tree vibrations *does* yield a constant value of the damping ratio.

Until now, analysis of tree bending and vibration has not been sufficiently sophisticated to take account of branches as anything other than added mass at each level. More elaborate theories are perfectly possible, but these will require more detailed information about the spatial distributions of mass and stiffness among the branches themselves.

7.3.4 The Gardiner Average Tree

This discussion requires physical dimensions with mass and stiffness distributions for a typical tree. The data source used here is Gardiner (1989) and the tree is described in Figs. 7.7 and 7.8.

The stem mass values in Fig. 7.7 have been calculated from the taper measurements in Fig. 7.8, with a small adjustment to the nominal timber density to maintain a self-consistent set of values. Thus with a wood density, assumed constant, of 918 kg m^{-3}, the reference mass per metre at breast height (0.09 h) is 19.63 kg m^{-1} on Gardiner's reference diameter of 0.165 m. The stem mass distribution then integrates over a 15 m height to give Gardiner's total stem mass of 162.3 kg. The branch mass distribution in Fig. 7.7 is from Gardiner's (personal communication, 1988) measurements on a typical Sitka spruce tree. The curve integrates to agree with Gardiner's total of 49.5 kg.

Fig. 7.8 shows stem taper measurements for all of the 10 trees. Like the mass values, the diameters are expressed as ratios with a reference value for

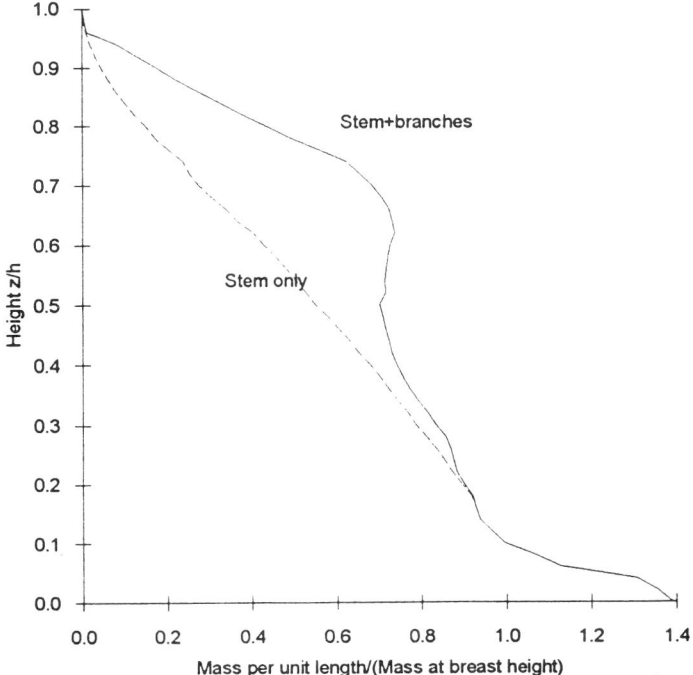

Fig. 7.7. Mass distribution for Gardiner Average Tree.

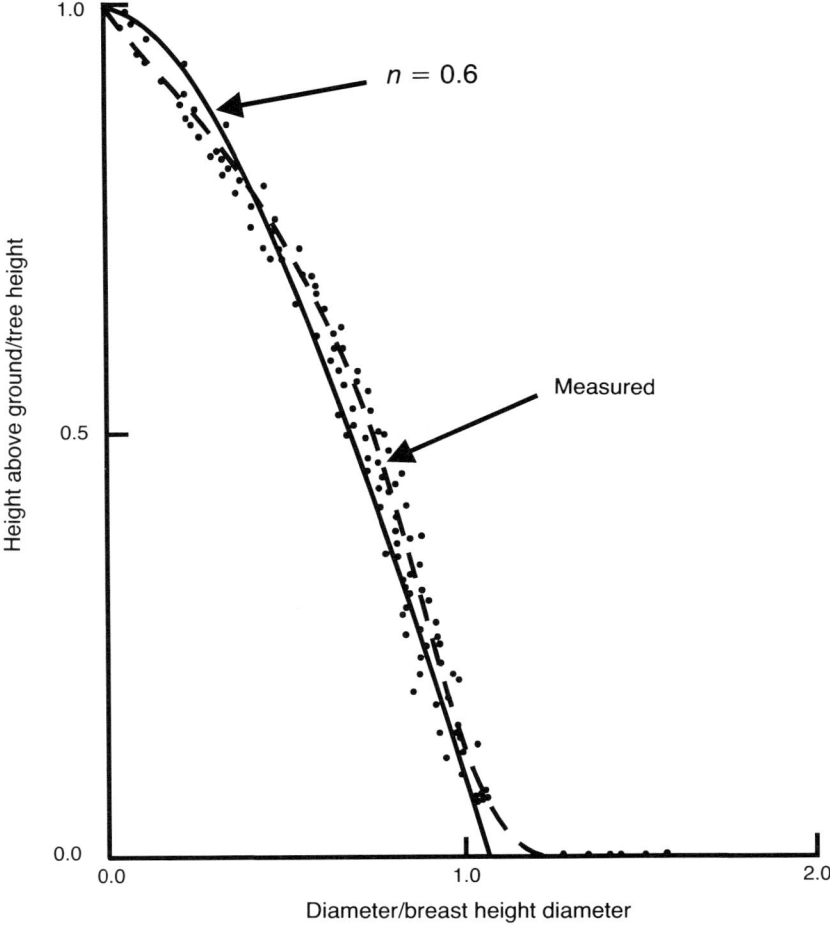

Fig. 7.8. Diameter measurements from Gardiner (1989).

each tree at breast height (approximately 1.3 m). Gardiner chose a 0.6 power law curve to represent these data. This allowed some easy analysis but the term d^4 in Eq. (7.2) magnifies errors and it is arguable that such an approximation may obscure some important characteristics. The alternative curve fits the data better. It is superimposed by the present author, and measured to produce a numerical description of the 'Gardiner Standard Tree', as distinct from the 'Gardiner Power Law Tree'.

7.4 Modelling at small scale

Tree models were tested in wind tunnels as long ago as 1963 (Walshe & Fraser, 1963), but Papesch (1984) was probably the first to construct aeroelastic model trees in bulk. He used tapered brushes on twisted wire stems, and filmed their motion when planted in various patterns in the Oxford wind tunnel. The models vibrated at about 7 Hz, which was acceptable at 1 : 75 scale, but they were much too heavy and much too stiff and therefore the sway deflections were unrealistically small.

7.4.1 Aeroelastic tree models

Since then, Dr B. A. Gardiner of the Forestry Commission, and the Wind Engineering Research Group in Oxford have developed an improved aeroelastic model using injection-moulded nylon (Stacey *et al.*, 1994). Arrays of up to 12 000 of these models have been used to simulate plantation dynamics in the Oxford wind tunnel (Fig. 7.9).

The models were 200 mm tall and were designed as shown in Table 7.1 to model the stiffness, mass and drag of a typical 15 m Sitka spruce, described hereafter as the Gardiner Standard Tree. They were much lighter and more flexible than the Papesch bottle-brush trees, but they were still more heavy and stiff than they should have been. Nevertheless, both the vibration frequency and the still-air damping were right and this allowed some confidence in the measured overturning moments.

7.4.2 Modelling wind flow

It is not sufficient just to design the model trees correctly. The motion must be excited by a correctly scaled gusty wind blowing over the canopy. Figs. 7.10 and 7.11 show that the windspeed and turbulence profiles were correct and the gust frequencies were realistic, all at a scale of 1 : 75. The target curves are from the full-scale data correlations of ESDU International Ltd (ESDU, 1982, 1985). Fig. 7.12 shows the devices used in the wind tunnel to generate this wind simulation.

7.4.3 Measuring overturning moments

One specimen tree was mounted on a two-component strain gauge balance (Fig. 7.13). Positioned at chosen points in the model plantation, this measured the fluctuating overturning moment at ground level (Fig. 7.14). By computer

Fig. 7.9. Aeroelastic tree models in wind tunnel.

sampling and analysis, mean values or the probability distribution for the largest moment could be examined.

As an example, Fig. 7.15 shows how the mean and extreme moments decay with distance from the upwind edge of a model plantation.

Table 7.1. *Dynamical similarity of aeroelastic tree models*

Name of parameter	Symbol	Full-scale value	Model value
Tree characteristics			
Mass	$M/\rho h^3$	0.051	0.102
Natural sway frequency	$fh/\bar{u}_R$	0.165	0.163
Still-air damping ratio	δ	0.048	0.048
Drag in uniform flow at speed U	$D/(\tfrac{1}{2}\rho U^2 h^2)$	Eq. (7.10)	0.018
Description of the wind			
Mean windspeed ratio at stated height	$\bar{u}/\bar{u}_R$	Fig. 7.11	Fig. 7.11
Turbulence intensity at stated height	$\sigma_u/\bar{u}$	Fig. 7.11	Fig. 7.11
Gust frequency at stated height	$nh/\bar{u}$	Fig. 7.12	Fig. 7.12
Nominal Reynolds number	$\rho \bar{u}_R h/\mu$	3.1×10^7	8.2×10^4
Result of experiment			
Bending moment at base of tree	$Q/(\tfrac{1}{2}\rho \bar{u}_R^2 h^3)$	As model	Measured

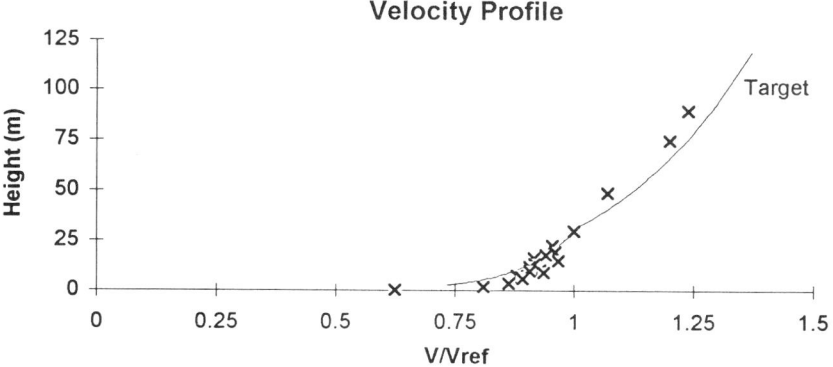

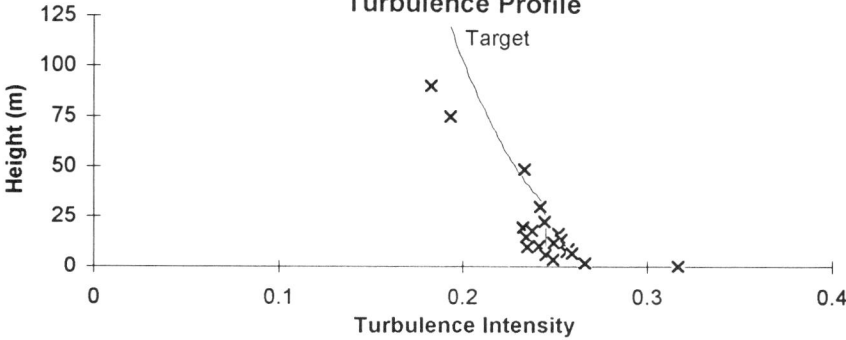

Fig. 7.10. Mean velocity (*a*) and turbulence (*b*) profiles in wind tunnel.

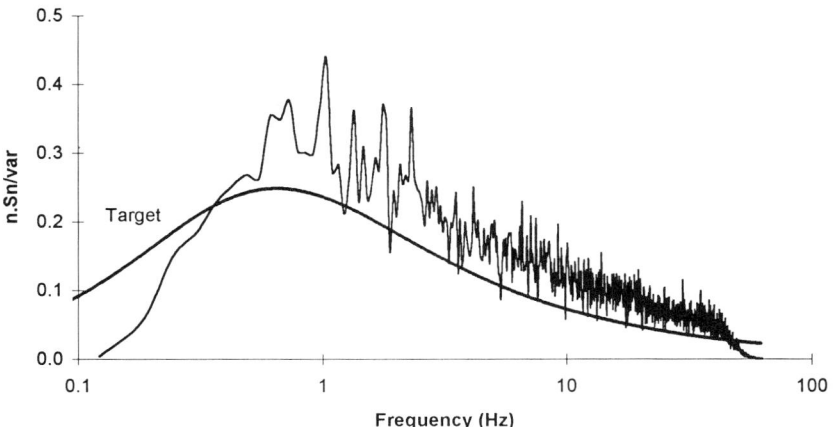

Fig. 7.11. Gust frequency spectrum in wind tunnel.

Fig. 7.12. Flow control devices in wind tunnel.

Fig. 7.13. Miniature balance.

7.5 Wind forces on a single tree

7.5.1 Drag coefficients

The link between the windspeed and the force on a tree is made by defining a drag coefficient C_D which defines the drag D for any size of tree (h) and any value of the reference windspeed U with air density ρ:

$$C_D = \frac{D}{0.5\rho U^2 h^2} \tag{7.9}$$

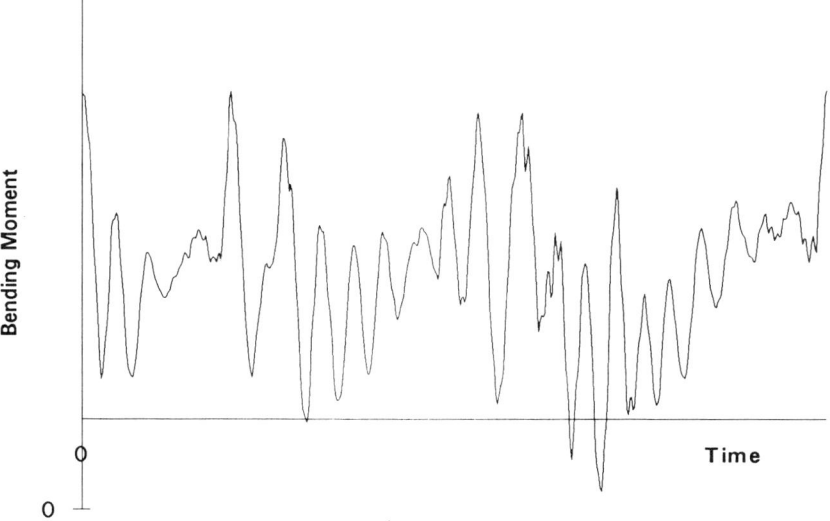

Fig. 7.14. Overturning moment time series.

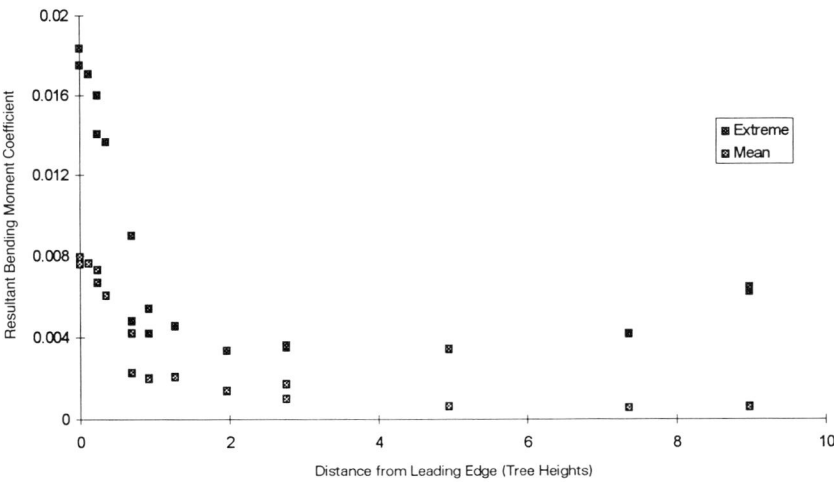

Fig. 7.15. Variation of overturning moments with distance from forest edge.

Several investigators have measured aerodynamic drag of trees or parts of trees. Mayhead *et al.* (1975) put whole small trees and the top parts of larger trees in a wind tunnel and measured the drag directly in a uniform steady wind. For Sitka spruce, the empirical formula below gives the drag D in Newtons:

$$D = 0.4352\, U^2 M^{0.667} \exp(-0.0009779\, U^2)$$

Assuming the standard sea level air density ρ of 1.225 kg m^{-3}, the drag coefficient (Eq. 9.7) is:

$$C_D = 0.7105 \left(\frac{M}{h^3}\right)^{0.667} \exp(-0.0009779\, U^2)$$

The Gardiner Standard Tree has a height h of 15 m and the live branch mass M is 49.5 kg, so that

$$C_D = 0.0426 \exp(-0.0009779\, U^2) \tag{7.10}$$

The exponential term in Eq. (7.10) emulates the effect of branch and foliage deflection. This streamlines the tree, so that the drag coefficient is halved between 0 and 27 m s^{-1}. Thus, over this speed range, D increases approximately in proportion to $U^{1.8}$ rather than U^2.

The difficulty with the Mayhead drag formula is that U is the speed of a test flow which is uniform over the height of the tree. For a tree in a forest, the corresponding speed is that of the retarded flow deep among the trees. Fig. 7.16 shows that this is not uniform, so there is no definable value for U.

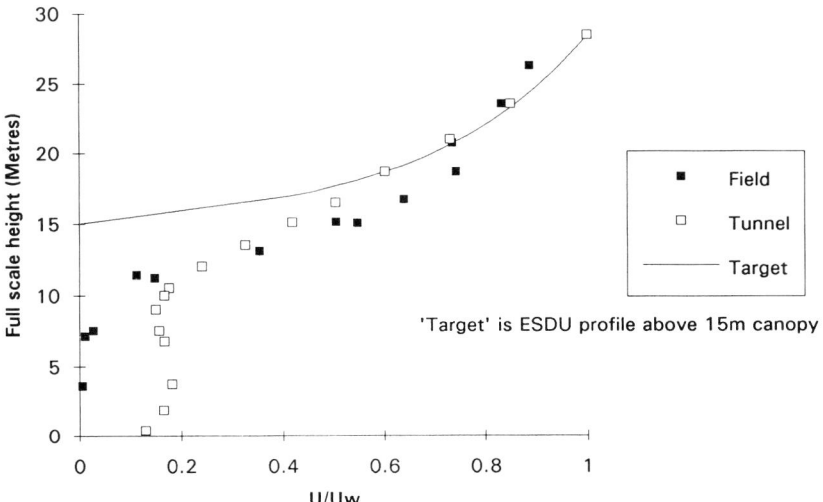

Fig. 7.16. Mean velocity profiles above and below canopy height.

To investigate how the drag of a tree is modified by canopy shelter, Stacey et al. (1994b) measured the bending moment coefficient C_Q of a tree model, first in uniform flow to match the Mayhead data, and then among a large array of similar tree models. For the canopy-sheltered test, the reference windspeed was taken at $1.9h$ (28.5 m for the Gardiner 15 m tree). This corresponds to the reference height where $U/U_W = 1$ in Fig. 7.16. The values of C_Q were 0.0101 in uniform flow and 0.00061 when sheltered. From actual mid-forest tree deflections Gardiner (1992) estimated C_Q to be 0.00085. These similar results confirm that the mean drag on a tree sheltered by many others, amounts to only 6–8% of the fully exposed value in the same windspeed.

7.5.2 Canopy drag and Reynolds stress

An alternative way to assess the mean drag forces on trees is to measure the Reynolds shear stress in the flow above the canopy.

The wind above the forest canopy has mean velocity components U parallel to the ground and W vertically upward, together with superimposed fluctuating components u, v and w in streamwise, lateral and vertical directions respectively. The product $\rho(W+w)$ describes an instantaneous upward mass flow per unit area which, when multiplied by the simultaneous instantaneous streamwise velocity $(U+u)$, gives an instantaneous upward flux of streamwise momentum per unit area. When time-averaged, the mean values of u and w are zero and the mean upward momentum flux becomes equal to $\rho UW + \overline{\rho uw}$. This momentum flux may be defined at the top boundary of a control volume of height H and streamwise length dx, while the drag on individual trees provides a horizontal shear stress τ_C on the lower boundary, as shown in Fig. 7.17.

In general, there will be unequal momentum flux terms at the upstream and downstream vertical boundaries of the control volume, so the momentum balance equation is

$$U_H W_H + \overline{u_H w_H} + \frac{\tau_c}{\rho} + \frac{d}{dx}\int_0^H \overline{(U+u)^2}\, dz = 0 \tag{7.11}$$

Within the control volume, mass is conserved, so the time-averaged continuity equation relates the vertical mean velocity W_H across the upper boundary to the flow through the vertical faces:

$$W_H + \frac{d}{dx}\int_0^H U\, dz = 0 \tag{7.12}$$

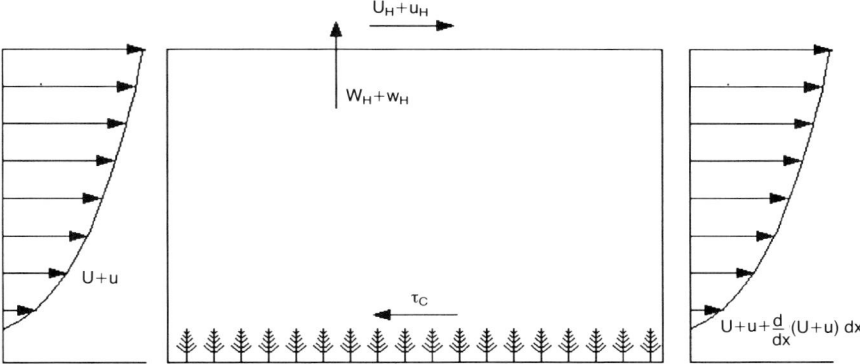

Fig. 7.17. Diagram for Reynolds stress.

In the special case where the mean velocity profile does not vary in the streamwise direction, Eq. (7.12) shows that W_H is zero. If in addition, the *turbulence* profile does not vary in the stream direction, then Eq. (7.11) reduces to the familiar form

$$\tau_c = -\rho \overline{u_H w_H} \qquad (7.13)$$

The analysis is set out here in order to emphasise that the validity of Eq. (7.13) depends upon spatial uniformity of the flow, which is also indicated by an invariance of Reynolds stress with height. If these conditions are fulfilled, then the Reynolds stress at any height H above the canopy becomes a direct measure of the mean wind force on the trees below, per unit ground area.

In a few of Gardiner's (1992) field measurements these conditions were satisfied and he was able to estimate in one case a Reynolds stress of 2.1 N m^{-2}. In the same wind, the mean overturning moment coefficient (C_Q = 0.00085: see Section 7.5.1) was estimated independently from tree deflections, and this corresponded to a mean canopy stress of 3 N m^{-2}.

7.5.3 Honami gusts

Interesting though Reynolds stresses and mean forces may be, they are of little practical importance when extremes of the fluctuating forces are much larger. Nevertheless, the Reynolds stress mechanism explains physically how fluid transferred *downward* by negative 'w' gust components will frequently carry an *excess of horizontal 'u' velocity*. This mechanism includes the penetration of the canopy by gusts of high-speed wind, causing sudden high wind forces on a few trees. These gust penetration events have been observed in the field and in the wind tunnel and have been called 'Honami gusts' (Gardiner, 1992). It seems likely that they are a significant source of damage and therefore it is important to consider tree sway as a response to gust events.

7.5.4 Gust excitation and sway response in the frequency domain

Recognising the importance of gust penetration events in driving tree sway, both I. Turnbull *et al.* (unpublished data) and C. J. Baker (unpublished data) advocate a transfer-function approach. This defines in the frequency domain, the fluctuating response of a tree in terms of the spectrum of a defined gust event. In each case the aerodynamic admittance part of the transfer function incorporates the drag, not as a mean force related to a mean windspeed as in Section 7.1, but as a fluctuating force which is related empirically to the fluctuations of the chosen wind descriptor.

Deciding the measure of gust flow which influences most closely the fluctuating force on a tree remains a matter of debate. Baker uses the longitudinal velocity near the tree crown. In contrast, Turnbull *et al.*, seeking to represent the physical Honami gust mechanism, use the instantaneous product $u'w'$ of longitudinal and vertical velocities close above the canopy. In either case, the definition demands a measurement, and the measurements are not yet made.

7.6 Pull and sway computations

The computations in this section have been programmed using a commercial PC spreadsheet (Microsoft Excel for Windows). In all cases the Gardiner Average Tree, as defined in Figs. 7.7 and 7.8, is taken to have a height h of 15 m. At breast height $(0.09\,h)$ it has a reference stem mass m_r of 19.63 kg m^{-1} and a reference diameter d_r of 0.165 m. Taken with a nominal value of 6.57×10^9 Pa for Young's modulus E, this reference diameter gives a reference stem stiffness EI_r of 2.39×10^5 N m^2.

7.6.1 Stem bending due to a steady point force

A tree of height h and mass per unit length $m(z)$ is pulled by the tension P of a rope as shown in Fig. 7.18. The rope is attached at height λh and inclined downward at an angle θ. The force bends the tree to a curved shape $x(z)$, the

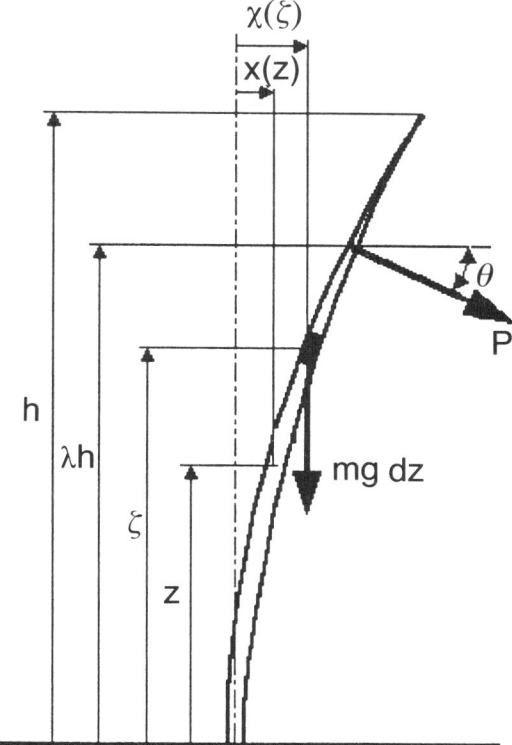

Fig. 7.18. Diagram for tree pulling equations.

overhanging weight increases the bending effect and the bending moment $M(\zeta)$ at height ζ is given by

$$M(\zeta) = P \cos(\theta)(\lambda h - \zeta) + P \sin(\theta)[x(\lambda h) - x(\zeta)] + \int_{\zeta}^{h} mg[x(z) - x(\zeta)] \, dz \quad (7.14)$$

The Moment Area Theorem (Eq. 7.5 and Eq. 7.4) gives the deflection $x(z)$ in terms of the bending moment distribution $M(\zeta)$ and the distribution of bending stiffness $EI(\zeta)$:

$$x(z) = \int_{0}^{z} \frac{M(\zeta)(z-\zeta)}{EI(\zeta)} \, d\zeta$$

Solving this pair of equations requires two numerical integrations and is easy on a modern spreadsheet. Data required are the distributions of mass $m(z)$ and bending stiffness $EI(z)$ together with an initial guess for the deflection curve $x(\zeta)$ in Eq. (7.14). The improved deflection curve $x(z)$ is copied back to start the next iteration.

7.6.2 Some computed pull-deflection results

Fig. 7.19 shows the surface stress distributions in the outer fibres, calculated by Eqs. (7.14) and (7.3) for the Gardiner Average Tree and the 0.6 Power Law Tree (see Section 7.3.4). Each tree is pulled with a horizontal force at $0.8h$. The effect of overhanging weight is included, and the load P is chosen arbitrarily to be $0.1\ EI_r/h^2$.

For the Average Tree, the stress is approximately constant over the whole height of the stem below the load point. Similar uniformity has been observed by Milne (1991). This suggests that the Gardiner Average Tree has a stem shape that is well adapted to this loading.

In contrast, the stress distribution computed for the 0.6 power law shape is quite different. This difference suggests strongly that it is not sufficiently

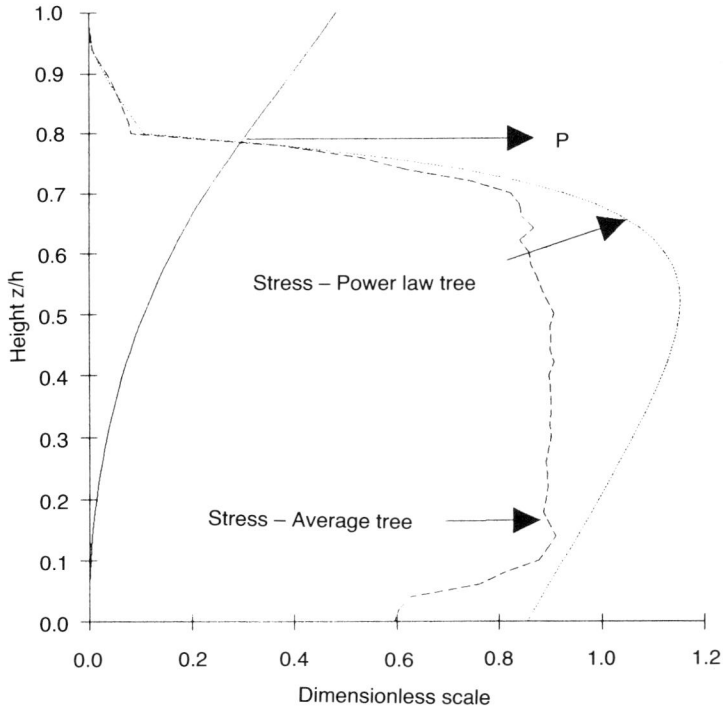

Fig. 7.19. Stress distributions for average and power law trees, pulled at $0.8h$.

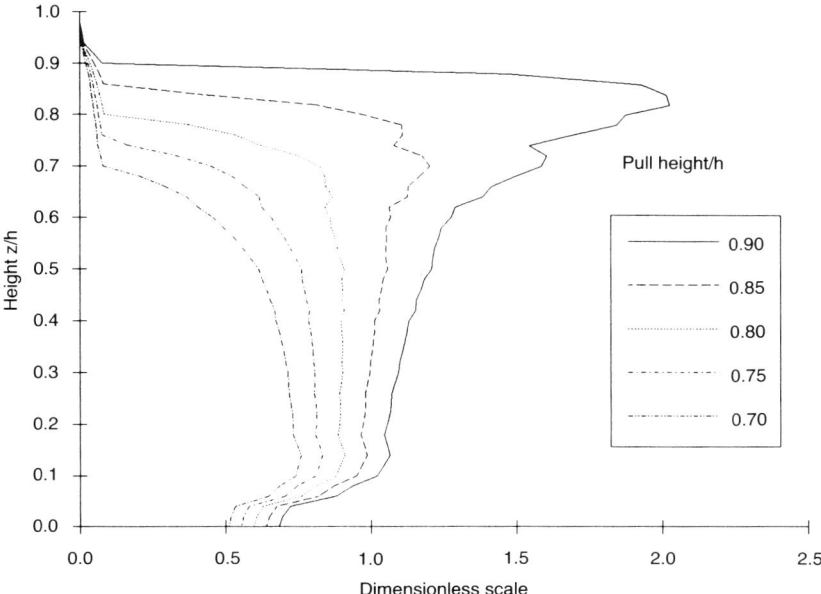

Fig. 7.20. Effect of pull height on stress distribution for Gardiner Average Tree.

accurate to represent the data by an analytical function. Discarding the Power Law Tree and returning to the Average Tree, Fig. 7.20 shows how the stress distribution changes with the loading point. A tree pulled from too high suffers excessive stress at the top. Pulled from too low, the stress is largest at the bottom. The uniformity of stress for a pull height of $0.8\,h$ suggests that the Gardiner Average Tree probably has a growth-experience of wind loading with a centre of pressure at about this level.

7.6.3 Analysis of sway vibrations by Stodola's method

Mathematical models for sway vibrations of trees are improving rapidly. Rayleigh's approximate method with guessed modal deflection curves (e.g. Gardiner, 1989; Milne, 1991) is being superseded by exact solutions for realistic distributions of mass and stiffness. Morgan is working on a vibration analysis as part of his large deflection computer package (Morgan 1989, personal communication 1993) and Burden (1993) has completed an exact prediction based on the Gardiner Standard Power Law Tree.

The calculations below use Stodola's method (den Hartog, 1956). This originated as a graphical method in 1905, but is well suited to the modern

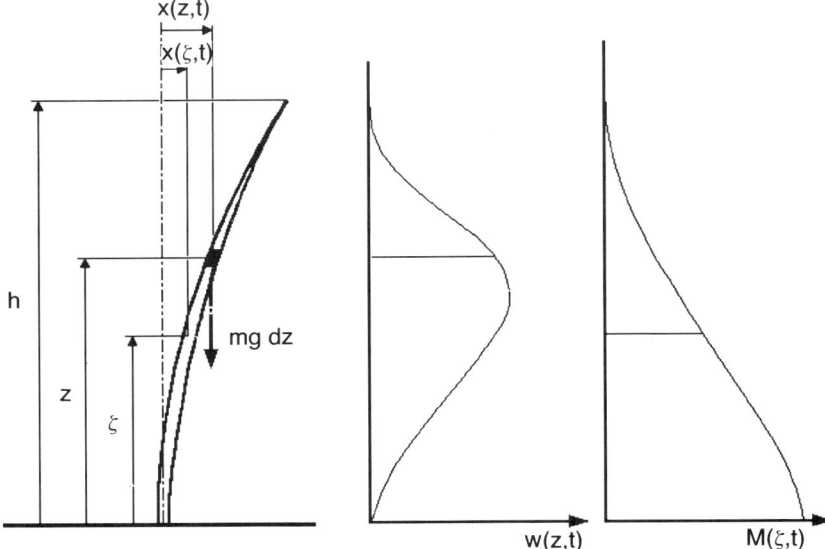

Fig. 7.21. Diagram giving notation for Stodola analysis.

PC spreadsheet (Fig. 7.21). The theory is valid for small deflections, and starts with the assumption that the tree stem sways with a natural frequency of ω radians per second and an amplitude $a(z)$. The function $a(z)$ describes the physical displacement curve of the tree stem at maximum deflection:

$$x(z,t) = a(z)\sin(\omega t) \qquad (7.15)$$

Differentiating $x(z,t)$ twice with respect to time yields the acceleration $\ddot{x}(z,t)$ of a stem element at height z:

$$\ddot{x}(z,t) = -a(z)\,\omega^2\,\sin(\omega t) \qquad (7.16)$$

This stem element has a mass $m(z)\,\mathrm{d}z$ and so the acceleration $\ddot{x}(z,t)$ requires a force $w(z,t)\,\mathrm{d}z$ by the elastic stem itself to produce it. Alternatively, $w(z,t)$ may be thought of as an externally applied 'inertia' load per unit length, distributed over the height of the stem and acting in a direction opposite to the direction of the acceleration $\ddot{x}(z,t)$:

$$w(z,t) = -m(z)\,\ddot{x}(z,t) \qquad (7.17)$$

Any distributed load $w(z,t)$ on the stem causes a corresponding distribution of bending moment. $M(\zeta,t)$. This is found by summing the bending moment at any point ζ due to all load elements $w(z,t)$ between ζ and the top of the tree ($\zeta < z < h$). These may include not only the inertia load distribution

$w(z)dz$ with a moment arm $z - \zeta$, but also the bending moment due to the weight $m(z) g \, dz$ of stem elements above ζ which overhangs by a horizontal distance $x(z,t) - x(\zeta,t)$:

$$M(\zeta,t) = \int_\zeta^h w(z,t)(z-\zeta)\,dz + \int_\zeta^h m(z) g \, [x(z,t)-x(\zeta,t)]\,dz \quad (7.18)$$

The computation loop is closed by calculating an improved deflection curve $x(z, t)$ from the bending moment $M(z,t)$ for the elastic tree stem. This is again found most easily by the second moment-area theorem (Eq. 7.5 with 7.4):

$$x(z,t) = \int_0^z \frac{M(\zeta,t)(z-\zeta)}{EI(\zeta)}\,d\zeta$$

To start the procedure, any guess will do for the mode shape $x(z,t)$ ... Thereafter, the bending moment distribution $M(z,t)$ is calculated by Eqs. (7.15)–(7.18) and an improved estimate of $x(z,t)$ is then found by Eqs. (7.4) and (7.5). This is copied back as the starting shape for a repeat calculation. The method converges very powerfully for the first mode. The interaction for the second mode requires a computational adjustment (den Hartog, 1956) to prevent the first mode reappearing.

7.6.4 Surface stress distributions for computed sway vibration

Stodola's method has been used to compute first- and second-mode vibrations for the Gardiner Average Tree. The method calculates the exact deflected shape, from which the surface stress distribution may be deduced. Fig. 7.22 shows the first mode and it is to be noted that, unlike the pull calculation which yields uniform stress, the vibration gives a maximum stress near the bottom of the tree.

It is known that trees do not always uproot but that they sometimes break part-way up the stem. The question has often arisen as to whether this breakage might be the result of a second-mode resonant vibration. Fig. 7.23 shows the Stodola computation of the second mode. It has a surface stress maximum at 80% tree height, whereas stem breakage usually occurs at about 30% tree height. This suggests that whatever it is that breaks trees part-way up, it is *not* second-mode vibration.

7.6.5 Conclusions from pull and sway computations

1. It is impossible to design a tree stem to give uniform stress under both direct wind forces and sway vibration. The two types of loading require different stem taper profiles.

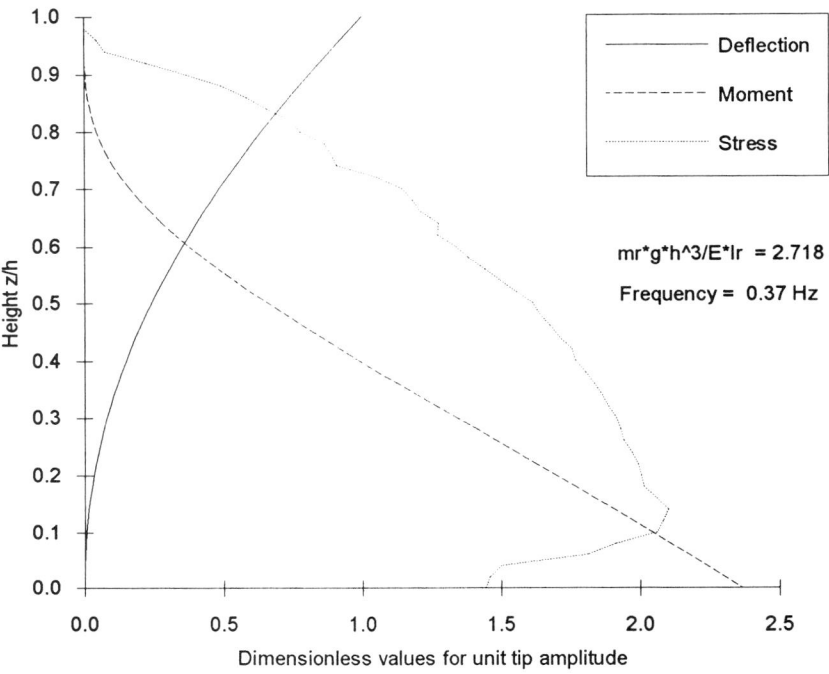

Fig. 7.22. Gardiner Average Tree: first mode sway vibration.

2. The Gardiner Average Tree appears to be optimally adapted to direct wind forces at approximately 80% tree height, but not to sway vibrations.
3. When the Gardiner Average Tree vibrates in the first mode, the maximum stress is at ground level. Therefore sway vibrations may be expected to cause uprooting rather than stem breakage at mid-height.
4. A tree grown adaptively to have uniform stress under sway vibration would be thicker near the ground and thinner higher up than the Gardiner Average Tree.
5. Such a sway-adapted tree, if subjected to a direct wind force at 80% height, would show a maximum stress at mid-stem and might be expected to break there. It would be of interest to measure the *taper* of some trees that have snapped instead of uprooting.)

7.7 Dynamic response and damping

Having drawn these conclusions, some of them are immediately called in question by one final observation about aerodynamic damping.

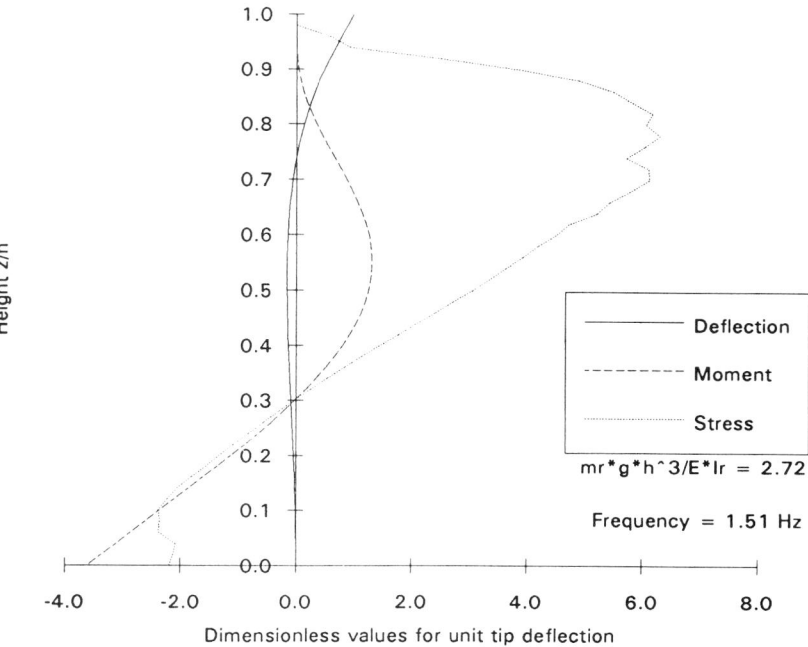

Fig. 7.23. Gardiner Average Tree: second mode sway vibration.

7.7.1 Damping in steady-state forced vibrations

Fig. 7.24 illustrates the fact that a tree subjected to a continuously fluctuating force will respond differently, depending upon the *frequency* of the force. The diagram describes the solution of Eq. (7.7) with the zero on the right-hand side replaced by a time-varying force $F(t)$:

$$\frac{d^2x}{dt^2} + 2\delta\omega_n \frac{dx}{dt} + \omega_n^2 x = F(t) \tag{7.19}$$

1. If the frequency is very low ($\omega/\omega_n < 1$), the tree will bend in proportion to the instantaneous wind force, and the maximum deflection and stress will correspond to the largest wind force, just as though that force were steady.
2. If the frequency is very high ($\omega/\omega_n \gg 1$), the tree will not respond at all to the fluctuations. The deflection and stresses will be related only to the mean force.
3. The worst situation is resonance, which occurs if the excitation frequency is close to the natural frequency of sway vibrations. With inadequate

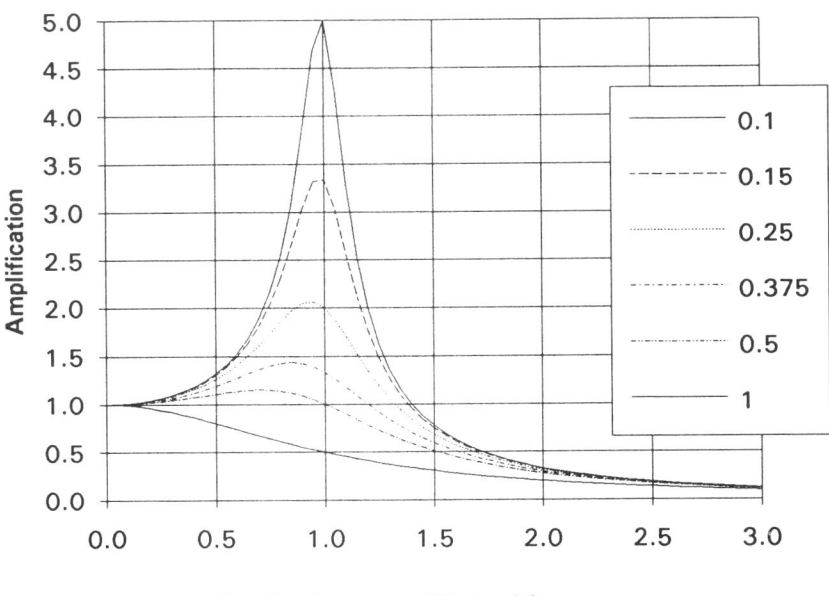

Fig. 7.24. Steady-state vibration response (Eq. 7.14).

damping, the resonant sway deflections and stresses can be alarmingly large. However, damping reduces this response, and with critical damping ($\delta = 1$) there is no resonance at all.

7.7.2 Transient response

Section 7.7.1 refers to steady-state response, but transients are equally important. The simplest example is the motion following a step change in the wind force, representing the arrival of a sudden gust. This particular solution to Eq. (7.19) is shown in Fig. 7.25. The diagram shows that an undamped system overshoots the new equilibrium to produce a maximum in the deflection or stress which is exactly twice the equilibrium value. However, as with resonance, damping attenuates the response until, with critical damping ($\delta = 1$), there is no overshoot at all.

7.7.3 Effect of strong wind on aerodynamic damping

The two examples in Sections 7.7.1 and 7.7.2 illustrate the way in which damping protects systems from the damaging effects of resonant or transient

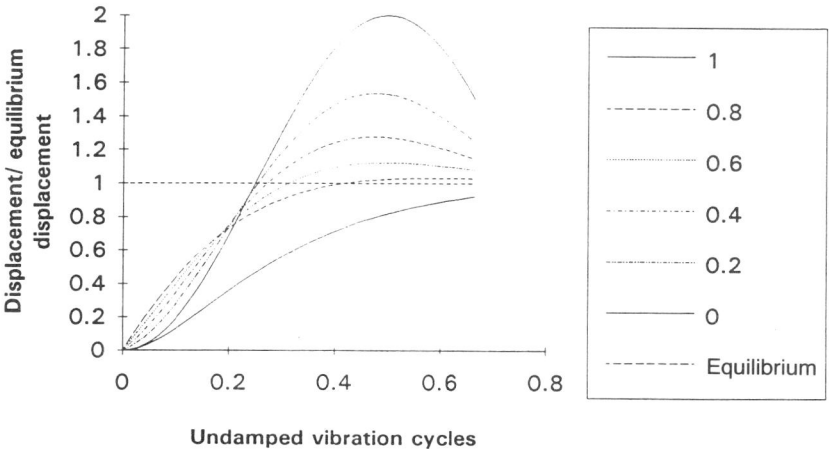

Fig. 7.25. Transient response to a step change in force (Eq. 7.14).

response. Table 7.1 gives the damping ratio of a 15 m Sitka spruce as 0.048. This is deduced from vibration decay in calm conditions and Milne (1991) suggests that perhaps 20% of this is hysteresis in the timber and soil, so 0.04 might be an appropriate aerodynamic contribution. However, tree sway is of concern only when there is a substantial windspeed V superimposed upon the oscillatory sway speed v. Defining the latter as positive in the downstream direction, the drag may be written

$$D = K(V - v)^n \tag{7.20}$$

If the branches and foliage remain rigid, the exponent n is probably 2. If they do deflect, n is reduced (Mayhead *et al.* 1975) and a value of 1.8 gives a reasonable approximation to Eq. (7.10) in the range 0–30 m s^{-1}. For small v/V, a binomial expansion of Eq. (7.20) yields the following dominant terms in the series:

$$D = KV^n - nKV^{(n-1)} v - \ldots$$

The second term is proportional to v and is the primary contribution to the aerodynamic damping force. It is actually *greater*, by the factor $n(V/v)^{(n-1)}$, than the still-air drag Kv^n. For example, if $n = 1.8$, $V = 30$ m s^{-1} and $v = 1$ m s^{-1} the factor is 25.8. Thus the still-air damping ratio of 0.04 will increase to 1.03 in a 30 m s^{-1} wind.

This is not rigorous, but it suggests strongly that the damping ratio increases with windspeed and may approach or even exceed the critical value above which resonance cannot occur. If this is true, then the focus of our research should shift away from vibration analysis to a more careful examination of direct wind loading.

7.8 Conclusions

In our concern to prevent wind damage to trees, the adaptive growth hypothesis has a profound effect in maintaining a proper research perspective. The hypothesis states that a tree will grow only sufficiently strong to resist the forces that have occurred during its growth history. Thus we expect trees spaced more widely to grow stronger in proportion to the increased exposure, but this may not improve the chances of surviving a catastrophic storm. Adaptive growth leaves trees vulnerable to events that are unprecedented in their growth history. Thus damage is most likely when felling, disease or a local accident exposes previously sheltered trees to the unobstructed wind, and it is in these situations that remedial measures are most likely to be useful.

Recent research has focused on the mechanism of wind damage. Why do some trees break when most uproot? Is damage caused by direct wind loading or by resonant vibration or by some transient combination of gust impulses? By keeping adaptive growth in mind, engineering analysis can be interpreted to yield some clues.

This chapter re-examines the taper of a representative selection of Sitka spruce stems, and shows that under direct wind loading they have uniform stress, whereas vibration loading causes a stress maximum near the base. Therefore, if this type of tree fails as a result of vibration resonance, it will fail by uprooting. Conversely a tree, adaptively grown under conditions of vibration but which fails due to direct wind loading, may break part-way up the stem. But above all, these engineering computations, interpreted with adaptive growth in mind, suggest that direct wind loading rather than resonance is the *normal* growth experience. Furthermore, a normally tapered stem under direct wind loading should exhibit no preferred failure mode. Quite separately, simple aerodynamics suggests that tree sway vibrations may be so heavily damped in strong winds that resonance cannot occur. This reinforces the expectation that direct wind loading is the norm in strong winds.

Faced with this, and the associated expectation of no preferred mode of failure, we ask: Why then do the majority of Sitka spruce trees in the United Kingdom uproot rather than break? And why are they increasingly vulnerable as they grow older and taller? Here the adaptive growth mechanism may yield one final clue, this time not by its relevance but curiously by its irrelevance. Tree roots grow adaptively, as Fig. 7.1 shows and as Stokes *et al.* (this volume) have demonstrated, but soil itself does not adapt. The root plate of a tree may extend in size and weight, but the shear strength of the soil is not involved in the biomechanical response system that monitors stress. Therefore, as illustrated by Matthek (this volume), any engineering assessment of tree resistance to windthrow must include not only the adaptive growth of the tree itself, but also the non-adaptation of the soil in which it grows.

Acknowledgements

The author acknowledges, with thanks, permission to use the following illustrations from other publications, detailed in the list of references below: Fig. 7.2 from Matthek (1989), Fig. 7.4 from Morgan (1989), Fig. 7.5 from Gardiner (1989), Fig. 7.7 from Milne (1991).

References

Bisshopp, K. E. & Drucker, D. C. (1945). Large deflections of cantilever beams. *Quarterly Journal of Applied Mathematics*, **3**, 372–5.

Burden, T. R. G. (1993). Dynamics of trees in strong winds. Dissertation, Dept of Theoretical Mechanics, University of Nottingham.

Cannell, M. G. R. & Morgan, J. (1988). Support costs of different branch designs. *Tree Physiology*, **4**, 303–13.

den Hartog, J. P. (1956). *Mechanical Vibrations*. McGraw-Hill, New York.

ESDU (1982) *Strong Winds in the Atmospheric Boundary Layer*. Wind Engineering Series, Data Item 82026. ESDU International Ltd, London.

ESDU (1985). *Characteristics of Atmospheric Turbulence Near the Ground*. Wind Engineering Series, Data Item 85020. ESDU International Ltd, London.

Gardiner, B. A. (1989) *Mechanical Characteristics of Sitka Spruce*. Forestry Commission Occasional Paper 24. Forestry Commission, Edinburgh.

Gardiner, B. A. (1992). Wind and wind forces in a plantation spruce forest. *Boundary-Layer Meterology*, **67**, 161–86.

Mattheck, C. (1989). Engineering components grow like trees. Kernforschungszentrum Karlsruhe, KfK 4648.

Mayhead, G. J., Gardiner, J. B. H. & Durrant, D. W. (1975). Physical properties of conifers in relation to plantation stability. Unpublished report. Forestry Commission, Edinburgh.

Milne, R. (1991). Dynamics of swaying of *Picea sitchensis*. *Tree Physiology* **9**, 383–99.

Milne, R. & Blackburn, P. (1989). The elasticity and vertical distribution of stress within stems of *Picea sitchensis*. *Tree Physiology*, **5**, 195–205.

Morgan, J. (1989). Analysis of beams subject to large deflections *Aeronautical Journal*, Nov., 356–60.

Morgan, J. & Cannell, M. G. R. (1990). Theoretical study of variables affecting the export of assimilates from branches of *Picea*. *Tree Physiology* **6**, 257–66.

Papesch, A. (1984). Wind and its effects on (Canterbury) forests. PhD Thesis, University of Canterbury, N.Z.

Stacey, G. R., Belcher, R. E., Wood, C. J. & Gardiner, B. A. (1994). Wind flow and forces in a model spruce forest. *Boundary-Layer Meteorology*, **69**, 311–34.

Stacey, G. R., Belcher, R. E., Wood, C. J. & Gardiner, B. A. (1995). Minimising wind forces in a spruce forest: wind tunnel and field study on thinning and respacing. *Boundary-Layer Meteorology*, in press.

Walshe, D. E. & Fraser, A. I. (1963). *Wind Tunnel Tests on a Model Forest*. National Physical Laboratory Aeronautical Report no. 1078.

8
Modelling mechanical stresses in living Sitka spruce stems

R. MILNE

Abstract

Mathematical models of the transfer of wind momentum into forest canopies and the subsequent bending stresses in individual tree stems are presented. The windspeed profiles within and above canopies are predicted for conditions where the extreme windspeed likely to occur in a 50 year period in south-west Scotland is used as a reference. The stresses predicted by the models in these conditions are compared for Sitka spruce (*Picea sitchensis* (Bong.) Carr.) growing in an unthinned stand (3800 stems ha^{-1}) and in a stand recently thinned by removing half the trees. The mechanical model of the trees includes the responses to dynamic as well as static bending forces. The calculated variation of stress along the stems of six trees of different dimensions is compared and found to have a maximum near the stem base in some trees but a nearly constant pattern for several metres in other trees. The maximum value of stress is about 10 MPa for trees in the unthinned model and about 20 MPa for the recently thinned model. These stresses are compared with the estimates from published studies of the strengths of the root systems and of the stem wood. The increases in stress with thinning are shown to reflect qualitative observations in British plantations that wind damage is more likely and that this will be in the form of overturning rather than stem snap. It is concluded that the models should be a good basis for further work on estimating the risk of wind damage to plantation trees.

8.1 Introduction

Great Britain is one of the windiest islands in the world and extensive new plantations of Sitka spruce (*Picea sitchensis* (Bong.) Carr.) have been established over the last 40 years. Normal winter gales cause major damage on shallow soils in the form of uprooting and occasionally stem snap. The mechanical stresses generated in the stem, and hence also in the roots, are influenced by windspeed and its variation above the canopy, the way in which the canopy acts as drag elements transmitting wind force to the individual tree and the mechanical properties of the individual tree. The maximum bending of a tree stem will depend on the dynamics of movement as well as a static component which could be considered to be caused by the mean windspeed. Gusts of wind will interact with the movement of the tree in a way controlled by the variation of windspeed and the elastic properties of the tree.

Sitka spruce has been planted at densities as great as 4000 ha^{-1}, although 2500 ha^{-1} is now normal, and in such densities there is a strong economic incentive to thin the stands when an appropriate height is reached (8–15 m depending on density, etc.). It is, however, well known that thinning tends to promote the onset of wind damage. Managers have responded by designating existing forests as thin or no-thin areas and by reducing the planting density in new forests. The British Forestry Commission has developed a Windthrow Hazard Classification to assist in the designation of areas for different thinning practice by defining the height of the forest at which wind damage is likely to occur for different locations. This method is, however, empirical and more basic work is needed.

Lohmander & Helles (1987) have assessed the risk of wind damage to forest trees in Denmark by statistical methods, while more mechanistic models of the bending of tree stems and of the stresses generated have been published by Mayer (1987), Galinski (1989), Blackburn et al. (1988) and Blackburn & Petty (1988). In particular Blackburn et al. (1988) included the effect of canopy structure on wind profiles and drag forces using an exponential profile of windspeed within the canopy and included a simple description of the magnifying effect of tree swaying to predict maximum bending moments at the base of Sitka spruce stems. Previous work has also described the bending (Milne & Blackburn, 1989) and the dynamic properties (Milne, 1991) of swaying Sitka spruce trees, while descriptions of the occurrence of extreme windspeeds used by the building industry have been shown to apply to forest sites (Milne, 1992).

When a plantation is thinned two mechanisms lead to the onset of damage: (1) greater forces on the trees as the windspeeds are enhanced within the more open canopy, and (2) less damping of tree swaying as interaction with neighbours is reduced. In this chapter a model is presented which describes the tree as standard engineering system. It is used to estimate the relative importance of these two mechanisms to the mechanical stresses in individual trees in a typical unthinned stand and a hypothetical stand where alternate trees are removed to increase the windspeeds within the canopy and allow individual trees to sway freely. The structure of this thinned stand will emphasise the effects occurring immediately after thinning but will not describe the situation several years later when faster growth will have altered the stem and canopy dimensions of the trees. Six individual Sitka spruce trees with the elastic properties estimated by Milne (1991) are considered. Wind profiles, and hence drag forces, in the canopy are predicted from reference windspeeds far above the forest using a first-order closure model of momentum transport (Li et al., 1985; Miller et al., 1991). The reference windspeeds themselves are estimated by the building industry methods described in Milne (1992). The model calculates the bending moment, and mechanical stress due to this bending, at 1 m intervals up an individual tree stem. The canopy structure used is based on a survey of a 26-year-old (top height approximately 15 m) Sitka spruce plantation in south-west Scotland. A modified version of this structure is used to model the situation when alternate trees are missing. The bending moments and stresses for the six tree sizes in the unthinned and thinned model forests are compared with values of the stress in stems estimated from published measurements on Sitka spruce overturned manually (Coutts, 1986).

8.2 Methods

8.2.1 Botanical characteristics of forest

Two different situations are modelled; a typical Scottish unthinned Sitka spruce plantation, and the same forest thinned by removing alternate trees but assuming that the tree size distribution would be unaltered. The unthinned stand's characteristics were taken from measurements on an area of Moffat Forest in south-west Scotland (UK National Grid Ref. NT 019044) of top height 15 m and General Yield Class 20. This area was planted in 1962 in a peaty gley soil, sloping at 6°, with the trees at 1.52 m × 1.52 m spacing and with double mouldboard ploughing. Measurements of the vertical distribution of biomass on individual trees in this forest have been made and related to stem diameter at breast height. From this relationship, and the measured stem diameter distribution, a description of vertical canopy biomass distribution in the unthinned case has been published (Milne & Brown, 1990; Milne, 1991, 1992). The model canopy characteristics for the thinned stand were equal to those of the unthinned stand except that the total needle area was halved to simulate the removal of alternate trees.

8.2.2 Windspeeds

The UK Meteorological Office (Shellard, 1968; Hardman et al., 1973; Caton, 1976), the Engineering Sciences Data Unit (ESDU, 1984a) and the UK

Building Research Station (Cook, 1985) have mapped extreme hourly-mean windspeeds at sea level for the United Kingdom. These speeds refer to winds occurring during normal depressions and similar climatic conditions, but exclude unusual events which would happen less than once in every 100 years. A correction for site elevations above sea level is also available (Cook, 1985). Windspeeds with a return period of 50 years are given, but only for standard conditions, i.e. smooth agricultural countryside. Extreme windspeed over forests can be estimated from the standard open countryside values if the roughness length of the forest canopy is known (Milne, 1992).

Profiles of windspeed above and within the canopies to be modelled were estimated using a description of momentum transport into a canopy published by Li *et al.* (1985) and Miller *et al.* (1991) which uses first-order closure and an empirical estimate for large-scale injections and ejections of momentum. This model uses a profile of mixing length from ground level through the canopy and up to some reference point far above the forest canopy. The canopy top height is taken in this chapter as the height at which needle density becomes negligible. The mixing length profile used in the lower canopy is that suggested by Miller *et al.* (1991), but above the height of maximum canopy needle density a different approach is taken. In order to link the hourly-mean windspeed profile at the forest with the mesoscale wind conditions, a high-level reference at 10 times the canopy top height is used. At this point the value of mixing length was estimated from the wind profile model of Harris & Deaves (1981) as used in ESDU (1984a), i.e.

For standard smooth surface conditions:

$$V_M = \frac{u_{*M} \ln(z_M/z_{0M})}{k} \tag{8.1}$$

For the forest conditions:

$$u_{*F} = \frac{u_{*M} \ln(10^5/z_{0M})}{\ln(10^5/z_{0F})} \tag{8.2}$$

$$H_F = \frac{u_{*F}}{\beta f} \tag{8.3}$$

$$\tau_F = \rho u_{*F}^2 (1 - z/H_F)^2 \tag{8.4}$$

$$V_F = \frac{u_{*F} \ln(z/z_{0F}) + 34.5 fz}{k} \tag{8.5}$$

where:

f is the Coriolis parameter (1.1×10^{-4} s^{-1} for Scotland)
H_F is gradient height at forest (m)
k is von Karman's constant
u_{*F} and u_{*M} are friction velocities (m s^{-1})
V_F is reference forest wind speed (m s^{-1})
V_M is reference wind speed for standard conditions (m s^{-1})
z is reference height at forest (m)
z_M is reference height for standard conditions (10 m)
z_{0F} is roughness length of forest (m)
z_{0M} is roughness length for standard smooth surface conditions (0.03 m)
β is a constant of value 5.75
ρ is density of air (kg m^{-3})
τ_F is momentum flux at forest (N m^{-2})

The mixing length (l) at the reference height above the forest is then found from

$$\tau_F = \rho l^2 \left|\frac{dV}{dz}\right| \frac{dV}{dz} \qquad (8.6)$$

The mixing lengths between the value at the height of maximum needle density as used by Miller *et al.* (1991) and the value at $10h$ (h is canopy top height) from Eqs. (8.1)–(8.6) are then estimated by interpolation. The mixing length at $10h$ calculated by this method is dependent on the reference windspeed, in contrast to the fixed value used by Miller *et al.* (1991). The equation of momentum exchange used is:

$$\frac{d}{dz}\left(l^2 \left|\frac{dV}{dz}\right|\frac{dV}{dz} + C_q(V_r - V)V_r \frac{z}{h}\right) - C_d A|V|V = 0 \qquad (8.7)$$

where

A is the projected needle area of the canopy layer at height z
C_d is the drag coefficient of a canopy layer
C_q is a momentum dispersion flux constant
l is mixing length (m) at height z
V is hourly-mean windspeed at height z (m s^{-1})
V_r is hourly-mean windspeed at $2h$ (m s^{-1})

For further information on these parameters see Miller *et al.* (1991).

Given the estimated profile of mixing length and the profile of canopy needle density Eqs. (8.1)–(8.6) were used to solve the momentum equation (8.7) by also assuming (1) that at the reference height u_*, and hence the slope of windspeed with change in height, was as defined from the Harris & Deaves (1981) profile, (2) the windspeed at the ground was zero and (3) between $2h$ and $10h$ the large-scale dispersion term did not apply. This solution effectively involved finding a value of the forest roughness length (z_{0F}) which made the windspeed at the ground zero in the momentum equation. The windspeed at the reference height for this solution was found to be within 10% of the directly calculated Harris & Deaves value. The 50 year return period hourly-mean windspeed profile for each canopy type was calculated using the above method, taking the standard condition reference windspeed to be the extreme windspeed from the map of Cook (1985) at the Moffat location.

The roughness length for each forest type found from the profile of speeds during extreme wind conditions was used to estimate the horizontal turbulence intensity (I_u) at the top of each canopy using the following estimate of Harris & Deaves (1981) and ESDU (1984b)

$$I_u = \frac{\sigma_u}{u_{*F}} \cdot \frac{u_{*F}}{V_h} = \frac{7.5\eta(0.538 + 0.09\ln[(h-d)/z_{0F}])^{\eta^{16}}}{1 + 0.156 \ln(u_{*F}/fz_{0F})} \cdot \frac{u_{*F}}{V_h} \quad (8.8)$$

where

$$u_{*F} = \frac{kV_h}{\ln[(h-d)/z_{0F}]}$$

and

$$\eta = 1 - (6fz_{0F}/u_{*F})$$

where V_h is the hourly-mean wind speed at the canopy top height of h.

It was found necessary to use a zero plane displacement (d) within this equation and this was chosen to be 0.8 of median tree height as presented by Jarvis et al. (1976), Gardiner (1994) and others.

Ideally the variation of turbulence intensity with height within the canopy is needed but here, due to the lack of a readily available description, the turbulence intensity was assumed to be uniform down through the canopy.

The horizontal turbulence scale length, and hence the shape of the spectral distribution of horizontal turbulence, was estimated from the equations for these variables published by ESDU (1984c) with the apparent roughness length at the forest reference height and the zero plane displacement as parameters. ESDU (1987) have published more detailed estimates of turbulence spectra which take into account the effects of different windspeeds but, given the approximations inherent in the coupling of the Miller et al. (1991) profile to that of Harris & Deaves (1981), this additional complication was not considered appropriate.

8.2.3 Static stem bending

A computer model of the mechanical structure of Sitka spruce (Milne & Blackburn, 1989) was used to estimate the bending of six sample trees (Milne, 1991) due to steady hourly-mean winds at the extreme speed. The variation in wind drag down through the canopy was taken into account using the previously calculated profiles. The drag on 1 m deep layers of foliage on each tree was calculated using the drag coefficients for Sitka spruce published by Landsberg & Jarvis (1973).

8.2.4 Dynamic stem bending

Wind is gusting in nature and dynamic effects increase the bending of stems and hence the load on root systems. The additional dynamic effect can be split into, firstly, the resonant component due to gusts near the tree's natural sway frequency which will cause amplified sway compared with a uniform wind of the same speed and, secondly, gusts at other frequencies, the background component, which are assumed to cause additional but unamplified sway compared with uniform windspeeds. The natural frequency and damping of sway, with and without the interference of neighbours' branches, were measured by Milne (1991) for six Sitka spruce trees in Moffat Forest. These parameters were measured by manual swaying and were found to be related to the size and mass distribution of an individual (Milne 1991).

The damping (i.e. slowing down) of tree motion, which is of importance in determining the resonant component of dynamic forces, comes from three sources: interference between the branches of neighbours; aerodynamic drag on the branches moving through the air; and mechanical forces in the stem and roots. For closely spaced trees the first of these sources is as important as the other two sources combined, i.e. in a stand of 3800 stems ha^{-1} the movement of the trees is significantly restricted by branch interference. Damping with branch interference is used in this chapter for the unthinned case and for the recently thinned case it was assumed that no branch interference would occur. The natural frequency of an individual is assumed to be the same in unthinned and thinned models.

The background component of dynamic forces is determined by the turbulence energy as measured by the frequency spectrum of the variation in horizontal windspeed. ESDU (1984c) has published methods of estimating the

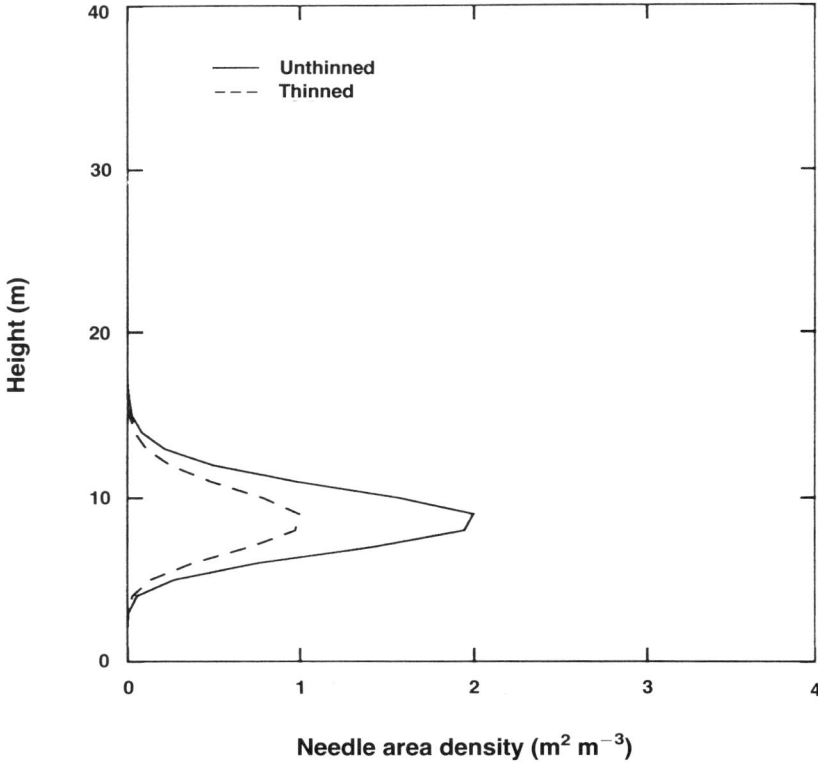

Fig. 8.1. Estimated projected needle area density distributions in unthinned (3800 stems ha^{-1}) and simulated thinned (1900 stems ha^{-1}) Sitka spruce plantations.

resonant and background components of movement and stress in structures where the turbulence characteristics of the wind are known as well as the elastic parameters of the structure. In this chapter the tree is treated as an elastic line-like structure similar to an electricity or lighting tower. As turbulence is a stochastic process with considerable variation this is also taken into account in the ESDU (1984c) methods to estimate the extreme values of movement and stress for given hourly-mean windspeeds. Here an estimate is made of the peak or extreme stress which would occur in an hour when wind with the 50 year *extreme* hourly-mean speed was blowing.

Table 8.1. *Characteristics of six 26-year-old Sitka spruce trees in the unthinned Moffat Forest southwest Scotland (Milne, 1991)*

Tree no.	Diameter at 1.3 m, D (cm)	Height, H (m)	H/D	Projected needle area (m²)	Natural frequency (Hz)	Damping ratio With interference	Damping ratio No interference
7	10.6	13.6	128	7.7	0.26	0.107	0.066
2	11.5	13.0	113	17.4	0.31	0.110	0.073
12	14.0	13.8	99	15.7	0.40	0.189	0.058
10	15.5	14.3	92	38.4	0.40	0.116	0.060
11	16.9	14.8	88	41.2	0.40	0.117	0.066
5	18.3	15.5	85	47.8	0.38	0.099	0.065

The damping ratios were measured with and without the interference of neighbours' branches.

8.3 Results

8.3.1 Tree characteristics and vertical distribution of needle density in canopy

The vertical distributions of needle area density estimated for the unthinned and thinned forest cases are shown in Fig. 8.1.

The characteristics of the six sample trees are presented in Table 8.1 (Milne, 1991). The trees are all of similar height but nos. 7, 2 and 12 have smaller canopies and narrower stems than nos. 10, 11 and 5.

8.3.2 Vertical variation of hourly-mean windspeed

The modelled windspeed variation with height within and above the unthinned and thinned forest models for a reference hourly-mean windspeed of 23.5 m s⁻¹ at a standard smooth site is shown in Fig. 8.2. This extreme speed is the 50 year return period wind as read from the map of Cook (1985) at Moffat Forest for a surface with standard site conditions but adjusted for elevation. The windspeeds calculated were found to be greater within, and for a considerable height above, the canopy of the thinned model compared with the unthinned model (Fig. 8.2). The apparent roughness length at $10h$ was found to be 1.13 m for the unthinned case and 1.01 m for the thinned case. The shapes of the modelled windspeed profiles were found not to vary with standard site windspeeds between 15 m s⁻¹ and 23.5 m s⁻¹.

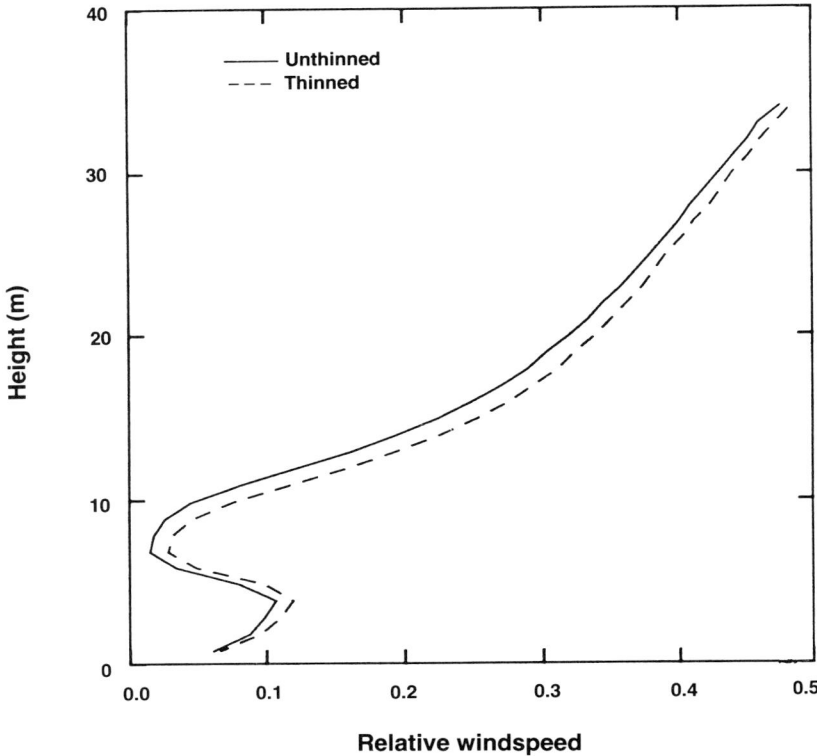

Fig. 8.2. Calculated profile of hourly-mean windspeed calculated from momentum transport model for unthinned and recently thinned Sitka spruce plantation models. The windspeed is shown relative to the speed at a height of $10h$ where h is the canopy top height.

The wind speed profile for the unthinned forest was calculated for a standard site condition hourly speed of 10 m s^{-1} and compares well with published measurements (Gardiner, 1994) in and above the canopy at Moffat Forest. The model, however, predicts an increase in speeds below the canopy which was not observed at Moffat Forest (Fig. 8.3).

8.3.3 Peak stresses on tree stems

The modelled bending stresses along each of the six sample tree stems, where they were assumed to be either within the unthinned or the thinned forest,

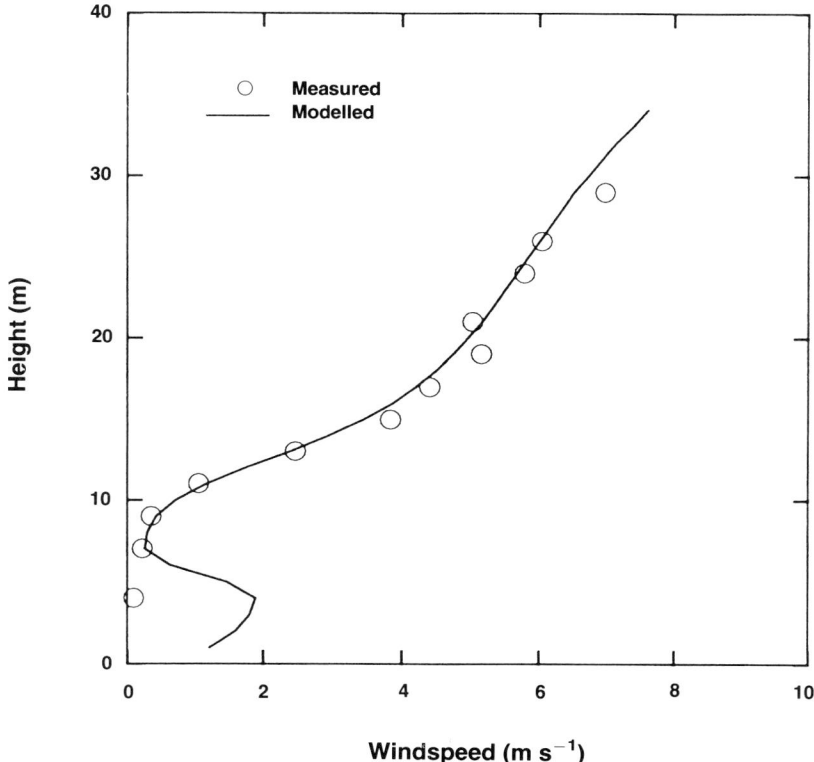

Fig. 8.3. Profile of hourly-mean windspeed in unthinned Sitka spruce stand at Moffat Forest, south-west Scotland, estimated by momentum transport model compared with field measurements of Gardiner (1994).

are shown in Fig. 8.4. Stress varies along a stem in patterns which vary from tree to tree. In some stems a notable maximum in peak stress occurs at about 6 m from the ground. Other stems have a considerable distance along their length where stress remains constant. Peak stress at these positions and at the maximum in the other stems has a value of about 10 MPa for the unthinned model. In the thinned case the pattern of stress variation along a given stem does not change significantly compared with the unthinned case but the value of maximum stress increases to around 20 MPa. Most of this increase is due to the effect of increased windspeeds in the recently thinned forest model and only in a single tree (tree 12) does the contribution from the decreased damping without branch interference approach 50% of the additional stress.

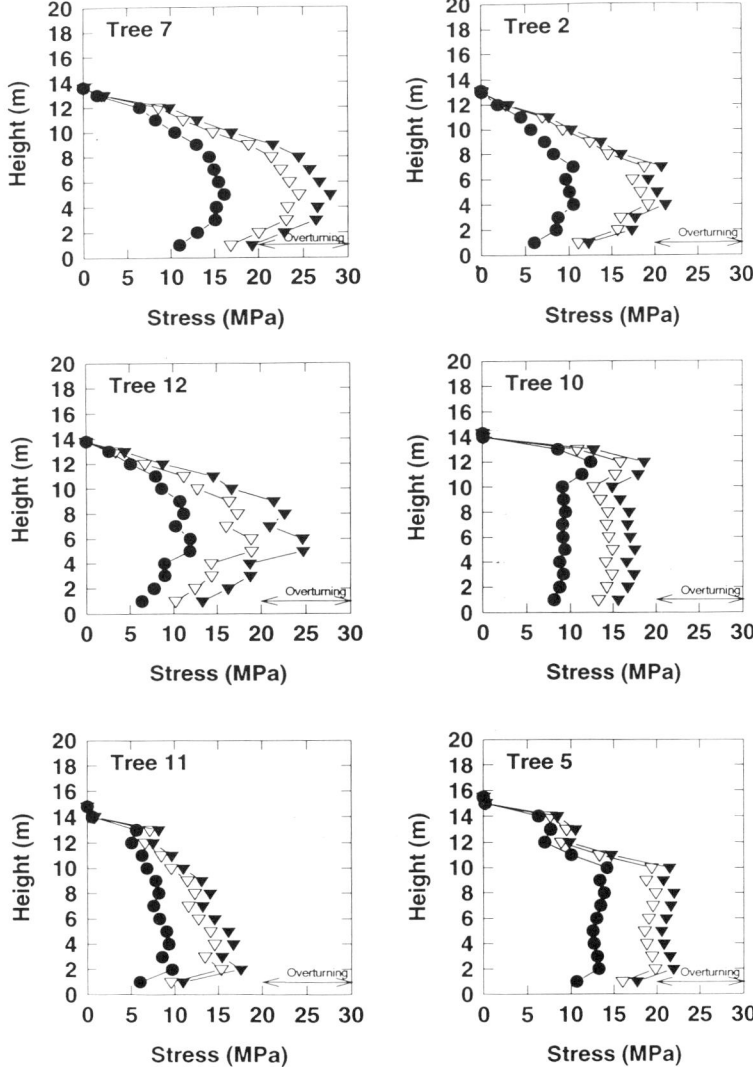

Fig. 8.4. Modelled stress in the outer layers of the stem of the six sample trees in unthinned and recently thinned Sitka spruce plantation models when subjected to a peak gust of wind (as occurs when the hourly-mean windspeed is at its 50 years return period maximum). Filled circles show stress from the unthinned forest model. Inverted triangles show stress from the thinned model. The hollow inverted triangles show the stress when damping is assumed to be the same in the thinned model as in the unthinned model, i.e. branch interference occurs. The filled inverted triangles show stress when lesser sway damping is assumed in the thinned model, i.e. branch interference does not occur. The basal stress estimated, from published studies of tree overturning by winch, to occur at uprooting is indicated.

8.4 Discussion

The momentum transfer model (Eq. 8.7) for estimating hourly-mean windspeed profiles proved to be useful but had some drawbacks. It estimated the profile above the canopy and in the upper canopy, where most drag occurs, much better than a logarithmic shape would have achieved (Fig. 8.3). A logarithmic profile with a zero plane displacement of around 10 m would have predicted zero windspeed near this height, which is not as measured (Fig. 8.3) for the unthinned forest (Gardiner, 1994). The model used is less empirical than a logarithmic profile and hence can be used to predict the profile for the simulated thinned forest where no windspeed data exist to which a logarithmic profile could be fitted. At the reference height of $10h$ the coupling to the profile methods of Harris & Deaves (1981) proved adequate for the situation here but could not be used where accuracy to better than 10% for the coupling of windspeed and its slope with height was required.

The roughness length parameter of Harris & Deaves (1981) was found to decrease from 1.13 m in the model of unthinned forest to 1.01 m in the model of thinned forest. This is not perhaps intuitive, for with a standard log profile over the apparently rougher thinned canopy surface it might have been expected that roughness length would be greater and zero plane displacement less than for the unthinned surface. However, the thinned forest model presents less momentum-absorbing drag elements, and therefore higher windspeeds would be expected at a given height around the top of the canopy, and this is as predicted by the Miller *et al.* (1991) model used in Eq. (8.7). However, roughness length is the only controlling parameter in the Harris & Deaves (1981) profile above $10h$ and has to have a smaller value for the thinned situation to produce the higher speeds.

The main failing of the momentum transport model occurs beneath the canopy, where it predicts an unobserved increase in windspeed. Gardiner (1994) has also noted this discrepancy for a different implementation of the same transport equations. This apparent error in the predicted windspeed profile is unlikely to affect the predictions of bending stress as there is only a small amount of momentum-absorbing material at these heights.

The extreme bending stress in the outer layer of the wood of the sample tree stems predicted to occur within an hour where the 50 year return windspeed is blowing (Fig. 8.4) were found to double from about 10 MPa in the model of the unthinned forest to 20 MPa in the model of a recently thinned forest. The contribution to this increase due to the effect of the stronger winds in the thinned canopy model was separated by initially keeping the damping ratio the same as in an unthinned stand model. The full effect of thinning was then calculated by reducing the damping ratio for each tree in the thinned model to that measured by Milne (1991) where branch interference was eliminated. In all except tree 12 the contribution of increased swaying due to reduced damping was about 10% of the total increase in bending stress. In tree 12 about 30% of the predicted increase in stress was due to reduced damping of sway (Fig. 8.4). This is directly related to the large recorded change in damping ratio (Table 8.1) for this tree when its neighbours' branch interference was removed in the swaying studies of Milne (1991).

The greatest stress was estimated to occur in tree 7 which had the narrowest stem. Trees 2 and 12 were also small and had large stress values, particularly in the thinned situation. There does not appear to be a consistent relationship between tree size and the stresses generated at the stem base (Fig. 8.4). Tree 7 (the smallest) and tree 5 (the largest) are each estimated to have a stress at the stem base approaching 20 MPa in the recently thinned case. Milne & Brown (1990) discussed published studies where the static forces required to uproot forest conifers were measured by using a winch and cable to pull the trees over. The results of such studies are difficult to generalise but indicate that, in typical peaty gley forest soils (where damage is commonest), the stress at the surface of the lower stem is 20–30 MPa, on average, when a tree overturns. In deeper-rooting sites the stem stress is about 40 MPa at overturning. These values are indicative of the strength of the soil/root system in normal and wind-stable sites respectively. If the trees in the winching studies were not seriously wind damaged before the work commenced then it can be assumed that transient stresses above these values would cause progressive damage. It can therefore be predicted, from the calculations presented here which include static and dynamic forces, that unthinned stands will in general be more stable because the peak stress of 10 MPa in the lower stem during extreme wind conditions is only about 30% of the stress which occurs under static overturning. However, in a recently thinned forest the predicted peak stress rises for many trees to be close to that observed for static overturning. Assuming that root strength is directly related to the stress at static overturning, all trees with root strengths of less than 20–30 MPa in peaty gleys and 40 MPa in better soils will be at risk of the onset of accumulating damage.

The variation of bending stress up the stems was calculated to have a pronounced bell shape in the smaller trees whilst in larger trees the stress remained constant over several metres (Fig. 8.4). Morgan & Cannell (1994) have discussed hypotheses which propose that tree stems will grow more quickly at positions along their length where mechanical stress is greater. This process results in a shape of stem such that stress along the stem is constant where wind conditions are equal to the average condition occurring over the life history of the tree. Morgan & Cannell (1994) calculated that a tree which had developed under low windspeeds would not have constant stress along its stem during a later period when the windspeed was much greater than the previous average. The stress would be greater in the lower stem. The six trees studied here have stress patterns in the stem of three types: trees 7, 2 and 12, peak in the lower stem; trees 10 and 5, stress constant over several metres; tree 11, pattern intermediate between the previous two (Fig. 8.4). The predicted stresses given here are for extreme windspeeds and as trees 7, 2 and 12 are smaller they may have developed lower in the canopy with lesser wind speeds on average and hence, as suggested by Morgan & Cannell (1994), would have more stress in the lower stem for the extreme speeds used here. Trees 10 and 5 may have developed in the upper canopy and hence have a shape producing more constant stress pattern along their length even for strong winds.

The maximum values of stress in the stem above the base were in the range 10–15 MPa for trees in the unthinned case and 25–30 MPa for the thinned case. As the breaking strength of the wood in the stem is known to be 40–45 MPa, the risk of stem breakage due to the wind should be small in both unthinned and thinned Sitka spruce forests. Trees 7, 2 and 12 were those with the greatest stress values in the stem above ground level and hence the greatest risk of damage from snap, particularly after thinning. Subdominant trees in a stand develop in more sheltered conditions, and hence in lower windspeeds, and may therefore fall into the category of being at risk to stem snap after thinning. These quantitative modelling predictions are in agreement with qualitative observations in British plantations, namely that thinning significantly increases the likelihood of wind damage and secondly that stem breakage is not unknown, but this mode of damage is usually associated with snow or stem defects.

8.5 Conclusions

The modelling methods presented provide a useful starting point for further work on quantitatively assessing the effects of forest management on the risk of wind damage to Sitka spruce. The estimated stresses in typical tree stems in models of differently spaced forests compare favourably with the limited information from published field measurements. Measurements of canopy structure and individual tree dynamic parameters under different silvicultures are required before the models presented can be more widely used.

References

Blackburn, P. & Petty, J. A. (1988). Theoretical calculations of the influence of spacing on stand stability. *Forestry*, **61**, 235–44.

Blackburn, P., Petty, J. A. & Miller, K. F. (1988). An assessment of the static and dynamic factors involved in windthrow. *Forestry*, **61**, 29–43.

Caton, P. G. F. (1976). *Maps of Hourly-Mean Wind Speed over the United Kingdom, 1965–1973*. Climatological Memorandum 79. Meteorological Office, Bracknell.

Cook, N. J. (1985). *The Designer's Guide to Wind Loading of Building Structures*, part 1. Butterworth, London.

Coutts, M. P. (1986). Components of tree stability in Sitka spruce on peaty gley soil. *Forestry*, **59**, 173–97.

ESDU (1984a). *Strong Winds in the Atmospheric Boundary Layer*, part 1, *Mean-Hourly Wind Speeds*. Data Item 82026. Engineering Sciences Data Unit, London.

ESDU (1984b). *Strong Winds in the Atmospheric Boundary Layer*, part 2, *Discrete Gust Speeds*. Data Item 83045. Engineering Sciences Data Unit, London.

ESDU (1984c). *The Responses of Flexible Structures to Atmospheric Turbulence*. Data Item 76001. Engineering Sciences Data Unit, London.

ESDU (1987). *Calculation Methods for Along-Wind Loading*, part 2, *Response of Line-Like Structures to Atmospheric Turbulence*. Data Item 87035 (amended March 1989). Engineering Sciences Data Unit, London.

Galinski, W. (1989). A windthrow-risk estimation for coniferous trees. *Forestry*, **62**, 139–46.

Gardiner, B. A. (1994). Wind and wind forces in a plantation spruce forest. *Boundary-Layer Meteorology*, **67**, 161–86.

Hardman, C. E., Helliwell, N. C. & Hopkins, J. S. (1973). *Extreme Winds over the United Kingdom for Periods Ending 1971*. Climatological Memorandum 50A. Meteorological Office, Bracknell.

Harris, R. I. & Deaves, D. M. (1981). The structure of strong winds. In *Proceedings of CIRIA Conference on Wind Engineering in the Eighties*, November 1980. Construction Industry Research and Information Association, London.

Jarvis, P. G., James, G. B. & Landsberg, J. J. (1976). Coniferous forest. In *Vegetation and the Atmosphere*, vol. 2, *Case Studies*, ed. J. L. Monteith, pp. 171–240. Academic Press, London.

Landsberg, J. J. & Jarvis, P. G. (1973). A numerical investigation of the momentum balance of spruce forest. *Journal of Applied Ecology*, **10**, 645–55.

Li, Z. J., Miller, D. R. & Lin, J. D. (1985). A first-order closure scheme to describe counter-gradient momentum transport in plant canopies. *Boundary-Layer Meteorology*, **33**, 77–83.

Lohmander, P. & Helles, F. (1987). Windthrow probability as a function of stand characteristics and shelter. *Scandinavian Journal of Forest Research*, **2**, 227–38.

Mayer, H. (1987). Wind-induced tree sways. *Trees*, **1**, 195–206.

Miller, D. R., Lin, J. D. & Lu, Z. N. (1991). Airflow across an alpine forest clearing: a model and field measurements. *Agricultural and Forest Meteorology*, **56**, 209–25.

Milne, R. (1991). Dynamics of swaying of *Picea sitchensis*. *Tree Physiology*, **9**, 383–99.

Milne, R. (1992). Extreme wind speeds over a Sitka spruce plantation in Scotland. *Agricultural and Forest Meteorology*, **61**, 39–53.

Milne, R. & Blackburn, P. (1989). The elasticity and vertical distribution of stress within stems of *Picea sitchensis*. *Tree Physiology*, **5**, 195–205.

Milne, R. & Brown, T. A. (1990). *Tree Stability and Form*. Final Report to the CEC, Contract MA-0061-UK(BA). Commission of the European Communities, Brussels.

Morgan, J. & Cannell, M. G. R. (1994). Shape of tree stems: a re-examination of the uniform stress hypothesis. *Tree Physiology*, **14**, 63–74.

Shellard, H. C. (1968). *Tables of Surface Wind Speed and Direction over the United Kingdom*. Meteorological Office, Met. O.792. HMSO, London.

9
Experimental analysis and mechanical modelling of wind-induced tree sways

D. G. E. GUITARD and P. CASTERA

Abstract

Investigations have been carried out for the purpose of understanding and predicting different types of storm damage to forest, such as stem, stock or root breaks. The tree sway mechanical model used enables the determination of the modal parameters, natural frequency and shape deflection of the stem, for the two or three first oscillation modes of the structure. Inputs are specific dendrometric data, including mass distribution and conicity of the stem, orientation and biomass of branches. The purpose of this model is to allow the estimation of the fundamental oscillation mode (responsible for stock or root breaks) and the first harmonic (probably responsible for stem breaks) and to simulate the influence on those parameters of silvicultural practices such as pruning and pollarding. For the moment, in our description, the tree stem is considered as a cantilever beam of circular section, rigidly embedded in the soil. Experimentally, the signal delivered by an extensometer plugged under the bark on the tree stem is analysed using a dual-channel signal analyser. The modal parameters of the dynamic behaviour of the tree are identified, including resonance frequencies and associated damping factors. The methodology is first applied to a laboratory tree scale model and a good fit of the experimental results to predicted values is shown. Preliminary experimental results from a young budding red oak indicate the sensitivity of the dynamic spectrum to such a biological change.

9.1 Introduction

For about 4 years the research team in wood mechanics of the Laboratory of Wood Rheology of Bordeaux (LRBB) has been conducting an analysis of the mechanical behaviour of tree structure. In preliminary investigations the mechanical behaviour of the structure when submitted to static bending was analysed. Experimental measurements of the rigidity in bending of a tree stem have been discussed by Langbour (1989), as related to the main geometric parameters of the stem and the elastic modulus in the longitudinal direction of the constitutive material. Other examples are given by the work by Cinotti (1989) on frost cracks in logs, and by Fournier (1989) on internal stress development in growing trees.

In previous papers (Guitard, 1990; Guitard *et al.*, 1991) the background of a method for calculating the first vibration modes in dynamic bending of the tree structure, using biomass distribution, was proposed. The method has been used to simulate the response of a particular species, maritime pine (*Pinus pinaster* Ait.), and to analyse the influence of silvicultural practices such as pruning and pollarding on the resonance frequencies of the first two swaying modes.

The aim of the present chapter is, first, to illustrate the use of the model for interpretation of the experimental frequency spectrum of a tree structure. In this particular case the studied structure is a scale model of a tree. A second approach shows the sensitivity of the dynamic spectrum to biological changes in a young, budding red oak (*Quercus rubra* L.).

9.2 Mechanical measurement and modelling of tree sway under dynamic bending

A first step in the model consists in establishing schematic representations of the geometry of the structure, and weight distributions within the stem. A significant level of complexity must be introduced in these representations, so that particularities of the structure can be taken into account. On the other hand the complexity of the model must be restricted given the limitation of practical field measurements.

9.2.1 Tree structure

The morphology of the structure is described as follows (Fig. 9.1). The stem is divided into n logs, the ith log being situated between the whorls of height z_{i-1} and z_i, respectively, and characterised by the area $S(z)$ of any cross-section and the bending inertia $I(z)$ of this section at the position z. The material of this log has a longitudinal modulus of elasticity E_L and a specific gravity ρ.

Branches at the whorl of height z_i are considered through the resulting weight M_i of all branches at this height and the moment of inertia I_i, calculated around the bending axis. Note that the proposed modelling will not account for the stiffness of branches themselves.

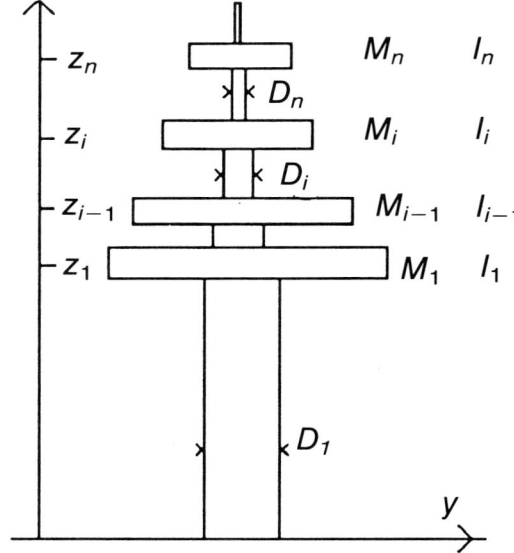

Fig. 9.1. Idealised morphology of the tree structure.

9.2.2 Equations of motion and approximate solution method

The following developments are, *a priori*, restricted to the bending of the stem in a plane (torsion is not yet considered). When the structure is moved away from the equilibrium state, free sways take place and the structure has a periodic deflection $U_x(z,t)$. Under the assumption of small and non-dissipative displacements, the resulting movement is the superposition of different modes, a fundamental sway and harmonic sways:

$$U_x(z,t) = \sum_{p=1}^{\infty} U_p(z) \sin\left(\frac{2\pi p}{T} t + \varphi_p\right) \qquad (9.1)$$

where $U_p(z)$, $\omega_p = (2\pi p)/T$ and φ_p are, respectively, the deflection, frequency and phase of the sway mode of order p.

It is therefore correct to analyse the harmonic sway of the structure using a restriction of Eq. (9.1) such as:

$$U_x(z,t) = U(z) \sin(\omega t) \qquad (9.2)$$

The kinetic energy $W_c(t)$ of the structure is given in Eq. (9.3) as the superposition of three terms: the first for translation related to the distributed weight between z_{i-1} and z_i for each log; the second related to the mass M_i of all branches of the whorl of order i; and finally a rotation term for all branches of the ith whorl, characterised by inertia I_i, around the rotational axis being considered:

$$W_c(t) = \omega^2 C(U) \cos^2(\omega t)$$

$$= \omega^2 \sum_{i=1}^{n} \left\{ \int_{z_{i-1}}^{z_i} \rho \frac{S(z)}{2} [U(z)]^2 \, dz + \frac{M_i}{2} [U(z_i)]^2 + \frac{I_i}{2} \left(\frac{\partial U(z_i)}{\partial z} \right)^2 \right\} \cos^2(\omega t) \quad (9.3)$$

The elastic potential energy $W_e(t)$ of the structure, when only bending effects are considered, is given by:

$$W_e(t) = P(U) \sin^2(\omega t) = \sum_{i=1}^{n} \left\{ \int_{z_{i-1}}^{z_i} \frac{E_L I(z)}{2} \left[\frac{\partial^2 U(z)}{\partial z^2} \right]^2 dz \right\} \sin^2(\omega t) \quad (9.4)$$

where $P(U)$ represents the maximum elastic potential during one cycle, E_L is the longitudinal modulus of elasticity of wood, and $I(z)$ is the inertial momentum of any cross-section with regard to the bending axis.

Under the energy conservation assumption, the first derivative of the total energy of the system $W_T(t)$ must be zero. Therefore

$$\frac{dW_T}{dt} = \frac{d[W_c(t) + W_e(t)]}{dt} = 0 = [P(U) - \omega^2 C(U)] 2\omega \sin(\omega t) \cos(\omega t) = 0 \quad (9.5)$$

The corresponding sway frequency ω is determined by solving:

$$P(U) - \omega^2 C(U) = 0 \quad (9.6)$$

In the following sections it will be assumed that the junction of the tree at the soil is perfectly rigid (different boundary conditions could be considered due to alternative information about root behaviour). Then the boundary conditions for $U(z)$ are:

$$U(0) = 0 \quad \text{and} \quad \frac{dU(0)}{dz} = 0 \quad (9.7)$$

The objective is to estimate the fundamental and the first harmonic sways which induce natural deformations of the structure as illustrated in Fig. 9.2. The displacement function $U(z)$ is approximated by a five-parameter polynomial form $U^*(z)$ of order 6, satisfying the kinematic conditions:

$$U(z) = U^*(z) = a_2 z^2 + a_3 z^3 + a_4 z^4 + a_5 z^5 + a_6 z^6 \quad (9.8)$$

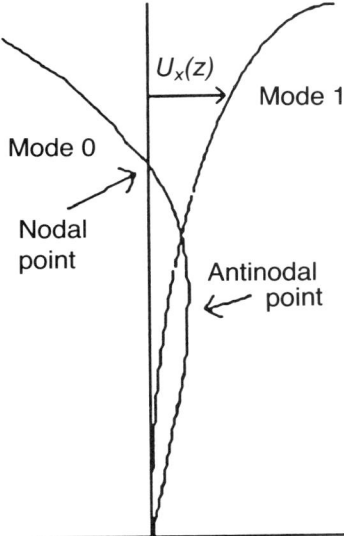

Fig. 9.2. Schematic displacements for the two first bending modes.

Each $U^*(z)$ is characterised by a vector A of components $[a_i]$ with i varying from 2 to 6. Using the polynomial of Eq. (9.8) in Eq. (9.6) leads to an eigenvalue problem:

$$[a_i]^T \left[[K] - \omega^2[M]\right][a_i] = 0 \tag{9.9}$$

$$\det\left[[K] - \omega^2[M]\right] = 0 \tag{9.10}$$

$[K]$ is a matrix of rigidity taking into account the bending stiffness of the logs in the stem; $[M]$ is a matrix containing the weights of logs, branches and the moments of inertia in rotation of the same branches.

The roots in ω^2 of the characteristic Eq. (9.10) provide five pulsations. However, in the following sections only the displacement functions of the fundamental and the first harmonic modes will be analysed, corresponding to the eigenvectors of the two lowest frequencies. The modes of higher range are not considered because they are not estimated with sufficient accuracy.

9.2.3 Free frequencies and damping measurements

In order to measure sway frequency in sample trees, an extensometer is plugged under the bark on the tree stem (Fig. 9.3). When the tree structure is subjected to random wind oscillations, the signal is registered by the FFT

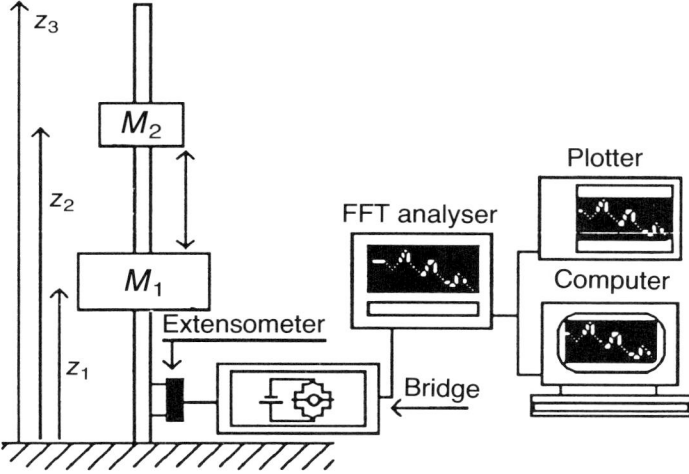

Fig. 9.3. Experimental equipment, around a tree scale model (M_1 and M_2).

analyser (Brüel & Kjær T2034). The frequency spectrum obtained can be plotted or stored in a computer.

The remaining problem is the identification of the resonance frequency N_i of each oscillating mode of rank i; N_1 for the fundamental mode and N_2 for the first harmonic (Fig. 9.4).

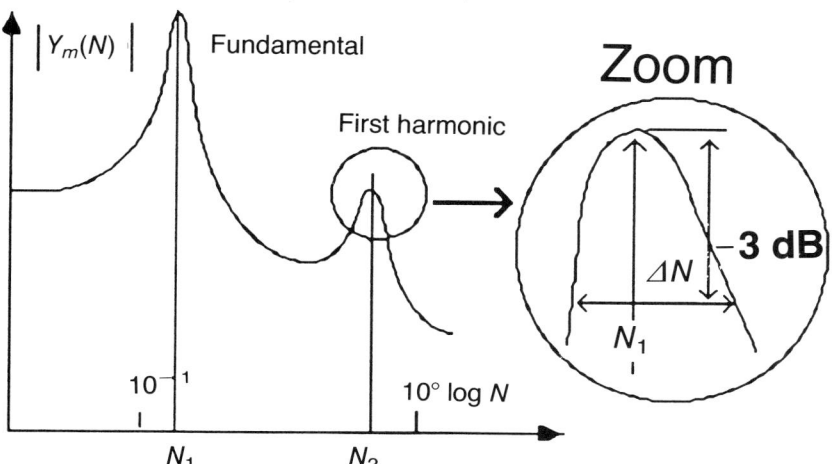

Fig. 9.4. Frequency spectrum, resonance maxima, damping.

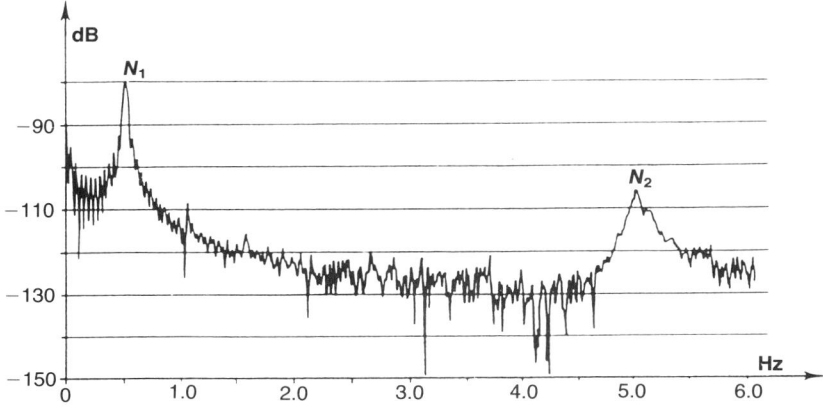

Fig. 9.5 Frequency spectrum for oscillations of the tree scale model.

The experiment gives access to complementary information concerning the damping of motion by energy loss during swaying, but this is not yet taken into account by the modelling. The structural damping is estimated by the bandwidth ΔN_i, at -3 decibels around a given resonance frequency N_i.

In such conditions, a given tree structure can be studied by the modelling, taking into account biomass distribution and the mechanical properties of wood, and can also be analysed by the experimental approach.

9.3 Tree scale model behaviour

To improve the methodology, a preliminary study has been conducted on a scale model of a tree (Fig. 9.3). The tree stem is simulated by a wooden beam of z_3 1.78 m long, with a diameter D of 14.1 mm. The inertia of the branches is simulated by two masses: M_1 of 1.3 kg at the level $z_1 = 1$ m, and M_2 of 1.2 kg at the level $z_2 = 1.5$ m. The longitudinal Young's modulus of the dry wood is estimated experimentally ($E_L = 18\,500$ MPa) and specific gravity (ρ) is 0.57 g cm^{-3}.

Fig. 9.5 gives the frequency spectrum from 0 to 6.3 Hz. Two resonance frequencies are identified: for the fundamental mode $N_1 = 0.51$ Hz, for the first harmonic $N_2 = 5.1$ Hz. These experimental data are compared with calculations from the model in Table 9.1. The agreement between experimental and calculated frequencies is good enough to allow the use of the model for identification, without ambiguity, of the correct frequency mode N_i among the multiplicity of frequencies that can occur in a given spectrum. Note that the modelling results are useful to identify the resonance maxima in the spectrum (Fig. 9.5); this is not so simple in the case of a real tree.

Table 9.1 *Experimental and modelling results for the tree scale model*

	N_1	ΔN_1	N_2	ΔN_2
Experimental	0.51	0.035	5.05	0.075
Modelling	0.06	—	5.3	—

9.4 Simulation of silvicultural practices

A realistic set of data, obtained from dendrometric measurements on a 23-year-old Maritime pine tree, has been used in the following simulations (see details in Guitard *et al.*, 1991; Radi, 1992). The tree chosen is in an experimental stand of the INRA (Domaine de l'Hermitage, INRA Pierroton).

The two first sways of the tree in bending are characterised by the two natural frequencies N_1 and N_2, and the corresponding displacement functions that are schematically represented in Fig. 9.2. It is of interest to make a comparison with the results of Wood (1990) who used the experimental work of Gardiner (1989) on a Standard Sitka spruce (Table 9.2). Note that the two trees exhibit similar characteristics. The Standard Sitka spruce is a little smaller than the Maritime pine and has a lighter crown also. The first natural frequency N_1 is also smaller.

A first simulation consisted in analysing the effect of pollarding the tree, more or less severely, on the two first sways of the structure. Reduction of the height of the tree from H to h increases the natural frequencies n_1 and n_2. The effect is greater on n_2. The curves shown in Fig. 9.6 indicate that the variation of n_1 is roughly proportional to power 3 of the ratio of the total residual mass m to the total initial mass M of the tree.

A second simulation consisted in pruning the tree more or less severely, from the base to the top. The effect of pruning on n_1 is significant, as indicated in Fig. 9.7. Pruning has a smoother effect on the frequency n_2. This is due to the fact that when pruning occurs at the level of the nodal point of the first harmonic sway, the evolution becomes less significant. Fig. 9.8 illustrates

Table 9.2. *Comparison between the Maritime pine tree and the standard Sitka spruce*

	Standard Sitka spruce (experimental)	*Pinus pinaster* (calculated)
Height (m)	15	17.5
Mass of the stem (kg)	162.3	268.7
Mass of living branches (kg)	49.5	94.7
First natural frequency N_1 (Hz)	0.33	0.395
Second natural frequency N_2 (Hz)	—	1.510

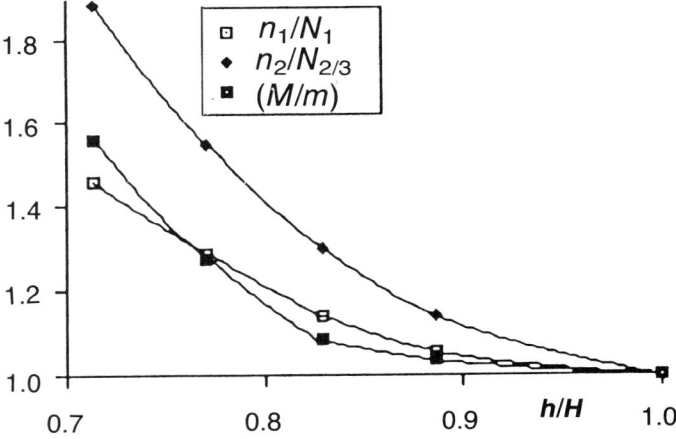

Fig. 9.6. Effect of pollarding on natural frequencies n_1 and n_2. M, H, N_1, N_2 are respectively the mass, height and the two first natural frequencies of the tree; m, h, n_1, n_2 are the mass, height and the two first natural frequencies of the pollarded tree.

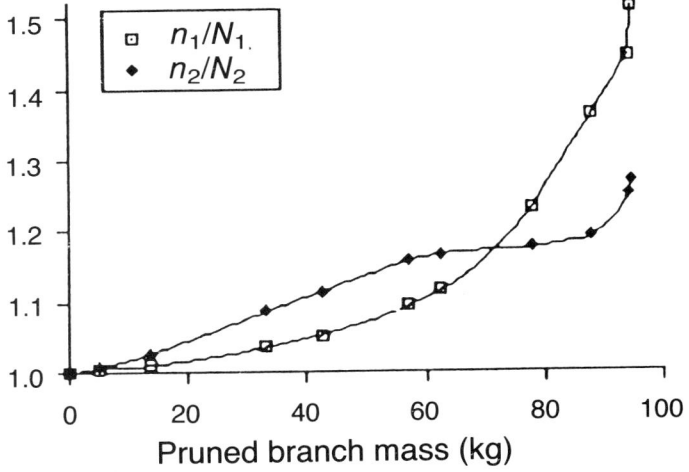

Fig. 9.7. Effect of pruning on natural frequencies.

the evolution of n_1 plotted against the percentage of pruned branch mass (ratio of the mass of pruned branches $(M_b - m_b)$ to the total mass of branches M_b) at the power 3.

The evolution of the curvatures C along the stem provides additional information on the deflections. Normalised curvatures C/C_{max} corresponding to the whole tree are shown in Fig. 9.9. For the first mode it is observed that the maximal curvature occurs near $0.15H$. This result must be related to the

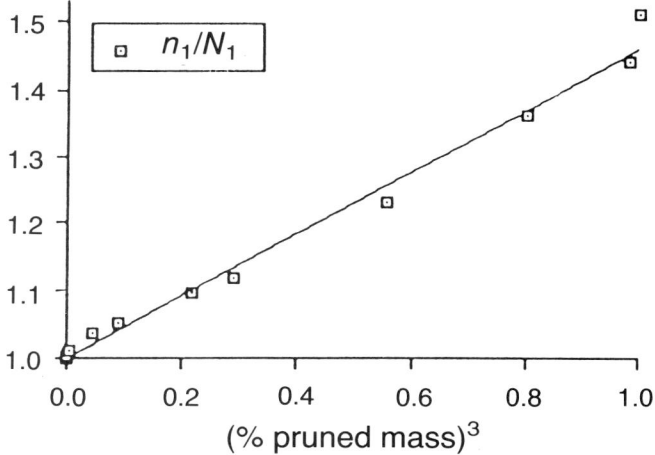

Fig. 9.8. First natural frequency versus percentage of pruned mass.

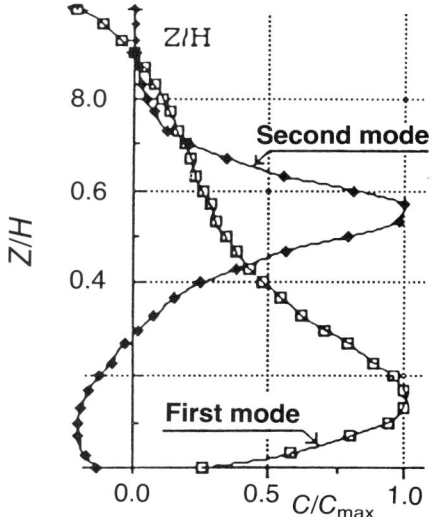

Fig. 9.9. Curvature along the pine stem.

conicity of the first log near the roots. In the case of a large resonance movement of the tree due to wind (assuming that the root system resists), for the first mode rupture of the stem may occur near the roots at the predicted level. With regard to the deflection of the second sway mode, the maximum curvature occurs near $z = 0.55H$, close to the antinodal point as observed earlier.

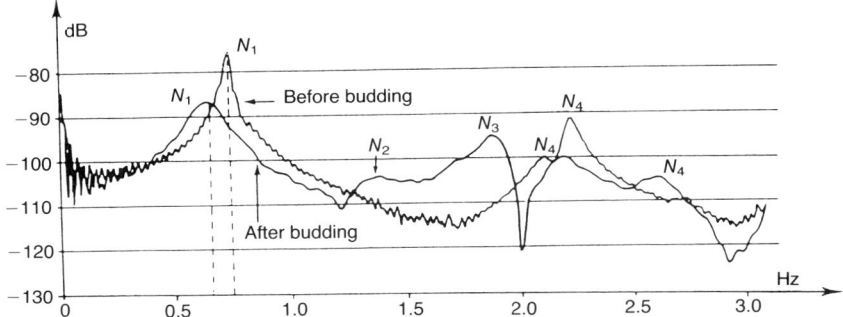

Fig. 9.10. Dynamic spectrum analysis of a red oak before and after budding.

Again, in the case of large resonance movement on the second mode, rupture may occur near the mid-height of the stem.

These examples illustrate the possibilities afforded by the mechanical model for analysing tree sway and its consequences.

9.5 Field experiments on a young red oak

Fig. 9.10 represents some preliminary results obtained from the dynamic spectrum analysis of a 6.2 m tall young red oak. It is the superposition of two spectra obtained at two different periods of growth: a winter measurement when the tree had no leaves, and a second one recorded in spring, just at the beginning of budding. It was soon found that spectra recorded in field conditions are more complicated than those obtained on the scale model, and consequently more difficult to analyse.

Looking at the lowest frequency peak, in each spectrum, it seems that comparison of winter measurement ($N_1 = 0.75$ Hz and $\Delta N_1 = 0.04$ Hz) with the spring measurement ($N_1 = 0.66$ Hz and $\Delta N_1 = 0.08$ Hz) shows a decrease in N_1 associated with an increase in ΔN_1. The decrease in frequency can be interpreted as the result of increased biomass (primary and secondary growth and mass of new leaves) associated with a quasi-constant rigidity of the stem, before the lignification of the radial cellular increment. The increase in damping may be associated with the aerodynamic drag due to the presence of many new leaves.

The identification of the second mode is more difficult because the expected maximum seems not to be unique. At least two, and even three maxima can be detected in the range 1.3–3.0 Hz (Table 9.3).

In fact, a real tree does not have the symmetrical properties of the tree scale model. Langbour (1989) also showed that the static bending stiffness

Table 9.3. *Observed resonance frequencies (Hz) on a young red oak*

	N_1	N_2	N_3	N_4
Before budding	0.75	2.08	2.15	2.6
After budding	0.66	1.35	1.85	2.2

of a given tree is not a constant but varies according to the orientation of the considered flexure plane. This work, although related to the second mode of oscillation, remains valid for the first mode. We did find during the experiments that a bending oscillation of the tree in one plane often gives free oscillations out of the plane by a combination of different modes.

Finally we must recognise that twisting modes can occur, and should be included in the spectra. They are not considered at present by the modelling and thus cannot be predicted.

9.6 Concluding remarks

Numerical and experimental tools have been developed to study and characterise wind-induced tree sways. The preliminary results discussed here have shown the interest and complementarity of the experimental approach by spectrum analysis, and the theoretical approach using mechanical modelling of the tree structure that takes into account biomass distribution and rheological properties of wood.

The modelling enabled prediction of qualitative and even quantitative effects of silvicultural treatments on the sway modes of the tree. As shown in a previous paper, we must bear in mind that the use of models requires non-conventional morphological measurements.

The last example gives evidence of the difficulties remaining for a full understanding of experimental results. Whatever the quality of the tools, the user must be able to read and interpret frequency spectra.

References

Cinotti, B. (1989). La gélivure des chênes: front de gel source de contraintes internes, incidence des propriétés anatomiques et mécano-physiques. Doctoral Thesis, INP de Lorraine, Nancy, France.

Fournier, M. (1989). Mécanique de l'arbre sur pied: maturation, poids propre, contraintes climatiques dans la tige standard. Doctoral Thesis, INP de Lorraine, Nancy, France.

Guitard, D. (1990). L'arbre soumis aux aléas climatiques de courte durée, vents. Deuxième séminaire 'Architecture, structure, mécanique de l'arbre'. B. Thibaut, Montpellier. France.

Guitard, D. (1992). Mechanical modelling of tree sways: probability of wind damage. IUFRO A11 – Division 5 Conference, 'Forest Products' Proceedings 1, pp. 186–7. Nancy, 23–28 August. A. Arbolor, Nancy, France.

Guitard, D. & Fournier, M. (1989). Eléments de mécanique des solides déformables en vue d'une application à la mécanique de l'arbre sur pied. Premier séminaire 'Architecture, structure, mécanique de l'arbre'. B. Thibaut, Montpellier, France.

Guitard, D., Castera, P., Fournier, M. & Radi, M. (1991). Comportement en flexion dynamique d'un arbre: modélisation du Pin maritime. Troisième séminaire 'Architecture, structure, méanique de l'arbre. B. Thibaut editor. Montpellier, France.

Langbour, P. (1989). La rigidité de l'arbre sur pied comme indicateur du module d'élasticité longitudinal du bois. Doctoral Thesis, INP de Lorraine, Nancy, France.

Radi, M. (1992). Analyse morphologique de l'arbre en vue de sa modélisation mécanique. Doctoral Thesis, Université Bordeaux I, France.

Wood, J. C. (1990). *Aeroelastic Modelling of Trees Susceptible to Windthrow Damage.* University of Oxford. Final Report, Contract CEE ref: MA1B-0063-UK (BA).

10
Failure modes of trees and related failure criteria

C. MATTHECK, K. BETHGE and W. ALBRECHT

Abstract

Three failure modes of trees are described, selected from a more extensive study, because of their practical importance. These are: failure of hollow trees, axial splitting of hazard beams, and windthrow.

10.1 Failure of hollow trees by cross-sectional flattening

The failure mechanism observed in nature in hundreds of hollow trees is shown in Fig. 10.1. Hollow trees will fail when a certain ratio of wall thickness t to stem radius R is reached. The failure mechanism by cross-sectional flattening is due to lateral forces which increase with the curvature of the 'pipe' due to bending. At a certain degree of flattening the hoop stresses exceed the value of circumferential strength resulting in axial splitting. When the hollow tree collapses into individual timber boards the stiffness is dramatically reduced and overall breakage will normally happen. It has been shown in a field study (Mattheck et al., 1993) with more than 700 trees that failure can start if 68–70% of the stem radius is hollow or decayed wood (Figs. 10.1, 10.2). This failure has to be expected if the crown of the tree is not reduced and therefore the full canopy (sail) area is under wind loading. If, on the other hand, the crown volume is reduced, it is possible that trees with even much smaller t/R ratios will resist the wind, with a much smaller canopy.

10.2 Axial splitting of hazard beams

Lateral forces are also responsible for axial splitting of hazard beams. However, they act in the opposite direction, pulling laterally away from the centre of a curved branch or stem, if it is straightened by bending. Hazard beams

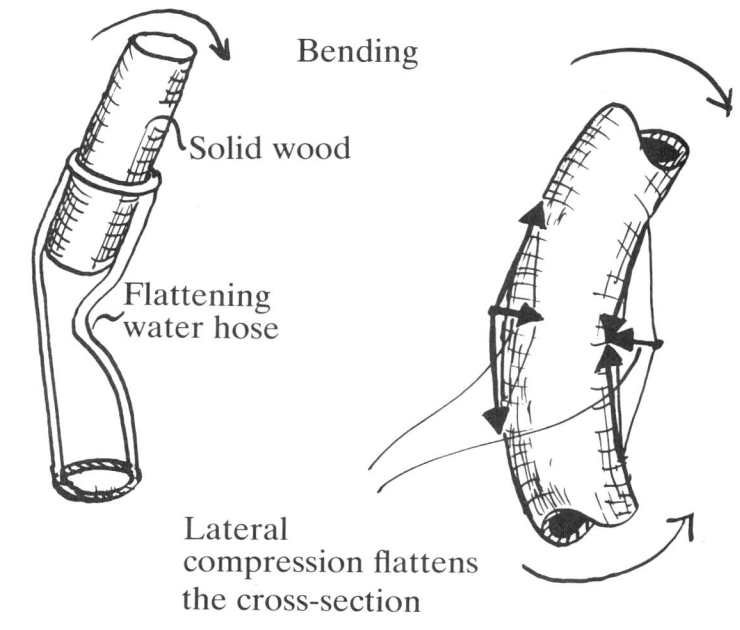
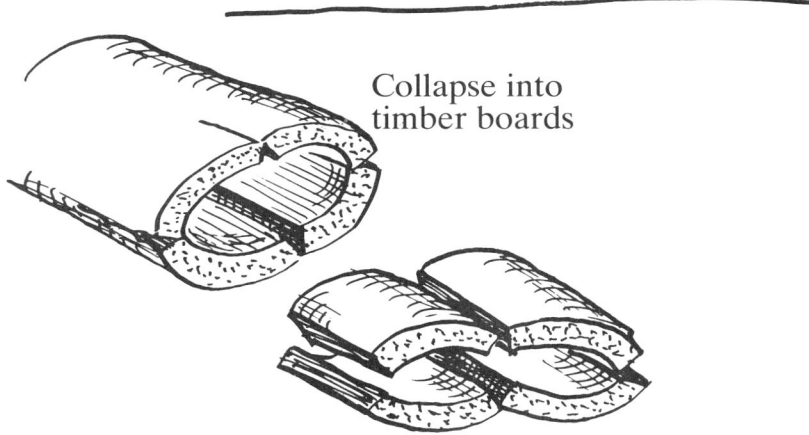

Fig. 10.1. A hollow tube of any material will flatten in cross-section under bending load. Due to lateral weakness wood will split axially into timber boards which finally will break individually.

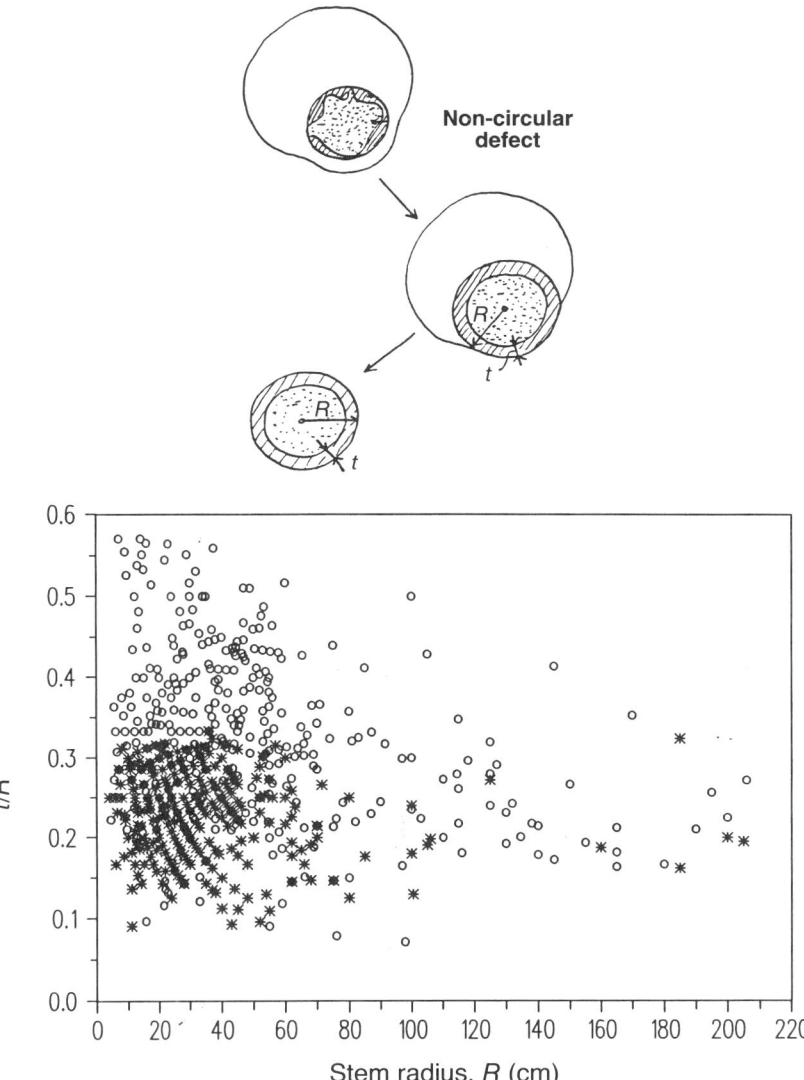

Fig. 10.2. Result of a field study on about 800 hollow standing (open circles) and broken (asterisks) trees (non-symmetric hollowness has to be measured as shown in the figure).

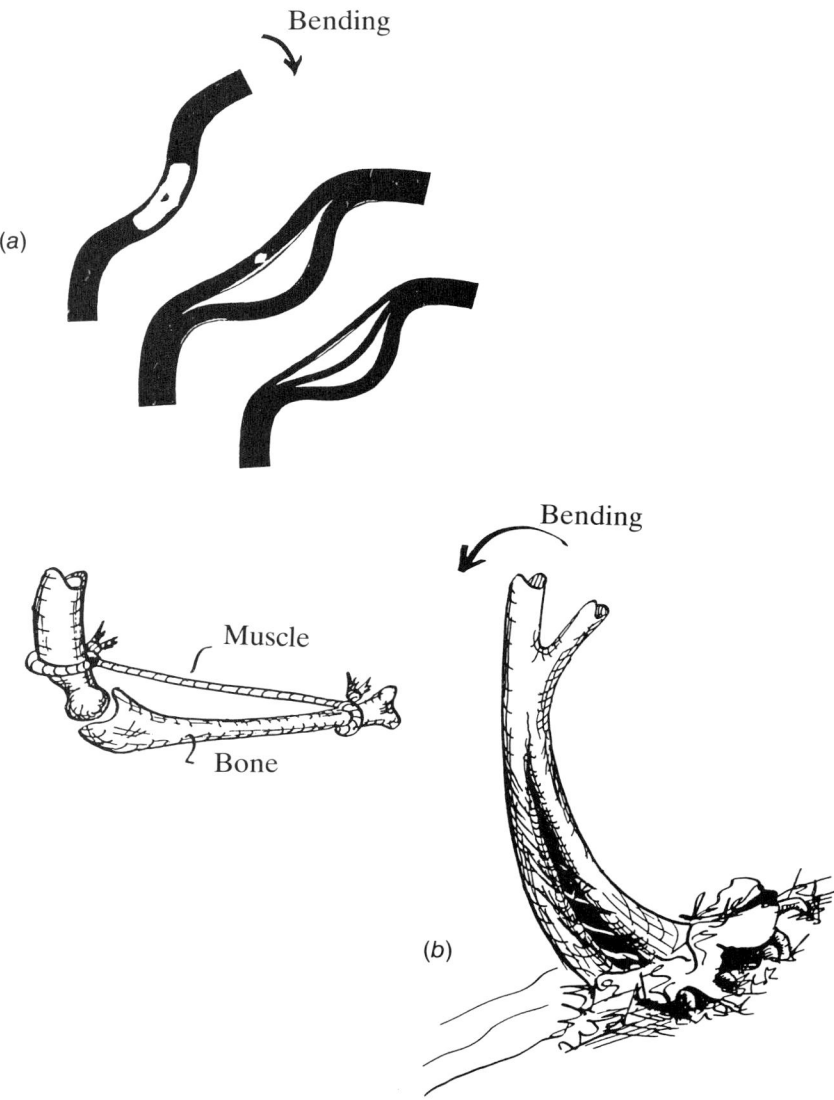

are presented in Fig. 10.3 (Mattheck, 1991; Mattheck & Bethge, 1991). The failure type in Fig. 10.3c is standard for spruce trees with their shallow rooting systems. The failure criterion is determined when lateral stress exceeds lateral strength. The trees seem to perceive the stress and respond by increasing lateral strength according to the distribution of lateral stresses in order to avoid failure. This has been proved up to now only for three ash trees

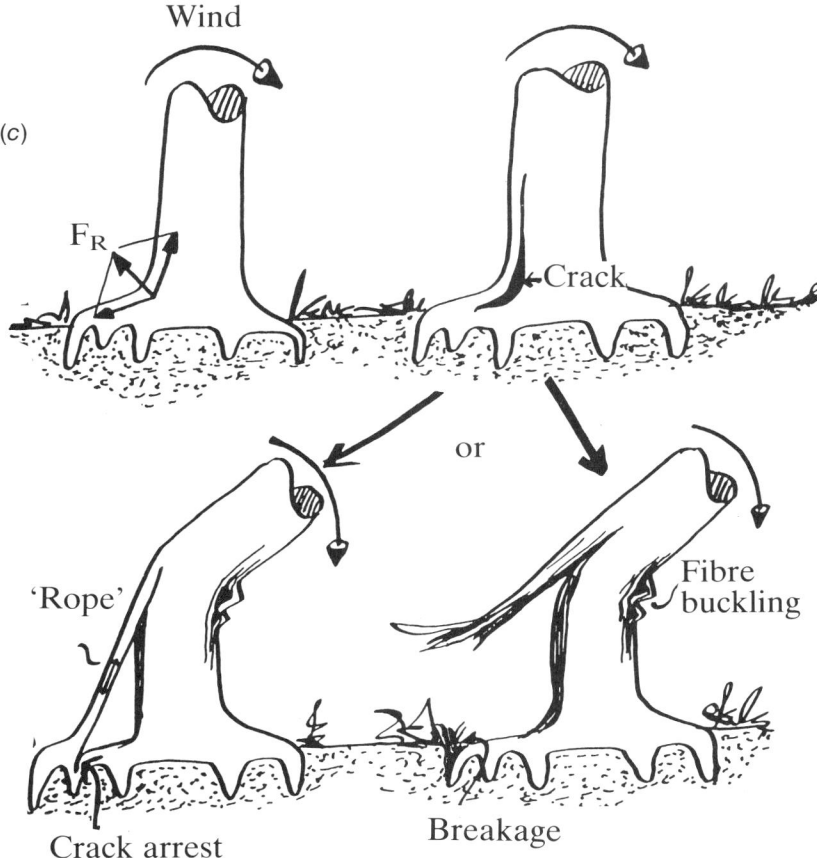

Fig. 10.3. Hazard beams in trees: (*a*) multiple splitting in a branch; (*b*) stable tree with split; (*c*) wind breakage, typical for spruce, starting by root delamination.

and a general statement regarding other species cannot be given yet, although it is plausible. A model for one of the ash trees is presented in Fig. 10.4. The bending strength measured by breaking samples taken with an increment borer is distributed locally just as the lateral stresses computed with the finite element method (FEM).

Before the stresses are plotted the 'fibres' in the FEM mesh are arranged along the force flow. This is achieved by adjusting the local orthotropy system of each element to the local force flow iteratively. The method used for this is called CAIO (computer-aided internal optimisation) (Mattheck, 1993). By local increase of lateral strength, the trees become safer against failure by axial splitting.

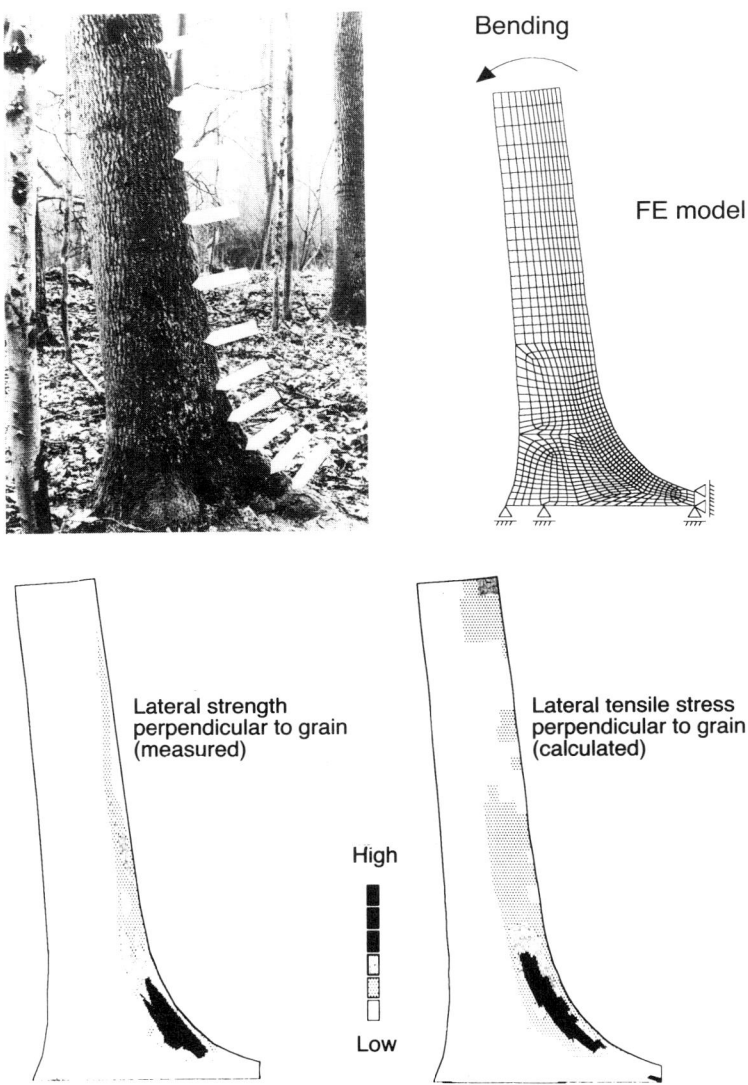

Fig. 10.4. Cylindrical cores, 5.3 mm in diameter, have been taken from the ash tree (arrows) and broken incrementally to determine the strength distribution. The calculated lateral stresses are also plotted. The highest lateral strength is found at the sites of highest computed lateral stresses.

Failure modes of trees

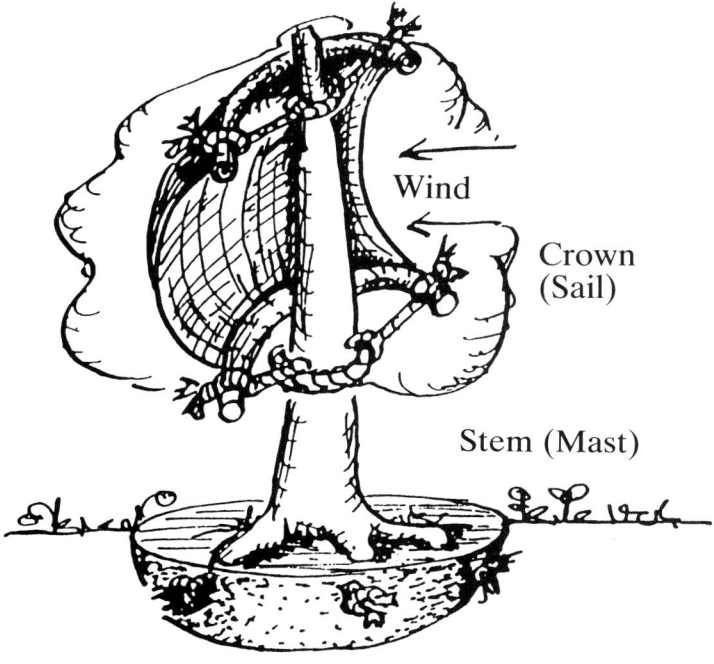

Roots as an anchor

Fig. 10.5. The tree has no weak spot. Crown (sail), stem and root plate fit mechanically with each other.

10.3 Failure by windthrow

In Putz *et al.* (1983) it is shown that windthrow happens mainly to trees with strong roots, whilst trees with weak or decayed roots tend to snap. However, healthy trees are a chain of equal links: crown sail, stem stiffness and root plate are well adjusted to each other (Fig. 10.5). Coutts (1983) also mentions that the bending moment at which the stem would break is often close to the bending moment of windthrow. A correlation between stem radius R and radius of the root plate R_W measured for wind-blown trees is therefore expected. Fig. 10.6 shows the result of a field study on about 1500 wind-blown trees (nearly all European hardwood and softwood species). Thinner, light-weight trees with lower density wood need a larger root plate in relation to their stem radius than trees with thick stems. The reason may be that critical shear strength of the soil increases with the normal pressure, because the soil below heavy trees is more shear resistant and a smaller shear zone is sufficient to anchor the tree.

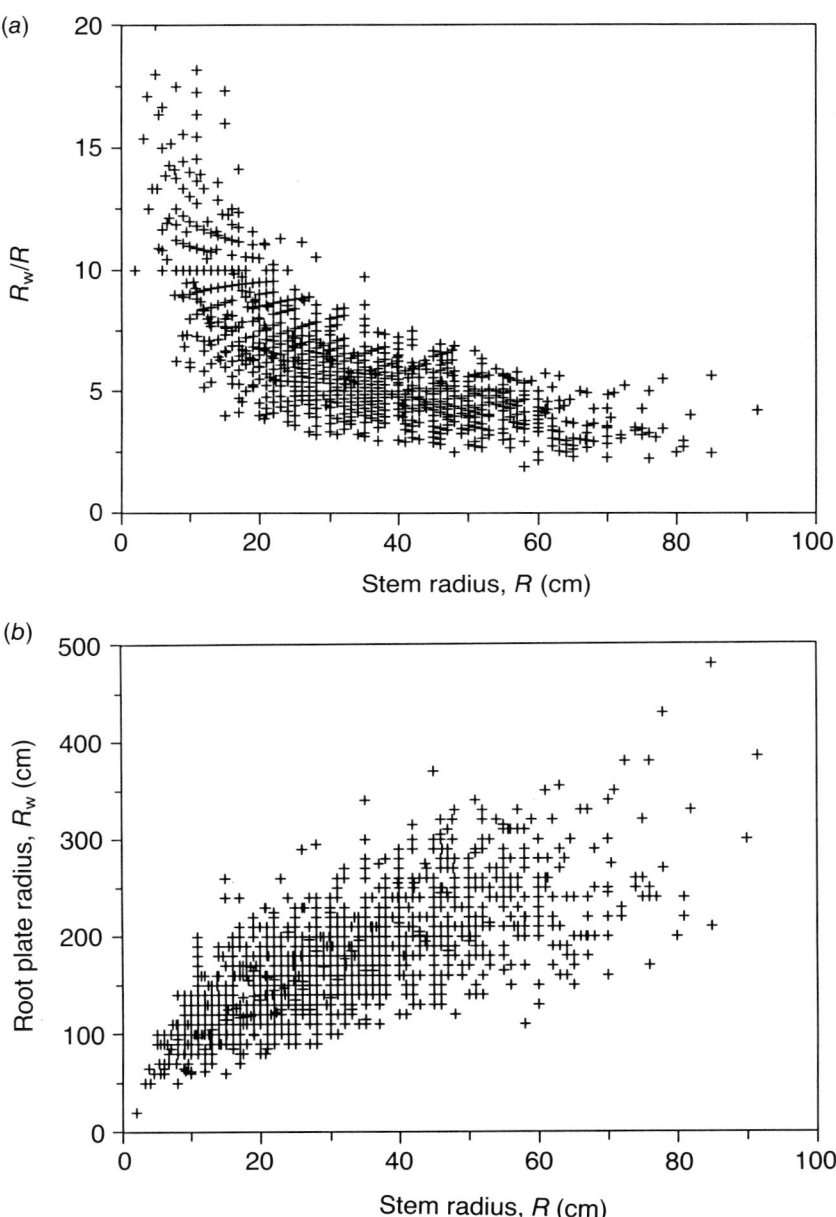

Fig. 10.6. Field study on 1500 windthrown trees: (a) root plate radius R_w normalised by the stem radius R and plotted versus the stem radius; (b) absolute R_w values versus stem radius R.

An upper envelope in Fig. 10.6a and b can be interpreted as a failure diagram for windthrow.

Acknowledgements

The authors are grateful to Professor Dr Hans Kübler, Mr Ted Green and the tree expert Helge Breloer who helped a great deal in the field studies.

References

Coutts, M. (1983). Root architecture and tree stability. *Plant and Soil*, **71**, 171–88.
Mattheck, C. (1991). *Trees: The Mechanical Design.* Springer-Verlag, Berlin.
Mattheck, C. (1993). *Design in der Natur.* Rombach-Verlag, Freiburg.
Mattheck, C. & Bethge, K. (1991). Failure of trees induced by delamination. *Aboricultural Journal*, **15**, 243–53.
Mattheck, C. & Breloer, H. (1993). *Handbuch der Schadenskunde von Bäumen- der Baumbruch in Mechanik und Rechtsprechung.* Rombach-Verlag, Freiburg.
Mattheck, C., Bethge, K. & Erb, D. (1993). Failure criteria for trees. *Arboricultural Journal*, **17**, 201–9.
Putz, F. E., Coley, P. D., Lu, K., Montalvo, A. & Aiello, A. (1983). Uprooting and snapping of trees: structural determinants and ecological consequences. *Canadian Journal of Forestry Research*, **13**, 1011–20.

11
An experimental investigation of the effects of dynamic loading on coniferous trees planted on wet mineral soils

M. RODGERS, A. CASEY, C. McMENAMIN
and E. HENDRICK

Abstract

Two field studies are presented on the *in situ* dynamic loading of mature coniferous trees planted on wet mineral soils. The studies consisted of examining the behaviour of a complete tree under natural wind loading and a number of trees with truncated stems under forced dynamic loading. The test site has a history of tree instability and the mineral soil under the rootplates consists of a clayey silty sand. The rootplates of the tested trees were shallow and the main roots had an asymmetrical radial distribution about the tree stem centre. The dynamic loading caused an increase in soil pore water pressure, and in the forced loading tests it led to hydraulic fracturing of the rootplate.

11.1 Introduction

Windthrow is a major source of economic loss in Irish and United Kingdom forests and crop instability imposes important restrictions on silviculture. This chapter presents details of two field studies (Rodgers *et al.*, 1990; McMenamin, 1992) on the dynamic loading of mature Sitka spruce trees planted on double mouldboard plough ridges. The objectives of the studies were:

1. To develop a mechanical rocking device and a high-speed data logging system which could be used to assess the stability of trees in the field.
2. To study the behaviour of trees and rootplates subjected to forced dynamic loading.
3. To study the behaviour of a tree under natural storm conditions.
4. To identify soil properties which are important for tree stability.
5. To investigate the stiffness values and Young's moduli of the trees.

6. To monitor the changes in damping that occur during forced dynamic loading.

11.2 Site characteristics and preparation

The test site, at Castledaly State Forest, is situated 40 km south east of Galway City in the west of Ireland near the village of Ardrahan. The site was planted with Sitka spruce in 1967 and has experienced extensive windthrow in the past. During the test periods the top mean height of the forest was about 14 m. The rootplates of the selected test trees did not show any signs of windthrow failure and were located more than 10 m in from the edge of the forest area.

The soils on the test site are typical surface water gleys. The topsoil is 300–400 mm thick consisting of a dark mineral soil with a high organic content and overlies a sandy loamy subsoil. The water table is at or near ground surface throughout the year.

The site was prepared for planting using a double mouldboard plough. Young trees were planted on ribbons of overturned soil formed during

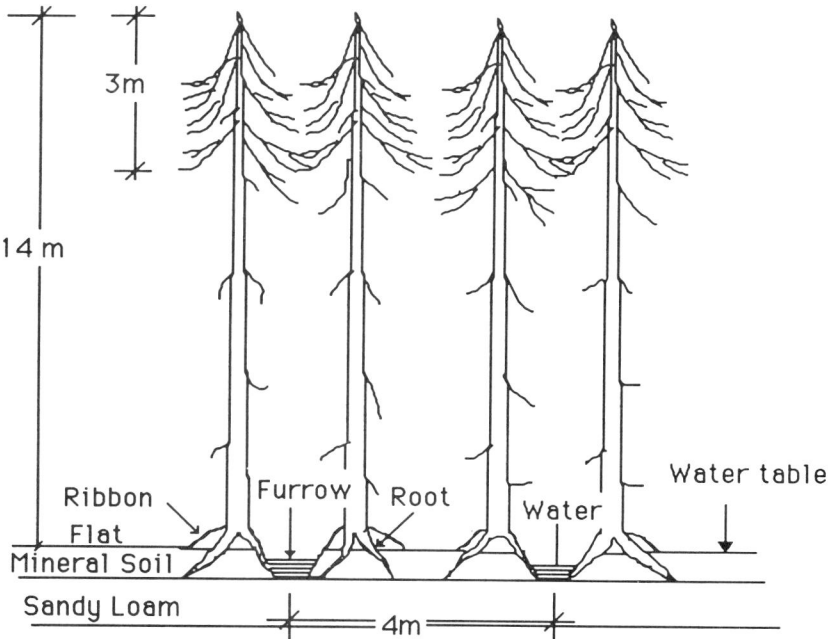

Fig. 11.1. Tree development on double mouldboard ploughed land.

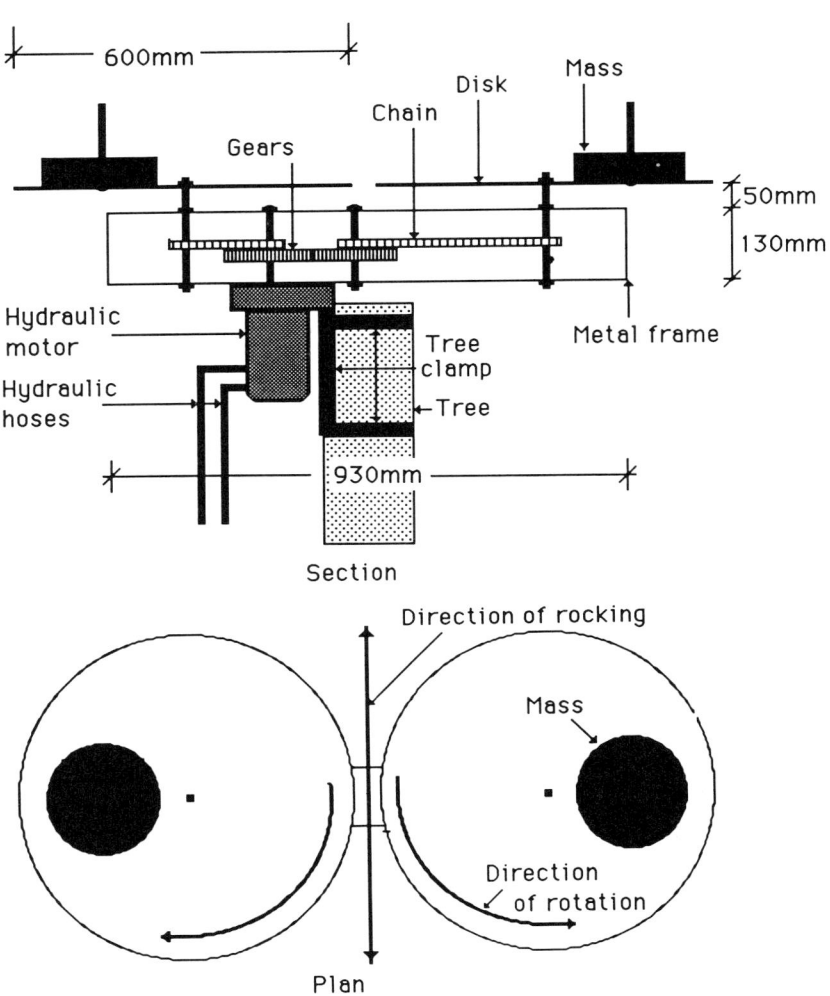

Fig. 11.2. Section and plan of tree rocker.

ploughing. The tree roots tend to follow the line of these ribbons with few roots growing at right angles to the ribbons (Fig. 11.1). The roots usually terminate near the bottom of the undisturbed aerated top soil.

11.3 Field equipment

In order to measure the effects of forced dynamic loading a mechanical device for rocking the test trees was developed. This device or rocker con-

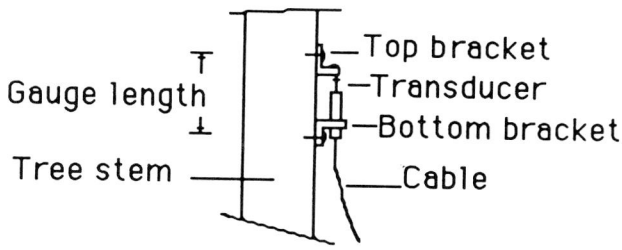

Fig. 11.3. Strain device.

sisted of an arrangement that included two discs loaded eccentrically with selected masses and rotated by a hydraulic motor through gears, chains and sprockets (Fig. 11.2). The motor was activated by a hydraulic pump that was driven by a petrol engine. The rocker enabled field testing to proceed independently of weather conditions and allowed each test tree to be rocked in a selected vertical plane. The rocker was mounted for the test duration, 6 m above ground surface, on top of the truncated stem of the tree.

Eight transducers were used to measure the effects of the dynamic loading on each tree. These included soil pore water pressure transducers, displacement transducers and strain measuring devices (Fig. 11.3). Displacement transducers were used to measure the lateral movement of the stem and the vertical movement of the rootplate and, along with dial gauges, to evaluate the strain in the outer fibres of the stem.

In the forced loading tests an Apple Macintosh II microcomputer was used to record the response of the transducers in the field. The signals from each of the eight transducers were sampled 20 times per second using a high-speed 12-bit analogue-to-digital converter in combination with LabView, a National Instruments software system. LabView enabled results to be plotted and checked when a test was in progress. A Mac SE microcomputer was used in the laboratory to store the transducer data in the wind loading test.

11.4 Field tests

11.4.1 Forced loading tests

11.4.1.1 Equipment installation

The ground in the vicinity of each test tree was flooded for at least 3 days prior to testing in order fully to saturate the soil in the rootplate. The pore water pressure transducers were inserted to a depth of about 400 mm below the original

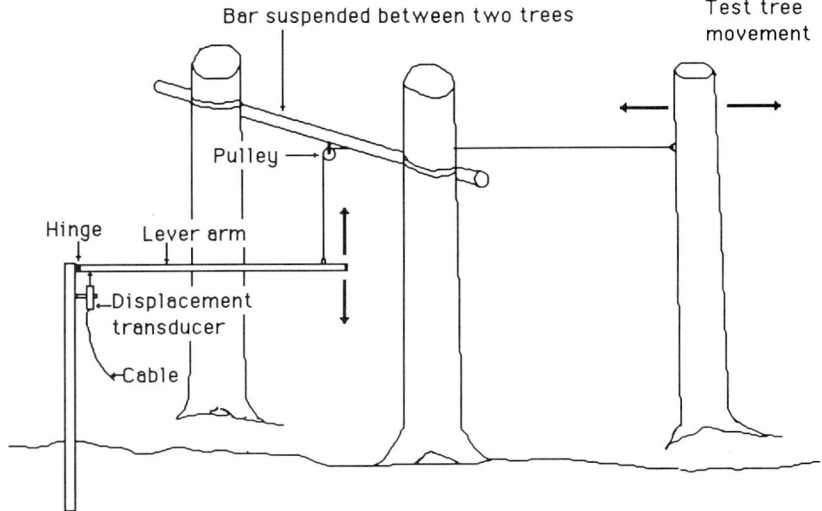

Fig. 11.4. Arrangement for measuring the horizontal displacement of the tree stem.

ground surface. Great care was taken to ensure that these transducers were saturated with de-aired water. The equipment was installed on the day before testing in order to allow soil pressures to equilibrate. Arrangements of a line, pulley, hinged lever arm and transducer were used to monitor the large horizontal displacements of the tree stem (Fig. 11.4). The vertical movement of the line, which was evaluated from the transducer reading and the geometry of the arrangement, was the same as the horizontal movement of the tree where the line was attached to the stem. The masses on the rocker and the frequency of rocking of the tree were varied during the forced loading test.

11.4.1.2 Stiffness value and Young's modulus

The stiffness value of a tree indicates its degree of inflexibility. This was measured by monotonically pulling the test tree with small forces that caused minimal disturbance to the rootplate (Fig. 11.5). Monotonic testing was carried out before and after the forced dynamic tests and at rest intervals during the tests. It was also carried out on four other trees that were overturned. Young's moduli for the trees were also obtained from these tests.

11.4.1.3 Damping

Damping of a system is measured by the rate of decay of oscillation. To measure damping, the truncated stem was first pulled to one side using a rope; the field microcomputer for recording the data was started and the rope

Effects of dynamic loading 209

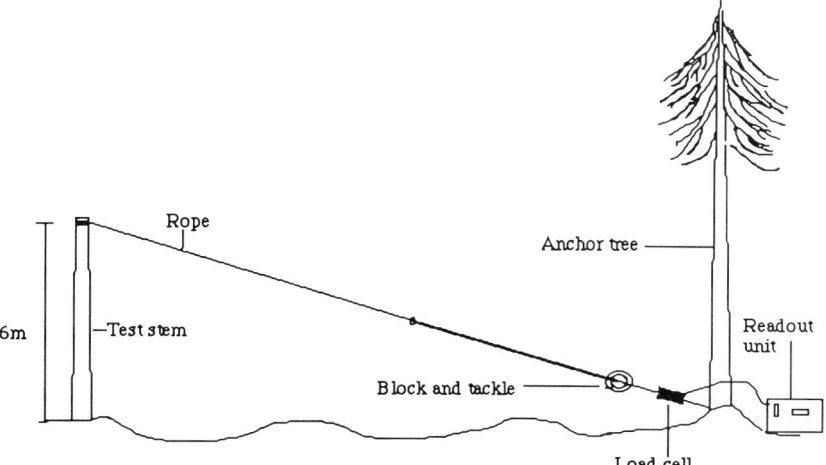

Fig. 11.5. Tree pulling arrangement.

was then released causing the tree to oscillate. The amplitude and frequency of oscillation were evaluated from readings from the horizontal displacement measuring devices. These tests were carried out immediately after the stiffness values tests.

11.4.2 Wind loading tests

The instrumentation used in the forced loading tests was also used in the wind loading tests. A battery-operated system for remote data logging and transmission was developed for these tests to monitor, record and download measurements from the field instrumentation. A three-cup anemometer was placed in the middle of the canopy at about 10 m above ground surface. The data logging and transfer system was activated when the anemometer measured a windspeed in excess of 5 m s^{-1}. The data from the instruments were first stored and then transmitted from the site by radio signals via a PacComm modem to the laboratory at University College Galway. A diagrammatic view of the system is shown in Fig. 11.6.

11.5 Results

11.5.1 Forced loading tests

The results from forced loading tests on three trees (A, B and C) are given here. Displacement, strain and pore water pressure data for tree A and strain

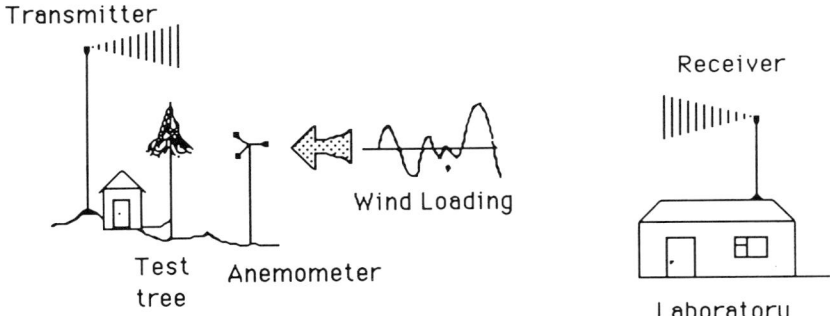

Fig. 11.6. Diagram of remote data logging and transfer system.

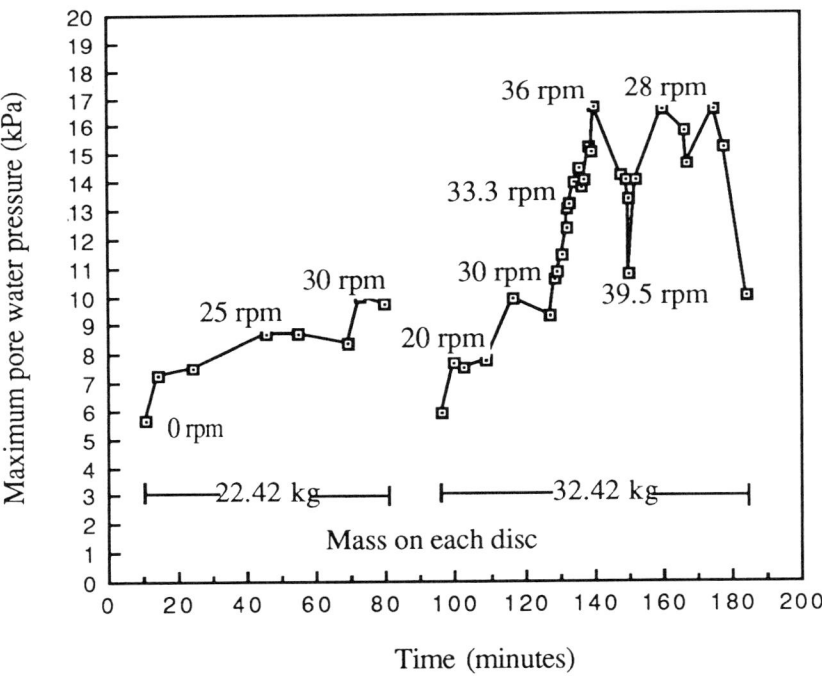

Fig. 11.7. Maximum pore water pressure behaviour under the rootplate of tree A.

data for tree B are presented. Pore water pressures, horizontal displacements, bending moment estimates and results of calibration tests are given for tree C. All the structural roots of trees A and C were contained in the ribbon and ran parallel to the plough furrow. All the structural roots of tree B were also contained in the ribbon with the exception of one root that was growing at right angles to the furrow in line with the direction of rocking of the tree.

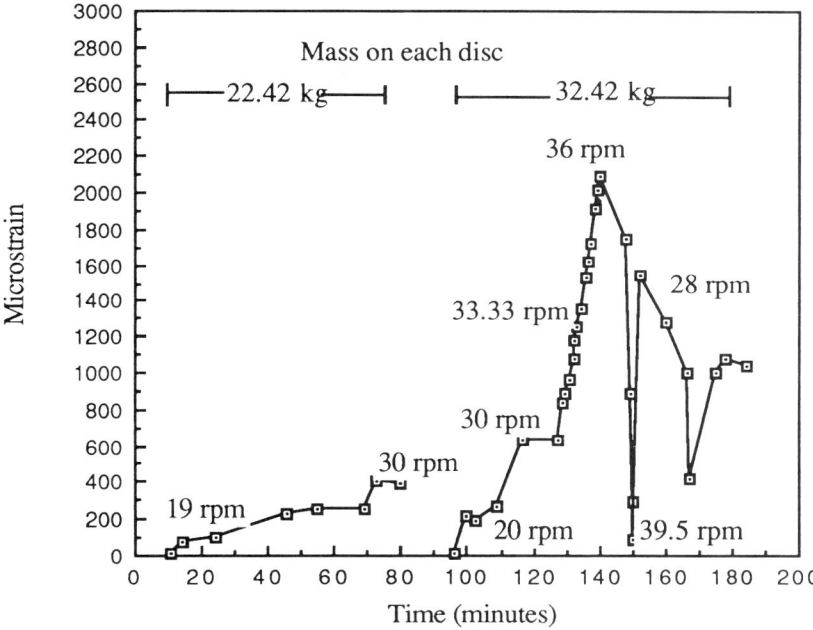

Fig. 11.8. Extension strain behaviour at 1.3 m above the root collar of tree A.

Fig. 11.7 illustrates the behaviour of the pore water pressure during the test at one transducer location under the rootplate of tree A. The masses used on each disc and some of the frequencies of rocking are shown on the figure. The maximum pore water pressure recorded was 17 kPa. This pressure caused extremely large hydraulic gradients in the soil and led to hydraulic fracture in the rootplate.

For tree A, from about 130 min onwards, the maximum extension strain in the stem at 1.3 m above ground surface and the maximum vertical upward movement of the rootplate increased as the maximum pore water pressure in the soil increased (Figs. 11.8 and 11.9). This strain increase was substantial even though the masses on the rocker remained the same and the frequency of rocking only changed from 30 to 36 revolutions per minute. It was caused by the increased inertial and eccentric loading effects of the rocker and tree resulting from the large displacement of their centres of gravity.

At about 140 min the motion of the tree changed from a rocking to a loop motion in plan, and this change was sensed by all the transducers and can be seen in Figs. 11.7, 11.8 and 11.9. This change occurred when hydraulic fracturing of the soil took place in the rootplate. The loop motion continued even

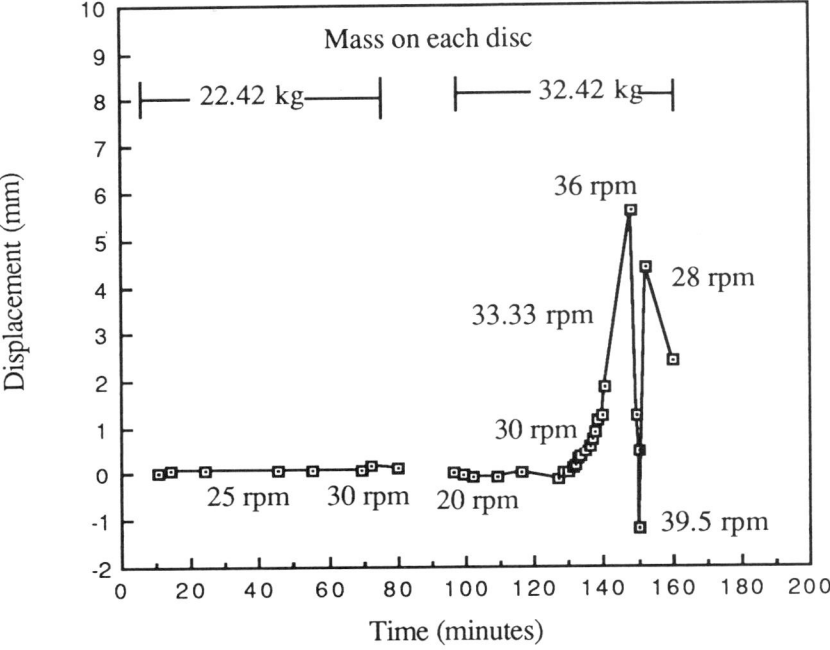

Fig. 11.9. Vertical upward movement of the rootplate of tree A.

when the frequency of rotation of the rocker was increased. The rocking motion with large strains recurred when the frequency was reduced below the frequency at which the rocking motion initially changed to the loop motion. It was decided not to overturn the tree using the rocker because damage to the test equipment would have occurred. The above observations suggest how windthrow of a tree might occur under storm conditions: a strong wind could cause hydraulic fracturing or failure of the soil in the rootplate and the tree could be subsequently overturned by winds of lesser magnitude.

Test tree B had a single root in line with the direction of rocking and it was necessary to apply larger masses and frequencies to tree B than to tree A to cause similar magnitudes of rocking movement (Table 11.1). The diameters of tree A and tree B at 1.3 m above ground surface were 172.8 mm and 185 mm respectively. Fig. 11.10 illustrates the behaviour of the strain measurement device for tree B. Greater inertial and eccentric loading effects on tree A caused its maximum strains to be similar to those of tree B.

Non-destructive monotonic pulling tests, using small loads, were carried out on tree C and on the storm tree. For each load increment, horizontal displacements at 3 m and 6 m above ground surface were recorded and strains at three

Table 11.1. *Comparison of horizontal displacements of trees A and B*

Tree	Mass per disc (kg)	Rev/min	Maximum horizontal displacement (mm) at three heights above the ground surface		
			3 m	5 m	6 m
A	32.45	30	32.5	65	107.5
B	41.46	35	20	66.5	94.5

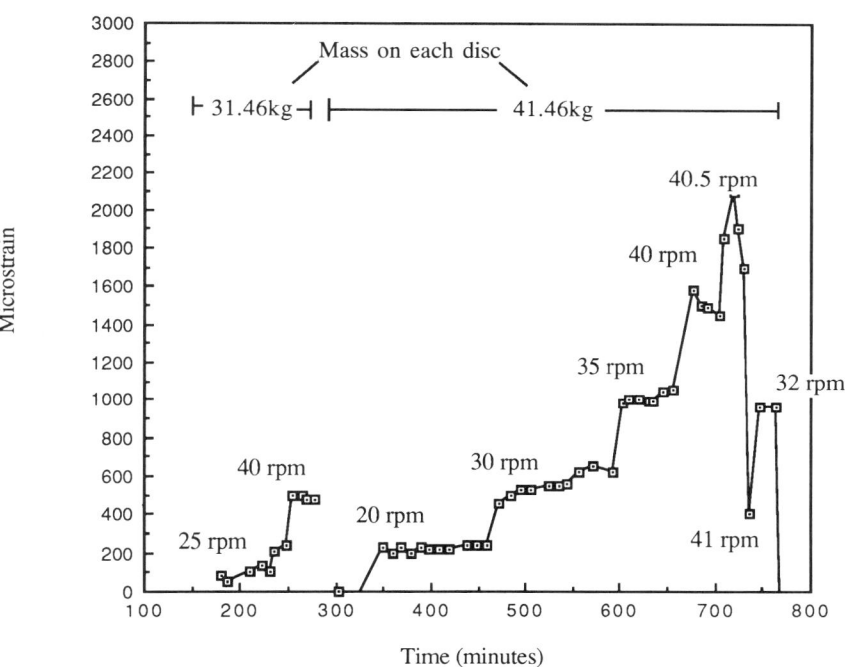

Fig. 11.10. Extension strain behaviour at 1.3 m above the root collar of tree B.

heights on the stem were evaluated from the strain device measurements. Tree C was subjected to four sets of non-destructive tests: one before commencement of dynamic loading, one at the end of the test and two at intermediate stages. The storm tree was subjected to one pulling test before monitoring began. The bending moment that caused each strain during the pulling tests was calculated by multiplying the value of the applied load by its normal distance to the rootplate. Results for the monotonic pulling test on tree C are given in Fig. 11.11. The results from these tests were subsequently used, as calibrations, to estimate the overturning moments at the base of the trees from strain values obtained during the forced and wind loading on the trees.

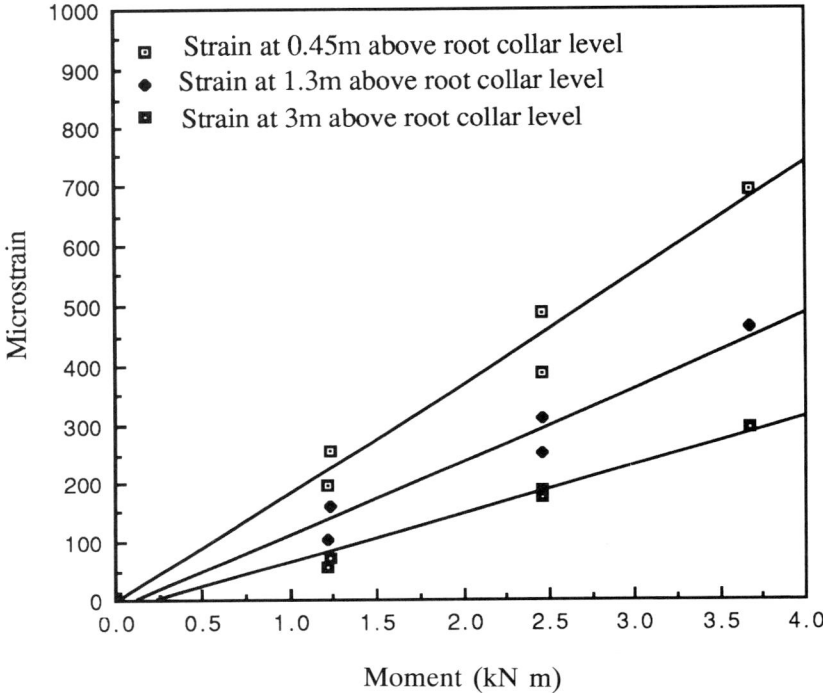

Fig. 11.11. Results of pulling tests on tree C.

The measured pore water pressures and horizontal displacements and the estimated bending moments for the forced loading test on tree C are presented in Fig. 11.12. This shows that the pore water pressure started to rise dramatically at about 100 min after the commencement of the test when the estimated bending moment was about 7 kN m. The pore water pressure increased to a maximum value of about 16 kPa at 135 min with a bending moment of about 7.5 kN m and then decreased rapidly when hydraulic fracturing of the soil occurred. The rocking of the tree was continued using the same masses on the discs but with a reduced frequency. The displacements of the tree stem increased as a result of the hydraulic fracturing and consequent soil loosening in the rootplate. The increased displacements caused greater inertial and eccentric loading effects resulting in bending moments of about 10 kN m. After the dynamic test, the tree was pulled over at an ultimate overturning moment of 12 kN m. This overturning moment had the same value as the average

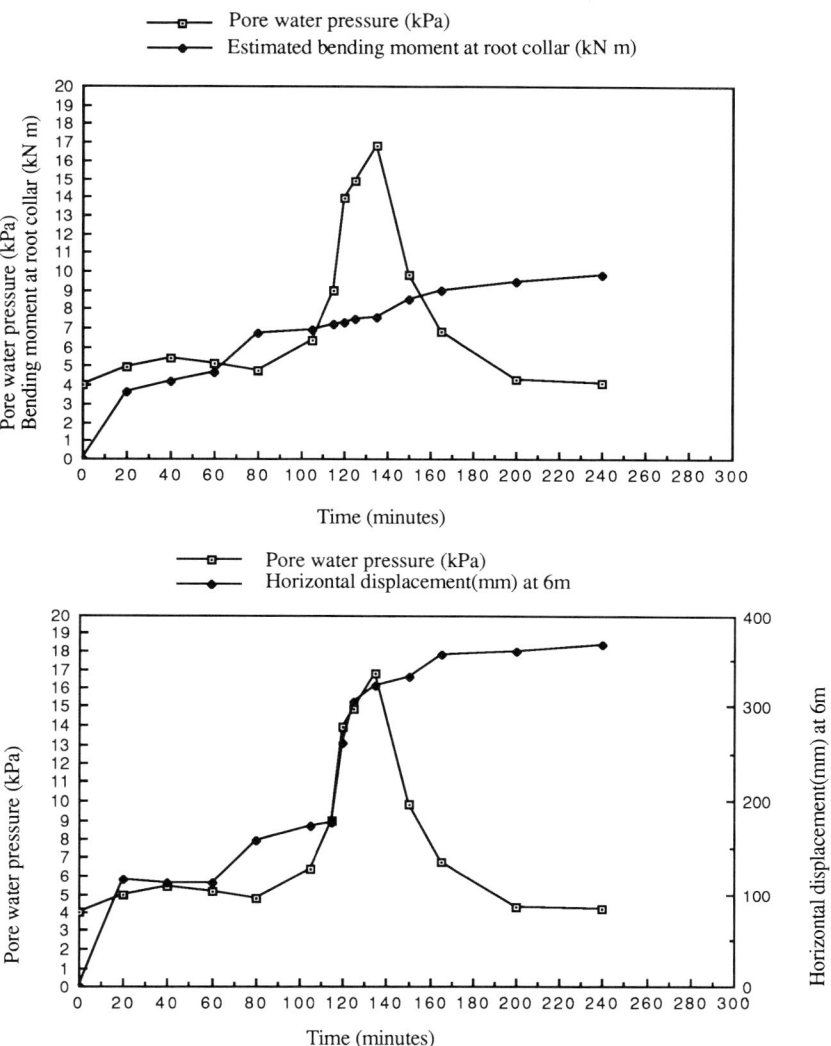

Fig. 11.12. Results of forced loading test on tree C.

ultimate overturning moment obtained for four other test trees that were not dynamically loaded by the rocker. It would seem, from the above, that rootplate failure in wet mineral soils can be initiated by a dynamic bending moment that is substantially less than the ultimate overturning moment of the tree.

Table 11.2. *Stiffness, Young's modulus and damping results for tree C*

Tree	Time (min)	Stiffness (kN m^{-1})	Young's modulus at 3 m (GPa)	Damping ratio (%)
C	0	7.81	5.26	4.1
C	135	5.85	5.41	7.6
C	175	5.71	5.65	9.9
C	237	5.17	5.92	11.5

Average gravimetric moisture content of stem wood at 1.3 m above root collar = 190%.

11.5.2 Stiffness value and Young's modulus

Estimates of the stiffness of the tree system were calculated from each monotonic non-destructive test as follows:

Stiffness (kN m^{-1}) = Applied force/horizontal deflection at 6 m

Estimates of Young's modulus for the stem at the strain device position were obtained from each monotonic pulling test as follows:

$$\text{Young's modulus} = My/I\epsilon$$

where M is the bending moment at the strain measurement position (kN m), y is the distance from the tree centre to the centre of the strain measuring device (m), I is the second moment of area of the stem assumed to be circular (m^4) and ϵ is the strain at the transducer position.

The stiffness and Young's modulus for tree C are presented in Table 11.2. The stiffness of the tree system reduced as a result of the loosening that occurred in the rootplate.

11.5.3 Damping

The results from damping tests on tree C are also presented in Table 11.2. The damping ratio (τ) was estimated using the formula:

$$\tau = (1/2\pi n)\log_e(A_0/A_n)$$

where A_0 is the amplitude at time T_0, A_n is the amplitude at time T_n, and n is the number of cycles between T_0 and T_n.

As the test progressed the damping increased due to the loosening in the rootplate.

11.5.4 Storm monitoring

Measurements of windspeed, pore water pressure, strain and horizontal displacement were made at the test site during storms in April 1992. The experimental tree had a diameter of 195 mm at 1.3 m above ground surface. The transducers were scanned five times per second for 8 s over three scan periods during a small storm. The first scan was taken at the start of a storm and the second and third scans were taken 43 min and 86 min, respectively, after the first scan. Results are presented in Fig. 11.13 and indicate that average windspeeds greater than 8 m s^{-1} caused significant dynamic loading on the tree and the pore water pressure in the soil to rise. The bending moment, estimated from the strain device measurements and the storm tree monotonic calibration test, had values up to 6 kN m, which was just less than the moment of 7 kN m that initiated the substantial increase in pore water pressure leading to hydraulic fracture in the forced dynamic rocking test on tree C.

11.6 Conclusions

1. A method for dynamically loading trees in the field was successfully developed. The equipment included a versatile tree rocker and a computer data logging system for sampling transducer signals and recording measurements.

2. During the dynamic loading of the trees, high pore water pressures were generated in the soil causing hydraulic fracturing in the rootplate. The hydraulic fracturing of the soil took place at bending moments that were substantially less than the ultimate overturning moment of the tree. It was easier to rock the trees with large displacements once the hydraulic fracturing had occurred. The above observations suggest how windthrow of a tree might occur under storm conditions: a strong wind could cause hydraulic fracturing or failure of the soil in the rootplate, the tree could be subsequently rocked with large displacements by winds of lesser magnitude and these large displacements could lead to increased bending moments that could eventually overturn the tree.

3. All the trees tested had shallow rootplates. This was possibly due to the high water table level. Lowering the water table might increase rooting depth which in turn could lead to greater tree stability. The lowering of the water table might also reduce the build-up of pore water pressure.

4. Tree systems that had roots growing in the direction of rocking were more stable in that direction than those that had roots growing only at right angles to the direction of rocking. This indicates that a more uniform root spread could give a more stable tree system.

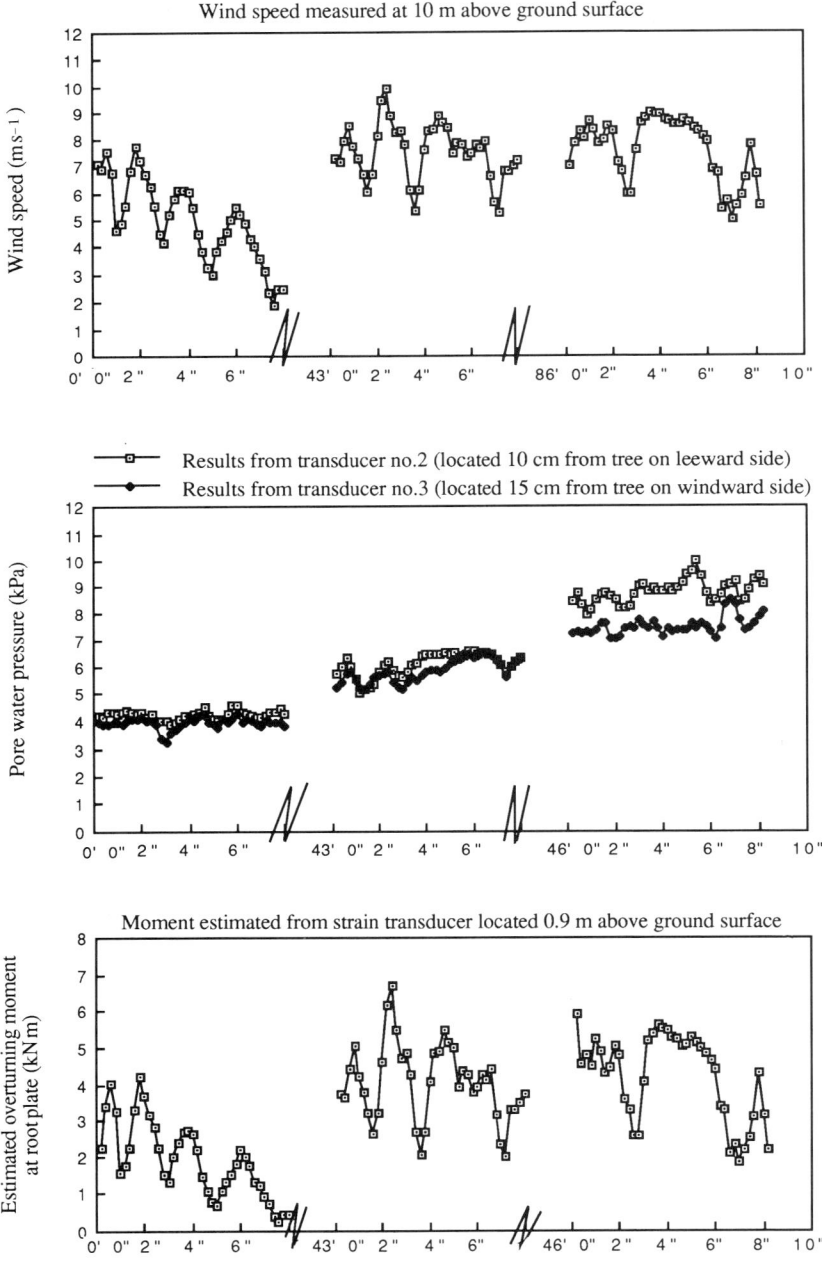

Fig. 11.13. Results of storm monitoring.

5. Monotonic tree pulling tests can give useful information about ultimate overturning moment, Young's modulus, damping and soil resistance. Strain devices greatly enhance the usefulness of these tests.

6. A system was developed for remote monitoring of the movement of a tree and the pore water pressure under its rootplate during a storm. Monitoring tests indicated that the pore water pressure increased in the soil during stormy conditions when the wind had an average speed of 8 m s^{-1}. These stormy conditions generated values of overturning moments which were just less than the moments that caused hydraulic fracturing in the rootplate of one of the test trees that was loaded using the rocker.

Conclusions 3 and 4 strongly suggest that site preparation methods which encourage roots to grow deeper and to have a radial symmetry about the stem centre should be adopted on surface water gley soils.

Acknowledgements

This project was supported by the European Community through its FOREST research programme. The authors also wish to acknowledge the help received from Coillte and The Irish Forest Service, which included financial assistance, technical assistance and the use of the test site at Castledaly.

References

McMenamin, C. (1992). An experimental investigation of the stability of Sitka spruce trees planted on a surface-water gley soil. Unpublished thesis submitted to the National University of Ireland for the degree of Master of Engineering Science.

Rodgers, M., Hendrick, E. & Casey, A. (1990). *Determination of the Effects of Site Preparation Treatments on the Stability of Sitka Spruce on Wet Mineral Soils.* Report to the European Commission.

12

Measurement of wind-induced tree-root stresses in New Zealand

A. J. WATSON

Abstract

A small load cell was developed to measure *in situ* stresses generated within a tree-root system as the surrounding soil mass was being subjected to an external force. The load cells appear to represent a viable method of monitoring tree-root stresses and could readily be modified to register root stresses generated in trees subjected to wind loading. The ability of these instruments to measure *in situ* both tension and compressional forces will give an opportunity to investigate relationships between various above-ground and below-ground tree components during storm conditions.

12.1 Introduction

Wind is a serious risk factor in many conifer plantations in New Zealand. Since the early 1940s windthrow has accounted for up to 45 000 ha of damaged trees. Wind damage also has important implications for management of indigenous forests.

Exotic forests on the Canterbury Plains, South Island (lat. 42° 45′ S, long. 172°45′ E), are often underlain by thin soils and compacted gravels, which tend to produce shallow plate-type root structures of usually less than 1 m total depth. Significant windthrow of exotic forest stands was recorded as early as 1914, with subsequent major damage occurring in 1945, 1964 and 1975 (Somerville, 1979). Genetic research over recent years has resulted in trees with improved form and wood-producing qualities. Whether there has been a corresponding improvement in below-ground qualities has yet to be investigated. However, the genetically improved stock has led to management practices tending towards lower stocking rates. This will increase wind turbulence within the forest and possibly increase windthrow, particularly after final thinning.

In the past the soil/root components of tree stability were investigated using static loads only. In those studies the overturning force was applied artificially by a winch and then related to the various components of root anchorage. Windthrow is a dynamic process. Research and subsequent models (Papesch, 1974; Coutts, 1986; Blackwell *et al.*, 1990) have analysed the interaction between wind, tree crown, stem deflection, tree vibration and their associated moments. Ideally, the dynamic components of root anchorage should be included, as the roots and soil are subjected to oscillating forces transmitted by the stem. The objective of this study was to attach load cells to radiata pine (*Pinus radiata* D. Don.) roots and monitor the wind-induced stresses within the tree-root system and relate these to the above-ground components of the overturning forces.

12.2 Load cell development

During 1991 a project supported by the US/NZ Cooperative Research Program of the US National Science Foundation was set up to shear a soil block containing a tree-root system from its base and monitor the forces generated within instrumented lateral and vertical roots. The information was used to calibrate two- and three-dimensional models of root geometry and soil–root interaction (Wu *et al.*, 1988*a*, *b*). As part of this research a load cell was developed to measure the tensional and compressional stresses generated within the tree-root network as it was being subjected to an external force.

The basic components of the load cell are two resistance strain gauges of the type used in engineering to measure the stresses generated by experimentally induced loadings in reinforced concrete columns and beams. The gauges consist of a grid of fine wire bonded to a thin insulating foil backing. The electrical resistance of the grid varies linearly with strain. The gauge is attached to the test specimen with a designer adhesive. When the specimen is loaded, the strain is transmitted to the wire grid through the adhesive and foil backing material. The strain in the specimen is found by measuring, via a Wheatstone bridge circuit, the change in the electrical resistance of the grid material. An increase in electrical resistance indicates tension; a decrease indicates compression.

With the tree-root experiment, problems arose in finding a suitable adhesive that would attach the gauge directly to moist root material and ensure good bonding between the gauge and the live root surface. This was difficult in the field as the contact surfaces had to remain clean. In addition radiata pine roots are capable of 10–20 mm elongation if tensioned to breaking point. Elongations of this magnitude were sufficient to rupture the strain

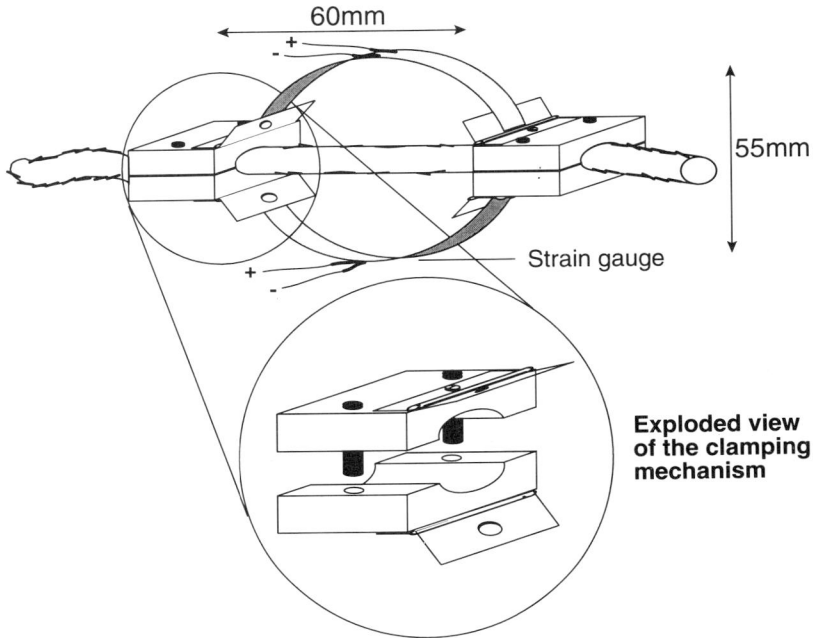

Fig. 12.1. Schematic diagram of a load cell and clamping mechanism.

gauges. These difficulties were overcome by attaching the strain gauge to a thin strip of metal that could be clamped onto the selected root. The strip was bowed and the strain gauge glued to the apex of its concave face. This created a 'soft' system that deformed with the root, so that only a small portion of the total stress was transferred directly to the gauge. This method eliminated the problem of adhesives and simplified field work as the gauge/ strips could be put together in the laboratory. It also meant the gauges could be re-used.

In practice it was unlikely that the root specimens being tested would be straight or free of defects. Consequently as the root straightened under load, one portion could be in tension and another in compression. To be aware of when this was happening and to monitor the consequences, two gauge/strips were used in tandem, positioned directly opposite each other on either side of the root to form a load cell (Fig. 12.1).

12.3 Construction

The basic construction was as follows: thin strips of stainless steel were bent symmetrically so that the arms were at approximately 90°. A 120 ohm, copper–nickel alloy wire strain gauge was glued to the apex of the concave face and the immediate area sealed by applying two or three coats of waterproofing cement. Each end of the strip was attached, via a small brass hinge, to an aluminium block, which formed one-half of the root clamping mechanism. The two small complementary blocks were designed to fit around the specimen root and were held together by two 2 mm diameter metal pins (Fig. 12.1). The pins prevented the two halves of the load cell from moving relative to each other as the root was stressed.

To attach the load cells, a short section of the specimen root was excavated using hand tools. The length of root exposed need be only a little greater than the length of the load cell, approximately 60 mm. The two halves of the clamping mechanism were fitted around the root specimen and tightly strapped together with a plastic electrical cable tie. A rigid covering can be placed around the instrument and the hole backfilled with soil to give protection from the weather.

12.4 Calibration

A linear relationship between changes in strain gauge length and electrical resistance for each load cell was obtained using a manual extensometer. The regressions were compared using a single sample t-test. Neither the slopes nor intercepts were found to be statistically different ($p<0.05$). A mean regression was calculated, and used in both field and laboratory to measure the changes in lengths of the instrumented roots as they were being subjected to an external force.

After the field test, a section of root containing the load cell was cut from the specimen root, taken to the laboratory and calibrated in tension on a Floor Model 1195 Instron Universal Testing Machine. The details of the testing equipment and technique are outlined in O'Loughlin & Watson (1979). The Instron measured the change in root length (mm) and the stress (kN) required to produce that change and hence provided an extension/stress relationship for each root/load cell combination.

12.5 Preliminary results

The field data generated two traces, one for each half of the load cell (Fig. 12.2). If the traces followed similar trends the root was considered to be in either tension or compression. When they revealed opposite trends, one-half of the load cell was in tension and the other in compression, i.e. diverging traces, indicating either root bending or straightening. The two traces were

224   A. J. Watson

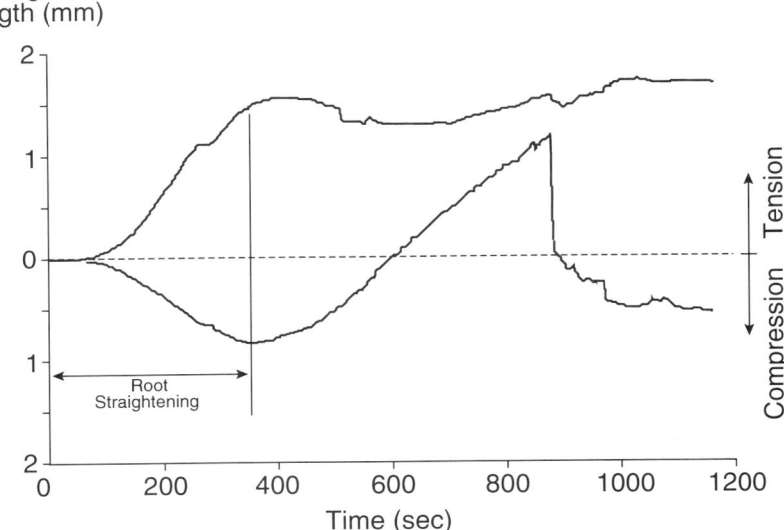

Fig. 12.2. Field measurements of the rates of change in length when a root was subjected to an external force.

averaged to give a single trace of change in root length (mm) against time (s).

The laboratory data generated on the Instron produced linear relationships ($r^2 > 0.95$) over the data range, of stress (kPa; kN divided by cross-sectional area of the root) against changes in root length (mm). These linear regressions were used, with the field data on root extension, to determine the stresses in the roots as an external force was applied to the surrounding soil mass (Fig. 12.3).

12.6 Discussion

The load cells appear to represent a viable method of monitoring *in situ* tensile and compressional stresses generated within selected tree roots as they are being subjected to external forces. The operation and performance of the load cells showed they were robust, reliable and easily installed. They could readily be modified to monitor the root stresses generated in trees subjected to wind loadings. The modifications envisaged are a more robust clamping mechanism that would allow investigation of roots up to 30 mm in diameter and an alteration to the field programmes to allow the data loggers to act as

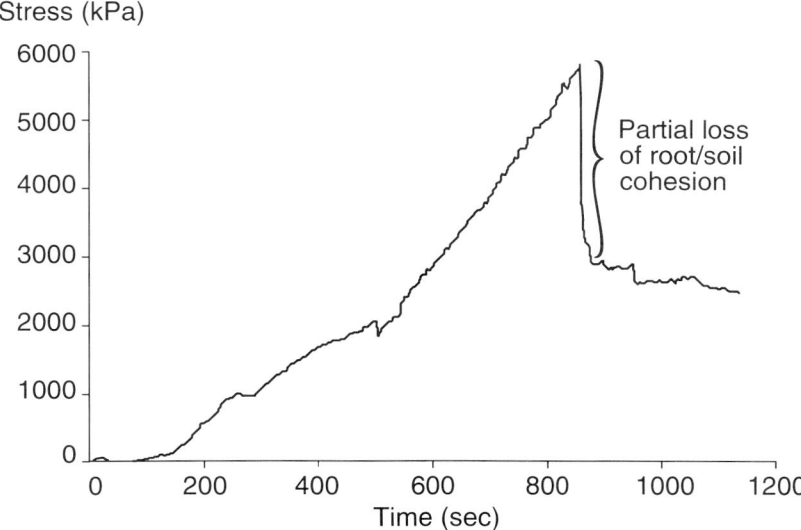

Fig. 12.3. Variation in root stress generated by an external force derived from field and laboratory measurements.

event recorders, whereby only specific events above some threshold limit would be monitored.

The ability of the load cells to measure *in situ* stresses will give an opportunity to investigate changes in relationships of various above- and below-ground parameters during potentially hazardous wind conditions. Initially, the components of the overturning forces that would be continually monitored could include wind velocity and direction and stem deflection. The components of anchorage would be tension/compression in selected windward and leeward roots, and soil properties such as soil moisture. Possible objectives would be to determine the wind-induced stresses within the soil/root system and relate these to more readily available above-ground parameters, thereby enabling the prediction of critical wind velocities that could cause crop damage.

References

Blackwell, P. G., Rennolls, K. & Coutts, M. P. (1990). A root anchorage model for shallowly rooted Sitka spruce. *Forestry*, **63**, 73–91.

Coutts, M. P. (1986). Components of tree stability in Sitka spruce on peaty gley soil. *Forestry*, **59**, 173–97.

O'Loughlin, C. L. & Watson, A. J. (1979). Root wood strength deterioration in radiata pine after clearfelling. *New Zealand Journal of Forestry Science*, **9**, 284–93.

Papesch, A. J. G. (1974). A simplified theoretical analysis of the factors that influence wind throw of trees. In *Fifth Australian Conference on Hydraulic and Fluid Mechanics*, University of Canterbury, New Zealand, pp. 235–42.

Somerville, A. (1979). Root anchorage and root morphology of *Pinus radiata* on a range of ripping treatments. *New Zealand Journal of Forestry Science*, **9**, 294–315.

Wu, T. H., Bettadapura, D. & Beal, P. E. (1988a). A stochastic model of root geometry. *Forest Science*, **34**, 980–97.

Wu, T. H., McOmber, R. M., Erb, R. T. & Beal, P. E. (1988b). A study of soil–root interaction. *Journal of Geotechnical Engineering*, ASCE **114**, 1351–75.

13

New methods for the assessment of wood quality in standing trees

K. BETHGE and C. MATTHECK

Abstract

Two instruments are described for assessing the trunks of standing trees for defects which affect their strength. One, the Metriguard Stress Wave Timer, measures the time taken for a sound wave to travel through the trunk; defects reduce the velocity of the wave. The other, called the Fractometer, measures the stiffness, strength and static fracture energy of a core removed with an increment borer.

13.1 Introduction

Two devices for the assessment of wood quality in standing trees are described. The first is a modification and an adaptation of an existing device, namely a Metriguard Stress Wave Timer, and the second is a new tool for measuring the strength of core samples. These two devices support a method of assessment known as Visual Tree Assessment (VTA) which was developed at the Karlsruhe Nuclear Research Centre (Mattheck & Breloer, 1993). VTA is based on the observation that a tree which contains a defect (crack, hollow, etc.) will repair itself by attachment of more wood at the weakened cross-section in order to restore the state of even load distribution (Mattheck, 1991). Figure 13.1 shows some typical symptoms and related defects. If these symptoms are visible in trees that could pose a hazard to the safety of the public, further investigation is justified. Also, in commercial forestry it may be necessary to check the wood quality of a stand as the basis of management decisions. This requires an assessment of the size of the internal defect, and a measure of the strength of the affected wood – tasks that can be achieved with the devices described below.

Fig. 13.1. Defects in trees and related symptoms.

13.2 The Metriguard Stress Wave Timer

An acoustic device has been in use in the United States for the detection of decay in poles and timber constructions (Hoyle & Pellerin, 1978). It consists of a hammer, a sensor and a clock. The hammer induces a sound wave on one side of the test sample and starts a clock. When the wave has travelled through the wood it reaches the sensor, which stops the clock (Fig. 13.2). An apparent velocity v can be defined as:

v = distance from hammer to sensor/travelling time measured

For the application of the method to trees, it is necessary to tap into the

Table 13.1. *Radial stress wave velocity for different species of green, healthy and standing hard- and softwoods*

Species	Radial stress wave velocity (m s^{-1}) for increasing stem radii
Hardwoods	
Maple	1006, 1082, 1103, 1136, 1426
Birch	967 , 1023, 1026, 1077, 1150
Sweet chestnut	1215, 1375
Oak	1382, 1416, 1430, 1430, 1450, 1495, 1500, 1610
Ash	1162, 1210, 1210, 1218, 1214, 1227, 1379
Lime	940, 1008, 1061, 1073, 1183
Plane	950, 1033
Pine poplar	967, 1090, 1144
Black locust	934, 1088, 1100, 1184, 1463
Red beech	1206, 1228, 1286, 1365, 1371, 1377, 1410, 1412
Horse chestnut	873, 921, 1103, 1146
Black poplar	869, 943, 1034, 1042, 1048, 1057
Silver poplar	821, 950, 1016, 1108
Willow	912, 1028, 1155, 1216, 1333
Softwoods	
Douglas fir	950, 1013, 1030, 1034, 1091, 1209, 1295, 1323
Spruce	931, 972, 1040, 1048, 1056, 1085
Pine	1066, 1073, 1122, 1132, 1146
Larch	1023, 1159, 1238, 1338
Fir	910, 983, 996, 1100, 1166

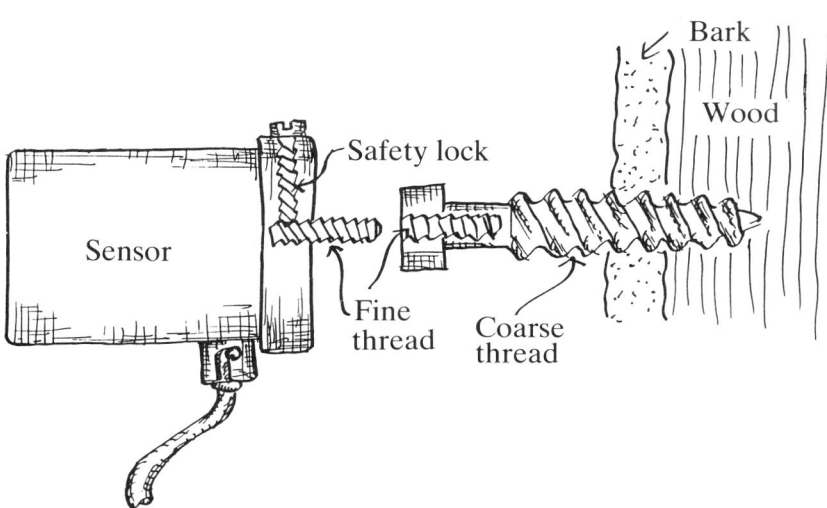

Fig. 13.2. How the sensor is attached to the screw.

Table 13.2. *Minimum and maximum values for fracture moment and angle in fractometer units for different species (population size: 50 per species) of green hard- and softwoods*

Species	Min. and max. values for fracture moment in fractometer units	Min. and max. values for fracture angle in degrees
Hardwoods		
Maple	35, 135	12, 25
Birch	8, 70	13, 30
Copper beech	86, 130	15, 20
Sweet chestnut	10, 100	10, 26
Oak	19, 130	7, 20
Alder	6, 63	12, 16
Ash	7, 120	12, 22
Hornbeam	37, 120	13, 16
Lime	2, 80	15, 37
Plane	88, 120	10, 23
Pine poplar	2, 25	15, 20
Black locust	55, 123	18, 25
Red beech	10, 150	12, 22
Horse chestnut	18, 110	12, 30
Black alder	6, 63	13, 15
Black poplar	2, 29	13, 23
Silver poplar	3, 29	12, 25
Elm	23, 78	10, 22
Willow	6, 25	12, 30
Softwoods		
Douglas fir	5, 9	12, 24
Yew	60, 92	11, 17
Spruce	2, 34	5, 30
Pine	2, 27	12, 22
Larch	5, 25	11, 30
Fir	18, 28	13, 25
Juniper	14, 37	13, 15

xylem to avoid acoustic absorption and scattering by the bark. Using steel screws at both sides (Fig. 13.2) reproducible results could be achieved. Table 13.1 shows velocities measured for some healthy trees. The table is useful as a reference for the assessment of defects. Fig. 13.3 shows the reduction of sound velocities due to the presence of defects when compared with data for healthy wood in Table 13.1. Both cracks and hollows reduce the velocity dramatically and enable a definitive diagnosis of the internal condition. The sound velocities are related to the modulus of elasticity E and the wood density ρ by $v = \sqrt{E/\rho}$. As the modulus of elasticity usually changes much faster than the density (Wilcox, 1978), decay can be detected quite well.

Assessment of wood quality in standing trees 231

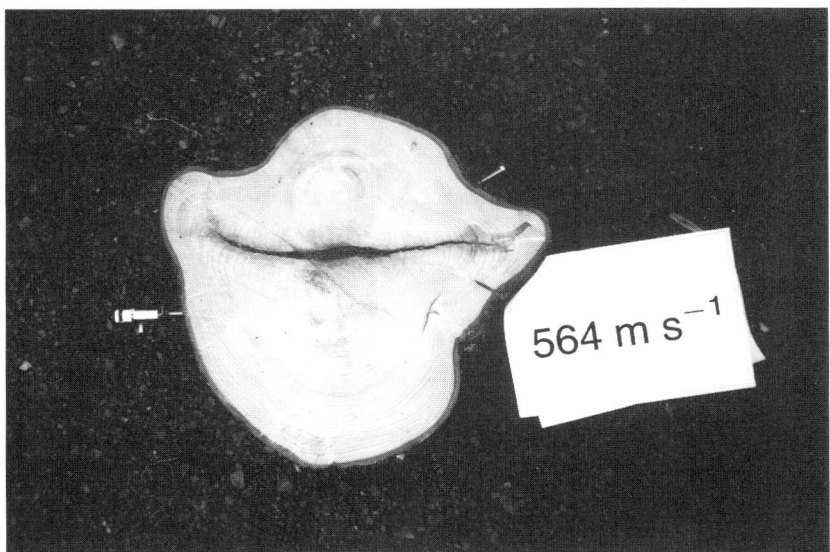

Fig. 13.3. Some examples showing dramatic reductions in sound velocities due to the presence of defects inside the tree.

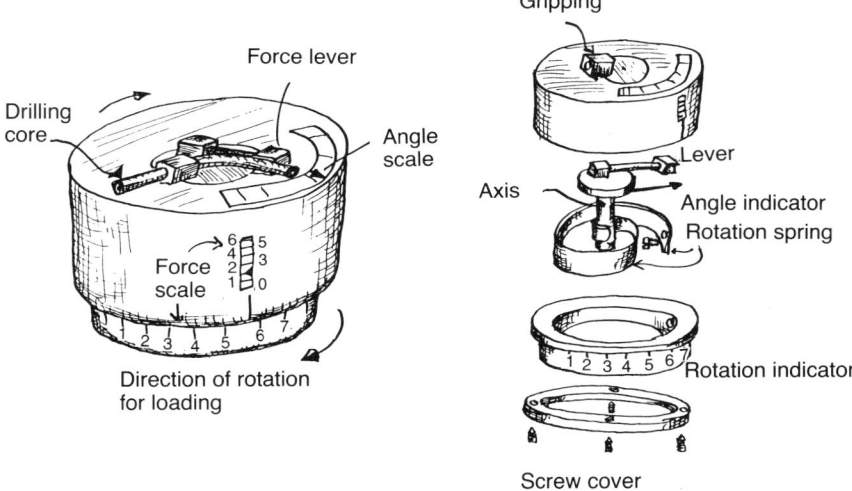

Fig. 13.4. The major elements of the Fractometer design.

However, in these cases it is always the stiffness that is measured and not the strength of the wood. To detect the strength of wood, a device was developed at the Karlsruhe Nuclear Research Centre which enables the user to determine stiffness, strength and static fracture energy of a cylindrical core removed with an increment borer from a standing tree. This device is known as a Fractometer.

13.3 The Fractometer

Fig. 13.4 shows the principle of the Fractometer and a photograph. In order to make tree assessment under field conditions easier a study was made of European trees (Table 13.2). The values can be directly compared with the values obtained with the Fractometer. Using a standard drilling core of 5.3 mm in diameter, the following calibration was found:

$$100 \text{ Fractometer units} = 22.6 \text{ MPa}$$

The Fractometer (Mattheck & Bethge, 1993) also allows the assessment of decayed wood. Large bending angles combined with still quite normal fracture stresses indicate preferential lignin destruction by fungus. On the other hand a preferential destruction of cellulose fibres (as for example in early stages of brown rot) is indicated by brittle fracture at small bending angles and low fracture stresses. The device is not a high-precision instrument but a reliable tool which will make decision making easier on a quantitative basis, even under rough field conditions. The procedure of VTA for the assessment of city trees consists of the three following steps:

1. Searching for defect symptoms.
2. Confirmation of a defect behind a symptom by use of the Metriguard hammer.
3. If the sound velocities are alarmingly low, removing a core from the tree with an increment borer in order to determine the remaining fracture strength.

References

Hoyle, R. J. & Pellerin, R. F. (1978). Stress wave inspection of a wood structure. In *Proceedings of the Fourth International Nondestructive Testing of Wood Symposium*, Vancouver, WA, USA, pp. 33–45.
Mattheck, C. (1991). *Trees: the Mechanical Design*. Springer-Verlag, Berlin.
Mattheck, C. & Bethge, K. (1993). Ein Prüfgerät für Holz im Taschenformat: das Fractometer. *Allgemeine Forst Zeitschrift*, **3**, 2–3.
Mattheck, C. & Breloer, H. (1993). *Handbuch der Schadenskunde von Bäumen: der Baumbruch in Mechanik und Rechtsprechung*. Rombach-Verlag, Freiburg.
Wilcox, W. W. (1978). Review of literature on the early stages of decay on wood strength. *Wood Fibre*, **9**, 252–7.

Part III
Tree physiological responses

14

Wind-induced physiological and developmental responses in trees

F. W. TELEWSKI

Abstract

The influence of wind on tree growth and development is interpreted within the context of stress and strain relationships. The primary stress is the force of the wind applied to the tree. The fluttering of leaves and branches, the back and forth swaying motion of the stem, the displacement or wind-induced lean of the stem and failure of the stem or roots, resulting in windthrow, are the viable, mechanical strains manifest by the tree. Secondary stresses include the influence of gravity due to displacement, and changes in the atmospheric conditions around leaves. As the magnitude of the stress (windspeed) increases, so do the resulting strains, resulting in a cascade of physiological strain responses. The physiological responses range from rapid changes in transpiration and photosynthesis at the foliar level, to reduced translocation, callose formation and ethylene production in the phloem and cambial zone. Long-term developmental and structural changes occur in canopy architecture, leaf, stem and root morphology, and modifications of cell structure and biomechanical properties of the xylem. The interaction between acute and chronic wind stress, and dynamic and static loading stresses, with their influence on physiology and development is discussed.

14.1 Introduction

Since Metzger (1893) first proposed wind as the most significant or 'Massgebender' factor affecting growth of trees, the literature on wind stress has been intertwined within the fields of ecology, physiology and forestry. The ecological literature has focused on canopy and leaf responses, originally interpreting wind-formed trees as a pathological condition of induced nutrient and dehydration stress (assumed as increased transpiration), leading to mechanical damage (Schimper, 1903; Shreve, 1914; Daubenmire, 1959; Odum,

1970; Weaver *et al.*, 1973). Historically, wind-induced developmental responses of the stem, specifically reaction wood formation and the gravitropic response, have been emphasised in the physiological and forestry literature (Timell, 1986*b*, *c*). More recently, the theory of wind-induced increase in transpiration, leading to dehydration stress, has been dispelled (Grace, 1977, 1981; Dixon & Grace, 1984) and a wind-induced mechanical (motion) effect on stem development has been positively identified (Table 14.1).

In his treatise on plant responses to stress, Levitt (1980*a*) applied the stress and strain terminology of mechanics to plant physiology. This is an appropriate approach that can be used to interpret the stress and strain relationships within trees exposed to wind stress (Levitt, 1980*b*). Stress is defined as the external factor acting on an organism, whereas the strain is the resulting response or deformation (physical or chemical change) in the organism. The resulting strain can produce secondary stress, resulting in secondary strains within the system.

The strains can be elastic or plastic. Elastic strains are defined as reversible physical or chemical changes. The swaying of a tree in response to wind, returning to the original, vertical position is considered an elastic strain. Plastic strains are defined as irreversible physical or chemical changes. The permanent displacement of a tree from its original orientation is considered a plastic strain on the tree. Plastic strains are considered injurious to the organism. However, the organism may survive and recover from the plastic strain. The formation of reaction wood in a displaced tree would be an example of recovery. Jaffe (1973, 1980) defined the strain and associated physiological and morphological responses of plants to wind and other mechanical stresses as thigmomorphogenesis.

The motion induced by wind within the foliage and vascular support tissues does affect physiology and development. Wind also induces several secondary stresses within a tree, such as the influence of gravity in a displaced tree, or changes in the atmospheric conditions around leaves. For this reason, it is critical to realise that the type and magnitude of the physiological and developmental response can be dependent on the amplitude and frequency of the stress applied to the tree (Jaffe *et al.*, 1980), in concert with seasonality and phenology. The force applied to a tree will be determined by the characteristics of the wind for a given environment. The wind can be manifest as a semi-continuous mostly turbulent flow, or it can occur as highly turbulent gusts. Seasonality and frequency of cyclonic storm systems also define the mechanical stress conditions for an environment. However, to clarify the stress as applied to a tree in this chapter, wind will be defined as chronic or acute, resulting in a dynamic and/or static displacement of the tree.

Table 14.1. *The effect of various mechanical perturbations (MP) on the height and diameter of different woody species*

Species	Height	Diameter	MP[a]	Reference
Pinus radiata	−	+	WF	Jacobs (1954)
Larix laricina	−	+	WL	Larson (1965)
Liquidambar styraciflua	−	+	WF	Neel (1967)
Zelkova serrata	−	+	WF	Neel (1967)
Myoporum laetum	−	+	WF	Harris & Hamilton (1969)
Picea glauca	ND	+	WF	
Pinus contorta	ND	+	WF	Bannan & Bindra (1970)
Pinus strobus	ND	+	WF	
Pseudotsuga menziesii	−	+	F	Reich & Ching (1970)
Liquidambar styraciflua	−	+	S	Neel & Harris (1971)
Betula verrucosa	−	+	WF	
Eucalyptus polyanthemos	−	+	WF	
Eucalyptus sideroxylon	−	+	WF	
Fraxinus uhedei	0	+	WF	
Grevillea robusta	−	+	WF	Leiser et al. (1972)
Liquidambar styraciflua	0	+	WF	
Pistachia chinensis	−	+	WF	
Quercus ilex	−	+	WF	
Schinus terebinthifolius	−	0	WF	
Pinus taeda	−	+	WF	Burton & Smith (1972)
Mimosa pudica	−	ND	R	Jaffe (1973)
Juglans nigra	0	0	WL	Heiligmann & Schneider (1974)
Juglans nigra	−	−	WF	Heiligmann & Schneider (1975)
Pinus resinosa	−	+	T	Quirk et al. (1975)
Pinus resinosa	−	0	V	Quirk & Freese (1976a)
Pinus resinosa	0	−	C	Quirk & Freese (1976b)
Pseudotsuga menziesii	−	+	F	Carlton (1976)
Pseudotsuga menziesii	−	0	S	Kellogg & Steucek (1977)
Acer saccharinum	−	0	S	
Juglans nigra	−	0	S	Ashby et al. (1979)
Liquidambar styraciflua	−	0	S	
Pinus contorta	−	0	WL	Rees & Grace (1980a)
Pinus contorta	−	0	S	Rees & Grace (1980b)
Pinus taeda	−	+	F	Telewski & Jaffe (1980)
Didymopanax pittieri	−	+	WF	Lawton (1982)
Abies fraseri	−	+	F	Telewski & Jaffe (1986a)
Abies fraseri	−	+	WF	Telewski & Jaffe (1986a)
Pinus taeda	−	+	F	Telewski & Jaffe (1986b)
Liquidambar styraciflua	−	0	WF	Holbrook & Putz (1989)
Pinus taeda	−	+	F	Telewski (1990)

+, an increase in growth; −, a decrease in growth; 0, no change; ND, no data available.
[a] MP treatments: C, compression; F, flexing; R, rubbing; S, shaking; V, vibration; WF, wind in the field; WL, wind in the laboratory.

14.2 Chronic versus acute wind stress

Wind is a ubiquitous component of the environment, differing in frequency, amplitude and direction from one habitat to another, and within a habitat. As a tree grows through a surface- or canopy-boundary layer, it experiences an increased loading due to increased wind velocities above the boundary layer. The growth process incorporates both physiological and biomechanical acclimation to the site-specific, chronic wind conditions within the development of foliage, canopy and vascular support tissues. The developmental acclimation prevents the tree from buckling under the existing wind loading conditions by reducing drag and increasing mechanical strength (Telewski & Jaffe, 1986a). Developmental acclimation is consistent with McMahon's (1973) biomechanical analysis of tree stems functioning as self-supporting columns under loading conditions. The work of McMahon has since been supported and the formulae used to describe the tree as a self-supporting structure has been reviewed and modified to account for taper, canopy and stand structure (McMahon & Kronauer, 1976; King & Loucks, 1978; King, 1981, 1986; Morgan & Cannell, 1987; Holbrook & Putz, 1989).

The biomechanical relationship between the height and diameter of a tree is determined by the constant surface strain hypothesis (Wilson & Archer, 1979). The possible mechanism for maintaining a balance in structural design has been proposed by Mattheck (1990, 1991, 1993) and Ennos in this volume, where loading stresses are distributed to the surface of all load-carrying parts of the tree consistent with the constant surface strain hypothesis. Regions of the tree experiencing higher strains induced by the loading stress will be strengthened by adaptive growth, resulting in a reduction in the surface strain.

However, acute wind stress, usually associated with a cyclonic storm or frontal system, can exceed the biomechanical acclimation (design) limits for a tree. If the design limits are exceeded, the tree will fail, resulting in foliar damage, defoliation and branch loss at the canopy level, and stem displacement and possibly windthrow at the whole tree level. Wood properties were reported to be the most important factor determining the type of wind-induced death in trees. In tropical moist forests, trees with lower wood density and low stem taper tend to snap, whereas trees of higher wood density and increased taper tend to uproot (Putz *et al.*, 1983). Coniferous trees of low stem taper are more liable to experience stem failure (Petty & Swain, 1985). The structural acclimation can also be compromised by pathogens weakening the structural tissues of the tree, such as *Armillaria* root rot (Shaw & Taes, 1977) or *Fomes applanatus* butt rot (Landis & Evans, 1974). In these circumstances stem failure can occur at wind loading pressures considered as chronic for the given environment.

What is not presently understood is the level of acclimation that occurs in the biomechanical design of a tree for existing wind conditions for a specific environment. Trees appear to be 'over designed' as free-standing columns, with their buckling limits well above the theoretical buckling limit determined by McMahon (1973) and others. Only under conditions where wind sway is inhibited by artificial mechanical constraints or by trees growing in dense stands will trees approach or exceed the calculated buckling limit (King, 1981; Holbrook & Putz, 1989).

McMahon (1973) suggests 'overdesign' is a response to the mechanical loading conditions imposed on the tree by the environment. It is not clear whether the structural acclimation in trees is optimised for an environmentally dependent average windspeed and vector and, if so, to what extent the tree is overdesigned to withstand the greater than average or acute windspeeds.

14.3 Physiological and developmental responses

Physiological responses to wind have been studied in leaves and in relation to the growth of woody tissues. Foliar studies have focused on heat transfer, transpiration, photosynthesis, desiccation, and leaf growth and expansion. Meristematic studies have focused on apical growth and stem elongation and on differentiation of the cambium to form xylem and phloem, the role of plant growth regulators, and changes in anatomy and morphology of the xylem cell elements. The interactions between the different responses and stresses associated with wind are outlined for foliar tissues (Fig. 14.1) and for vascular tissues (Fig. 14.2).

14.3.1 Foliar and canopy responses

Wind stress is separated into primary direct mechanical movement of the foliage and secondary altered local atmospheric conditions around the individual leaves and within the canopy (Fig. 14.1).

14.3.1.1 Primary, wind-induced mechanical stress

The direct elastic effect of the actual mechanical motion of leaves can influence foliar physiology. Ferree & Hall (1981) reported a decrease in net photosynthesis and transpiration in response to the rubbing of leaves as a simulation of the effect of wind. In soybean plants, mechanical shaking induced

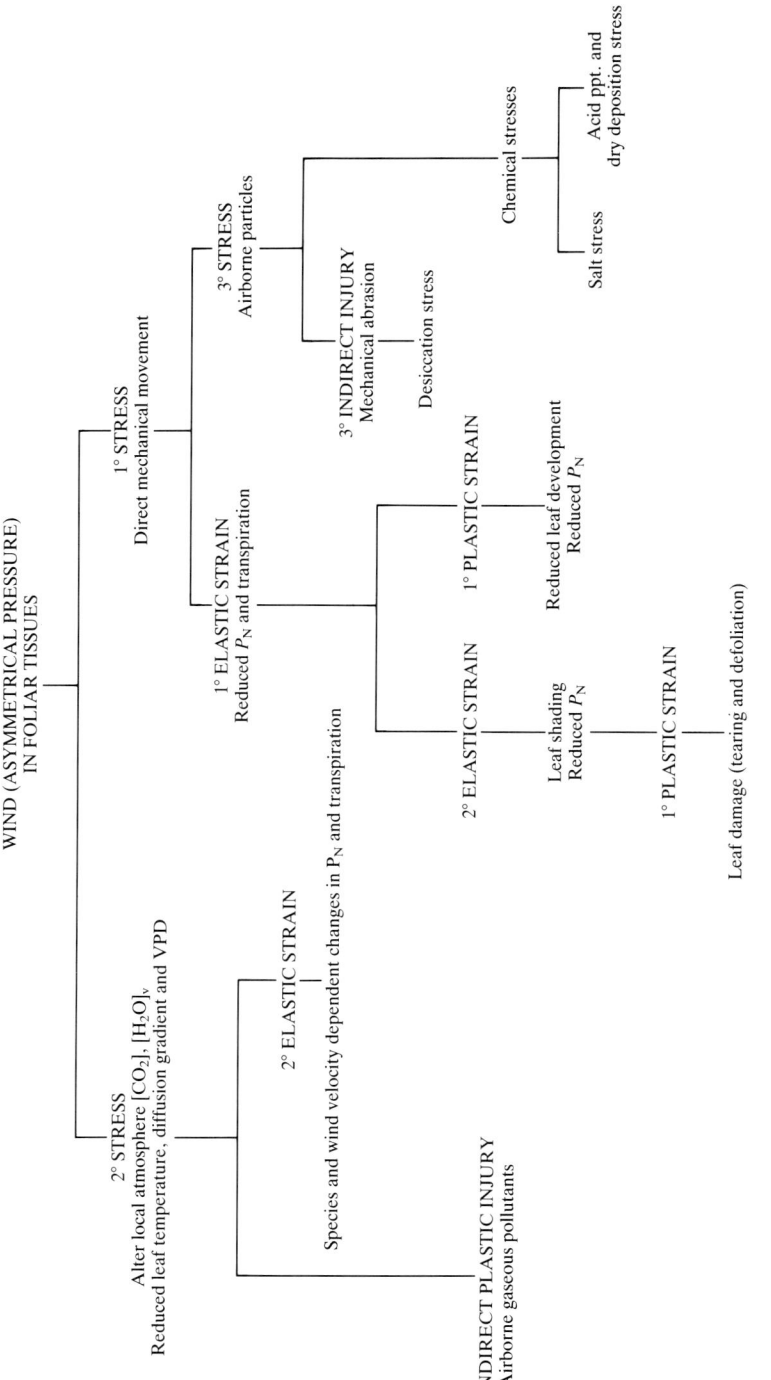

Fig. 14.1. Stress, strain and injury induced by wind in foliar tissues. $[H_2O]_v$, water vapour; VPD, vapour pressure deficit; P_N, net photosynthesis; ppt., precipitation.

Wind-induced physiological responses

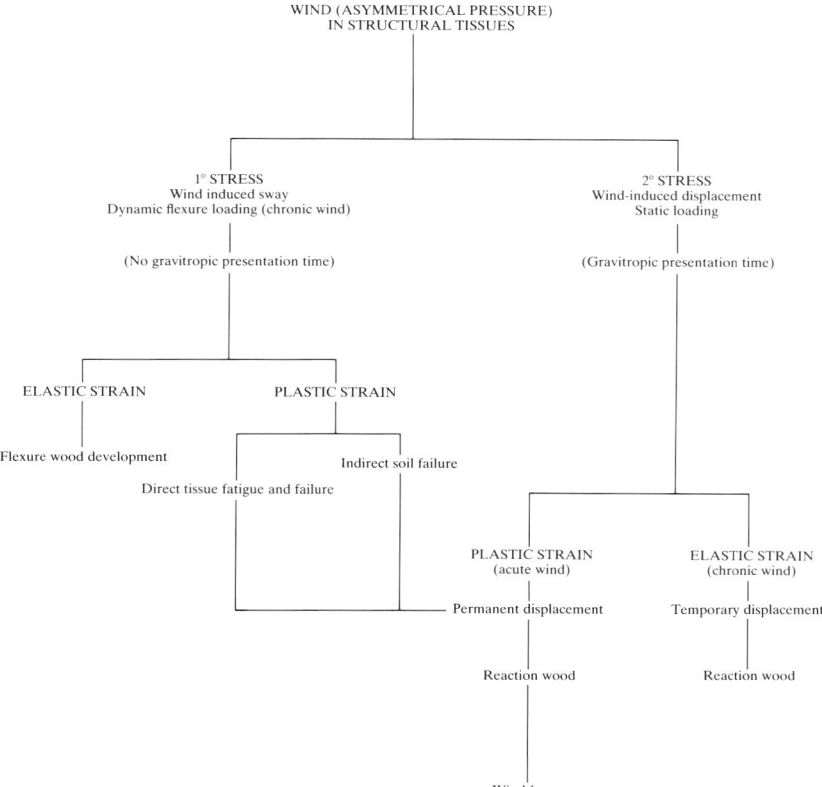

Fig. 14.2. Stress, strain and injury induced by wind in negatively orthogravitropic axial stem tissues.

stomatal closure, reduced transpiration and increased leaf water potential within minutes after the treatment (Pappas & Mitchell, 1985).

Net photosynthesis can be reduced as a secondary direct elastic strain by reductions in effective photosynthetic area due to clustering of leaves (Vogel 1989) and changes in leaf display to available irradiation (Caldwell, 1970). A wind-induced reduction in leaf area, due to retarded leaf expansion in young, developing foliage (Heiligmann & Schneider, 1974; Flückiger et al., 1978; Rees & Grace, 1980a) also reduces the effective photosynthetic area. The reduced foliar expansion is considered part of the primary direct

Fig. 14.3. Short-shoot needles and the juvenile leaves to the left of each letter represent the control specimen of the pair. (*a*) Short-shoot needles from the basal region of the stem. (*b*) Short-shoot needles from the apical region of the stem. (*c*) Juvenile leaves from the mid section of the stem.

mechanical stress since mechanical shaking of trees can mimic the wind-induced reduction in leaf enlargement. Mechanical shaking resulted in smaller leaves in *Juglans nigra* and *Liquidambar styraciflua* with developing foliage (Ashby *et al.*, 1979), and shorter needles in *Pinus contorta* (Rees & Grace 1980*b*), and in *Pinus taeda* and *Abies fraseri* seedlings (Telewski & Jaffe, 1981, 1986*a*) (Fig. 14.3). Primary direct plastic strains at the foliar level are often manifest by tearing, shredding, lesions, holes, distortions and abscission of leaves (Wilson 1980, 1984; Rushton & Toner, 1989).

Wind-blown, airborne particulates, such as dust, sand and ice crystals, may create an indirect plastic, mechanical abrasion strain (Tranquillini, 1979). Plastic strain at the foliar level is manifest by winter desiccation due to wind-induced cuticular abrasion (Hadley & Smith, 1983, 1986). Chemical stresses, such as salt stress, can be induced by airborne particulates creating the complications associated with salt spray along coastal regions (Moss, 1940; Rutter & Edwards, 1968; Oosting & Billings, 1942). Wind can also increase the deposition rate of pollutants, such as acid fog, on foliar surfaces (Lovett *et al.*, 1982; Lovett, 1984).

14.3.1.2 Secondary, wind-induced atmospheric stress

At the foliar level, wind may have the greatest elastic strain by its influence on the vapour pressure gradient and the thickness of the boundary layer around the leaf (Grace, 1977, 1981; Dixon & Grace, 1984). The alteration of the gradient and layer results in several physical changes in the atmosphere immediately surrounding the leaf. The degree to which leaf temperature and hydration are affected varies with the windspeed, tree species and the vapour pressure deficit. At low wind speeds with high irradiance an increase in transpiration may be observed as a result of the high leaf temperatures and consequent high leaf-to-air gradient in vapour pressure. As the wind speed increases, leaf temperature falls, the vapour pressure difference decreases and a reduced rate of transpiration is often observed (Dixon & Grace, 1984). Mansfield & Davies (1985) also suggested a role for changes in the carbon dioxide (CO_2) concentration. At zero to low speeds of wind, the CO_2 level around the leaf decreases due to demand by photosynthesis and the rate of diffusion of CO_2 from the atmosphere into the leaf. As windspeed increases, there is a higher concentration of CO_2 available to the leaf due to a breakdown of the diffusion gradient. This suggests a role for the CO_2-sensing mechanism of the guard cells, resulting in closure of the stomata at higher levels of mesophyll CO_2 (Mansfield & Davies, 1985).

The effect of leaf cooling in response to low windspeeds on the photosynthetic rate will depend on whether photosynthetic enzymes have a lower or higher optimal temperature for the photosynthetic reaction at ambient conditions. This assumption is based on leaf temperature to net photosynthetic studies, and the existence of plants with lower and higher temperature optima (Mooney & West, 1964; Mooney & Shropshire, 1967; Neilson et al., 1972; Smith & Hadley, 1974; Slatyer & Ferrar, 1977). If the leaf temperature is greater than optimal and is decreased by wind to a level closer to the optimal temperature for enzyme activity, photosynthesis should increase, and vice versa. This relationship, in addition to an increased diffusion rate of CO_2, may explain the initial increase in the photosynthetic rate observed by Tranquillini (1979) in *Larix decidua* and *Pinus cembra* at wind velocity less than 10 m s^{-1}.

Leaf and canopy responses will directly affect stem development via changes in transpiration and photosynthetic rates and the net photosynthate (P_N) that will be made available to the meristematic zones.

14.3.2 Woody tissue response

Dynamic, flexural loading, resulting in alternating compressional and tensional forces in the stem, will be considered as the primary stress. The displacement of the stem, resulting in static compressional and tensional forces within the stem, will be considered a secondary, gravitropic response. The arguments for and against the role of compressional and tensional forces in reaction wood formation have been presented and reviewed by Wilson & Archer (1977), Boyd (1977, 1985), Timell (1986b) and Telewski (1989). For the purposes of this chapter an exhaustive review of the gravitropic literature will not be presented; however, the reader may wish to consult the volumes by Timell (1986a,b,c).

14.3.2.1 Dynamic and static loading stresses

The use of dynamic and static loading will be applied to developmental changes induced in the vascular cambium and vascular tissue. A flow chart of wind stress (asymmetrical pressure) and induced strain relationships for a tree stem is presented in Fig. 14.2. The first dichotomy in wind-induced tree stress in structural tissues separates wind-induced sway, as a primary stress, from wind-induced displacement, as a secondary stress.

Wind-induced sway, or flexure, initiates a dynamic loading in stem tissues under turbulent or gusty conditions. The stem moves back and forth through the vertical position and returns to the vertical position after the stress has been removed. Under these conditions the tree is experiencing only dynamic bending, alternating between compression and tension, as a back and forth motion. The stem is never displaced in one direction long enough to meet the requirements for the presentation time necessary to induce a gravitropic response. (The presentation time is the minimum stimulation time required to elicit a response.) The amount of time required to induce compression wood formation has been estimated to be within the range of 0.6–76 min by Westing (1965) and may require as much as 12 h (Jaccard, 1919) to 15 h (Burns, 1920) of static loading. Several factors appear to influence the presentation time, including stage of cambial development, tissue age, angle of displacement and temperature (Timell, 1986b).

Both direct and indirect plastic strains can be initiated under wind-induced sway. The direct plastic strain is expressed as tissue fatigue and failure, especially in roots, due to prolonged exposure to sway. The indirect plastic strain results from soil failure around the roots of the tree (Coutts, 1986). Both plastic strains will result in stem displacement, initiating the secondary stress response due to stem displacement and static loading.

Static loading due to a strong, constant wind flow, resulting in a prolonged displacement of the stem, is considered as a secondary wind-induced stress. Under these conditions the stem may sway but never moves back through the vertical before the necessary presentation time required to induce a gravitropic response. Once the asymmetric pressure of the wind is removed, the stem may return to the vertical (elastic strain). These conditions would stimulate the formation of reaction wood in a vertical stem as reported by Larson (1965).

If structural failure occurs (plastic strain), the stem will remain displaced, with windthrow the extreme condition. Static loading in the permanently displaced stem is now induced by the continuous influence of gravity. The displaced stem will initially experience continuous compression on the lower side of the stem and continuous tension on the upper side of the stem. The force of compression and tension may vary as sway of the displaced tree can continue under windy conditions.

The strain induced by displacement of the stem to gravity stimulates eccentric growth and mostly the formation of reaction wood, eventually altering the internal compressional and tensional strains within the displaced stem (Timell, 1986b; Wilson & Archer, 1977). The formation of reaction wood in the stem will function to return the apical meristem to the vertical position; however, in most trees the lower stem remains in the displaced position giving the appearance of sweep.

14.3.2.2 Primary, wind-induced sway (dynamic flexure loading)

Developmental responses. From low to high windspeeds, the stem and branches will respond to the applied stress by a swaying motion. The wind loading induces compressive and tensional forces within the tissues which resist the effect of the wind, and return the tree to its original position. Most often the tree will overshoot the original orientation, creating an oscillation. Without further external applied force the oscillation will dampen out, with the tree returning to its original position.

The total effect of flexure stress on tree growth and development (thigmomorphogenesis) results in a more compact growth form, with greater stem taper, shorter branches and smaller leaves. Stem taper is altered by either a reduction in stem elongation (Fig. 14.4) and/or an increase in radial growth (Table 14.1). The ultimate effect is to reduce speed-specific drag of the crown (Telewski & Jaffe, 1986a) and maintain elastic or geometric similarity between the stem diameter and height growth, producing a greater margin of safety against mechanical failure under windy conditions (Leiser *et al.*, 1972; McMahon, 1973; McMahon & Kronauer, 1976; King & Loucks, 1978; King, 1981; Long *et al.*, 1981; Lawton, 1982; King, 1986; Rich *et al.*, 1986; Telewski & Jaffe, 1986a, b; Morgan & Cannell, 1987; Holbrook & Putz, 1989).

At the tissue level, the increase in radial growth in coniferous species is usually asymmetrical (Fig. 14.5) and results from an increase in the number of tracheids in the direction of flexure, or on the leeward side of the stem (Larson, 1965; Bannan & Bindra, 1970; Burton & Smith, 1972; Quirk *et al.*, 1975; Telewski & Jaffe, 1981, 1986a; Telewski, 1989). The tracheids that develop in this region are shorter than tracheids in other portions of the stem

248 F. W. Telewski

Fig. 14.4. Six-month-old seedlings of *Pinus taeda*. L. exhibiting reduced height and shorter leaf and needle growth in response to mechanical stimulation induced as flexing (Stim.). Each division on the scale represents 10 mm.

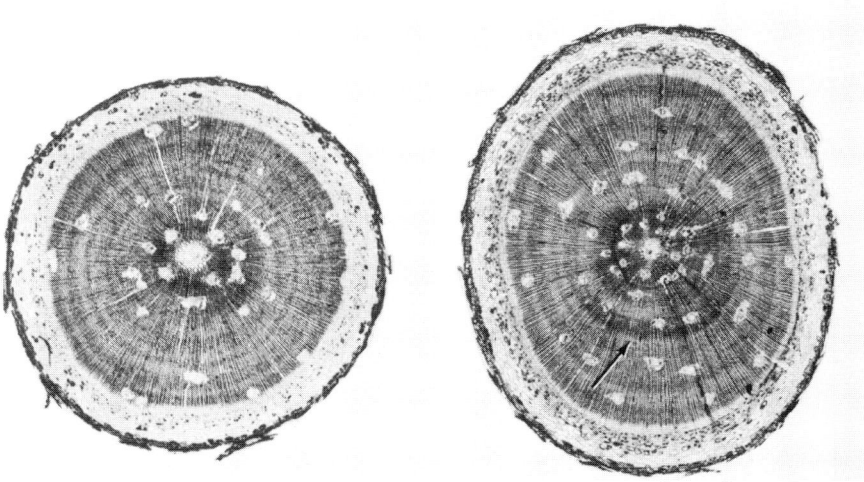

Fig. 14.5. Hypocotyl transverse sections of six-month-old *Pinus taeda* seedlings. *Left*: Control seedling with equal number of tracheids along all radial axes. *Right*: Mechanically perturbed seedling showing eccentric radial growth due to increased tracheid production along the long axis. The arrow points to an arc of thick walled tracheids perpendicular to the direction of perturbation and may have been induced by stem displacement to the gravitational vector rather than the actual flexural motion. Reproduced from Telewski & Jaffe (1981) with the permission of the publisher (National Research Council Canada)

develop in this region are shorter than tracheids in other portions of the stem or in non-stressed trees (Bannan & Bindra, 1970; Telewski & Jaffe, 1986*a*, *b*). Increased growth is also observed at branch bases and stem at the branch nodes, a focal point of flexure for branches (Telewski & Jaffe, 1986*a*).

Although a wind-induced mechanical response has been reported in a number of woody angiosperm species (Table 14.1), documentation of anatomical changes has been lacking. Neel & Harris (1971) observed a reduction in vessel element length and diameter in shaken *Liquidambar styraciflua* as compared with the vessel elements of untreated trees.

Mechanical stress can induce changes in the grain angle. Mechanical stress, applied as torque, induced a change in grain angle from the vertical from 5° to 17.5° in *Pinus radiata* (Quirk *et al.*, 1975). Telewski & Jaffe (1986*a*) observed an increase in grain angle of *Abies fraseri* growing on the windward side of the ridge, and also as a result of flexure stress induced under laboratory conditions. The application of ethrel, which gives off ethylene, to the stems of *Pinus taeda* seedlings induced an increase in grain angle (Telewski & Jaffe, 1986*c*).

In *Abies fraseri*, flexure stress also stimulated an increase in average growth ring density expressed both as grams per cubic centimetre and as the ratio cell wall area to cell lumen area when compared with non-stressed controls (Telewski, 1989). Increased density (g cm^{-3}) and radial growth was also reported in flexed *Pinus taeda* (Telewski, 1990) and non-constrained (free swaying) *Liquidambar styraciflua* (Holbrook & Putz, 1989). This translates into a greater mass and volume of xylem produced per square centimetre of cambial surface area in response to flexure.

At the cellular level, tracheids of *Abies fraseri* which differentiated under the influence of flexure exhibit increased microfibril angles that approach the less extreme values found in compression wood (Telewski, 1989). There were no significant differences in lignification in *Pinus taeda* in response to mechanical stress (Telewski & Jaffe, 1981).

The observed morphological and anatomical changes also influence the biomechanical properties of the woody stem. Flexural stiffness (EI) is the product of the elastic modulus (E) and the second moment of cross-sectional area (I). The increased radial growth produced in response to flexure will increase the second moment of area (I); however, there is a decrease in the elastic modulus (E) in the stems of *Abies fraseri* and *Pinus taeda* exposed to flexure. The increase in radial growth has a greater influence on the calculation of EI than the decreased elastic modulus, since I scales with stem radius to the fourth power, resulting in an overall stem of greater flexural stiffness (Telewski & Jaffe, 1986*a*, *b*). Different results were reported in a shaded, or allowed to sway freely under field conditions (Holbrook & Putz, 1989). Free swaying trees were shorter and had an apparently lower value of E and higher value for I (expressed as diameter growth) suggesting an increase in flexural stiffness. However, the whole-tree flexibility, measured as deflection/force (radian N^{-1}), of free swaying trees was not significantly different from that of constrained conspecifics. Trees that received lateral shade and were constrained were significantly taller and thinner, with increased whole-tree flexibility (decreased stiffness) compared with constrained or free swaying trees.

The reduction of wind-induced sway in *Pinus sylvestris* also modified secondary growth in lateral roots. The reduced motion resulted in a decreased radial growth and a decrease in the amount of compression wood in exposed parts of horizontal roots (Fayle, 1976). A recent study on the influence of wind on root development is presented by Stokes *et al.* in this volume.

Physiological responses. The dynamic, elastic strain induced by mechanical flexure stress stimulates a cascade of physiological responses in the stem. In studies conducted on *Phaseolus vulgaris*, the most immediate response to mechanical flexure is a reduction in the rate of translocation of photosynthate in the phloem within 10 min and occurring almost simultaneously with the formation of callose plugs in the phloem (Jaffe & Telewski, 1984). Although phloem transport in response to flexing was not measured in woody plants, callose deposited in the phloem of *Pinus taeda* in response to flexure parallels the response of *Phaseolus vulgaris*. Callose deposition in *Pinus taeda* peaks at 9 h and is re-absorbed within 25 h after the stress is applied (Jaffe & Telewski, 1984).

The plant growth regulator ethylene is produced by trees in response to wind, flexure and bending. Leopold *et al.* (1972) reported increased levels of ethylene in branches of *Pinus strobus*, *Pyrus malus* and *Prunus persica* after bending. Leopold & Brown (unpublished report cited in Telewski & Jaffe, 1986c) recorded a correlation between wind storms and extra-high levels of ethylene production in field-grown *Pinus strobus*. The release of ethylene peaked 15 h after flexure in greenhouse-grown *Pinus taeda* seedlings that had never previously been exposed to a significant mechanical stress (Jaffe & Telewski, 1984; Telewski & Jaffe, 1986c). Seedlings preconditioned to flexure stress produced a peak of ethylene 9 h after flexure. More ethylene was produced by non-preconditioned seedlings than by the preconditioned seedlings, suggesting a physiological hardening to continuous presentation to flexure. Similar results were reported for *Abies fraseri* growing under natural field conditions (Telewski & Jaffe, 1986c). Trees within the same species but of different genetic origin can produce significantly different levels of ethylene in response to flexure (Telewski & Jaffe, 1986c; Telewski, 1990).

The treatment of trees in the genus *Pinus* with ethrel, a compound which releases ethylene when neutralised, can mimic many observed growth responses in mechanically stressed trees (Telewski, 1990). In general, ethrel stimulates radial growth, cambial division and tracheid formation, tracheid radial cell wall thickness, increased proportions of ray parenchyma cells, resin canals, and wood density. Ethrel inhibits tracheid extension, cell enlargement (Brown & Leopold, 1973; Wolter, 1977; Barker, 1979; Telewski *et al.*, 1983; Telewski & Jaffe, 1986c; Yamamoto & Kozlowski, 1987) and inhibits height growth (Telewski *et al.*, 1983; Telewski & Jaffe, 1986c). Biomechanical properties of the stem treated with ethrel also mimic the response to flexure stress (Telewski & Jaffe, 1986c).

14.3.2.3 Secondary, wind-induced displacement (static loading)

If the wind is continuous and unidirectional, the tree can be displaced and maintained in the displaced position for several hours without returning to the vertical position. Under these conditions the stem and branches are maintained in a state of static loading, with continuous compression on the leeward side of the structure and tension on the windward side. The degree of static loading may vary due to some gust- or turbulence-induced sway; however, we may assume that the tree is not returning to the vertical condition, or beyond, where the internal strain forces would be zero, or reversed. Besides being under static loading due to the wind, the tree also experiences a complicated interaction with gravity, as it is now displaced from its original position. Therefore, the tree will be subject to gravitropic responses, beyond any mechanical, flexural responses induced by the wind.

The induction of a gravitropic response – here, the formation of reaction wood (see Timell, 1986a, for detailed anatomical definition) – requires exposure to the gravitational field for a period of time known as the presentation time. If the stem is returned to the vertical position before the required exposure time, no reaction wood will develop. If, however, a stem is displaced for a time equal to, or greater than, the required presentation time, reaction wood will develop, even if the stem is returned to the vertical position (Timell, 1986b).

Therefore, under field or experimental conditions it is possible to induce reaction wood formation in vertical stems. This is accomplished by the application of a constant, asymmetrical pressure, such as wind, which is capable of maintaining the displacement of the stem for a period of time equal to or greater than the required presentation time. However, it is not possible to isolate a static loading response in stem development from a gravitropic response and reaction wood formation within the influence of a gravitational field. Because of the complex relationship between wind, flexure, static loading and gravity effects on tree growth, Telewski (1989) considered the exposure to a gravitational field to be a secondary wind-induced stress and the formation of reaction wood to be a secondary wind-induced strain.

Chronic and acute winds that exceed the design limits for a tree are capable of inflicting a plastic strain, defined as permanent lean or displacement or, in the extreme, windthrow. Chronic winds with force not exceeding the design limit of the tree can induce a plastic strain by weakening the mechanical strength of fine roots or inducing material fatigue. The failure of roots or stem tissue, or the soil holding the roots, will result in the tree's permanent displacement with respect to the gravitational field. As a result of this displacement, the tree will experience long-term and possibly continuous static loading. Rarely is the tree ever returned to the vertical position by the formation of reaction wood, although the formation of this type of wood will significantly alter the internal static forces imposed on the stem by the force of gravity (Timell, 1986b). The changes in internal forces induced by reaction wood formation are known as internal growth strains. Functionally, the internal growth strains return the tree to a vertical orientation. However, the resulting development in the stem produces a distinctive sweep in the bole, which effectively returns the apical meristem to the vertical orientation, not the entire tree.

Under conditions where the displacement and/or deflected stem mass is relatively small, the formation of reaction wood may return the stem to the vertical position with negligible sweep (Timell, 1986b). This was the suggested function of spiral compression wood formation in young stems of *Pinus taeda* and *Abies concolor* and in mature stems of *Picea* (Telewski, 1988).

Developmental response. Other agents besides wind that can displace a tree and induce lean and static loading include snow, soil creep, erosion, landslides, earthquakes and tree falls (Timell, 1986b,c; Robertson, 1990). Therefore, static loading is not a wind-specific stress. The greatest singular response of a tree to static loading induced by displacement with respect to gravity is the formation of reaction wood.

Vertical stems grow in a direction opposite to that of gravity and are negatively orthogravitropic. Lateral branches, however, assume an equilibrium position at a given angle to gravity and are plagiogravitropic. The mechanism of perception of the gravistimulus in plants is still being investigated. The overwhelming evidence supports the role of statoliths in sensing gravity, although they cannot account for all incidents of graviperception (Sack, 1991). Amyloplasts, which are plastids containing starch, are believed to be the primary statoliths, although it has been suggested that other organelles can function in this capacity. Although the statolith theory appears valid for most herbaceous species, statoliths are not clearly involved in the perception of gravity in the induction of reaction wood development. The presence of statoliths in woody plants has rarely been reported and never firmly established (Timell, 1986b). Timell (1986b) states that statoliths have never been observed in mature gymnosperms and, therefore, their location is unknown.

However, Hawker (1932) observed the presence of statoliths in a region consisting of two to four layers surrounding the central vascular strand in the hypocotyl in seedlings of 12 conifer species. Statoliths disappeared as the seedlings aged, with a concomitant loss of gravisensitivity (Hawker, 1932). From the available data on statoliths, it appears that they are correlated with primary and possibly early secondary growth (metacambium), in a region close to the apical meristem. Therefore, they are associated with the early primary growth responses to a gravistimulus. As a woody plant matures, and the vascular cambium develops, the role of statoliths appears to diminish as the required presentation time increases, leaving open the question of a mechanism for gravistimulus perception.

In coniferous species, reaction wood is compression wood that develops on the lower side of tilted stems and on the underside of branches. In dicotyledonous angiosperm species, reaction wood is referred to as tension wood that develops on the upper side of tilted stems and on the upper side of branches.

The xylem of compression wood is characterised by tracheids with greatly thickened and lignified secondary cell walls, with a circular cross-section. The cells of compression wood also contain a $(1-3)$-β-D-glucan, laricinan (Hoffmann & Timell, 1970, 1972), similar to another glucan, callose. The circular nature of the cells creates gaps or air spaces in the middle lamella between tracheids. The tracheids are shorter than in normal wood with increased angle of microfibrils within the secondary cell walls. Compression wood is also known to have a greater density and lower modulus of elasticity than normal wood (Timell, 1986c).

Within the wood of dicotyledonous angiosperms, tension wood is characterised by an increase in the number of thick-walled, gelatinous fibres, and fewer and smaller vessels. The gelatinous fibres have either an unlignified

S_2 layer, unlignified S_3 layer or a separate gelatinous $S_4(G)$ layer in the cell wall (Wardrop, 1964). Therefore, tension wood is less lignified than normal wood.

Physiological response. The formation of reaction wood in response to static loading and displacement with respect to gravity is considered a gravitropic response. Current theories regarding the perception of the gravitropic stimulus discount the involvement of any static compressive or tensional loading in inducing the response.

The involvement of compressive and tensional forces in reaction wood development within negatively orthogravitropic stems was dismissed after researchers conducted a series of looping experiments in which compression wood and tension wood failed to develop in regions of the stem under continuous compression or tension for conifers and dicotyledonous angiosperms respectively (Hartmann, 1942; Sinnott, 1952; Wardrop, 1965). For example, if the formation of compression wood was consistent with righting the looped tree stem to the original vertical position, compression wood should develop in the region under compression. This was not observed. Compression wood formed only on the lower side of the upper and lower portion of the loop. The forces created internally would be in conflict with each other if the function of compression wood development were to return the entire tree to the vertical position. Instead, the development of reaction wood in negatively orthogravitropic stems appears to be a response to gravity, creating conflicting internal growth strains that would not necessarily facilitate the righting of the entire tree.

However, compression wood can develop opposite or perpendicular to the gravitational vector, in a fashion consistent with a compressive force-induced response in plagiogravitropic branches (Timell, 1986*b*; Wilson *et al.*, 1989). The development of compression wood in lateral branches is a consistent mechanism for righting the position of the branch with respect to the stem and other branches within the canopy of the tree (Timell, 1986*b*).

The deposition of callose in phloem tissue has also been observed to occur as an early response to gravity and the gravitropic response in the herbaceous species *Zea mays* and *Pisum sativum* (Jaffe & Leopold, 1984). As mentioned above, a (1–3)-β-D-glucan, laricinan, which is similar to callose, is a component of compression wood cell walls (Hoffmann & Timell, 1970, 1972). Although it is quite possible for callose to have a role in the gravitropic response in woody plants, no studies have been conducted to determine whether callose will develop in phloem tissue of woody plants under the influence of gravity. Timell (1986*b*) reviewed the physiological responses associated with compression wood formation. As mentioned above, bending does induce ethylene production in woody tissues. The precursor of ethylene, 1-aminocyclopropane-1-carboxylic acid (ACC), was observed in the compression wood vascular cambium of *Pinus contorta* (Savidge *et al.*, 1983). The application of ethrel stimulated the formation of compression-wood-like tracheids (Barker, 1979); however, Timell (1986*b*) questions whether the observed tissues contained true compression wood.

The induction of compression wood formation after the application of supraoptimal concentrations of indole-3-acetic acid and other auxins was first demonstrated by Onaka (1940), and has since been demonstrated by several authors (Timell, 1986*b*). In contrast, the formation of tension wood appears to be stimulated by a deficiency of auxin (Timell, 1986*b*). Ethylene is a known block of basipetal auxin transport, and may function to alter the local concentration of auxin in the cambial zone during graviperception.

14.4 Synthesis

The influence of wind on tree growth and development is very complex and it is doubtful whether any one stimulus is responsible for all the observed responses, especially in the cases of secondary stress responses. Within the context of this chapter, the influence wind has on a tree has been separated into a primary mechanical stress and a secondary stress due to alteration of the local atmosphere. The primary mechanical stress has subsequently been divided into a wind-induced dynamic flexure stress and a gravitational static loading stress. The relationship between gravity and flexure stimulus perception and the resulting developmental responses that these two apparently different stimuli impart to structural tissues needs to be addressed further.

It would appear that the two developmental responses to flexure and gravity are different – a response to different stimuli. However, both gravity and flexure create internal mechanical compressive and tensional strains. In the case of gravity the strains are static, whereas in flexure the strains are dynamic. From the perspective of design principles for biological structural systems, compression wood and flexure wood must share some structural similarities (Telewski, 1989). Trees experience complex loading patterns involving tension, compression, torque, shear and bending, whether they are vertical to gravity or displaced with respect to gravity. Purely tensile elements resist tensile forces only, whereas bending and compression elements also resist tension and shear (Wainwright *et al.*, 1976). Compression elements need to be shorter and larger in external diameter, increasing I, in order to withstand compression (Wainwright *et al.*, 1976). Compression wood tracheids are consistent with this requirement. Trees experience a complexity of loading patterns under windy conditions, including compression, tension, torque, shear and bending. Therefore, flexure wood should, and does, have some characteristics of compression elements (Telewski, 1989).

What is not clear is whether there are different mechanisms of perception for the two stresses, resulting in similar physiological, developmental and mechanical responses, or whether there is a single biomechanically controlled stimulus perception (Boyd, 1985; Edwards & Pickard, 1987). Is the formation of reaction wood purely a perception of the gravitational field via some type of statolith model functioning at the level of the vascular cambium, without a mechanical loading component of compressive and tensional forces within the tissues of a stem? Despite over 100 years of research in the field, including the application of the statolith model to reaction wood development, the perceptive mechanism for the gravitropic response in the secondary growth of woody plants is still elusive (Timell, 1986b; Sack, 1991). However, information gathered from the literature on primary growth responses in herbaceous plants may answer these questions. Edwards & Pickard (1987) described stretch-activated ion channels in protoplasts of cultured cells derived from tobacco stem pith that facilitate the transport of calcium ions across cell membranes, establishing electrochemical gradients, and function as a mechanoreceptor system. They go on to suggest that mechanotransductive channels with sensitivity to stretching forces 'are responsible for the wide variety of responses plants make to mechanical stimuli', including wind and gravity.

This hypothesis has since been supported by other observations. Within the first second of wind-induced motion of *Nicotiana plumbaginifolia*, cytosolic calcium levels increased (Knight *et al.*, 1992). Wind, rain and touch induced the expression of calmodulin, a calcium binding protein, and calmodulin-related genes within 30 min in *Arabidopsis*, with a return to basal levels between 1 and 2 h (Braam & Davis, 1990). Increased cytosolic calcium is associated with callose formation (Kauss, 1987), an increased occurrence of exocytotic voltage transients (Pickard, 1984) and increased ethylene biosynthesis (Pickard, 1984).

The existence and function of a system of stretch-activated channels could account for a series of pressure-induced regulation systems in plants, including turgor sensing and osmoregulation, alignment of cell division, response to flexure and graviperception (Edwards & Pickard, 1987; Ding & Pickard, 1993). A mechanotransductive ion channel system not only accounts for flexure and static bending responses in structural tissues, but could explain the reduction in leaf enlargement and, possibly, mechanosensitive calcium channel transduction leading to stomatal closure (Blatt, 1991) associated with wind-induced mechanical movement and fluxes in leaves. Such a system could also explain the sensitivity of the vascular cambium to gravity in the absence of statoliths.

The soft tissues of the vascular and cork cambial zones, phloem and cortical tissues, located near the region of surface strain transduction proposed by Wilson & Archer (1979) and Mattheck (1990, 1991, 1993), could be potential sites for the perception of any mechanical signal. Zones of increased surface strain may respond via the mechanotransductive ion channel system. The perceptive mechanism would respond to wind-induced sway and wind-induced displacement, controlling morphogenesis to facilitate the growth and development of a structure which would redistribute internal loading strains. The modified structure would be capable of withstanding the complex loading requirements within the immediate environment of the tree.

References

Ashby, W. C., Kolar, C. A., Hendricks, T. R. & Phares, R. E. (1979). Effect of shaking and shading on growth of three hardwood species. *Forest Science*, **25**, 212–16.

Bannan, M. W. & Bindra, M. (1970). The influence of wind on ring width and cell length in conifer stems. *Canadian Journal of Botany*, **48**, 255–9.

Barker, J. E. (1979). Growth and wood properties of *Pinus radiata* D. Don in relation to fertilization, bending stress, and crown growth. *New Zealand Journal of Forest Research*, **10**, 15–19.

Blatt, M. R. (1991). Ion channel gating in plants: physiological implications and integration for stomatal function. *Journal of Membrane Biology*, **124**, 95–112

Boyd, J. D. (1977). Basic cause of differentiation of tension wood and compression wood. *Australian Forest Research*, **7**, 121–43.

Boyd, J. D. (1985). *Biophysical Control of Microfibril Orientation in Plant Cell Walls*. Nijhoff/Junk, Dordrecht.

Braam, J. & Davis, R. W. (1990). Rain-, wind-, and touch-induced expression of calmodulin and calmodulin-related genes in *Arabidopsis*. *Cell*, **60**, 357–65.

Brown, K. M. & Leopold, A. C. (1973). Ethylene and the regulation of growth in pine. *Canadian Journal of Forest Research*, **3**, 143–5.

Burns, G. P. (1920). *Eccentric Growth and Formation of Redwood in the Main Stem of Conifers*. Vermont Agricultural Experiment Station Bulletin 219.

Burton, J. D. & Smith, D. M. (1972). *Guying to Prevent Wind Sway Influences Loblolly Pine Growth and Wood Properties*. US Forest Service Research Paper SO-80, Atlanta.

Caldwell, M. M. (1970). Plant gas exchange at high wind speeds. *Plant Physiology*, **46**, 535–7.

Carlton, R. R. (1976). Influences of the duration of periodic sway on the stem form development of four-year-old Douglas fir, *Pseudotsuga menziesii*. Unpublished Master's Thesis, Oregon State University.

Coutts, M. P. (1986). Components of tree stability in Sitka spruce on peaty gley soil. *Forestry*, **59**, 173–97.

Daubenmire, R. F. (1959). *Plants and the Environment: A Textbook of Plant Autecology*. Wiley, New York.

Ding, J. P. & Pickard, B. G. (1993). Mechanosensory calcium-selective cation channels in epidermal cells. *Plant Journal*, **3**, 83–110.

Dixon, M. & Grace, J. (1984). Effect of wind on the transpiration of young trees. *Annals of Botany*, **53**, 811–19.

Edwards, K. L. & Pickard, B. G. (1987). Detection and transduction of physical stimuli in plants. In *The Cell Surface in Signal Transduction*, ed. E. Wagner, H. Greppin & B. Millet, pp. 41–66. Springer-Verlag, Berlin.

Fayle, D. C. F. (1976). Stem sway affects ring width and compression wood formation in exposed root bases. *Forest Science*, **22**, 193–4.

Ferree, D. C. & Hall, F. R. (1981). Influence of physical stress on photosynthesis and transpiration of apple leaves. *Journal of the American Society for Horticultural Science*, **106**, 348–51.

Flückiger, W., Oertli, J. J. & Flückiger-Keller, H. (1978). The effect of wind gusts on leaf and growth and foliar water relations of aspen. *Oecologia*, **34**, 101–6.

Grace, J. (1977). *Plant Response to Wind*, Academic Press, London.

Grace, J. (1981). Some effects of wind on plants. In *Plants and Their Atmospheric Environment*, eds. J. Grace, E. D. Ford & P. G. Jarvis, pp. 31–56. Blackwell Scientific, Oxford.

Hadley, J. L. & Smith, W. K. (1983). Influence of wind exposure on needle desiccation and mortality for timberline conifers in Wyoming, USA. *Arctic and Alpine Research*, **15**, 127–35.

Hadley, J. L. & Smith, W. K. (1986). Wind effects on needles of timberline conifers: seasonal influence on mortality. *Ecology*, **67**, 12–19.

Harris, R. W. & Hamilton, W. D. (1969). Staking and pruning young *Myoporum laetum* trees. *Journal of the American Society for Horticultural Science*, **94**, 359–61.

Hartmann, F. (1942). Das statische Wuchgesetz bei Nadel- und Laubbäumen. In *Neue Erkenntnisse über Ursachen, Gesetzmässigkeit und Sinn des Reaktionsholzes*, p. 111. Springer-Verlag, Vienna.

Hawker, L. E. (1932). Quantitative study of the geotropism of seedlings with special reference to the nature and development of their statolith apparatus. *Annals of Botany*, **47**, 121–57.

Heiligmann, R. & Schneider, G. (1974). Effects of wind and soil moisture on black walnut seedlings. *Forest Science*, **20**, 331–5.

Heiligmann, R. & Schneider, G. (1975). Black walnut seedlings growth in wind protected microenvironments. *Forest Science*, **21**, 293–7.

Hoffmann, G. C. & Timell, T. E. (1970). Isolation of a β-1,3-glucan (laricinan) from compression wood of *Larix laricina*. *Wood Science and Technology*, **4**, 159–62.

Hoffmann, G. C. & Timell, T. E. (1972). Polysaccharides in compression wood of tamarack (*Larix laricina*). I. Isolation and characterization of laricinan, an acidic glucan. *Svensk Papperstidn*, **75**, 135–42.

Holbrook, N. M. & Putz, F. E. (1989). Influence of neighbors on tree form: effects of lateral shade and prevention of sway on the allometry of *Liquidambar styraciflua* (Sweet gum). *American Journal of Botany*, **76**, 1740–9.

Jaccard, P. (1919). *Nouvelles recherches sur l'accroissement en épaisseur des arbres*. Foundation Schnyder von Wartensee Zürich, Switzerland.

Jacobs, M. R. (1954). The effect of wind-sway on the form and development of *Pinus radiata* D. Don. *Australian Journal of Botany*, **2**, 35–51.

Jaffe, M. J. (1973). Thigmomorphogenesis: the response of plant growth and development to mechanical stimulation. *Planta*, **114**, 143–57.

Jaffe, M. J. (1980). Morphogenetic responses of plants to mechanical stimuli or stress. *Bioscience*, **30**, 239–43.

Jaffe, M. J. & Leopold, A. C. (1984). Callose deposition during gravitropism of *Zea mays* and *Pisum sativum* and its inhibition by 2-deoxy-D-glucose. *Planta*, **161**, 20–6.

Jaffe, M. J. & Telewski, F. W. (1984). Thigmomorphogenesis: callose and ethylene in the hardening of mechanically stressed plants. In *Phytochemical Adaptations to Stress*, ed. B. N. Timmermann, C. Steelink & F. A. Loewus, pp. 79–95. Plenum Press, New York.

Jaffe, M. J., Biro, R. L. & Bridle, K. (1980). Thigmomorphogenesis: calibration of the parameters of the sensory function in beans. *Physiologia Plantarum*, **49**, 410–16.

Kauss, H. (1987). Some aspects of calcium-dependent regulation in plants. *Annual Review of Plant Physiology*, **38**, 47–204.

Kellogg, R. M. & Steucek, G. L. (1977). Motion-induced growth effects in Douglas fir. *Canadian Journal of Forest Research*, **7**, 94–9.

King, D. (1981). Tree dimensions: maximizing the rate of height growth in dense stands. *Oecologia*, **51**, 351–6.

King, D. (1986). Tree form, height growth, and susceptiblity to wind damage in *Acer saccharum*. *Ecology*, **67**, 980–90.

King, D. & Loucks, O. L. (1978). The theory of tree bole and branch form. *Radiation and Environmental Biophysics*, **15**, 141–65.

Knight, M. R., Smith, S. M. & Trewavas, A. J. (1992). Wind-induced plant motion immediately increases cytosolic calcium. *Proceedings of the National Academy of Sciences, USA*, **89**, 4967–71.

Landis, T. D. & Evans, A. K. (1974). A relationship between *Fomes applanatus* and aspen windthrow. *Plant Disease Report*, **58**, 110–13.

Larson, P. R. (1965). Stem form of young *Larix* as influenced by wind and pruning. *Forest Science*, **11**, 212–42.

Lawton, R. O. (1982). Wind stress and elfin stature in a montane rain forest tree: an adaptive explanation. *American Journal of Botany*, **69**, 1224–30.

Leiser, A. T., Harris, R. W., Neel, P. L., Long, D., Slice, N. W. & Maire, R. G. (1972). Staking and pruning influence trunk development of young trees. *Journal of the American Society for Horticultural Science*, **97**, 498-503.

Levitt, J. (1980a). *Response of Plants to Environmental Stresses*, vol. I. Academic Press, New York.

Levitt, J. (1980b). *Response of Plants to Environmental Stresses*, vol. II. Academic Press, New York.

Leopold, A. C., Brown, K. M. & Emerson, F. H. (1972). Ethylene in the wood of stressed trees. *Hortscience*, **7**, 175.

Long, J. N., Smith, F. W. & Scott, D. R. M. (1981). The role of Douglas-fir stem sapwood and heartwood in the mechanical and physiological support of crowns and development of stem form. *Canadian Journal of Forest Research*, **11**, 459–64.

Lovett, G. M. (1984). Rates and mechanism of cloud water deposition to a subalpine balsam fir forest. *Atmospheric Environment*, **18**, 361–71.

Lovett, G. M., Reiners, R. A. & Olson, R. K. (1982). Cloud droplet deposition in subalpine balsam fir forest. *Science*, **218**, 1303–5.

Mansfield, T. A. & Davies, W. J. (1985). Mechanisms for leaf control of gas exchange. *BioScience*, **46**, 158–64.

Mattheck, C. (1990). Why they grow, how they grow: the mechanics of trees. *Arboricultural Journal*, **14**, 1–17.

Mattheck, C. (1991). *Trees: The Mechanical Design*. Springer-Verlag, Berlin.

Mattheck, C. (1993). *Handbuch der Schadenskunde von Bäumen*. Rombach-Verlag, Freiburg.

McMahon, T. (1973). Size and shape in biology. *Science*, **179**, 1201–2.

McMahon, T. A. & Kronauer, R. A. (1976). Tree structures: deducing the principles of mechanical design. *Journal of Theoretical Biology*, **59**, 443–66.

Metzger, A. (1893). Der Wind als massgebender Faktor für das Wachsthum der Bäume. *Mündener Forstl*, **3**, 35–86.

Mooney, H. A. & Shropshire, F. (1967). Population variability in temperature related photosynthetic acclimation. *Oecoplogia Plantarium*, **2**, 1–13.

Mooney, H. A. & West, M. (1964). Photosynthetic acclimation of plants of diverse origin. *American Journal of Botany*, **51**, 825–7.

Morgan, J. & Cannell, M. G. R. (1987). Structural analysis of tree trunks and branches: tapered cantilever beams subject to large deflections under complex loading. *Tree Physiology*, **3**, 365–74.

Moss, A. E. (1940). Effect of wind-driven salt water. *Journal of Forestry*, **38**, 421–5.

Neel, P. L. (1967). Factors influencing tree trunk development. *Proceedings of the International Shade Tree Conference*, **43**, 293–303.

Neel, P. L. & Harris, R. W. (1971). Motion-induced inhibition of elongation and induction of dormancy in *Liquidambar*. *Science*, **173**, 58–9.

Neilson, R. E., Ludlow, M. M. & Jarvis, P. G. (1972). Photosynthesis in Sitka spruce (*Picea sitchensis* Bong. (Carr.). II. Response to temperature. *Journal of Applied Ecology*, **9**, 721–45.

Odum, H. T. (1970). Rain forest structure and mineral cycling homeostasis. In *A Tropical Rain Forest*, ed. H. T. Odum, pp. 3–52. United States Atomic Energy Commission, Division of Technology Information, Washington, DC.

Onaka, F. (1940). On the influence of heteroauxin on the radial growth and especially the formation of compression wood in trees (in Japanese). *Journal of the Japanese Forestry Society*, **22**, 573–80.

Oosting, H. J. & Billings, D. W. (1942). Factors affecting vegetational zonation on coastal dunes. *Ecology*, **23**, 131–42.

Pappas, T. & Mitchell, C. A. (1985). Influence of seismic stress on photosynthetic productivity, gas exchange, and leaf diffusive resistance of *Glycine max* (L.) Merrill cv. Wells, II. *Plant Physiology*, **79**, 285–9.

Petty, J. A. & Swain, C. (1985). Factors influencing stem breakage of conifers in high winds. *Forestry*, **58**, 75–84.

Pickard, B. L. (1984). Voltage transients elicited by sudden step-up of auxin. *Plant, Cell and Environment*, **7**, 171–8.

Putz, F. E., Coley, P. D., Lu, K., Montalvo, A. & Aiello, A. (1983). Uprooting and snapping of trees: structural determinants and ecological consequences. *Canadian Journal of Forest Research*, **13**, 1011–20.

Quirk, J. T. & Freese, F. (1976a). Effects of mechanical stress on growth and anatomical structure of red pine: compression stress. *Canadian Journal of Forest Research*, **6**, 196–202.

Quirk, J. T. & Freese, F. (1976b). Effects of mechanical stress on growth and anatomical structure of red pine: torque stress. *Canadian Journal of Forest Research*, **6**, 374–81.

Quirk, J. T., Smith, D. M. & Freese, F. (1975). Effects of mechanical stress on growth and anatomical structure of red pine (*Pinus resinosa* Ait): torque stress. *Canadian Journal of Forest Research*, **5**, 691–9.

Rees, D. J. & Grace, J. (1980*a*). The effects of wind on extension growth of *Pinus contorta* Douglas. *Forestry*, **53**, 145–53.

Rees, D. J. & Grace, J. (1980*b*). The effects of shaking on extension growth of *Pinus contorta* Douglas. *Forestry*, **53**, 155–65.

Reich, F., P. & Ching, K. K. (1970). Influence of bending stress on wood formation of young Douglas fir. *Holzforschung*, **24**, 68–70.

Rich, P. M., Helenurm, K., Kearns, D., Morse, S. R., Palmer, M. W. & Short, L. (1986). Height and stem diameter relationships for dicotyledonous trees and arborescent palms of Costa Rican tropical wet forest. *Bulletin of the Torrey Botanical Club*, **113**, 241–6.

Robertson, A. (1990). Directionality of compression wood in balsam fir wave forest trees. *Canadian Journal of Forest Research*, **20**, 1143–8.

Rushton, B. S. & Toner, A. E. (1989). Wind damage to leaves of sycamore (*Acer pseudoplatanus* L.) in coastal and noncoastal stands. *Forestry*, **62**, 67–88.

Rutter, N. & Edwards, R. S. (1968). Deposition of air-borne marine salt at different sites over the college farm, Aberystwyth (Wales), in relation to wind and weather. *Agricultural Meteorology*, **5**, 235–54.

Sack, F. D. (1991). Plant gravity sensing. *International Review of Cytology*, **127**, 193–253.

Savidge, R. A., Mutumba, G. M. C., Heald, J. K. & Wareing, P. F. (1983). Gas chromatography-mass spectroscopy identification of 1-aminocyclopropane-1-carboxylic acid in compression wood vascular cambium of *Pinus contorta* Dougl. *Plant Physiology*, **71** 434–6.

Schimper, A. F. W. (1903). *Plant Geography upon a Physiological Basis*. Oxford University Press, Oxford.

Shaw, C. G. III & Taes, E. H. A. (1977). Impact of *Dothistroma* needle blight and *Armillaria* root rot on diameter growth of *Pinus radiata*. *Phytopathology*, **66**, 1319–23.

Shreve, F. (1914). *A Montane Rain-forest: A Contribution to the Physiological Plant Geography of Jamaica*. Carnegie Institition of Washington, DC, Publication 109.

Sinnott, E. W. (1952). Reaction wood and the regulation of tree form. *American Journal of Botany*, **30**, 69–78.

Slatyer, R. O. & Ferrar, P. J. (1977). Altitudinal variation in the photosynthetic characteristics of snow gum, *Eucalyptus pauciflora* Sieb. ex Spreng. V. Rate of acclimation to an altered growth environment. *Australian Journal of Plant Physiology*, **4**, 595–609.

Smith, E. M. & Hadley, E. B. (1974). Photosynthetic and respiratory acclimation to temperature in *Ledum groenlandicum* populations. *Arctic and Alpine Research*, **6**, 13–27.

Telewski, F. W. (1988). Intra-annual spiral compression wood: a record of low-frequency gravitropic circumnutational movement in trees. *International Association of Wood Anatomists Bulletin*, n.s., **9**, 269–74.

Telewski, F. W. (1989). Structure and function of flexure wood in *Abies fraseri*. *Tree Physiology*, **5**, 113–22.

Telewski, F. W. (1990). Growth, wood density, and ethylene production in response to mechanical perturbation in *Pinus taeda*. *Canadian Journal of Forest Research*, **20**, 1277–82.

Telewski, F. W. & Jaffe, M. J. (1981). Thigmomorphogenesis: changes in the morphology, and chemical composition induced by mechanical perturbation of 6-month-old *Pinus taeda* seedlings. *Canadian Journal of Forest Research*, **11**, 380–7.

Telewski, F. W. & Jaffe, M. J. (1986a). Thigmomorphogenesis: field and laboratory studies of *Abies fraseri* in response to wind or mechanical perturbation. *Physiologia Plantarum*, **66**, 211–18.

Telewski, F. W. & Jaffe, M. J. (1986b). Thigmomorphogenesis: anatomical, morphological and mechanical analysis of genetically different sibs of *Pinus taeda* in response to mechanical perturbation. *Physiologia Plantarum*, **66**, 219–26.

Telewski, F. W. & Jaffe, M. J. (1986c). Thigmomorphogenesis: the role of ethylene in the response of *Pinus taeda* and *Abies fraseri* to mechanical perturbation. *Physiologia Plantarum*, **66**, 227–33.

Telewski, F. W., Wakefield, A. H. & Jaffe, M. J. (1983). Computer-assisted analysis of tissues of ethrel-treated *Pinus taeda* seedlings. *Plant Physiology*, **72**, 177–81.

Timell, T. E. (1986a). *Compression Wood in Gymnosperms*, vol. I. Springer-Verlag, Berlin.

Timell, T. E. (1986b). *Compression Wood in Gymnosperms*, vol. II. Springer-Verlag, Berlin.

Timell, T. E. (1986c). *Compression Wood in Gymnosperms*, vol. III. Springer-Verlag, Berlin.

Tranquillini, W. (1979). *Physiological Ecology of the Alpine Timberline*. Springer-Verlag, Berlin.

Vogel, S. (1989). Drag and reconfiguration of broad leaves in high winds. *Journal of Experimental Botany*, **40**, 941–8.

Wainwright, S. A., Biggs, W. D., Currey, J. D. & Gosline, J. M. (1976). *Mechanical Design in Organisms*. Princeton University Press, Princeton.

Wardrop, A. B. (1964). The reaction anatomy of arborescent angiosperms. In *The Formation of Wood in Forest Trees*, ed. M. H. Zimmermann, pp. 405–56. Academic Press, New York.

Wardrop, A. B. (1965). The formation and function of reaction wood. In *Cellular Ultrastructure of Woody Plants*, ed. W. A. Côté, Jr, pp. 373–90. Syracuse University Press, Syracuse.

Weaver, P. L., Bruck, D. L. & Byer, M. D. (1973). Transpiration rates in the Luquillo Mountains of Puerto Rico. *Biotropica*, **5**, 123–33.

Westing, A. H. (1965). Formation and function of compression wood in gymnosperms. *Botanical Review*, **31**, 381–480.

Wilson, B. F. & Archer, R. R. (1977). Reaction wood: induction and mechanical action. *Annual Review of Plant Physiology*, **28**, 23–43.

Wilson, B. F. & Archer, R. R. (1979). Tree design: some biological solutions to mechanical problems. *Bioscience*, **29**, 293–8.

Wilson, B. F., Chien, C.-T. & Zaerr, J. B. (1989). Distribution of endogenous indole-3-acetic acid and compression wood formation in reoriented branches of douglas-fir. *Plant Physiology*, **91**, 338–44.

Wilson, J. (1980). Macroscopic features of wind damage to leaves of *Acer pseudoplatanus* L. and its relationship with season, leaf age, and windspeed. *Annals of Botany*, **46**, 303–11.

Wilson, J. (1984). Microscopic features of wind damage to leaves of *Acer pseudoplatanus* L. *Annals of Botany*, **53**, 73–82.

Wolter, K. E. (1977). Ethylene-potential alternative to bipyridilium herbicides for inducing lightwood in red pine. In *Lightwood Research Coordinating Council*

Proceedings of the Annual Meeting 1977, ed. M. H. Esser, pp. 90–9. USDA Forest Service, Ashville.

Yamamoto, F. & Kozlowski, T. T. (1987). Effect of ethrel on growth and stem anatomy of *Pinus halepensis* seedlings. *International Association of Wood Anatomists Bulletin*, n.s., **8**, 11–20.

15

Responses of young trees to wind: effects on root growth

A. STOKES, A. H. FITTER
and M. P. COUTTS

Abstract

Two experiments were carried out with Sitka spruce and European larch grown under intermittent wind in tunnels. In the first, both species were grown to determine whether there are any specific effects of wind on root growth, as opposed to effects on growth in general. In the second experiment, larch seedlings had their tap root removed 20 mm below the soil surface to mimic the formation of a shallow root plate, such as develops when larch is planted on seasonally waterlogged peat in much of the United Kingdom. In both experiments, lateral roots were counted and their orientation relative to the tap root recorded. In the wind-stressed trees in both experiments there was an increase of almost 60% in the number of large roots on the windward side of the trees and of 45% on the leeward side of larch trees compared with the number of roots growing at right angles to the direction of wind. In the first experiment, the sum of the cross-sectional area (Σ CSA) of lateral root bases was greater on the windward side of the tree compared with the other sides in both species. In the second experiment, Σ CSA of lateral root bases of wind-stressed larches was greatest on the leeward side whereas the control plants had a larger Σ CSA in the regions perpendicular to the wind direction. It appears that wind action stimulates the diameter growth of those roots most important for anchorage, but has a smaller effect on root development than other factors such as uneven nutrient supply. Once the tree is established and has reached a point where it is more vulnerable to wind, even these small changes may, however, be important in determining resistance to windthrow. A prediction of the role wind plays in the development of young trees is important to the forest industry.

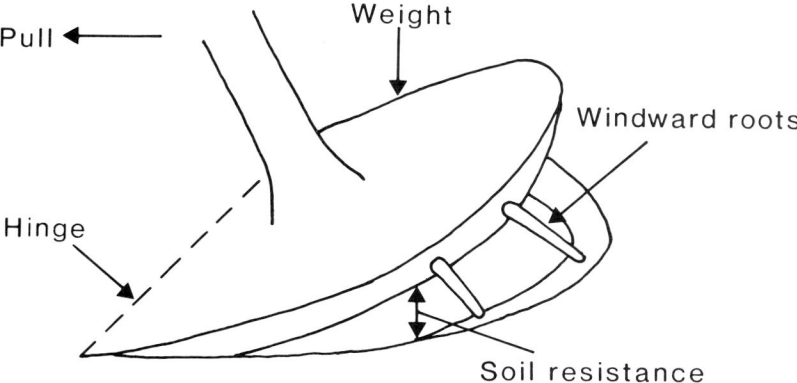

Fig. 15.1 Diagrammatic view of a shallow-rooted tree, showing four components of the anchorage which resist overturning (after Coutts, 1983a).

15.1 Introduction

Damage by wind is an important problem for forestry in the British uplands. Windthrow, the uprooting of trees, is much more common than stem breakage (Holtam, 1971). Extensive research has been carried out in attempts to predict and reduce wind damage (Booth, 1974; Somerville, 1980) but less has been done on the mechanics of how plants are uprooted (Coutts, 1983a, 1986; Ennos, 1989, 1991). Furthermore, only limited work has been carried out on the effects of wind action on root growth (Satoo, 1962; Heiligmann & Schneider, 1974), even though root form is the dominant influence on uprooting resistance (Ennos, 1993).

Coutts (1983a, 1986) quantified the forces involved in anchorage of shallow-rooted *Picea sitchensis* (Bong.) Carr. (Sitka spruce) by measuring the turning moment during the displacement of the soil–root plate when pulling mature trees with a winch attached to the stem. By sequentially cutting or breaking the roots and soil, he then divided the total resistive turning moment of the anchorage system into four components (Fig. 15.1): resistance to uprooting by soil underneath and at the sides of the plate, the weight of the plate, the resistance of leeward roots to bending and the resistance of windward roots in tension. When applied forces were maximal, the most important component in resisting uprooting was resistance from windward roots held in tension (54%). Other factors contributing to resistance included the weight of the root–soil plate (31%), the bending of roots at the leeward hinge (8%) and finally soil resistance underneath and at the edges of the plate (7%). In two non-woody species (*Impatiens glandulifera* and *Helianthus annuus*) Ennos *et al.* (1993) found that the windward roots again provided the most resistance to uprooting.

In order to achieve anchorage, roots must transfer into the soil forces which the shoot experiences. The stronger the soil and root–soil bond and the greater the root surface area, the larger the uprooting force that can be resisted, although the cost of constructing such a root system must be balanced against the benefits gained in anchorage (Fitter & Ennos, 1989; Ennos, 1993). For trees, the relevant uprooting force is when the wind or weight of the tree causes the plant to topple over from the base. This horizontal force will set up a moment about the base of the stem. The roots will be held in tension or compression and both forces will be transferred to the soil. In order to resist the turning moments transmitted by the stiff stem, trees need a rigid root system. This is achieved either by having a tap root from which horizontal lateral roots are attached, or by having a plate of lateral roots with sinkers growing downwards. If these sinkers quickly reach the water table and die, as happens for example when Sitka spruce is planted on peat in much of the United Kingdom, then anchorage strength is reduced.

The growth of woody roots, which form the essential structure for anchorage, is influenced by a number of internal and environmental factors (Coutts, 1987). Mechanical stresses from wind action must affect the activity of the root cambium, since there is an increase in diameter growth when roots are stimulated (Jacobs, 1954; Fayle, 1968; Wilson, 1975). Mechanical stimulation has been most thoroughly researched in stems, where thigmomorphogenesis (Jaffe, 1973; Telewski, this volume), in which stem elongation is reduced and radial growth increased, occurs and is intuitively adaptive. The forest industry needs to know whether comparable effects also take place underground, where the tree is anchored. Jacobs (1954) attached guy ropes to the stems of young *Pinus radiata* so that only the tops of the plants would sway. The diameter of the guyed stems was greater than the controls above the point of attachment, but smaller below. After 2 years the ropes were removed and, in the first high wind, all the stems broke or blew over. Jacobs also found that guyed trees had thinner woody roots. Fayle (1968) repeated this experiment on *Pinus sylvestris* saplings. After 2 years the lateral roots of free-standing trees showed a 75% increase in their annual ring widths when compared with the guyed trees.

Most research into thigmomorphogenesis has concentrated on flexing or shaking (Jacobs, 1939; Jaffe, 1973; Telewski, 1990), which is an artificial and often unrealistic way to mimic changes in a plant exposed to wind stresses. There are few studies in which young trees have been subjected to wind from an early age (Satoo, 1962; Heiligmann & Schneider, 1974; Rees & Grace, 1980; Telewski & Jaffe, 1986) and only two of these refer to root growth. Satoo (1962) exposed *Robinia pseudoacacia* seedlings to constant wind velocities of 3.6 m s^{-1} for 4 weeks and recorded reductions in root length and root and shoot dry weight compared with controls in still conditions. Heiligmann & Schneider (1974) found that shoot and root dry weight were decreased when *Juglans nigra* was grown at windspeeds of 2.8 m s^{-1} compared with plants grown at windspeeds of 0.1 m s^{-1}. In both experiments, exposure to wind resulted in a general reduction in growth, with no specific effects on roots recorded. However, both these wind tunnel studies used relatively fast continuous windspeeds, unlike field conditions where windspeeds can be low or zero for extended periods. Telewski (this volume) states that plants do not respond in the same way to continuous and intermittent wind action: far greater responses occur when plants are stressed periodically rather than continuously.

The present study, carried out in two wind tunnels, examines the changes in shoot and root growth of Sitka spruce and European larch seedlings under intermittent wind, in order to determine whether effects on plants occur under more realistic wind conditions; whether there are any specific effects of wind on root growth and whether any responses to wind could be viewed as adaptive. Ideally such experiments should be done on large trees, but it is difficult to work with mature trees. However, since the lateral roots of a mature tree develop early in its life (Coutts, 1983*b*), the roots laid down by a seedling tree will determine the structure of the mature system. Experimentation on young trees can therefore be a realistic experimental model.

15.2 Experimental tests

Wind tunnels were set up in glasshouses; each was constructed of dexion frames with polythene sides. In the first experiment, the tunnel was 3 m long, 1 m high and 1 m wide with three wind generators (electrically driven fans, 20 cm in diameter) positioned at one end. In the second experiment, two wind tunnels and two control tunnels (no wind) 0.25 m wide were set up, with one generator per wind tunnel. Twenty-four European larch (*Larix decidua* Mill.) and 24 Sitka spruce (*Picea sitchensis* (Bong.) Carr. were grown from seed in the first, but only European larch (14 plants in each tunnel) in the second experiment. In the second experiment the tap roots were cut 20 mm below the soil surface, using carefully inserted scissors, to mimic the formation of a shallow soil–root plate. The wind generators were operated for 6 h during the day and a further 6 h at night. The generators were positioned at soil level. Windspeeds experienced by plants ranged from 0.5 to 2.9 m s^{-1} in the first wind tunnel, 0.8 to 5.9 m s^{-1} in the second, with 'calm' in the control tunnels. The plants were harvested from the first and second experiments after 30 and 20 weeks, respectively. In the first experiment, after 30 weeks Sitka spruce reached a height of 40.2 ± 8.6 cm with a basal diameter of 9.9 ± 2.0 mm and larch were 58.3 ± 16.7 cm tall with a basal diameter of 12.2 ± 1.9 mm. In the second experiment, larch were 39.6 ± 13.5 cm tall with a basal diameter of 6.0 ± 1.0 mm after 20 weeks. The number, orientation, depth and diameter of lateral roots were recorded for all plants in both wind tunnels.

There was much variation in shoot growth between plants of both species in the first experiment although regressions of stem length, biomass and stem basal diameter on wind velocity were not significant; therefore variation was not due to wind action. In the second experiment, the regressions of larch shoot height and stem diameter on windspeed were significant: plants became shorter, and biomass and basal diameter were reduced with increasing windspeed ($R^2 = 0.30$, $p = 0.025$; $R^2 = 0.16$, $p = 0.035$; $R^2 = 0.16$, $p = 0.002$, respectively).

Root systems were examined in order to discover whether wind was affecting root growth and development. In the first experiment, roots >2 mm basal diameter in the top 50 mm of soil were studied as they were considered to be the most important for anchorage (lateral root basal diameter ranged from 0.2 to 10.5 mm in Sitka spruce and 0.2 to 13.5 mm in larch). The numbers of these lateral roots were calculated for each 15° sector around the stem for all the plants (Fig. 15.2). There was a large peak of roots in the direction of the wind (90°), for both species, where the number of roots had increased by 58% in Sitka spruce and 59% in larch, compared with the number of roots growing at right angles to the direction of the wind. The data for larch also suggest two lesser peaks of enhanced growth perpendicular to the wind direction (0°) and also on the leeward side (270°), where there was an increase of 41% in the number of roots compared with the number growing at right angles to the wind direction. Chi-squared analysis of the numbers of roots and their orientation for either Sitka spruce or larch shows that both distributions of roots are significantly different from uniform (Fig. 15.2, $p < 0.001$ in all cases).

For plants in the second wind tunnel, where the tap root had been cut, all lateral roots were included in the analysis, irrespective of size, because the plants were younger and few roots were larger than 2 mm in diameter. The numbers of roots were again calculated for each 15° sector around the stem. Two large peaks emerged on the histogram of treated plants (Fig. 15.3): one peak was towards the direction of wind (90°) and the other exactly opposite, i.e. on the lee side (270°). The numbers of windward and leeward roots had increased by 57% and 49%, respectively, compared with the numbers of roots

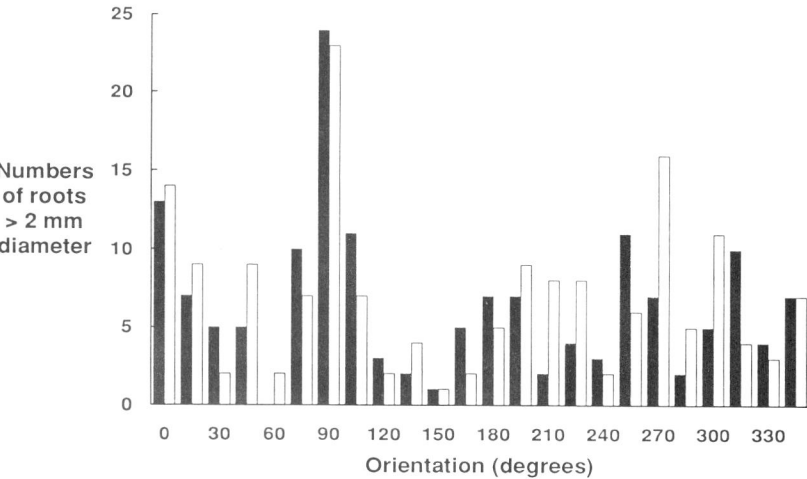

Fig. 15.2. Numbers of European larch and Sitka spruce lateral roots >2 mm diameter, for wind-stressed plants only. Both distributions differ significantly from uniform (Sitka spruce: $\chi^2_{23} = 92.65$, $p = 0.001$; European larch: $\chi^2_{23} = 88.6$, $p = 0.001$). Black bars, Sitka spruce; white bars, European larch.

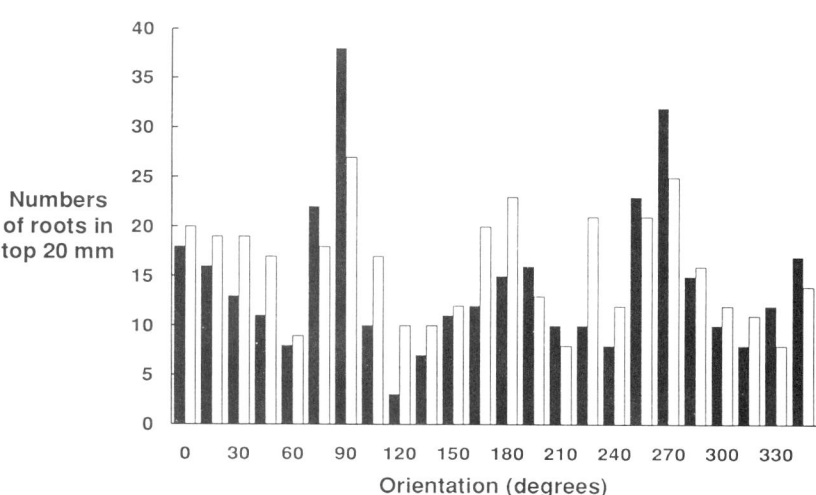

Fig. 15.3. Numbers of European larch lateral roots in the top 20 mm of soil for both control and wind-stressed plants. The two distributions differ significantly from each other ($\chi^2_{23} = 39.21$, $p = 0.02$). Black bars, treated plants; white bars, control plants.

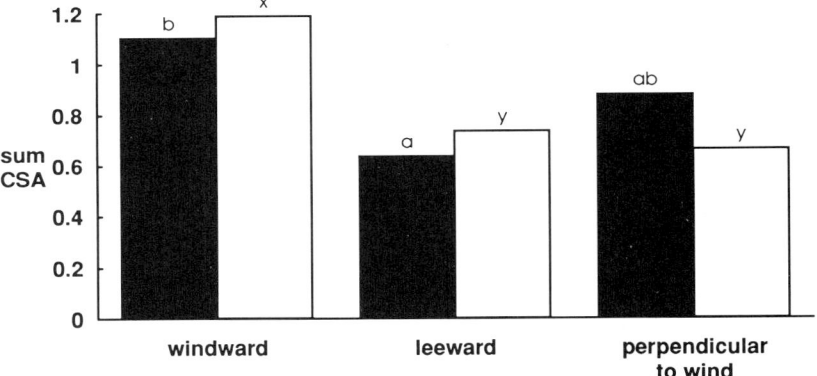

Fig. 15.4. Σ CSA of Sitka spruce and European larch root bases >2 mm diameter in three sectors around the plant's central axis. Windward roots have a significantly greater Σ CSA than leeward roots (Sitka spruce: $F_{2,92} = 3.53$, $p = 0.033$; larch: $F_{2,89} = 4.25$, $p = 0.017$; log transformation in both cases). Columns differently superscripted differ at $p = 0.05$. Black bars, Sitka spruce; white bars, larch.

perpendicular to the wind direction. In the control plants, three peaks appeared, all smaller than the two large peaks in the treated plants. One peak was in the direction of wind (90°), one opposite on the lee side (270°) and the third peak perpendicular to the flow of wind (180°). The distribution of lateral roots in wind-stressed plants was different from those in the control tunnel (Fig. 15.3, $p = 0.02$).

The increase in numbers of roots on the windward and to a lesser extent on the leeward side could imply a greater total root growth in these regions. The sum of the cross-sectional area (Σ CSA) of the bases of the roots growing on windward and leeward sides and perpendicular to these directions was calculated using 30° sectors. The two perpendicular sectors were combined as there should be no difference between them. In the first experiment, root bases of both Sitka spruce and larch had greater Σ CSA values in the windward sector than in the leeward sector (Fig. 15.4). However, there were no differences in mean CSA per root between any sector for either species, suggesting that the effect was not due to increased growth of a few roots. In the second experiment, wind-stressed larches had a significantly larger Σ CSA in the leeward sector than all other sectors (Fig. 15.5). The mean CSA per root was almost significantly larger in the leeward sector ($F_{2,132} = 2.48$, $p = 0.087$). The control plants had a significantly larger Σ CSA in the sectors perpendicular to the wind direction (Fig. 15.5). There were no differences in mean CSA between the sectors of the control plants.

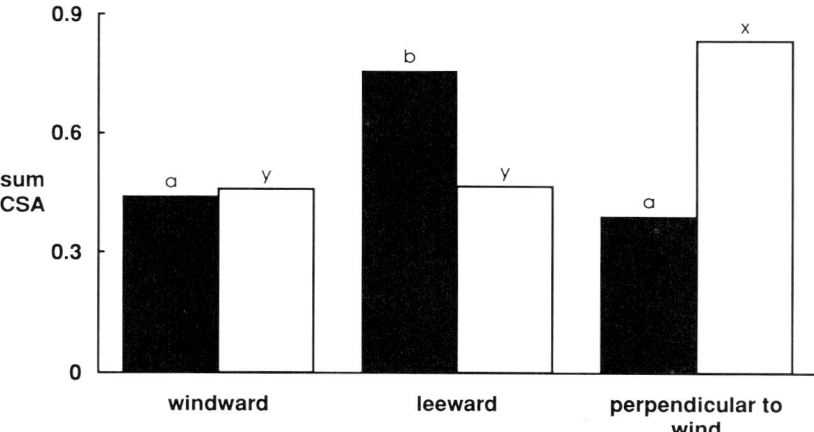

Fig. 15.5. Σ CSA of wind-stressed and control European larch root bases >2 mm diameter in sectors around the plant's central axis. Leeward roots of wind-stressed plants have a significantly greater Σ CSA than other roots ($F_{2,106} = 6.78$, $p = 0.002$, log transformation). Roots perpendicular to the direction of wind in control plants have a greater Σ CSA than other roots ($F_{2,102} = 8.51$, $p = <0.001$, log transformation). Columns differently superscripted differ at $p = 0.05$. Black bars, wind-stressed plants; white bars, control plants.

15.3 Discussion

Shoot growth in the first experiment was not altered by wind action, which may be due to the large variation noticed in the plant material or the low windspeeds experienced by the plants. In the second experiment, where windspeeds were greater, shoot growth was significantly reduced, although stem basal diameter did not increase with windspeed as reported earlier (Larson, 1965; Grace, 1977). However, in both experiments significant differences were found in root growth between wind-stressed and control plants, and between roots with different azimuths in wind-stressed trees, even at windspeeds which were too low to cause any large responses in shoot growth.

In the first experiment, both species had an increase of almost 60% in the number of larger roots growing towards the direction of the wind compared with roots growing perpendicular to the wind direction. As only roots >2 mm diameter at the root base were recorded, this implies that the numbers of larger roots had increased. Although the windward roots did not have a larger mean CSA of lateral root base per root, Σ CSA of root bases was significantly larger in this sector. In the second experiment, root number in wind-stressed larch was increased by 57% on windward and by 49% on leeward sides. The leeward roots had a larger mean CSA of root base per root than roots from the other sectors, and Σ CSA was also increased in the leeward sector. The control plants had more roots and a larger Σ CSA of root bases in the sectors perpendicular to the wind direction, where the smallest roots were found in the wind-stressed plants. It is striking that in both species there was a consistent increase in the numbers of roots important for anchorage on the windward side, and in larch on the leeward side also. This suggests that non-uniform distributions readily develop in Sitka spruce and larch and that wind can alter the distribution with very similar patterns of root growth forming under different windspeeds.

It appears that the distribution of resources within the root system has altered in response to wind so that those roots which provide the greatest contribution to anchorage receive more resources in both Sitka spruce and European larch. This was more pronounced in the second experiment where the tap root was removed. Instead of resources being sent to the tap root, which is an important component in resisting uprooting in young plants (Ennos, 1993), resources were diverted to windward and leeward lateral roots, which increased in number and cross-sectional area in the top 20 mm of soil. The increase in number of leeward and windward roots will mean a higher concentration of roots per unit area of soil, which in turn increases the tensile strength of the soil (Wu, 1976). Resistance to pullout will therefore be increased, especially on the windward side as these roots will be held in tension and are the most important component in resisting uprooting by wind (Coutts, 1983a, 1986). The possible increase in radial growth of root bases on the leeward side in the second experiment indicates that these roots have a higher resistance to bending (bending rigidity is proportional to the fourth power of radius). Roots perpendicular to the direction of wind are subject to torsion and will offer little resistance to uprooting. These species therefore respond to wind-induced stresses by enhanced growth of roots in positions most efficient in promoting anchorage, namely on the leeward and windward sides of the tree.

Wind action has a smaller effect on root architecture than does uneven nutrient (Drew et al., 1973) or water supply (Lesham, 1970; Coutts, 1982) and may be less important during this stage of plant growth. The results show an increase in allocation of resources to roots on the windward and leeward sides of the treated trees; however, the magnitude of the increase is small compared with the increases in root base radial growth seen in experiments where trees were staked in the field for 2–4 years (Fayle, 1968, 1976; Wilson, 1975). This could be due to the brevity of the laboratory experiments, which lasted only 20 and 30 weeks. There may also be a change in the importance of factors affecting root allocation once the tree is established and has reached a point where it is more vulnerable to wind. The responses to wind stress may be amplified as the tree matures.

It is very striking that these apparently adaptive changes in root growth occurred at windspeeds too low to exert large effects on the shoots. It is not clear, therefore, what signal was transmitted to the roots. Larson (1965), referring to young *Larix laricina*, suggested that resources are diverted from height to diameter growth under the influence of wind sway. He visualised the lower, stressed bole of a tree as a metabolic sink to which assimilates flow. Ethylene production in stems increases under mechanical stress and is thought to be the mediator of increased radial growth and reduced elongation (Telewski, this volume). It can be assumed that an increase in ethylene production at the root bases in the stressed areas stimulated the radial growth of more roots. The extra wood formation, especially in the leeward roots of wind-stressed larches in the second experiment, implies that these roots are receiving more assimilates, but probably at the expense of another part of the plant.

These results suggest that unidirectional wind may result in adaptive responses in tree root systems from a very early age. The changes which occur in trees exposed to wind loading are of an ecological advantage, in that the increase in number and size of windward and leeward lateral roots should result in a tree able to withstand greater stresses with less likelihood of uprooting.

It would be of interest to foresters to know how a tree which had reacted to wind from one direction would respond to wind stresses if the wind direction were changed. It is believed that most wind damage to a forest is caused when wind comes from an unusual direction, thereby suggesting that forest trees may also adapt to unidirectional winds. Further work where young trees in wind tunnels are exposed to different wind directions and durations will help to clarify the role wind plays in the development of root systems and hence anchorage strength.

Acknowledgements

This work was supported by a grant from the Natural Environmental Research Council with assistance from The Forestry Commission.

References

Booth, T. C. (1974). Silviculture and management of high-risk forests in Great Britain. *Irish Forestry*, **31**, 145–53.

Bowen, G. D. (1991). Soil temperature, root growth and plant function. In *Plant Roots: The Hidden Half*, ed. Y. Waisel, A. Eshel & U. Kafkafi. Dekker, New York.

Coutts, M. P. (1982). Growth of Sitka spruce seedlings with roots divided between soils of unequal matric potential. *New Phytologist*, **92**, 49–61.

Coutts, M. P. (1983). Root architecture and tree stability. *Plant and Soil*, **71**, 171–88.

Coutts, M. P. (1983*b*). Development of the structural root system of Sitka spruce. *Forestry*, **56**, 1–16.

Coutts, M. P. (1986). Components of tree stability in Sitka spruce on peaty, gley soil. *Forestry*, **59**, 173–97.

Coutts, M. P. (1987). Developmental processes in tree root systems. *Canadian Journal of Forest Research*, **17**, 761–7.

Drew, M. C., Saker, L. R. & Ashley, T. W. (1973). Nutrient supply and the growth of the seminal root system in barley. *Journal of Experimental Botany*, **24**, 1189–202.

Ennos, A. R. (1989). The mechanics of anchorage in seedlings of sunflower *Helianthus annuus*. *New Physologist*, **113**, 185–92.

Ennos, A. R. (1991). The mechanics of anchorage in wheat *Triticum aestivum* L. I. The anchorage of wheat seedlings. *Journal of Experimental Botany*, **42**, 1601–6.

Ennos, A. R. (1993). The scaling of root anchorage. *Journal of Theoretical Biology*, **161**, 61–75.

Ennos, A. R., Crook, M. J. & Grimshaw, C. (1993). A comparative study of the anchorage systems of Himalayan balsam *Impatiens glandulifera* and mature sunflower *Helianthus annuus*. *Journal of Experimental Botany*, **44**, 133–46.

Fayle, D. C. F. (1968). *Radial Growth in Tree Roots*. Technical Report no. 9. Faculty of Forestry, University of Toronto.

Fayle, D. C. F. (1976). Stem sway affects ring width and compression wood formation in exposed root bases. *Forest Science*, **22**, 193–4.

Fitter, A. H. & Ennos, A. R. (1989). Architectural constraints to root system function. *Aspects of Applied Biology*, **22**, 15–22.

Grace, J. (1977). *Plant Response to Wind*. Academic Press, New York & London.

Heiligmann, R. & Schneider, G. (1974). Effects of wind and soil moisture on black walnut seedlings. *Forest Science*, **20**, 331–5.

Holtam, B. W. (1971). *Windthrow of Scottish Forests in January 1968*. Forestry Commission Bulletin 45. HMSO, London.

Jacobs, M. R. (1939). *A Study of the Effect of Sway on Trees*. Australian Commonwealth Forestry Bureau Bulletin 26.

Jacobs, M. R. (1954). The effect of wind sway on the form and development of *Pinus radiata* D. Don. *Australian Journal of Botany*, **2**, 35–51.

Jaffe, M. J. (1973). Thigmomorphogenesis: the response of plant growth and development to mechanical stimulation – with special reference to *Bryonia dioica*. *Planta*, **114**, 143–57.

Larson, P. R. (1965). Stem form of young *Larix* as influenced by wind and pruning. *Forest Science*, **11**, 412–24.

Lesham, B. (1970). Resting roots of *Pinus halepensis*: structure, function and reaction to water stress. *Botanical Gazette*, **131**, 99–104.

Rees, D. J. & Grace, J. (1980). The effects of wind on the extension growth of *Pinus contorta* Douglas. *Forestry*, **53**, 145–53.

Satoo, T. (1962). Wind, transpiration and tree growth. In *Tree Growth*, ed. T. T. Kozlowski, pp. 299–310. Ronald Press, New York.

Somerville, A. (1980). Wind stability: forest layout and silviculture. *New Zealand Journal of Forest Science,* **10**, 476–501.

Telewski, F. W. (1990). Growth, wood density and ethylene production in response to mechanical perturbation in *Pinus taeda*. *Canadian Journal of Forest Research*, **20**, 1277–82.

Telewski, F. W. & Jaffe, M. J. (1986). Thigmomorphogenesis: field and laboratory studies of *Abies fraseri* in response to wind or mechanical perturbation. *Physiologia Plantarum*, **66**, 211–18.

Wilson, B. F. (1975). Distribution of secondary thickening in tree root systems. In *The Development and Function of Roots*, ed. J. G. Torrey & D. T. Clarkson, pp. 197–219. Academic Press, New York.

Wu, T. H. (1976). *Investigation of Landslides on Prince of Wales Island, Alaska*. Geotechnical Engineering Report 5. Department of Civil Engineering, Ohio State University.

16
Wind stability factors in tree selection: distribution of biomass within root systems of Sitka spruce clones

B. C. NICOLL, E. P. EASTON, A. D. MILNER,
C. WALKER and M. P. COUTTS

Abstract

The effects of tree improvement on factors likely to influence tree stability were investigated using clonal Sitka spruce trees that had been grown for 11 years on an unploughed nursery site. The distribution of biomass between root and shoot, and within the root systems of trees from five improved clones, was examined and compared with control trees grown from unimproved cuttings and transplants. The direction (azimuth) of growth and dimensions of the main woody roots were also measured. Differences between clones were found in allocation of biomass between root and shoot, and in root system architecture. Large differences were found between clones in proportions of below-ground biomass allocated to stumps and woody roots (which function for anchorage). These results indicate that root : shoot ratio can be a poor indicator of tree stability when the stump is included as part of the root biomass. The distribution of root origins around the stump showed no significant clumping but the allocation of biomass between roots was found to differ between tree types. On average, the improved clones had allocated biomass to fewer roots than the controls. The amount of branching in the proximal 45 cm of the root system also differed between clones. Distribution of root cross-sectional area around the tree was significantly asymmetric in two of the clones. Overall, root biomass was allocated more to the lee side of the prevailing wind direction. The substantial differences found in allocation between root and shoot, and within the root system, may have implications for the wind stability of trees and could present opportunities for improving stability by clonal selection.

16.1 Introduction

Damage by windthrow causes important economic losses to forestry in northern Europe. Anchorage of trees is provided by the structural root system, but a common feature of unstable crops on wind-exposed upland sites is the shallow rooting caused by unfavourable soil conditions at depth (Armstrong *et al.*, 1976). On these sites the structural root system must develop from lateral roots growing near the soil surface. Attempts to reduce windthrow usually involve treatment of the site – for example, by using drainage to increase the depth of rootable soil (Savill, 1976) – but selection and breeding of trees with characteristics likely to promote stability has not been attempted. Good heritability of root characteristics has been reported for trees within their first year of growth (see Nambiar, 1985). However, the heritability of root system characteristics relating to stability of older trees has not previously been examined.

Clones of Sitka spruce (*Picea sitchensis* (Bong.) Carr.) selected on the basis of improved shoot characteristics (including growth rate and stem straightness) are increasingly being used in British forestry plantations. It is not known whether these trees have superior growth rates due to more efficient assimilate production or if their faster shoot growth takes place at the expense of the root system. Any reduction of the root : shoot ratio would be expected to make trees more prone to windthrow. The distribution of biomass within the root system must also be considered since an asymmetric root system will make the tree vulnerable to wind from certain directions. Clones of Sitka spruce have been found to differ in the initiation (Deans *et al.*, 1992) and growth rate (Coutts & Nicoll, 1990) of their lateral roots, but development of the structural root systems of clonal trees has not been investigated. For adequate support on shallowly rooted soils a tree must allocate sufficient biomass to the structural roots, which must have fairly even development around the tree. In this study, these factors have been examined in a comparison between improved Sitka spruce clones, cuttings of unimproved trees, and normal transplants, all grown for 11 years on a flat and even site.

16.2 Methods

Nine replicate trees (ramets) from each of five improved clones, along with nine transplants and nine unimproved cuttings, were examined from a clonal demonstration plot planted 11 years earlier in Teindland Forest, north-east

Scotland. The site was previously a heathland nursery on a gentle slope at an elevation of 120 m. The soil consisted of shallowly cultivated podzol with an induration at between 40 and 60 cm depth, and was covered with a vegetation layer consisting of *Rumex acetosella* L. and various moss species. A strip 1 m wide had been rotavated to a depth of 15 cm to provide a weed-free planting position. The trees were planted in April 1981 at 2 m spacing in a randomised block design with three blocks. Each block contained a row of six trees of each of the five improved clones, unimproved cuttings, unimproved transplants, and five unimproved clones (not used in this study). Three trees of each type were selected at random from each block, giving a total of 63 trees.

The 'improved' clones were juvenile cuttings from different half-sibling trees grown from seed of one open pollinated 'plus-tree' of Queen Charlotte Islands (QCI) provenance. This plus-tree had been selected as part of a tree improvement programme on the basis of stem straightness, fine branching characteristics, and above-average growth rate. The 'unimproved' control cuttings were each propagated from a different transplant grown from randomly selected seed of QCI provenance and hence were all genetically different. All cuttings were propagated following the method described by Mason (1984). Transplants were produced using standard nursery practice (Aldhous, 1972) in which seedlings are grown for 2 years in a seedbed, transplanted, and then grown for a further year before lifting. Cuttings and transplants were produced as bare root stock, while clonal plants were all container grown. Unground rock phosphate was applied along the rotavated strips immediately after planting at a rate of 375 kg ha^{-1} giving 50 kg ha^{-1} of elemental phosphate. Urea was applied to all trees at a rate of 350 kg ha^{-1} in June 1987.

The trees were harvested 11 years after planting, when mean tree height was 6.17 m and mean diameter at 1.3 m was 9.4 cm. The root systems were removed from the ground by lifting vertically with a hydraulic forwarder grab, before removing soil using a high-pressure air jet from an 'air-knife' (Briggs Technology, Pittsburgh, USA). No attempt was made to remove fine roots from the soil, and only coarse woody roots are included in this study.

The angle (azimuth) and diameter were recorded for each root emerging from the stump. Root angles were measured between a reference point marked on the downhill side of each stump (320° from north) before extraction, and the middle of each major lateral root at a point 45 cm from the anatomical centre of the stump. Root diameters were measured in the vertical and horizontal planes, also at 45 cm from the stump centre. The trees were

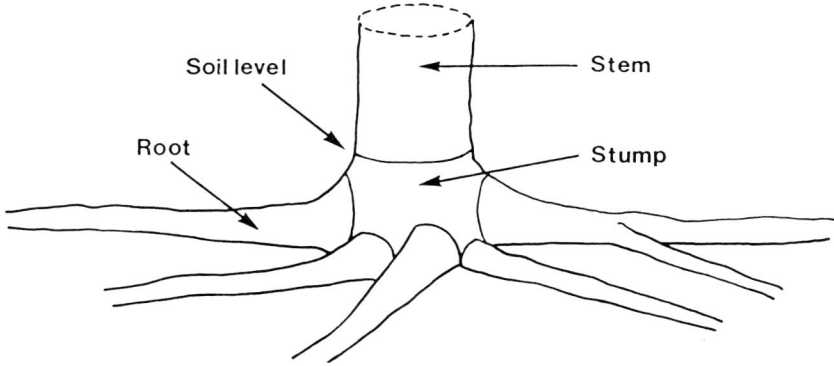

Fig. 16.1. The separation of a tree into root, stump and stem components.

cut into various components and dried to constant weight in a timber kiln at 70 °C. Dry weights were recorded for shoot components (not presented here in detail), and for stumps before and after removing lateral roots (Fig. 16.1). Root : shoot ratio was calculated in the normal way, with the stump included as part of the root biomass, and it was also calculated without the stump, where it is termed 'root only : shoot ratio'.

Data were analysed using analysis of variance (ANOVA) with percentages transformed to the arc-sine scale to satisfy the assumptions underlying this procedure, and differences were examined using the usual F-test. The outline ANOVA is shown in Table 16.1. Broad-sense clone-mean heritability (H_c^2) as described by Mullin & Park (1992) was calculated for each characteristic examined. This is a figure between 0 and 1 which indicates the proportion of the total variation among clones that is due to genotype, i.e. a value of 1 would indicate that the variation is all genotypic while

Table 16.1. *Outline analysis of variance*

Source of variation	Degrees of freedom
Block	2
Tree type	6
Control vs clone	1
Cutting vs transplants	1
Among clones	4
Error	54
Total	62

a value of 0 indicates that variation is entirely due to environmental factors. The uniformity of root emergence around the stump was examined by computing the L-statistic (Mardia, 1972) for each tree. This method examines the divergence of root angles from a model of uniformity where the angle between each root is the same for a given number of roots of equal length (Henderson et al., 1983). The smaller the L-statistic, the smaller the divergence from uniformity, while larger values indicate clustering. Clustering of roots was investigated using critical values of this statistic from Mardia (1972).

Curves were plotted for the mean distribution of cross-sectional area (CSA) among the 11 largest roots of each tree. Roots less than 10 mm in diameter at 45 cm from the stump centre were not measured but were assigned a diameter of 5 mm for the purpose of this analysis. A centre of root CSA was calculated for each tree to examine the directional allocation of biomass. The centre of CSA is similar to the centre of mass calculated by Coutts et al. (1990), and the centre of diameter used by Quine et al. (1991). The cartesian coordinates of the ith root (where $i = 1, \ldots, n$) weighted by CSA are:

$$X_i = D_i \sin \theta_i; \quad Y_i = D_i \cos \theta_i$$

where θ_i represents the angle of initiation,

$$D_i = \frac{d_i^2}{\sum_{i=1}^{n} d_i^2}$$

and d_i is the diameter of the ith root. Therefore the centre of CSA of a tree (i.e. its mean root position weighted by root CSA) has coordinates:

$$X = \sum_{i=1}^{n} X_i; \quad Y = \sum_{i=1}^{n} Y_i$$

The distance of the centre of CSA from the centre of the tree (R) and its orientation (ϕ) are given by: $R = (X^2 + Y^2)^{\frac{1}{2}}$; $\phi = \tan^{-1}(X/Y)$. Large values of R indicate that roots tend to cluster together in a preferred direction ϕ, while small values imply uniformity around the tree stump. Assuming that X_i and Y_i are independent normally distributed random variables with common variance, σ^2, the hypothesis of no clustering of roots can be tested using the statistic, $R^2/n\hat{\sigma}^2$ where $\hat{\sigma}^2$ is an estimate of the variance. Under the hypothesis of no clustering this statistic has an F-distribution with approximately 2, $2n - 1$ degrees of freedom. The hypothesis is rejected whenever $R^2/n\hat{\sigma}^2 > F(2, 2n - 1; a)$, when testing at the $a\%$ significance level.

16.3 Results

Mean total tree biomass ranged from 19.9 to 33.8 kg for clones and was 19.4 and 26.2 kg for transplants and cuttings respectively (Table 16.2). On average, clones had significantly greater total biomass than controls ($p < 0.01$), the mean being 29.3 kg for clones and 22.8 kg for controls. Differences amongst the five clones were also significant ($p < 0.01$).

Mean root : shoot ratios were between 0.27 and 0.39 for clones and were 0.32 and 0.30 for transplants and cuttings respectively. The mean root only (without stump) : shoot ratios were between 0.14 and 0.23 for clones and 0.20 and 0.16 for transplants and cuttings respectively. The comparison of mean ratios relative to the grand mean was altered considerably when the stump was ignored for some clones but very little for others (Fig. 16.2) because of large variation amongst clones in allocation between stump and roots (Fig. 16.3). For example, clone 77 had a larger root than stump biomass, whereas the reverse was true for clone 83. Root : shoot ratios calculated using both methods differed significantly amongst the seven tree types ($p < 0.05$), but appeared to be independent of total biomass. On average, clones had a significantly larger root : shoot ratio than controls ($p < 0.05$), but this was no longer the case when the stump component was removed (Table 16.2).

The main roots of all trees were predominantly in the surface 20 cm of the soil, and root depth was never greater than 60 cm. The mean number of main roots (>10 mm diameter) that emerged from the stump is shown in Table 16.2 with the percentage of these roots that had woody branches within 45 cm from the stump centre. The percentage of roots with branching differed significantly between clones ($p < 0.05$), with clone 74 having the least branching (10.4%) and clone 77 the greatest (35.8%), but on average there were no significant differences between controls and clones, or between the two controls. The mean number of main roots (>10 mm diameter) at the 45 cm measuring point (Table 16.2) also differed significantly between clones ($p < 0.01$), with clone 74 having the smallest number (6.1) and 77 the most (11.4). The broad-sense clone-mean heritabilities (H_c^2) for these characteristics are shown in Table 16.2. Horizontal and vertical diameters of roots measured 45 cm from the stump were similar, with the mean ratio (horizontal/vertical) ranging from 0.97 to 1.0.

Table 16.2. Mean total tree biomass, mean number of main (>10 mm diameter) roots that emerge from the stump, the percentage that have woody branches by 45 cm from the anatomical centre of the stump (mean percentages transformed on the arc-sine scale are shown with true means in brackets), mean number of main roots at 45 cm from the stump centre, mean total root : shoot ratio, mean root only : shoot ratio, the mean total cross-sectional area (CSA) of these roots, and the mean CSA per root

Tree type	Total tree biomass (kg)	No. of main roots	% with woody branches	No. of main roots at 45 cm	Total root : shoot ratio	Root only : shoot ratio	Mean accumulated CSA (cm^2)	Mean root CSA (cm^2)
Clone 74	19.92	5.6	12.8 (10.4)	6.1	0.39	0.21	52.7	9.4
Clone 77	29.50	8.3	36.1 (35.8)	11.4	0.37	0.23	77.0	6.3
Clone 78	31.70	7.6	29.8 (27.6)	11.3	0.33	0.19	76.8	8.3
Clone 80	31.68	6.4	28.4 (27.9)	7.8	0.27	0.14	42.4	5.3
Clone 83	33.76	7.1	16.8 (14.6)	7.3	0.39	0.17	57.9	7.8
Cuttings	26.22	6.2	34.9 (33.4)	9.0	0.30	0.16	43.8	4.9
Transplants	19.36	6.7	27.8 (24.9)	8.0	0.32	0.20	31.5	3.8
SED	3.70	0.93	7.17	1.5	0.03	0.03	14.4	1.7
H_c^2	0.77	0.62	0.72	0.81	0.81	0.73	0.55	0.44

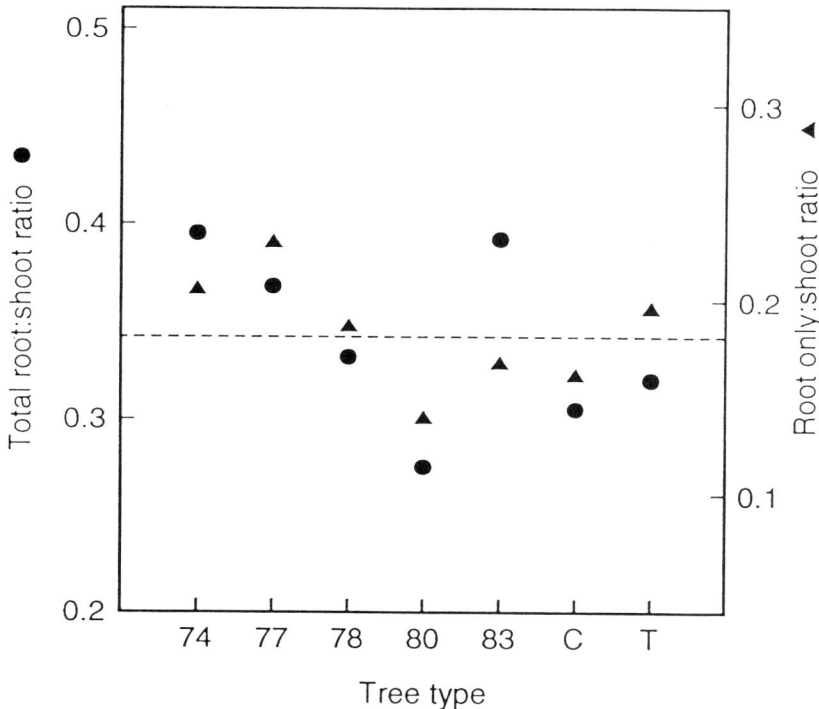

Fig. 16.2. Total root : shoot ratio and root only (without stump) : shoot ratio of 11-year-old improved clones (74–83), unimproved cuttings (C) and transplants (T), relative to the grand mean for each ratio (dashed line). Circles, total root : shoot ratio; triangles, root only : shoot ratio.

Mean accumulated cross-sectional areas (sum of CSA for all roots > 10 mm diameter) and mean root CSAs of main roots at 45 cm from the stump centre are shown in Table 16.2. On average, clonal trees had significantly greater mean accumulated CSA than controls ($p < 0.01$). There was no significant difference between cutting and transplant controls, or between individual clones. Fig. 16.4a shows the mean percentage of accumulated CSA contained in the i largest roots (where $i = 1, \ldots, 11$) for each of the seven tree types. Examples of the differences in allocation between roots described by the upper and lower curves in Fig. 16.4a (clones 74 and 77) are shown in Fig. 16.5. Clone 74 had 82% of CSA in the three largest roots while in clone 77 the three largest roots contained only 61%, and 85% was contained in six roots. Differences in the distribution of CSA between roots were significant between tree types. For example, a comparison of percentage CSA in the four largest roots

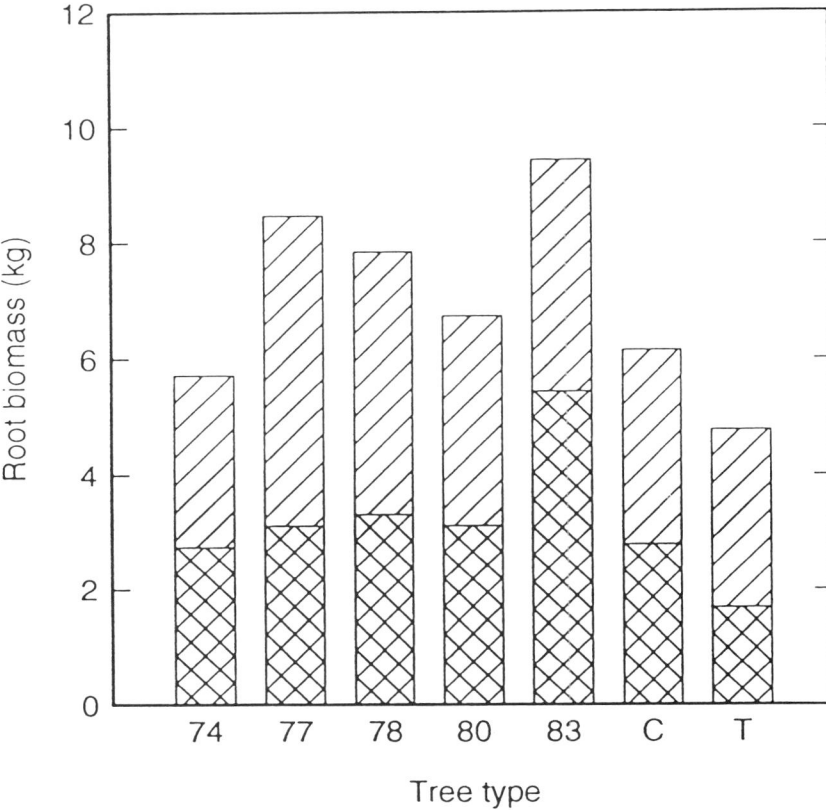

Fig. 16.3. Mean root biomass for each tree type – improved clones (74–83), unimproved cuttings (C) and transplants (T) – separated into lateral root (hatching) and stump component (cross hatching).

revealed significant variation between individual clones ($p < 0.01$) and, on average, clones had a significantly higher percentage than controls ($p < 0.01$), the average being 80.4% for clones and 73.2% for controls. Only one clone (77) had a smaller percentage of CSA in four roots than transplants or cuttings. The broad-sense clone-mean heritability (H_c^2) for percentage CSA in the four largest roots was 0.81. The log mean CSA for each of the 11 largest roots in clones 77 and 74 (the clones with extreme curves in Fig. 16.4a) are shown in Fig. 16.4b. The relationship between CSA and root number is described by a log regression line for both of these clones (Fig. 16.4b). The linear relationship is significant for both of the clones, and the two regression lines are significantly different from each other ($p < 0.05$).

No significant clustering of root origins around the stump was found on any of the trees using the L-statistic. There were, however, some differences between trees in the distribution of root biomass around the stump. The test statistic $R^2/n\hat{\sigma}^2$ was used to examine for clustering of root CSA on individual trees and for clustering of centres of CSA in each type. The hypothesis of no clustering was rejected for only one of the 63 trees (see Fig. 16.6a). Clones 77 and 83 both had centres of CSA that clustered in a preferred direction (Fig. 16.6b), although the directions for these clones were in opposite sectors. The centres of CSA of trees from clone 77 clustered in the direction 309° from the reference point, while for clone 83 the direction was 100°. The southern sector contained very few tree centres of CSA (Fig. 16.6a), and no mean centres of CSA (Fig. 16.6b). The mean centres of CSA tended to be directed downhill (Fig. 16.6b) and the mean centre of CSA for all tree types combined was 2° from North (42° from the reference point direction).

16.4 Discussion

There is no evidence from this study that the selection of trees for improved shoot characteristics results in reduced root : shoot ratios, and in fact four of the five clones had higher root : shoot ratios than the unimproved control trees. The improved clones had increased total biomass production rather than increased allocation to the shoot at the expense of the roots. When stump wood was removed from the ratio to produce a comparison between lateral root and shoot biomass, the relationship between clones was altered. As the stump makes no active contribution to stability in Sitka spruce, the comparison of the structurally important lateral roots with the above-ground components that they support is a more useful indicator of tree stability. Indeed, one clone (83) that looked the most promising from the point of view of growth rate and total root : shoot ratio (Fig. 16.2) allocated such a large proportion of below-ground biomass to the stump that root only : shoot ratio was below average and trees of this clone are likely to be more unstable.

Downward root growth in this experiment was restricted by an indurated layer in the soil. Consequently, trees developed shallow root systems that were comparable to those on sites where downward growth is restricted by a seasonally fluctuating water-table (see Coutts & Philipson, 1987). The use of a flat planting site in this study avoided the eccentric root development that may be associated with site cultivation (Savill, 1976). The structural root system develops through differences in the allocation of assimilates to individual roots undergoing secondary thickening (Fayle,

(a)

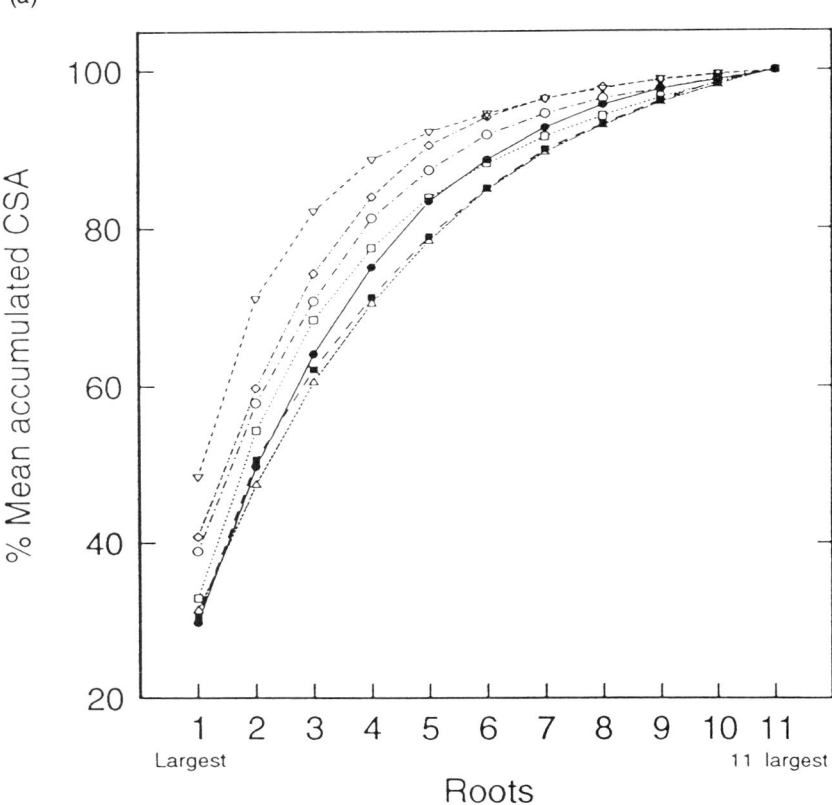

Fig. 16.4(a). For caption see opposite.

1975). Both the number and size of the structural roots are important, as is the distribution of biomass around the tree. If roots are considered to be circular beams their stiffness will be proportional to the fourth power of their diameter, and so a large number of thin roots will offer less resistance to bending than a few thick roots (Coutts, 1983a). However, where biomass is allocated predominantly to very few roots, a stable anchorage can be achieved only with an even root distribution around the stump (Coutts, 1983b). It is therefore expected that the clone with 85% of CSA in the six largest roots will have better stability than the clone with a similar percentage in three roots. On average, root systems of trees grown from improved clonal cuttings had a higher proportion of their

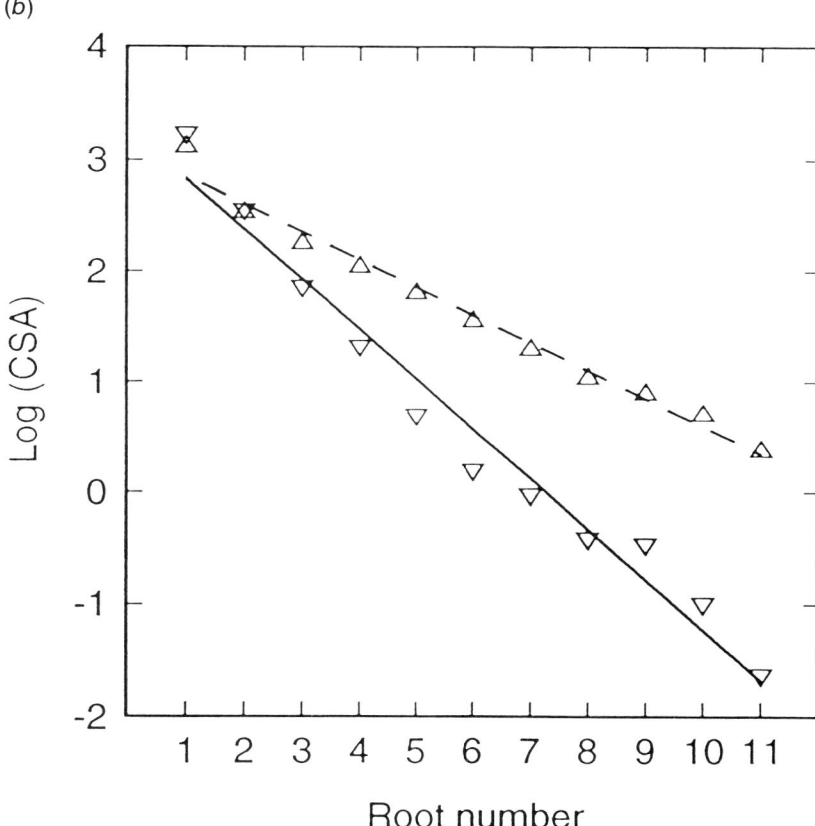

Fig. 16.4. Distribution of cross-sectional area (CSA) between roots. (*a*) Mean percentage of CSA in the i largest roots (where $i = 1, \ldots, 11$) for each of the tree types. Improved clones are shown as open symbols, and unimproved cuttings and transplants as filled symbols: ▽, clone 74; △, clone 77; □, clone 78, ○, clone 80; ◇, clone 83; ●, cuttings; ■, transplants. (*b*) Log mean CSA of each of the 11 largest roots of clone 74 (▽) and 77 (△). The regression line for clone 74 (continuous line) is described by Log(CSA) = 3.28 − 0.45Root number, and for clone 77 (dashed line) is described by Log(CSA) = 3.11 − 0.25Root number.

biomass allocated to the largest roots than that found in unimproved cuttings or transplants, and consequently these trees might be relatively less stable if their roots are unevenly distributed.

Henderson *et al.* (1983) found that in seven Sitka spruce trees, three had root origins evenly distributed around the stump, three were randomly distributed, and one was clustered. In the present study on a larger number

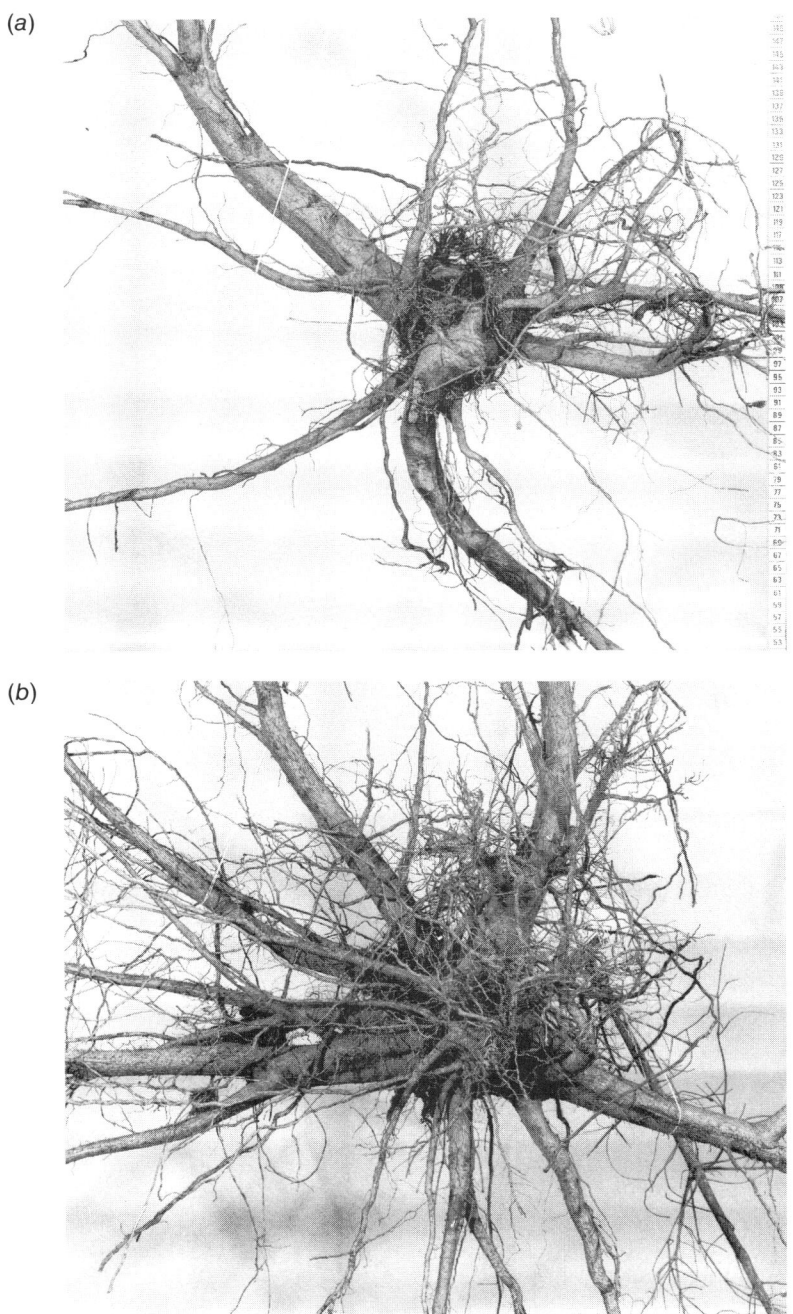

Fig. 16.5. Photographs of root systems described by the extreme lines in Fig. 16.4. (*a*) clone 74; (*b*) clone 77.

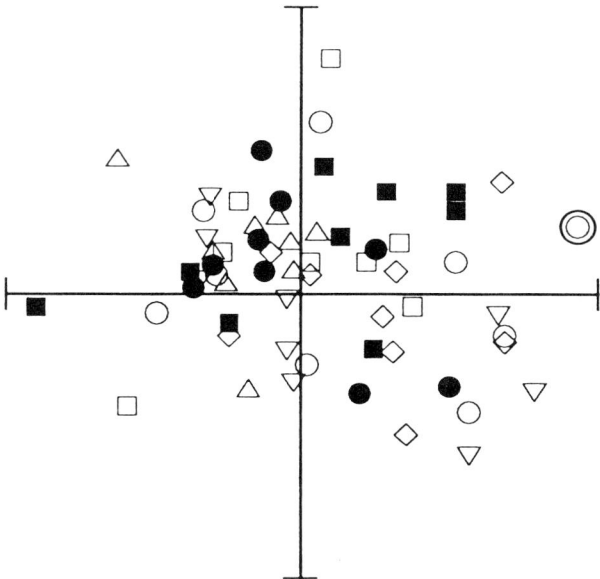

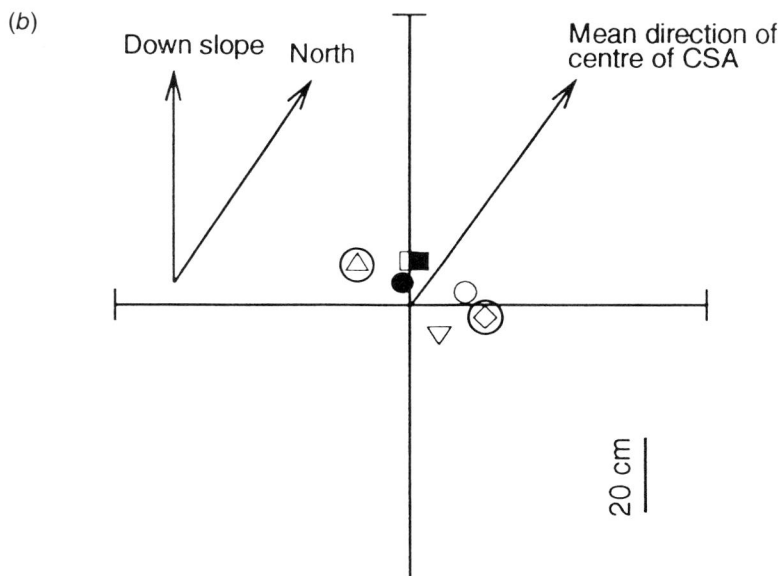

Fig. 16.6. Centres of root cross-sectional area (CSA): (*a*) all trees, (*b*) means for tree types. Improved clones are shown as open symbols, and unimproved cuttings and transplants as filled symbols. ▽, clone 74; △, clone 77; □, clone 78; ○, clone 80; ◇, clone 83; ●, unimproved cuttings; ■, transplants. Symbols surrounded by a circle indicate significant clustering of root CSA.

of trees, no clustering of initial direction was found, but uneven development evidently can occur as the roots undergo secondary thickening. The asymmetry of biomass allocation around trees in two of the clones, might indicate an inherent instability of these clones. Allocation of biomass did not differ between the two controls. Mason et al. (1986) found no differences between unimproved cuttings and transplants in root development after 3 years, and the present study indicates that this still holds at 11 years. Differences between clones in the branching of the structural roots could also be important for their stability, as branching of structural roots greatly reduces root stiffness (Coutts, 1983a). Clones varied between having 10% and 36% of their structural roots with major branches in the proximal 45 cm, indicating large variation in the support given by their root systems.

All tree types had a comparatively small amount of secondary thickening in the southern sector; biomass was concentrated downhill, and in a northerly direction. Similar observations of Sitka spruce root growth concentrated in a particular direction have been reported by Mason et al. (1986) and Coutts et al. (1990), but the reason for this growth pattern is not known. Robertson (1991) found tree ring widths to be greatest on the lee side of Balsam fir trees, and roots on the lee side might be expected to show correspondingly better growth. Wind was not recorded on this site, but measurements taken 12 km to the south-east at Rosarie Forest (B. A. Gardiner & A. Mackie, personal communication) show the wind to be predominantly from the south in both the growing and dormant seasons, with light winds commonly from the south-west. Thickening of roots in the present study therefore appears to have been concentrated in a direction away from the prevailing wind.

Despite the small number of clones examined in this study, the large genotypic variation we have found in characteristics likely to influence tree stability suggests an opportunity for improvement of stability by clonal selection. Although the usefulness of the clonal heritability values calculated here is limited by the experiment design (half-sibling clones planted on an even site), the high heritability values for root characteristics of these closely related clones suggest that further investigation would be worth while. The relationship between these characteristics and tree stability must be examined in practice, but the loss of trees to windthrow might be reduced by using clones which develop a high root (without stump) : shoot ratio, even biomass distribution between structural roots, and minimal structural root branching near the stem.

Acknowledgement

The work described forms part of a larger investigation supported by the EC FOREST programme. The clonal trial used for this study was designed by P. Biggin, established by G. R. Menzies, and managed by W. L. Mason. We are grateful to J. Davidson and other staff at Newton Nursery for their assistance with the field work.

References

Aldhous, J. R. (1972). *Nursery Practice*. Forestry Commission Bulletin 43. HMSO, London.
Armstrong, W., Booth, T. C., Priestley, P. & Read, D. J. (1976). The relationship between soil aeration, stability and growth of Sitka spruce (*Picea sitchensis* (Bong.) Carr.) on upland peaty gleys. *Journal of Applied Ecology*, 13, 585–91.
Coutts, M. P. (1983*a*). Root architecture and tree stability. *Plant and Soil*, 71, 171–88.
Coutts, M. P. (1983*b*). Development of the structural root system of Sitka spruce. *Forestry*, 56, 1–16.
Coutts, M. P. & Nicoll, B. C. (1990). Growth and survival of shoots, roots, and mycorrhizal mycelium in clonal Sitka spruce during the first growing season after planting. *Canadian Journal of Forest Research*, 20, 861–8.
Coutts, M. P. & Philipson, J. J. (1987). Structure and physiology of Sitka spruce roots. *Proceedings of the Royal Society of Edinburgh*, 93B, 131–44.
Coutts, M. P., Walker, C. & Burnand, A. C. (1990). Effects of establishment method on root form of Lodgepole pine and Sitka spruce and on the production of adventitious roots. *Forestry*, 63, 143–59.
Deans, J. D., Mason, W. L. & Harvey, F. J. (1992). Clonal differences in planting stock quality of Sitka spruce. *Forest Ecology and Management*, 49, 101–7.
Fayle, D. C. F. (1975). Distribution of radial growth during the development of red pine root systems. *Canadian Journal of Forest Research*, 5, 608–25.
Henderson, R., Ford, E. D., Renshaw, E. & Deans, J. D. (1983). Morphology of the structural root system of Sitka spruce. I. Analysis and quantitative description. *Forestry*, 56, 121–35.
Mardia, K. V. (1972). *Statistics of Directional Data*. London: Academic Press.
Mason, W. L. (1984). *Vegetative Propagation of Conifers using Stem Cuttings. I: Sitka Spruce*. Forestry Commission Research Information Note 90/84/SILN.
Mason, W. L., Mannaro, P. M. & White, I. M. S. (1986). Growth and root development in cuttings and transplants of Sitka spruce 3 years after planting. *Scottish Forestry*, 40, 276–84.
Mullin, T. J. & Park, Y. S. (1992). Estimating genetic gains from alternative breeding strategies for clonal forestry. *Canadian Journal of Forest Research*, 22, 14–23.
Nambiar, E. K. S. (1985). Increasing forest productivity through genetic improvement of nutritional characteristics. In *Forest Potentials, Productivity and Value*, ed. R. Ballard, P. Farnum, G. A. Ritchie & J. K. Winjum, pp. 191–215. Weyerhaeuser Co., Tacoma, WA.

Quine, C. P., Burnand, A. C., Coutts, M. P. & Reynard, B. R. (1991). Effects of mounds and stumps on the root architecture of Sitka spruce on a peaty gley restocking site. *Forestry*, **64**, 385–401.

Robertson, A. (1991). Centroid of wood density, bole eccentricity, and tree-ring width in relation to vector winds in wave forests. *Canadian Journal of Forest Research*, **21**, 73–82.

Savill, P. S. (1976). The effects of drainage and ploughing of surface water gleys on rooting and windthrow of Sitka spruce in Northern Ireland. *Forestry*, **49**, 133–41.

17

Development of buttresses in rainforest trees: the influence of mechanical stress

A. R. ENNOS

Abstract

An experimental study was performed to test the model of buttress development proposed by Claus Mattheck: that the high rates of growth occurring along the top of the junction between the trunk and the lateral roots are stimulated by the mechanical strains set up by wind forces which are concentrated in these regions. Strain gauges were attached around the base of the trunk and the lateral roots of young rainforest trees which were developing buttresses; strains were measured during simulated wind events when the trunk was bent over using a winch. Results supported Mattheck's theory: strains were concentrated along the top of developing buttresses and on the trunk above them – the places of maximum growth – and were negligible on the sides of the buttresses and at the base of the trunk where no growth was occurring. Buttress development seems to be controlled by the mechanical environment of the tree. Once the root system, in which lateral roots are attached to the subsoil well away from the trunk by sinker roots, has formed the growth of buttresses occurs automatically. This produces a mechanically efficient anchorage system in terms of the carbon investment. Trees which do not form buttresses instead transfer forces into the ground using a tap root or sinkers positioned close to the trunk.

17.1 Introduction

Root buttresses are a characteristic feature of the trees of lowland tropical rainforest. These huge plate-like structures which are formed around the base of the trunk have been the subject of speculation by generations of biologists who have attempted to discover what function they serve and what influences their development. However, even though it has generally been accepted that buttresses – as their name suggests – have a mechanical role (Richards, 1952;

Smith, 1972; Henwood, 1973) there has been no convincing account of exactly how they function. Nor has their pattern of development been explained: wood is progressively laid down along the tops of the lateral roots, most rapidly at their join with the trunk, until a plate-like structure is produced between the root and the trunk. At the same time radial growth of the rest of the lower trunk slows down or stops and it becomes tapered into an inverted cone from the top of the buttresses to the ground (Smith, 1972). Rather than carry out theoretical and experimental work on how buttresses help to support the trunk, most workers have attempted to deduce the function of buttresses by correlating their development with other morphological features of the trees and with the environmental conditions in which they grow (Baker, 1973; Richter, 1984; Lewis, 1988; Warren et al., 1988).

By far the greatest challenge to the anchorage systems of trees are wind forces which load the trunk in bending; the windward side is subjected to tension and the leeward side to compression. These forces must be transmitted into the ground to prevent uprooting (Coutts, 1983). Mattheck (1991, 1993) has suggested that, in trees which develop buttresses, this function is performed by sinker roots which grow out from superficial lateral roots and down into the subsoil well away from the trunk. Being far apart, they are well placed to anchor the tree, the windward roots resisting upward forces and the leeward roots downward forces. In such a system, the beneficial function of buttresses is clear. Without buttressing, the lateral roots would be subjected to large bending forces and might either snap or split at their junction with the trunk (Mattheck, 1991). Buttresses prevent such failure by bracing the roots to the trunk like angle brackets and smoothly transmitting tension to the windward sinkers and compression to the leeward sinkers (Fig. 17.1a). Buttresses therefore act both as props and as guy ropes to support the tree and, together with the sinker roots, make up an efficient anchorage system per unit of carbon invested.

Structural analysis of the anchorage system has also suggested how the characteristic growth pattern of the buttresses may be generated. The 'constant stress hypothesis', first put forward by Metzger (1893), proposes that the cells of the cambium can detect stretching and compression and respond by producing a greater number of thicker-walled wood cells; new wood is therefore added faster in areas which have experienced higher mechanical stresses and strains. The greater quantity of new wood laid down will increase the rigidity and so in turn reduce the stress and the amount of stretching. This pattern of feedback will result in a structure in which material is used optimally; the average stress experienced will be uniform all over the surface of the tree and there will be no weak points. The constant stress hypothesis

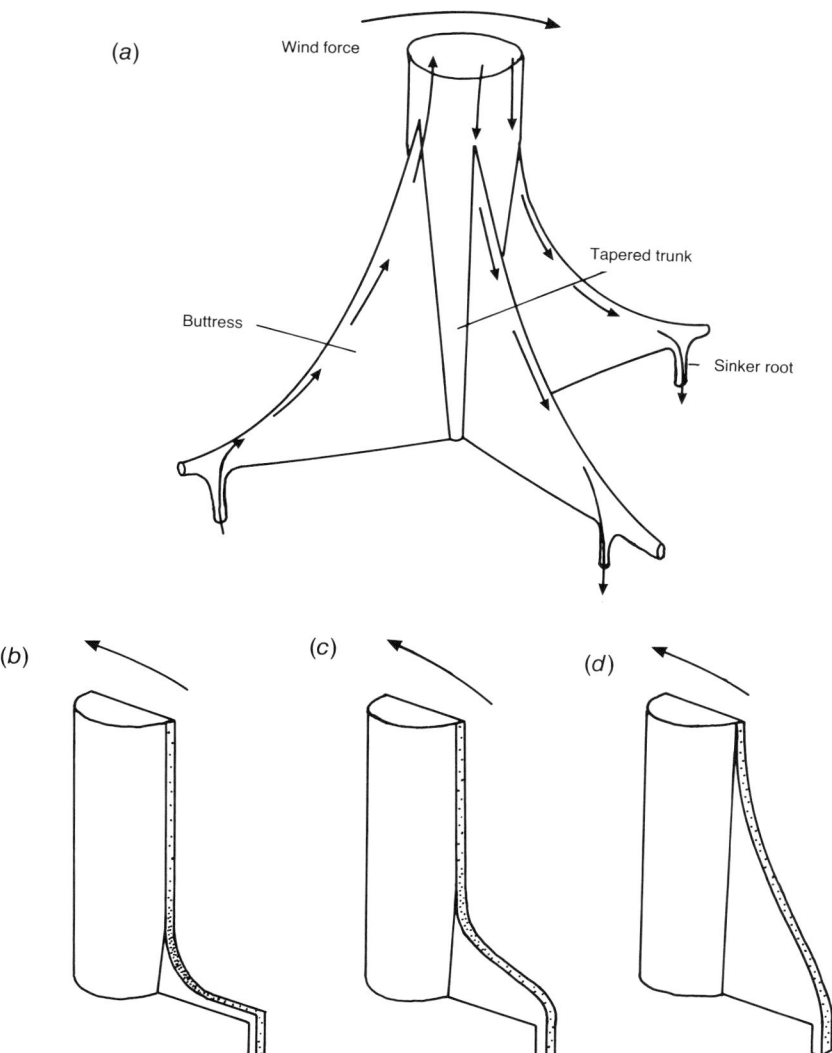

Fig. 17.1. Mattheck's model for the function and development of buttresses. If a tree is pushed over by the wind (a) the bending force is transmitted smoothly to lateral sinker roots by the buttresses. Windward sinkers resist upward forces and the buttress is put in tension while leeward sinkers resist downward forces and the buttresses are put in compression. (b), (c) and (d) show successive stages in Mattheck's simulation of buttress development. When the trunk is pulled over stress is concentrated at the top of the junction between the lateral root and the trunk (stippling). Growth in heavily stressed regions (b) results in the formation of buttresses (c and d) and a great reduction in stress concentrations.

can explain several features of gross tree morphology. First, the diameter of tree trunks tends to increase with the cube root of the distance below the crown, a pattern of growth which will result in equal stressing of the trunk by wind forces all the way down its length (Busgen & Munch, 1931; Larson, 1963). Second, trunk diameter varies with the degree of mechanical stimulation; both tethered trees (Jacobs, 1954; Holbrook & Putz, 1989) and those growing in sheltered sites (Larson, 1965) are taller and thinner than those grown in exposed sites. More recently, Mattheck (1991, 1993) has shown that the hypothesis can also explain many aspects of local growth. Using a finite element computer package to model mechanical stress distributions and to add material on to highly stressed areas, he simulated growth resulting in the expansion of the bases of branches where they join on to the trunk, the growth of calluses around wounds and the swelling of trees above supporting stakes.

When Mattheck modelled the stresses around the base of trees anchored by widely separated sinkers (Mattheck, 1993) he found that stresses would be concentrated at the top of the junction between the trunk and the lateral roots – just where growth is greatest. In contrast, stresses should be very low on the side of the lateral roots and around the base of the trunk. By adding material to highly stressed regions, therefore, he was able to simulate the growth of buttresses.

The aim of this study was to examine the pattern of mechanical stress around the base of young rainforest trees with developing buttresses and so test one aspect of Mattheck's theory of buttress development.

17.2 Materials and methods

Field work was carried out at the Danum Valley field centre, Sabah, Borneo, Malaysia, a 438 km^2 area of primary dipterocarp lowland rainforest. Four young trees of contrasting size and form were chosen for this preliminary study, of which two were specimens of *Ryparosa acuminata*, family Flacourtiaceae (locally named Gewei), a tree which produces pronounced buttresses. The younger of the trees had a basal diameter of 4 cm and had not yet started to develop buttresses on its lateral roots; the older, of basal diameter 17 cm, had well-developed buttresses which had both length and height of 30 cm. A third buttress-bearing tree, locally named Pisang-Pisang, from the family Annonaceae, was also examined. This tree had a basal diameter of 12 cm and had intermediate buttress development. Finally, one specimen of the genus *Eugenia*, family Myrtaceae, a tree which does not develop buttresses, was chosen for comparison. This tree had a basal diameter of 10 cm and the trunk was circular in section.

Plastic foil strain gauges of element length 5 mm (Kiyowa model KFG) were attached to selected points around the base of each tree oriented with the element parallel to the supposed direction of the stresses imposed on the tree by the wind – down the trunk and along the lateral roots. The sites were chosen to test whether stress did vary as predicted by Mattheck around the base of the trunk and the lateral roots. For this reason gauges were positioned in three areas: in a line down the trunk and out along the top of a lateral root; down the trunk beside the lateral root; and along the side of the lateral root or buttress. The exact position and orientation of each gauge is shown in Fig. 17.2a–d.

To attach the gauges, the outermost layer of moss and bark was first shaved away using a mortice chisel and the strain gauge was glued to the smooth exposed surface using 'Loctite 3' superglue. The gauge was then left to dry for over 15 min. To record strain, the wires of each gauge were attached to a battery-powered Vishay strain bridge which operated in quarter bridge mode. This apparatus allowed static readings to be taken of tensile and compressive strain of up to 10 000 μstrain, equal to extension and contraction of 1%.

Each tree was then subjected to a simulated wind force by pulling it sideways at a height of 2 m using a nylon rope wound around the trunk and anchored to the base of another tree 15 m away. Trees were pulled either towards or away from the side with strain gauges attached, which would result in the strain gauges being loaded in compression and tension respectively. Force was applied by hand, using a seven-pulley car engine winch (Halfords) which magnified applied forces by a factor of 7. The trees were loaded and unloaded by pulling on the winch with a spring balance capable of measuring up to 200 N and holding the rope taut in a cleat attached to the winch. The strains around each tree were measured a single gauge at a time as each tree was loaded and unloaded incrementally in five stages over a period of 5–10 min, to a maximum force of 910 N for the three thickest trees and 350 N for the young *Ryparosa* tree. These forces would create maximum bending moments about the base of each tree of 1820 N m and 700 N m, respectively.

17.3 Results

The system seemed to obey Hooke's law, each strain gauge recording a more-or-less linear response to an applied force. No hysteresis was observed and the strain returned to zero after unloading. The strains recorded for each tree decreased with the thickness and consequent rigidity of the trunk, ranging from a maximum strain of 0.22% recorded in the young *Ryparosa* tree to

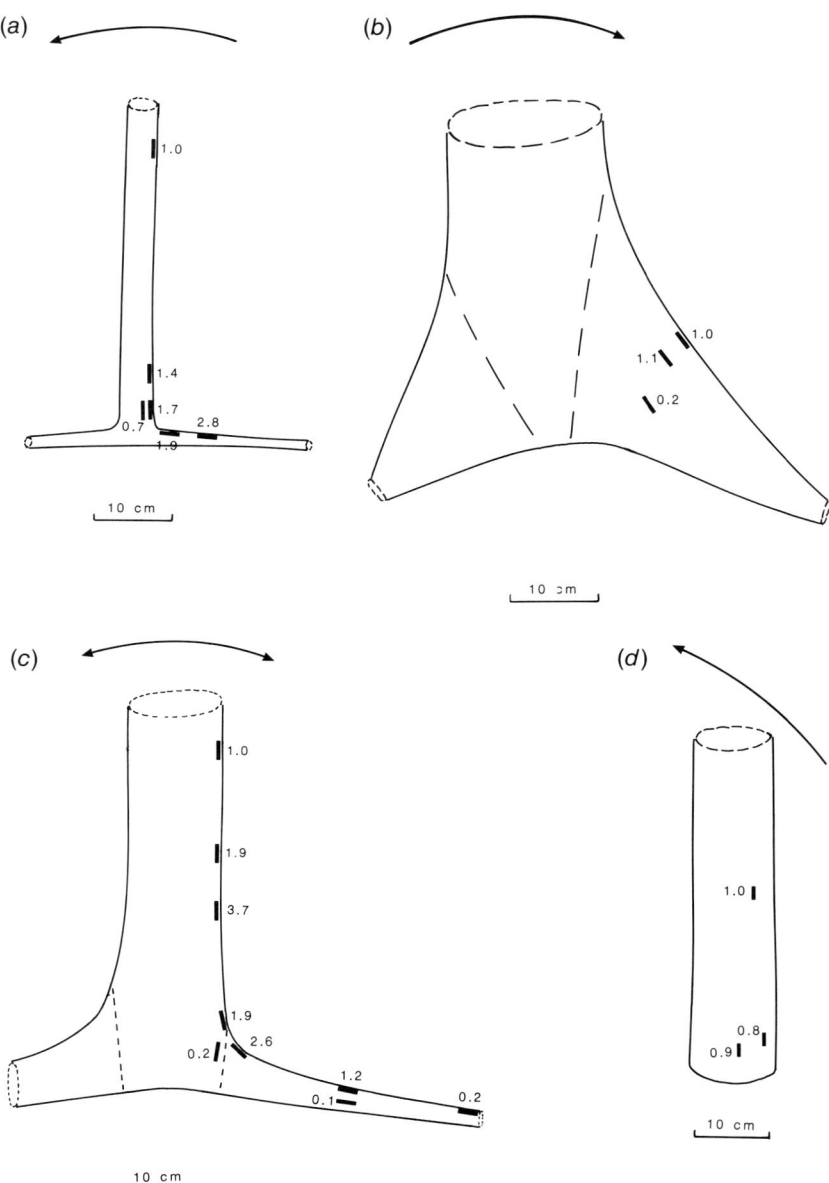

Fig. 17.2. Relative values for strain around the base of trees when pulled over in the direction shown by the arrows. (*a*) A young sapling of *Ryparosa acuminata*. (*b*) An older tree of the same species. (*c*) A specimen of Pisang-Pisang (family Annonaceae) and (*d*) *Eugenia*, a tree which does not develop buttresses. In trees (*a*)–(*c*), which develop buttresses strain is concentrated around the junction between the top of the lateral root and the trunk and along the top of the developing buttress. In tree (*d*), which does not develop buttresses, there is no such concentration of strain.

only 0.008% in the older one. Values for the strain around the base of each tree were similar but of opposite sign when it was pulled towards or away from the strain gauges.

The magnitudes of these strains are given in Fig. 17.2a–d, relative to a point up the trunk well above the lateral roots, which is given the value 1. The exception was the older *Ryparosa* tree on which gauges were placed only on the buttress (Fig. 17.2b). It can be seen that, in trees which are developing buttresses (Fig. 17.2a and c), the relative strain increased towards the junction between the trunk and the lateral root, just as Mattheck had predicted, and decreased at a distance from this junction, up the trunk and out along the lateral root. Strains were also very much lower in areas away from the predicted force flow: at the bottom of the trunk adjacent to the buttress and along the side of the developing buttress. The distribution of strain (and therefore applied stresses) resulting from wind forces was therefore qualitatively just what Mattheck had predicted, being highest in the areas which were growing fastest and lowest where growth was slow. A similar result was seen in the buttress formed by the older *Ryparosa* tree: strains were greatest around the top of the buttress where growth is fast and much lower down its sides where growth is very slow. In *Eugenia* (Fig. 17.2d), which was not developing buttresses, there was an even distribution of strain around the base of the stem.

17.4 Discussion

The strain gauge technique seemed to work well. It was easy to glue the gauges to these thin-barked trees and the readings of strain which were obtained were very stable. It may be possible to monitor strains due to wind forces over an extended period if a data logger is used, though changes in temperature might cause drifting of the zero reading. The main problems with the technique are the large numbers of strain gauges needed to obtain the overall strain distribution and the difficulty in deciding the principal strain axis. The former problem can be overcome to some extent by using a switching box to monitor each strain gauge in turn and the latter by using strain gauges with two orthogonal elements.

The results of the tests presented here showed excellent qualitative agreement with the predictions of Mattheck. In trees which were developing buttresses, strains (and hence presumably stresses) were concentrated around the top of the junction between the trunk and the developing buttresses, precisely where predicted. The distribution of strains also provides evidence supporting the tenets of the constant stress hypothesis; if growth continued to occur in areas of high strain buttresses would continue to develop. Hence this simple growth rule is capable of explaining even the apparently anomalous growth pattern of buttresses. It seems likely, therefore, that buttresses grow as a result of the early development and form of the root system. However, the cambium of different species may show *quantitatively* different sensitivity to mechanical stress: this may explain why buttress size and form vary even between species with similar root systems.

In one respect the results did differ form Mattheck's predictions. He suggested that on the leeward side of the tree the lateral roots would lean on the ground and so transmit compressive stresses to the ground close to the trunk. On this side of the tree the lateral roots should be much less stressed than those on the windward side. However, in this study there was no difference between the stresses on the two sides. This may be due to the soil conditions. Mattheck was simulating the growth of buttresses in poplars which grow on firm soils capable of resisting a strong downward force. In contrast, the lateral roots of these rainforest trees were growing through a very weak layer of humus which was easily compressed. Even on the leeward side, therefore, most stress would be transmitted out to the sinker roots. The difference in soil may be the reason why buttresses are much more asymmetrically arranged around poplars than around rainforest trees (Senn, 1923). In poplars, buttress growth will be stimulated only on the windward side of the prevailing wind, whereas in rainforest trees development should be stimulated on both windward and leeward sides.

The results of these tests point to two further possible lines of research. First, the measurement of strains may be combined with the local measurement of growth rates (preferably in temperate tree species which display prominent growth rings) both to provide a general test of the constant stress hypothesis and to calibrate the magnitude of the effect stress has on cambial growth. Second, a fuller study can be carried out to test Mattheck's theories on the function and development of buttresses. Results from strain gauge measurements on buttressed and unbuttressed trees can be combined with examination of root system morphology, the soil in which the trees are growing and, with the mechanics of anchorage, to determine why buttresses are confined to certain species growing in certain ecological conditions and geographical regions.

Acknowledgements

I would like to thank Ian Summers for his help in using strain gauges and the Department of Engineering, University of Manchester, for the loan of the

Vishay strain bridge. This work was supported by a travel grant from the Royal Society's South East Asia Fund.

References

Baker, H. G. (1973). Appendix. In A structural model of forces in buttressed tropical rain forest trees, by K. Henwood. *Biotropica*, **5**, 83–9.

Busgen, M. & Munch, E. (1931). *The Structure and Life of Forest Trees*. Wiley, New York.

Coutts, M. P. (1983). Root architecture and tree stability. *Plant and Soil*, **71**, 171–88.

Henwood, K. (1973). A structural model of forces in buttressed tropical rain forest trees. *Biotropica*, **5**, 83–9.

Holbrook, N. M. & Putz, F. E. (1989). Influence of neighbours on tree form: effects of lateral shade and prevention of sway on the allometry of *Liquidambar styraciflua* (sweet gum). *American Journal of Botany*, **76**, 1740–9.

Jacobs, M. R. (1954). The effect of wind sway on the form and development of *Pinus radiata* Don. *Australian Journal of Botany*, **2**, 35–51.

Larson, P. R. (1963). Stem form development of forest trees. *Forest Science*, Monograph 5.

Larson, P. R. (1965). Stem form of young *Larix* as influenced by wind and pruning. *Forest Science*, **11**, 412–24.

Lewis, A. R. (1988). Buttress arrangement in *Pterocarpus officinalis* (Fabaceae): effects of crown asymmetry and wind. *Biotropica*, **20**, 280–5.

Mattheck, C. (1991). *Trees: The Mechanical Design*. Springer-Verlag, Berlin.

Mattheck, C. (1993). *Design in der Natur: Der Baum als Lehrmeister*. Rombach-Verlag, Freiburg.

Metzger, K. (1893). Der Wind als massgeblicher Faktor für das Wachstum der Baume. *Mundener Forstliche Hefte*, **3**, 35–86.

Richards, P. W. (1952). *The Tropical Rainforest*. Cambridge University Press, Cambridge.

Richter, W. (1984). A structural approach to the function of buttresses of *Quararibea asterolepis*. *Ecology*, **65**, 1429–35.

Senn, G. (1923). Über die Ursachen der Brettwurzelbildung bei der Pyramiden-Pappel. *Verhandlungen der Naturforschenden Gesellschaft in Basel*, **35**, 405–35.

Smith, A. P. (1972). Buttressing of tropical trees: a descriptive model and new hypothesis. *American Naturalist*, **106**, 32–45.

Warren, S. D., Black, H. L., Eastmond, D. A. & Whaley, W. H. (1988). Structural function of buttresses of *Tachigalia versicolor*. *Ecology*, **69**, 532–6.

Part IV

Impacts of wind on forests and ecology

18

Hurricane disturbance regimes in temperate and tropical forest ecosystems

D. R. FOSTER and E. R. BOOSE

Abstract

We provide an overview of hurricane disturbance regimes in the north-eastern United States and the Caribbean, with a focus on ecological effects on temperate and tropical forests. Hurricanes in tropical regions occur with a higher frequency and reach higher intensity levels than in temperate regions. Slower movement of hurricanes in the tropics exposes forests to longer periods of damaging winds from a broader range of wind directions. At the regional to landscape level we are applying models of hurricane meteorology and topographic exposure to reconstruct wind conditions during historically important storms in order to compare these results with observations of forest damage, and to test two hypotheses: (1) regional gradients of hurricane frequency and intensity result from prevailing hurricane tracks and the configuration of coastlines and mountain ranges and (2) landscape gradients in wind exposure result from the interaction of peak wind directions and local topography. At the community level damage is controlled by windspeed, vegetation structure and composition, and site conditions. Damage patterns strongly control subsequent vegetation dynamics: (1) differential species damage determines the initial composition and structure of the vegetation, (2) leaf area establishment is controlled by damage type and forest composition, and (3) microenvironmental conditions and resource distribution are determined by the spatial pattern of residual vegetation. Field studies underline the offentimes low rate of initial mortality following catastrophic storms and the importance of releafing and sprouting in vegetation development. Changes in key ecosystem processes are often slight following even major hurricanes. Studies in Puerto Rico and New England document that nutrient retention was high, nutrient losses were minimal, soil moisture changed little, and minor changes in trace gas fluxes returned rapidly to pre-disturbance levels. Rapid recovery of biotic control of ecosystem processes results from the retention of organic matter on-site and high rates of survival and resprouting by tree species.

18.1 Introduction

Disturbance processes affect natural ecosystems in complex ways and across a range of ecological scales. One productive means of assessing the relative importance and role of a particular disturbance factor is through analysis of the disturbance regime (*sensu* Heinselman, 1973), including its: (1) physical basis and controlling environmental factors; (2) temporal and spatial distribution; (3) direct impact on the biota and site conditions; and (4) indirect effect on the microenvironmental, biological and soil processes controlling ecosystem response. In this review we seek to provide an overview of the physical nature, spatial and temporal scope, and ecological effects of hurricane winds on forest ecosystems, and to highlight areas in need of further research.

There are notable opportunities for increased dialogue between physical and biological scientists regarding the physical basis of hurricane damage. Although tropical storm meteorology is quite well understood, there has been little attempt to link this information with studies of airflow at the stand and landscape level or with field measurements of forest damage. As a result, ecological and forestry studies often make naive assumptions concerning the characteristics of hurricane winds and their potential to damage forests. A clear understanding of the general characteristics of tropical storms is critical to the interpretation of individual wind events and damage patterns at a landscape scale.

The assessment of temporal variation in wind disturbance is equally critical. For coastal and some marine environments meteorologists have compiled long-term analyses of storm activity (e.g. Lamb, 1991). However, with rare exceptions, such as Stephens' (1955) work in central Massachusetts, the temporal context for wind disturbance in inland regions is lacking. Most estimates by ecologists of the long-term importance of hurricanes are based either on extrapolation of short time series or on composites of hurricane tracks, which contain little information concerning storm intensity. We clearly need additional tools for estimating the return interval of tropical storms of varying intensity.

It is widely assumed that spatial variation in wind damage may explain vegetation patterns and ecosystem characteristics at a landscape to regional scale (Odum, 1970; Committee on Global Change, 1988). For example, sites that are exposed to severe winds as a result of aspect or physiographic position may differ in forest structure or composition from adjacent sheltered lowlands (Cline & Spurr, 1942; Foster, 1988a). Similarly, regional gradients in hurricane frequency may alter climatically controlled vegetation gradients (Smith, 1946). Despite the potential importance of variation in wind at these spatial scales the subject is seldom addressed in ecological studies.

In the following overview we do not presume to fill all of these gaps, but rather to explore the relevant information for tropical and temperate regions. We begin with an overview of hurricane meteorology, especially characteristics that are relevant to ecological studies and to the development of modelling approaches to explore the temporal and spatial distribution of these storms. We then address regional and landscape-level considerations and elaborate on preliminary results from simple meteorological and landscape models. In terms of biotic response we review recent information on community-level damage and vegetation development following catastrophic storms. This overview of stand dynamics provides the context for interpreting ecosystem responses. The coverage in this review is admittedly selective and incomplete, reflecting both the contents of other chapters in this volume and the current state of knowledge.

18.2 Hurricane meteorology

Hurricanes occur in all tropical oceans of the world except the South Atlantic and the eastern South Pacific. While the processes that control hurricane formation are not fully understood, meteorologists have identified several necessary conditions, including very warm sea surface temperatures (≥ 26 °C); an atmosphere free of vertical wind shear, temperature inversions, or dry layers that would inhibit developing cloud towers; and an external starting mechanism or energy source. Once formed, hurricanes often follow a parabolic path, drifting westward and slightly poleward with the tropical trade winds, then recurving eastward as they come under the influence of the prevailing westerlies of temperate latitudes (Fig. 18.1); however, there are many exceptions to this general pattern. Hurricanes that do follow a parabolic path over the oceans frequently become vigorous extratropical storms and may reach the polar oceans (Dunn & Miller, 1964; Simpson & Riehl, 1981; Anthes, 1982).

For the North Atlantic (including the Caribbean and Gulf of Mexico), sufficient data have been recorded to analyse the frequency, intensity and tracks of hurricanes over the last 120 years. As the hurricane season (June to November) progresses, there are regular shifts in the principal areas of storm formation and in the pattern of storm tracks. Hurricanes are most common

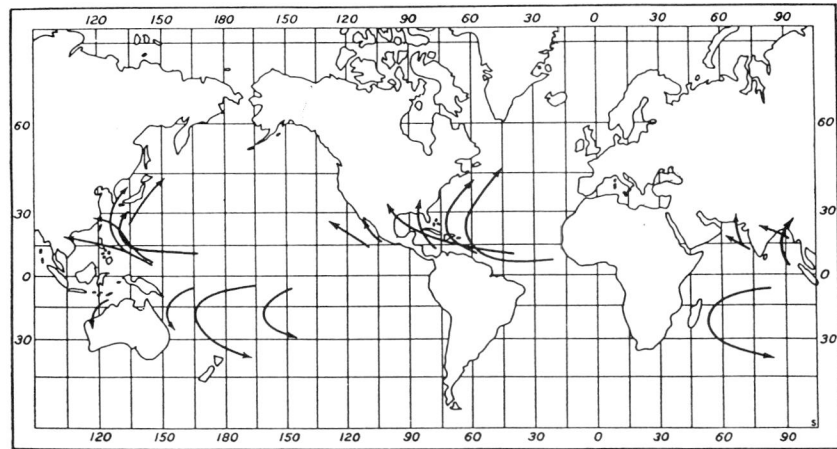

Fig. 18.1. Areas where hurricanes form. From Dunn & Miller (1964).

in September when ocean temperatures reach a maximum. Over the last 100 years (1886–1986) there have been an average of five hurricanes (sustained wind speed ≥ 32 m s^{-1}) per year (Neumann et al., 1987). The frequency of severe hurricanes (sustained wind speed ≥ 50 m s^{-1}) varies over multidecadal cycles, and has recently been shown to be closely linked to summer rainfall amounts in the Western Sahel region of West Africa; these rainfall amounts appear to be linked in turn to natural global-scale climate variation in both tropical and extratropical regions (Fig. 18.2; Gray, 1990). The potential effects of human-induced global warming on hurricanes may include increases in hurricane frequency and intensity, and expanded areas of hurricane formation and impact (Emanuel, 1987; Hobgood & Cerveny, 1988; O'Brien et al., 1992).

The mature hurricane is a great revolving vortex (counterclockwise in the northern hemisphere, clockwise in the southern) that reaches upward to an altitude of at least 7–8 km, and may extend outward to a radius of 1000 km. The eye at the centre of the hurricane is characterised by very low pressure and calm or light winds. Surrounding the eye, typically at a radius of 30 km (but variable from 10 to 80 km), is the eyewall, an area of intense convection where the highest windspeeds and heaviest precipitation normally occur. Sustained surface windspeeds (1 min average) as high as 88 m s^{-1} have been recorded, though in most hurricanes the maximum sustained speed is closer to 50 m s^{-1}. Outside the eyewall windspeeds normally decrease as a negative exponential function of the radius, so that sustained hurricane-force winds are rarely found beyond a radius of 100 km (Frank, 1977; Simpson & Riehl, 1981; Anthes, 1982).

Hurricane disturbance regimes in forests 309

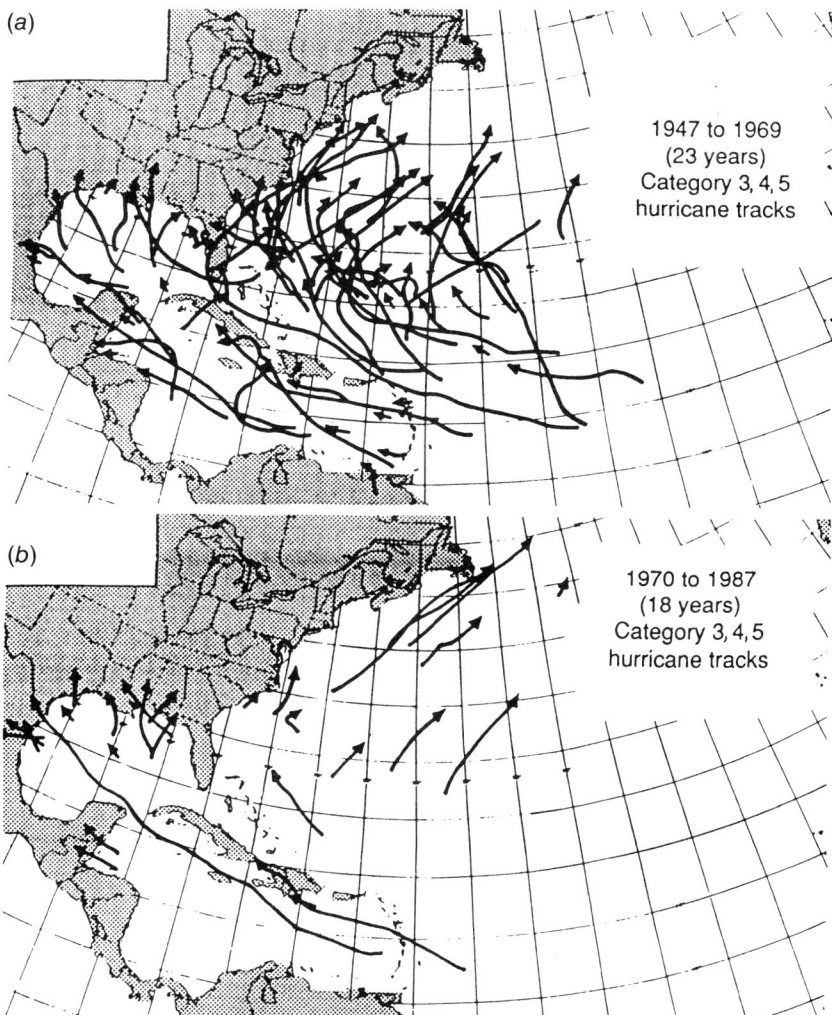

Fig. 18.2. Tracks of major hurricanes (sustained wind speed ≥ 50 m s^{-1}) in the North Atlantic for (*a*) the 23-year period 1947–69 when rainfall in West Africa was above average and (*b*) the 18-year period 1970–87 when it was below average. From Gray (1990).

On a regional scale, the area of widespread severe wind damage to forests rarely extends more than 100 km to either side of the storm track. Damage is often concentrated in areas directly impacted by the eyewall, and the worst damage normally occurs on the right side of the storm (in the northern hemisphere). On the right side the wind velocity relative to the ground is roughly equal to the sum of the rotational velocity of the wind around the centre and the forward velocity of the storm, while on the left it is roughly equal to the difference (Simpson & Riehl, 1981). In a fast-moving hurricane such as the 1938 New England hurricane (forward speed 25 m s^{-1}) virtually all severe damage is located to the right of the storm track (Boose et al., 1994). Note that the negative exponential decrease in sustained windspeed along radial lines means that the highest temporal and spatial gradients in windspeed and direction occur in the immediate vicinity of the eyewall. Similarly the highest spatial gradients in maximum windspeed during the storm's passage occur along the tracks of the right and left edges of the eyewall.

At a landscape or community level, finer patterns of damage may be created by smaller-scale meteorological processes. Hurricane winds within 500 m of the surface are characterised by gusts caused by turbulent eddies. The degree of gustiness is a function of height above the surface and of surface roughness. Peak gust speeds at standard height (10 m) over smooth ground are typically 30% higher than the sustained windspeed, and as much as 50% higher over rough terrain. Recent studies suggest that hurricane gusts typically recur at intervals of several minutes, accelerating for 5–10 s then slowing abruptly for 1–2 s (Atkinson & Holliday, 1977; Simpson & Riehl, 1981). It has been suggested that much of the damage to forests is caused by wind gusts, especially at higher windspeeds; slower winds may cause fatigue damage over time (Fujita, 1971). In addition, patches of extreme damage may be caused by convective phenomena such as intense thunderstorm cells or tornadoes embedded in the larger hurricane. Tornadoes are most likely to occur upon landfall; roughly 25% of the hurricanes that make landfall over the United States spawn tornadoes (Anthes, 1982).

Hurricane precipitation is most intense in the eyewall and in the spiral rainbands that normally extend outward from the eye. Rainfall rates are typically greater on the right side of the storm, and decrease logarithmically as one moves away from the centre; but there are many exceptions. In general rainfall patterns are much harder to predict than wind patterns. The actual rainfall at a particular site depends on the location relative to the storm track, the distribution of rainfall around the storm, the speed of storm movement, and surrounding topographic features. Peak rainfall for a moderate hurricane averages 35 cm (Simpson & Riehl, 1981; Anthes, 1982).

Hurricanes always weaken upon landfall, because the underlying energy source – warm ocean water – is removed. Such weakening is hastened in the vicinity of large mountain ranges because of increased surface friction. Upon landfall, windspeeds typically fall below hurricane levels in the first day. Rainfall rates may continue or even increase for several days. Hurricanes that pass over short stretches of land (e.g. small islands) frequently regain their former intensity as they move back out to sea. Storms that move over the colder temperate oceans weaken, again because the energy/heat source is removed. However, if conditions are right, the decaying hurricane in the temperate zone (over land or water) may regenerate as a vigorous extratropical cyclone (Simpson & Riehl, 1981).

There may be important differences in the hurricane disturbance regimes of tropical and temperate forests in hurricane-prone regions:

1. Hurricane frequency may be higher in the tropics. Most North Atlantic hurricanes, for example, never reach temperate land areas (Neumann *et al.*, 1987).
2. Hurricane intensity may reach a higher level in the tropics. Only there do storms attain the highest severity (sustained wind speed ≥ 70 m s^{-1}). Storms reaching temperate zones are generally moderated by colder ocean waters, an increase in the Coriolis force with latitude, and (frequently) the influx of cool, dry air into the storm (Byers, 1974; Simpson & Riehl, 1981).
3. Hurricanes in the tropics generally move more slowly than hurricanes in temperate zones. As a result, tropical sites may be subjected to longer periods of damaging winds and substantially greater rainfall from a single storm. In the faster-moving hurricanes of the temperate zone the area of highest winds is frequently shifted to the right side of the storm (see above), increasing the surface windspeed on that side but reducing the area of potential destruction and reducing the range of damaging wind directions.

18.3 Regional and landscape-level effects

The patterns of damage created by hurricane winds in a forested landscape are complex and may appear random or indecipherable (Shaw, 1983). The interaction of meteorological, physiographic and biotic factors that gives rise to such patterns has not been widely studied to date. At regional scales (~500 km) meteorologists have used damage to forests and to human structures as an indication of the strength and direction of the strongest surface winds for selected hurricanes at landfall (e.g. Fujita, 1971; Powell, 1982). At landscape scales (~10 km) ecologists have recognised the importance of hurricanes in establishing forest patterning and initiating vegetation dynamics (Naka, 1982; Denslow, 1985; Bellingham, 1991). Topographic exposure and forest structure and composition have been shown to explain much of the landscape pattern of damage in central New England in the 1938 hurricane

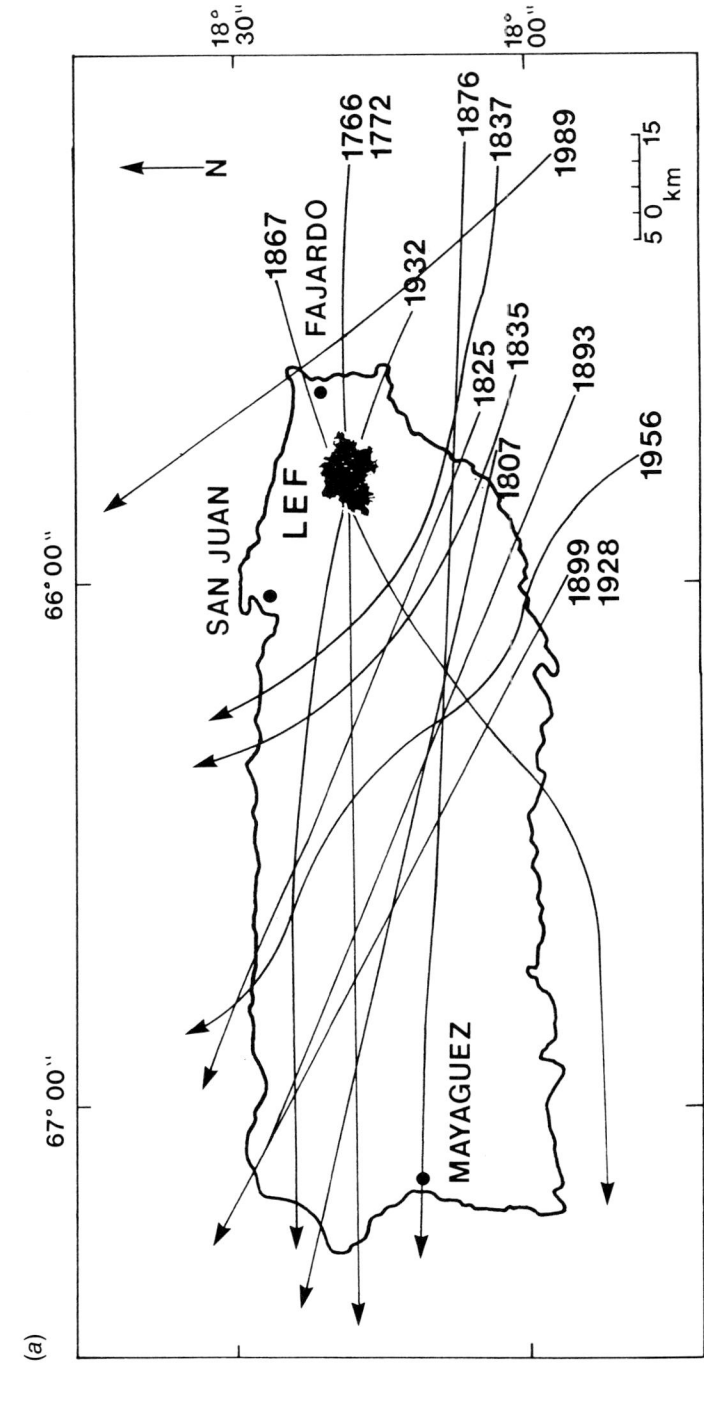

(a)

Hurricane disturbance regimes in forests 313

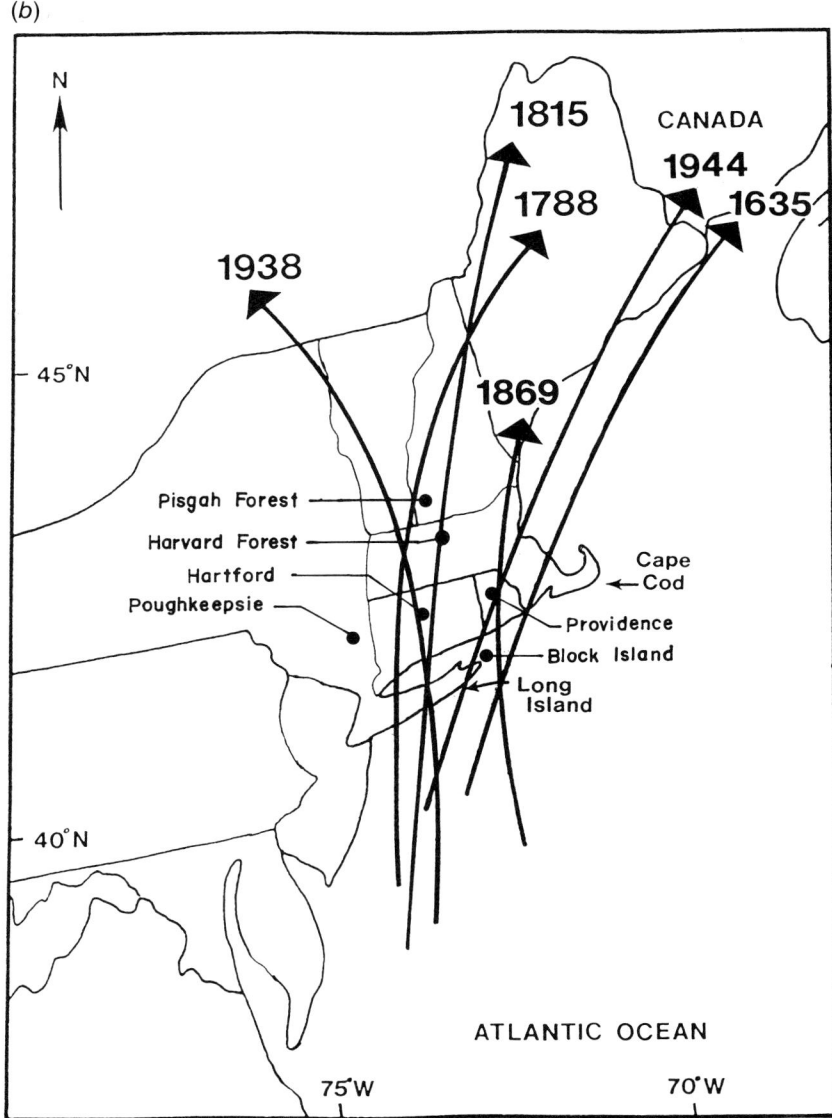

Fig. 18.3 (a) Tracks of severe hurricanes in Puerto Rico since 1700. LEF, Luquillo Experimental Forest; 1989, Hurricane Hugo. (b) Tracks of six severe hurricanes that caused significant forest damage in New England. From Boose et al. (1994).

(Foster & Boose, 1992). However, a deeper understanding of the processes that control patterns of forest damage will require an integration of hurricane meteorology, fluid dynamics, tree biomechanics and forest ecology.

An initial approach to this problem has been tested in a recent study of Hurricane Hugo (1989) in Puerto Rico and the 1938 New England hurricane (Fig. 18.3). The approach combines computer modelling of meteorological and topographic effects with assessment of actual forest damage (Boose *et al.*, 1994). The basic techniques and results to date are summarised below:

(1) Meteorological data are collected that describe the hurricane's location, size and intensity as a function of time. These data serve as input to a simple hurricane reconstruction model (HURRECON; Boose *et al.*, 1994) that reconstructs wind conditions during the storm as a function of location and time. The model approximates the broad-scale surface wind field shared by all hurricanes, and can generate estimates for forested sites where no meteorological data are available. Reconstructions of Hurricane Hugo and the 1938 hurricane were tested against actual meteorological observations, and found to be accurate enough for broad-scale studies of forest damage. Of course, smaller-scale effects such as intense thunderstorm cells or tornadoes that may produce severe local damage cannot be predicted by this technique.

(2) Peak wind directions from the meteorological model serve as input to a simple topographic exposure model (EXPOS; Boose *et al.*, 1994) that predicts areas protected from or exposed to the highest winds on a landscape scale. The exposure model requires an accurate digital elevation map, and classifies points as protected or exposed depending on whether or not they fall within the wind shadow cast by points upwind. The wind shadow is estimated by assuming that the air stream near the ground bends downward no more than a fixed inflection angle from the horizontal as it passes over a height of land. The model is a very simple approximation of a very difficult problem in fluid dynamics; nevertheless, exposure maps generated in this way were found to agree well with landscape-level patterns of damage in areas of very different topographic relief (see below).

(3) Regional-level evidence of damage is compared with predicted gradients in maximum windspeed. Predicted gradients for Hurricane Hugo agreed with scattered reports of damage to towns across eastern Puerto Rico, while predicted gradients for the 1938 hurricane matched detailed surveys of blowndown timber across New England (Fig. 18.4).

(4) Landscape-level patterns of damage for particular sites of interest are assessed through aerial photographs, archival records and/or field surveys. These patterns are analysed in the light of predicted gradients in maximum windspeed, predicted patterns of topographic exposure to the highest winds,

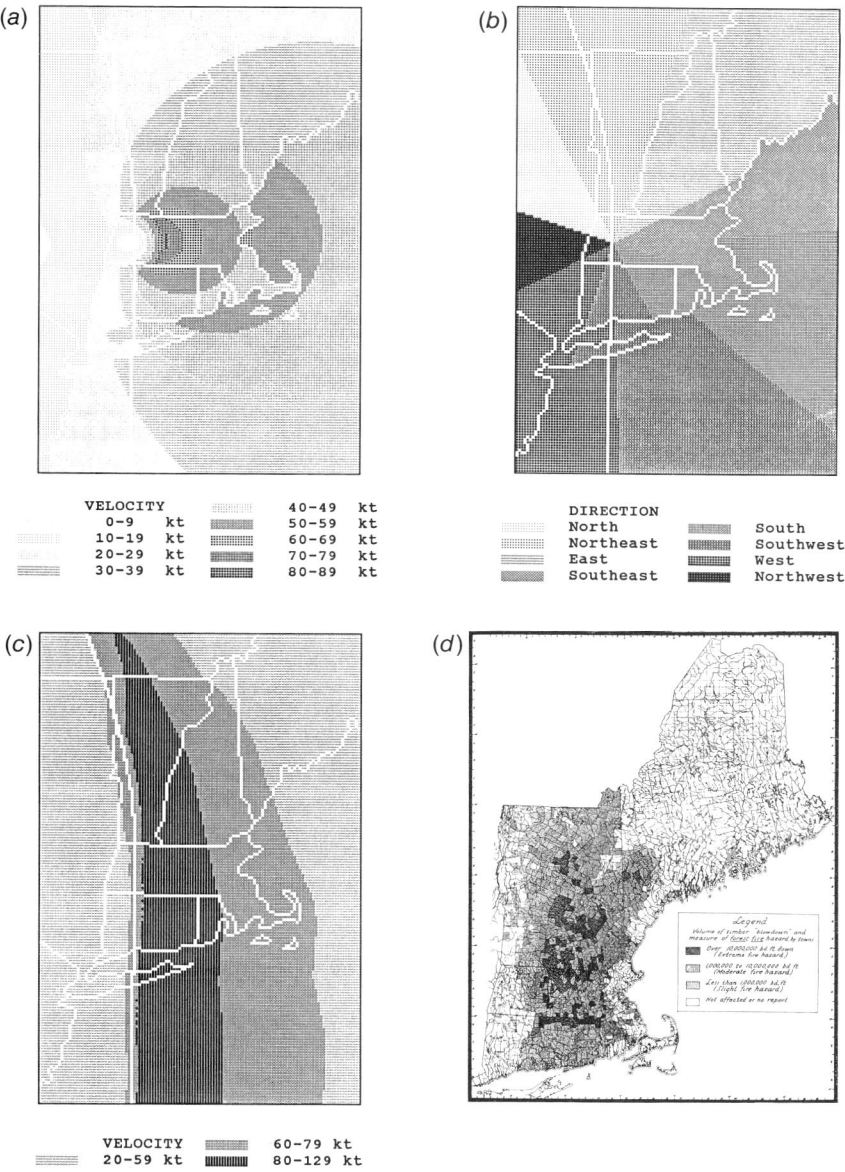

Fig. 18.4. Wind predicted by the HURRECON model and actual damage across New England from the 1938 hurricane. (*a*) Predicted sustained wind velocity and (*b*) predicted sustained wind direction at 1700 EST on 21 September (1 kt = 0.514 m s^{-1}). The model distinguishes land and water sites when calculating sustained wind values. (*c*) Predicted maximum gust velocity during the storm's passage. (*d*) Volume of timber blown down on a town by town basis, compiled by the Northeastern Timber Salvage Administration after the storm (1000 bd. ft. = 2.36 m^3). From Boose *et al.* (1994).

and available data on forest structure and composition. Results from the two hurricanes studied showed good agreement between predicted exposure and landscape-level patterns of damage (Fig. 18.5).

(5) The orientations of windblown trees at particular sites of interest are assessed through aerial photographs, archival records and/or field surveys. These orientations are analysed in the light of the modelled wind conditions for the site and local topographic features. In New England, treefall orientations closely matched predicted peak winds in the 1938 hurricane, while in Puerto Rico there was also fairly good agreement, with some apparent modification of wind direction by the mountainous terrain (Fig. 18.6).

Further analysis of other recent (twentieth-century) hurricanes for which good meteorological and forest damage data are available should improve this basic approach and refine the modelling techniques. An important ecological application of this approach will be to study long-term hurricane disturbance regimes at particular sites by extending the analysis to older storms (pre twentieth-century) to gain a better temporal perspective (even though data are less complete). Spatial variation in such regimes is anticipated. On a regional scale, we expect to find significant gradients of long-term hurricane frequency and intensity caused by prevailing historical hurricane tracks and the spatial configuration of coastlines and mountain ranges. On a landscape scale in hilly or mountainous areas, we expect to find significant gradients of long-term exposure to catastrophic hurricane winds caused by the interaction of prevailing peak wind directions and local topography. We expect that the range of peak wind directions at a particular site is constrained by historical storm tracks and storm velocity, and the location of coastlines and mountain ranges.

Work to date suggests that the following factors help to control regional scale (~500 km) hurricane wind damage to forests:

1. Regional wind velocity gradients resulting from hurricane size, intensity and proximity to the storm track. Note that hurricane size and intensity may change significantly during the storm's passage.
2. Large topographic features such as coastlines and mountain ranges which help to determine how much a hurricane will weaken before it reaches a particular site.
3. Regional vegetation zones resulting from differences in geology, climate and disturbance history. For example, high elevation forests, though often subject to higher wind speeds during a hurricane, may be better adapted to wind and suffer less damage than lower elevation forests.

At finer spatial scales, these regional factors may be modified by smaller-

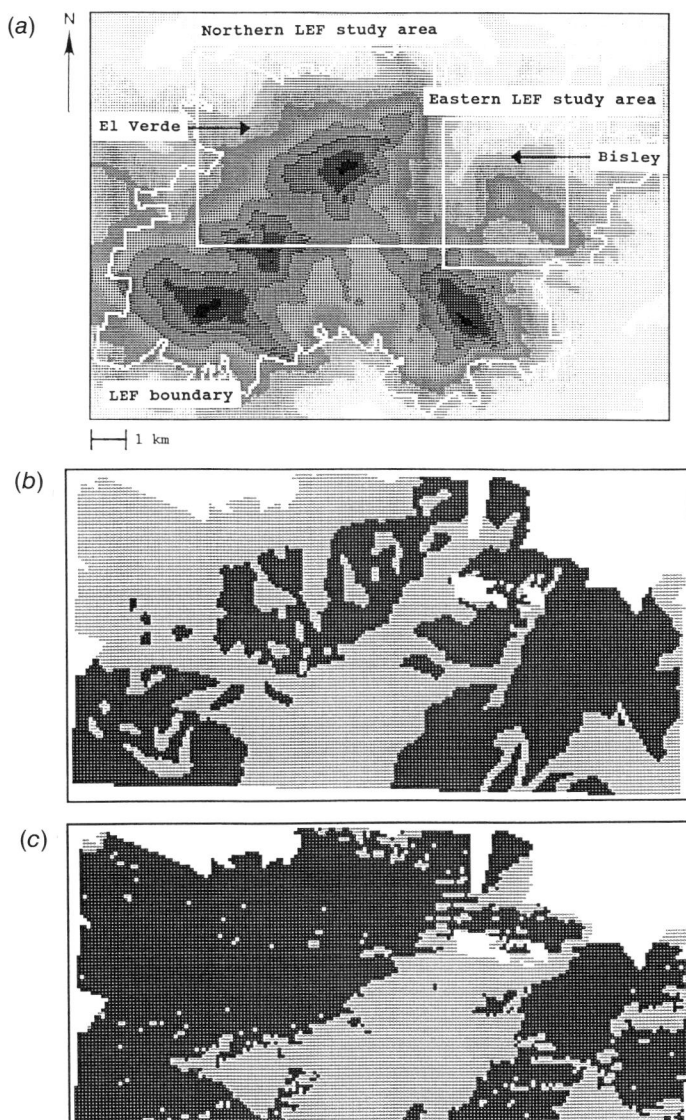

Fig. 18.5. Topographic exposure predicted by the EXPOS model and actual forest damage from Hurricane Hugo in the Luquillo Experimental Forest (LEF), Puerto Rico. (*a*) Study areas in the LEF. Elevation shown in 100 m contours ranges from 50 m to 1075 m a.s.l. (*b*) Forest damage in the Northern LEF study area. Light, <10% windthrow; dark, ≥10% windthrow. (*c*) Predicted exposure in the Northern LEF study area for the peak NNW wind predicted by the HURRECON model. Light, protected; dark, exposed. The lack of damage in exposed areas in the western portion of the study area parallels the predicted gradient in peak windspeed from east to west. From Boose *et al.* (1994).

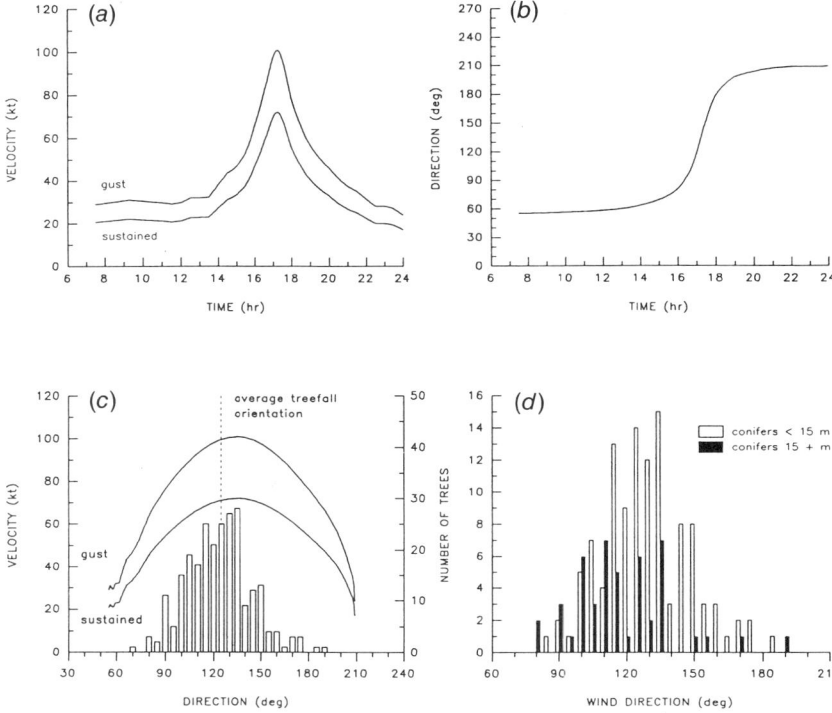

Fig. 18.6. Wind predicted by the HURRECON model and actual treefall orientation at the Harvard Forest in Petersham, Massachusetts, during the 1938 hurricane. (*a*) Predicted wind velocity and (*b*) predicted wind direction as a function of time (EST). (*c*) Lines indicate predicted wind velocity and bars indicate number of windthrown trees in 5-degree classes as a function of wind direction. (*d*) Number of windthrown conifers in 5-degree classes as a function of wind direction, showing taller and shorter stands separately. Average orientation angle for taller stands corresponds to wind directions earlier in the storm. From Boose *et al.* (1994).

scale processes. Factors controlling landscape-level (~10 km) hurricane wind damage to forests appear to include:

1. Regional-scale wind velocity gradients, which may be significant on a landscape scale as well, especially near the storm track. Intense thunderstorm cells or tornadoes may cause local patches or swaths of extreme damage, especially in the vicinity of the eyewall.
2. Variation in topographic exposure to wind. In hilly or mountainous areas topographic exposure may make the difference between little damage and complete destruction for the same forest type. The pattern and relative extent of protected and exposed areas vary with topography (Fig. 18.7).

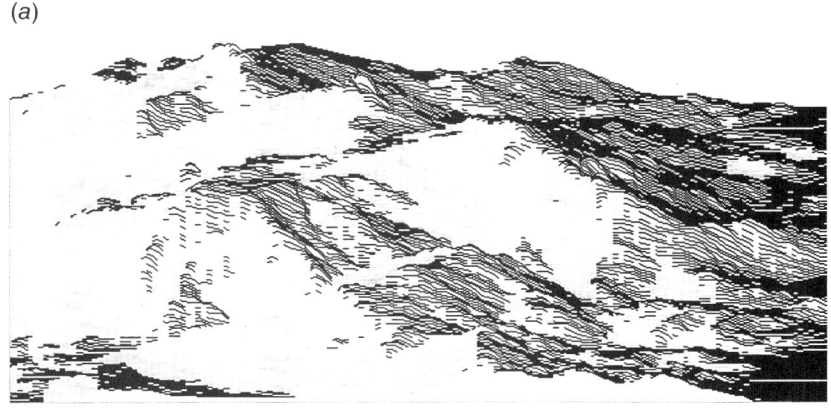

Fig. 18.7. Topographic exposure in contrasting landscapes predicted by the EXPOS model, using peak wind directions from the HURRECON model. Light, protected; dark, exposed. (*a*) Hurricane Hugo and mountainous terrain in the Luquillo Experimental Forest, Puerto Rico. (*b*) The 1938 hurricane and gently rolling terrain in the town of Petersham, Massachusetts. From Boose *et al.* (1994).

Other complex topographic effects at this scale include acceleration of the wind over ridges and summits, and channelling of the wind up valleys and around protuberances.

3. Differential response of individual stands to wind disturbance as a function of species composition and structure, and rooting conditions (see next section). These factors may in turn be strongly influenced by the land-use and natural disturbance history of the stand.

18.4 Community-level effects

Stand dynamics initiated by hurricane damage are dependent on two interrelated processes: initial physical impact of the wind and subsequent regrowth. Damage is determined by windspeed and duration, vegetation structure and composition and site conditions, especially soil characteristics and site stability. Forest regrowth is largely determined by the distribution of surviving vegetation and resulting microenvironmental conditions.

Variation in site exposure, storm conditions and vegetation result in complex damage patterns in forested landscapes. However, when adequate pre-hurricane data are available many of the relationships among these variables are remarkably straightforward (Foster, 1988b; Peterson & Pickett, 1991; Foster & Boose, 1992; Zimmerman et al., 1994).

The most complete effort to assess the relationships between forest characteristics and wind damage was that by Rowlands (1941) following the 1938 hurricane in New England. With samples stratified by wind exposure, soil drainage, stand age, structure and composition and with complete plot inventories before and after the storm it was possible to separate the relative importance of many factors. The results depict simple relationships between vegetation structure and composition and damage for exposed stands on well-drained soils (Foster, 1988b). Forest damage (percentage of windthrown trees) exhibited a positive relationship with stand height and age and negative relationship with density (Fig. 18.8). The slope of this relationship varied considerably with species composition: conifers (white pine, red pine, spruce) were much more susceptible to damage than hardwoods. In general, fast-growing and pioneer species were more susceptible than slower-growing, shade-tolerant species.

Within mixed forests of conifers and hardwoods the wind appeared to operate selectively on individuals according to their specific susceptibility. Thus, scattered white pine within an oak forest were preferentially windthrown (Rowlands, 1941; Foster, 1988b). The vertical distribution of damage also broadened with stand age. In young stands damage was largely confined to overstorey trees, whereas in older stands an increasing percentage of co-dominant, intermediate and understorey trees were damaged. Overall, uprooting of trees accounted for 75–90% of the windthrown stems.

These conclusions are supported by other research on the 1938 hurricane and different storms and forest types (Foster & Boose, 1992). New England studies suggest that fast-growing species are susceptible to damage as a result of tall stature, the concentration of foliage in the upper canopy and light wood (Jensen, 1941; Smith, 1946). Similar results have been reported from

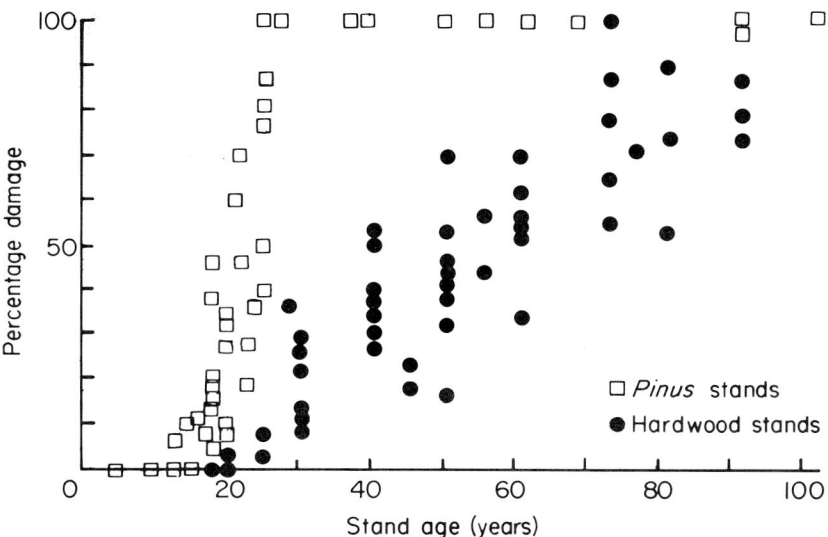

Fig. 18.8. Relationship between hurricane damage (measured as the percentage of individual trees damaged) and stand age for forests in central New England following the 1938 hurricane. Stand types are separated into conifer forests (primarily dominated by *Pinus strobus* and *P. resinosa*) and hardwood forest (consisting of *Quercus, Carya, Acer, Betula* spp.). From Foster (1988b).

tropical sites, although a straightforward relationship based on size or wood strength is not always apparent (Putz, 1983; Basnet, 1992).

The extent and type of damage affect subsequent vegetation dynamics in a number of important ways: (1) differential damage according to species, life history strategy and stratum will determine the initial composition and vertical arrangement of the vegetation; (2) re-establishment of leaf area through refoliation and sprouting is largely controlled by the type of damage suffered and species involved; (3) microenvironmental conditions and resource distribution are determined by the extent and structure of residual vegetation; and (4) site modification, including uprooting, controls many soil processes.

Differential species damage may produce abrupt changes in overstorey composition. For example, the 1938 hurricane transformed broad areas of New England from susceptible white pine forest to more resistant and more shade-tolerant hardwood stands comprised of oak, maple and birch (Clapp, 1938; Brake & Post, 1941). Similar responses occurred in old-growth white pine forests (Fig. 18.9; Henry & Swan, 1974). Additional studies suggest that selective damage to pioneer species by wind may actually advance succession (Jensen, 1941; Smith, 1946; Webb, 1986, 1989).

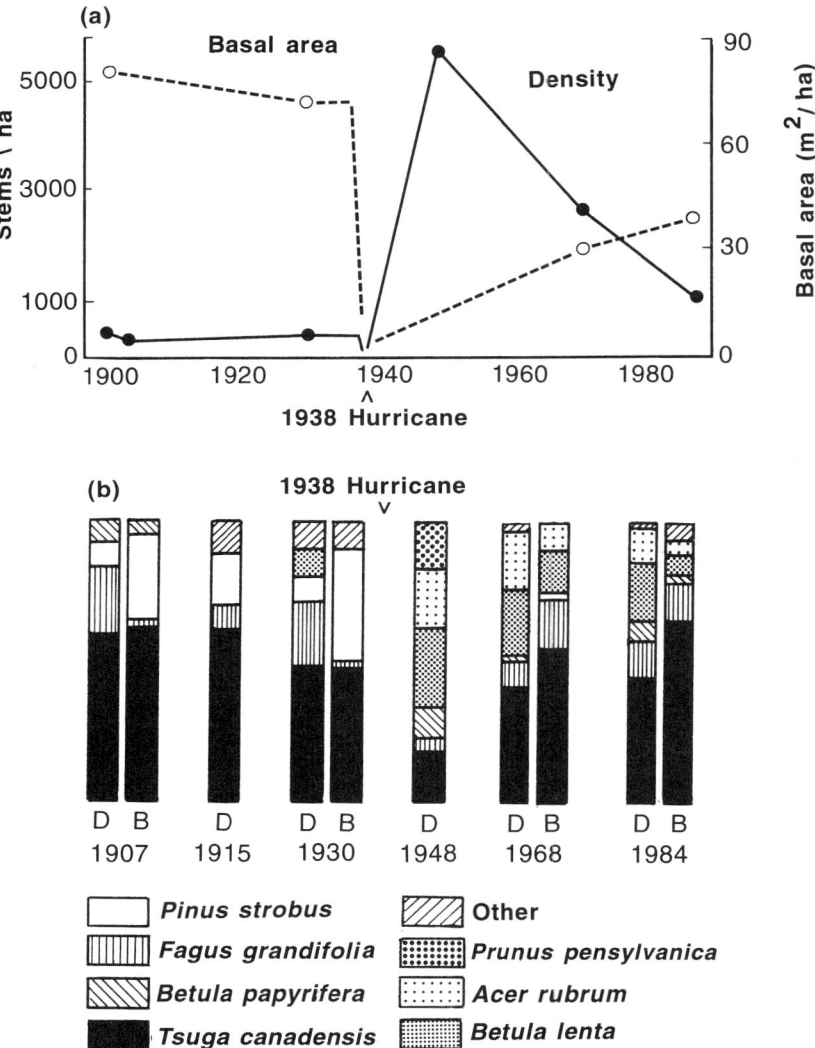

Fig. 18.9. Dynamics of the old growth white pine–hemlock forest on the Harvard Forest Pisgah tract, southwestern New Hampshire, following the 1938 hurricane. Changes from 1907 to 1984 are shown for (*a*) basal area and density and (*b*) relative density (D) and relative basal area (B). The forest stand was approximately 300 years old and undisturbed by human activity when it was completely windthrown by the hurricane. From Foster (1988*a*).

The extent of canopy removal and the type of damage strongly influence resource distribution and microenvironment heterogeneity (Figs. 18.10, 18.11). Of all microenvironmental factors affected, light is undoubtedly the most important in determining vegetation and ecosystem processes as it drives temperature and vapour pressure gradients and is critical to plant success (Sipe, 1990). Light quality, quantity and spatial pattern are determined by the size of canopy openings and rate of regrowth. For example, following Hurricane Hugo light at the seedling level increased from 23 to 404 μmol m^{-2} s^{-1} and was accompanied by a great increase in direct long beam radiation (Walker et al., 1992). However, this light environment was incredibly dynamic: regrowth from damaged stems and upgrowth from understorey vegetation rapidly reduced light levels near the ground. Light patterns are spatially and temporally complex, exhibiting tremendous variation across blowdowns and on a diurnal and seasonal basis (Chazdon & Fletcher, 1984; Walker et al., 1991; Carlton, 1993). As summarised by Sipe (1990) the light gradient in canopy gaps is also broad, spanning the range from full exposure to direct beam irradiance for much of the growing season to infrequent exposure to short sunflecks.

The mound and pit microtopography associated with uprooted trees forms important habitats in the post-hurricane landscape (Armson & Fessendon, 1973; Beatty, 1984; Denslow, 1985; Carlton, 1993). The extent of uprooting is strongly dependent on the long-term moisture status of the site; on wet sites root systems tend to be surficial and prone to uprooting. However, the effect of soil moisture on well-drained uplands is less clear and untested experimentally. For example, the extent of uprooting by an experimental hurricane created under relatively dry soil conditions was remarkably similar to that on similar sites following the 1938 hurricane, which was accompanied by over 20 cm of precipitation (Foster, 1988*b*; Lezberg & Foster, unpublished data). Other factors controlling uprooting include: tree characteristics (size, health, wood density and buttressing) and soil rockiness, drainage and depth (Brake & Post, 1941; J. E. Bertram, personal communication).

Windthrow microtopography may cover substantial ground in wind-prone forests. In New Brunswick, Canada, Lyford & MacLean (1966) identified 600 mounds and pits per acre covering approximately 50% of the surface. In Massachusetts, Stephens (1956) estimated that 15% of an old-growth forest site was windthrow microtopography; approximately 90% of this was produced by four severe hurricanes. Mounds averaged 6 m^2 in area and 60–100 cm in height, whereas pits were 30 cm deep and 2 m^2 in area.

Mound and pit topography increases site heterogeneity by creating soil microenvironments that vary in moisture, nutrients, temperature and stability

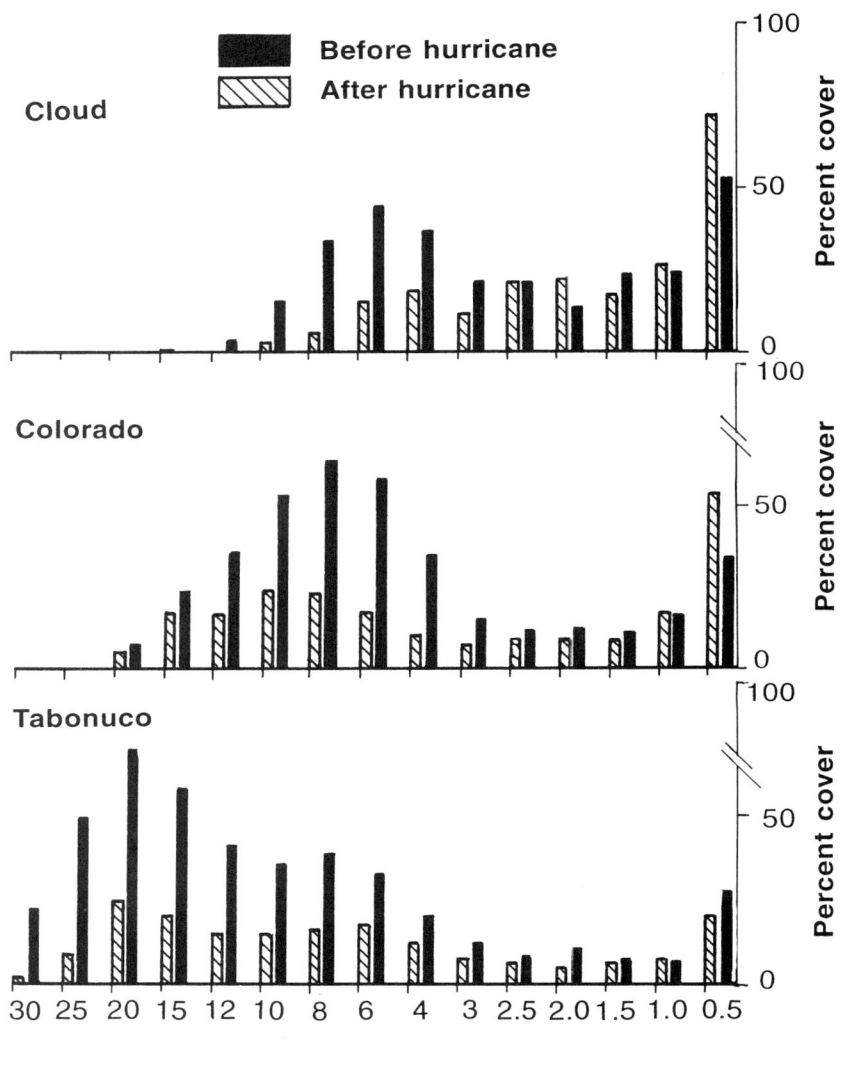

Fig. 18.10. Impact of Hurricane Hugo on the structure of forests in the Luquillo Experimental Forest, Puerto Rico. Vegetation height profiles taken before and after the hurricane for tabonuco (350 m elevation), colorado (750 m elevation) and cloud (1000 m elevation) forests. The horizontal scale shows the total points with cover as a percentage of the total number of grid points in each plot. The vertical scale shows the upper limit of each height interval. From Brokaw & Grear (1991).

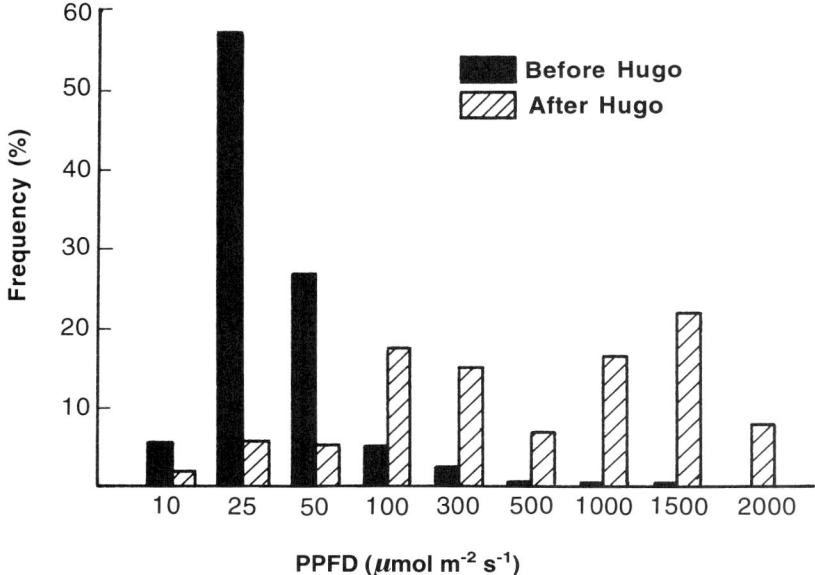

Fig. 18.11. The effect of hurricane damage on the light environment is illustrated by the frequency distribution of photosynthetic photon flux densities (PPFD; 1100–1300 hours local time) at Luquillo Experimental Forest, Puerto Rico, before ($n = 21\,660$ records) and after ($n = 34\,548$ records) Hurricane Hugo. From Walker et al. (1992).

(Peterson et al., 1993). Mounds provide seed beds of exposed soil that may reduce competition and stimulate the germination of some buried seeds (Beatty, 1984; Carlton, 1993). As a consequence, mounds are frequently occupied by pioneer species, such as birches, with small, wind-dispersed seeds.

The successional changes initiated by catastrophic blowdown are dependent on the initial vegetation, type and intensity of damage, and subsequent environmental conditions. Long-term studies following the 1938 hurricane document complex patterns of vegetation development in which advanced regeneration, buried seeds, seedling establishment and vegetative reproduction were variously important for different species at different times (Henry & Swan, 1974; Hibbs, 1983; Foster, 1988b; D. R. Foster, unpublished data). In old-growth white pine and hemlock forests and old-field white pine forests, pin cherry was the only arboreal species utilising the buried seed strategy. Its abundance varied considerably among sites, but was never as great as documented in the clearcutting studies at Hubbard Brook (Marks, 1974). Shade-tolerant saplings of hemlock and beech were released as advanced regeneration. In contrast, many species, including red maple, the birches (*Betula populifolia, B. lenta, B. papyrifera, B. alleghaniensis*) and oaks (*Quercus alba, Q. velutina, Q. borealis*) established as seedlings in the years before or just after the storm. The large numbers of multi-stemmed, bent and malformed trees in the modern landscape also indicate that regrowth and sprouting by survivors have been important processes. The resulting stands have a complex age structure of old stems, even-aged intolerant species and sprouts, and younger saplings of shade-tolerant species.

Recent studies after tornado damage in Pennsylvania, Hurricane Hugo in Puerto Rico and the hurricane experiment in Massachusetts highlight more of the mechanisms of regeneration. In particular, these studies underline the remarkably low mortality immediately following impact (Brokaw & Walker, 1991; Walker, 1991; Tanner et al., 1991; Lezberg & Foster, 1994) and document the importance of reiteration by damaged stems in the rapid recovery of forest structure.

In the hurricane experiment survival, re-leafing and sprouting of damaged trees produced a high leaf area index in the season following damage (Lezberg & Foster, 1994). Survival varied by species and by damage type and decreased through time from approximately 80% during the first growing season to 48% during the second (Fig. 18.12). Species differences are great: nearly 75% of the *Carya* and *Betula alleghaniensis* re-leafed during year 2, in comparison with 30% for *Quercus borealis* and *Pinus strobus*.

Sprouting frequency has increased steadily, with >50% of damaged stems sprouting in year 2 (Fig. 18.13). Again, considerably variation is noted between species; less than 30% of *Betula* stems sprout, in contrast to >60% of *Acer rubrum* and *Fraxinus americana* and nearly 95% of *Carya*. These observations on the experimental hurricane contrast with studies following the 1938 hurricane when pioneer, light-demanding species dominated. Differences may be due in part to variation in the forests studied. However, a plausible alternative is that timber salvage activities associated with the 1938 hurricane altered the trajectory of vegetation recovery (Brake & Post, 1941; NETSA, 1943). Logging following the hurricane was intended to reduce fire hazard and to recover the timber resource; it affected most of the impacted region and produced the largest timber salvage effort in United States history. In addition to cutting live and dead windthrown stems, loggers scarified the soil surface and concentrated slash into windrows for burnings. Collectively, these processes would have decreased the role of surviving vegetation and increased the role of light-seeded pioneer species that establish on bare mineral soil.

The importance of survival and reiteration affects interpretations of the role of hurricane impacts in controlling forest species composition and diversity.

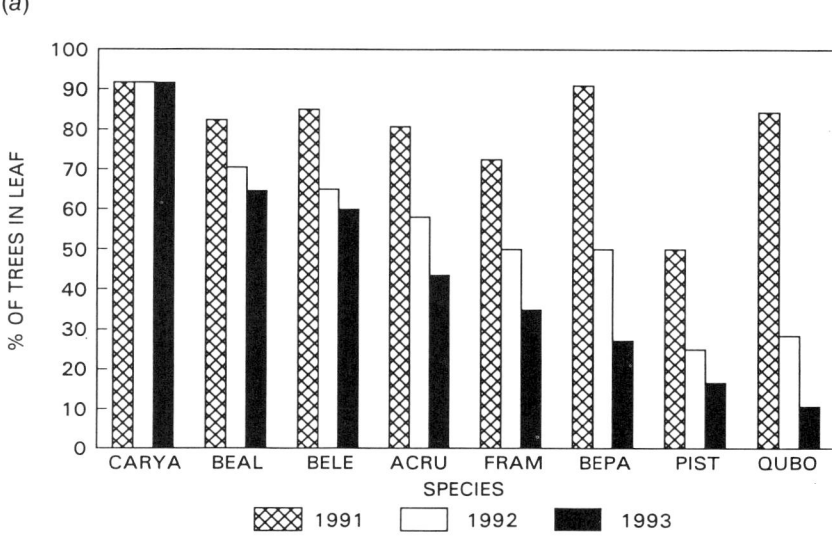

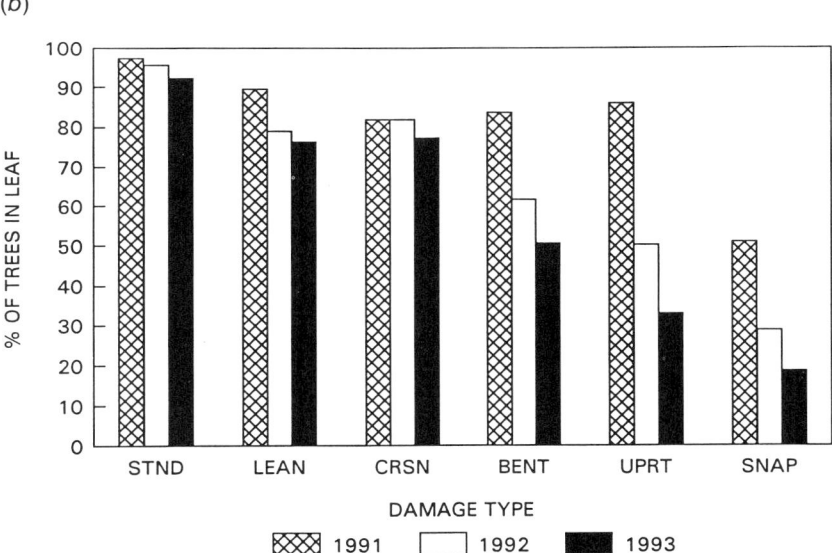

Fig. 18.12. Survival of trees following experimental hurricane on the Harvard Forest, Massachusetts. The percentage of live trees leafing out is shown by (*a*) species and (*b*) major class of damage for 3 years following the blowdown. Species acronyms are the first two letters of the genus and species: *Carya* spp., *Betula alleghaniensis*, *Betula lenta*, *Acer rubrum*, *Betula papyrifera*, *Fraxinus americana*, *Pinus strobus* and *Quercus borealis*. Damage classes: STND, standing; LEAN, leaning; CRSN, crown snap; UPRT, uproot.

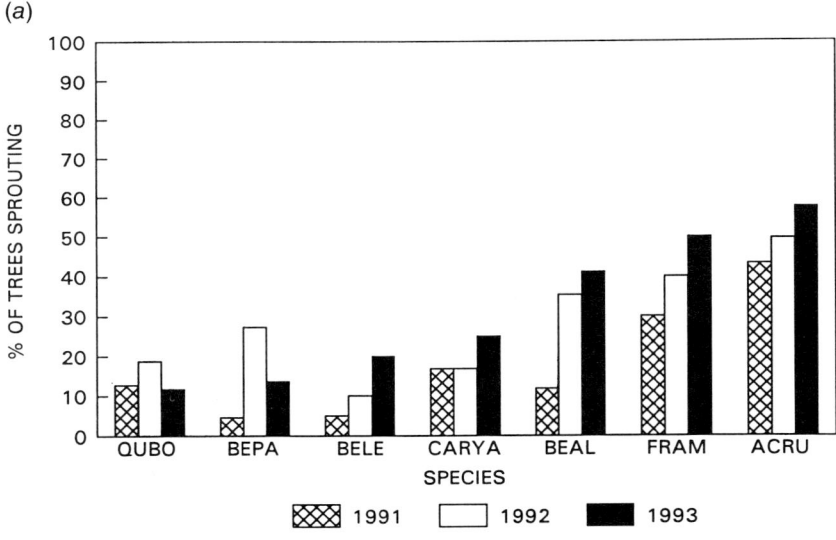
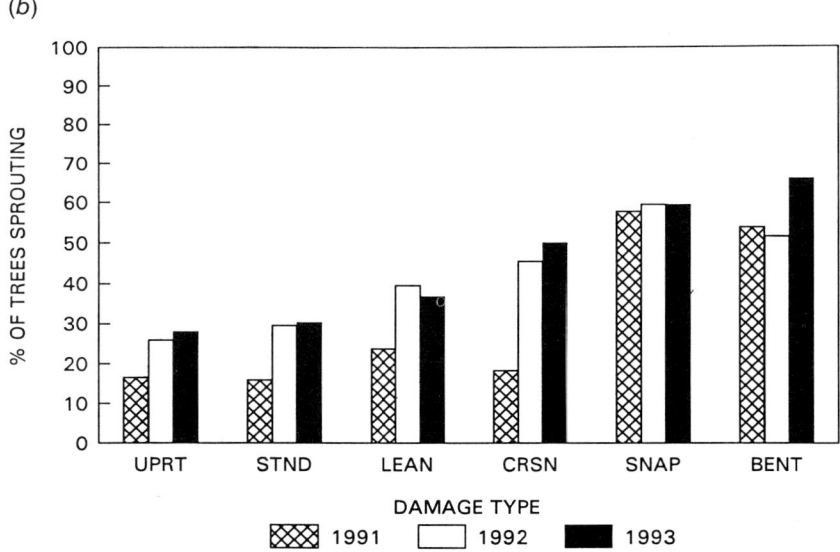

Fig. 18.13. The importance of vegetative reproduction in forest recovery following a hurricane is illustrated by the percentage of sprouts developing following experimental blowdown at the Harvard Forest, Massachusetts. The percentage of individuals developing sprouts is depicted by (*a*) species and (*b*) damage type for 3 years following the blowdown. Abbreviations as in Fig. 18.12.

Infrequent hurricane damage has been linked to the maintenance of high plant and animal diversity in some tropical forests (e.g. Lugo, 1988). This hypothesis is largely based on the intermediate disturbance hypothesis (Denslow, 1980) and is poorly tested. One attempt to assess the relationship between hurricane disturbance, floristic composition and diversity is a study in Puerto Rico. Using a stand development model (FORICO) Doyle (1981) demonstrated that dominance–diversity curves and the mixture of primary and pioneer species on forest plots matched model output based on an intermediate level of hurricane disturbance for eastern Puerto Rico (i.e. one hurricane every 9 years). This study has been cited as confirming the hurricane–diversity hypothesis (Lugo, 1988; Waide & Lugo, 1992; O'Brien et al., 1992). However, conflicting results are generated by models that use multiple tree gaps and spatial interactions among stands (O'Brien et al., 1992). There exists the real possibility that other disturbance factors (e.g. land-use) or plant–site relationships provide an alternative explanation of the forest composition.

In the absence of a definitive study, the relationship between hurricane disturbance and species diversity is speculative. The structural heterogeneity in standing vegetation, distribution of organic matter and microenvironments would appear to accommodate a broad range of species (Bormann et al., 1974; Beatty, 1984). However, many studies suggest that survival and vegetative reproduction are the major mode of regeneration and document very little compositional change following major windstorms (Brokaw & Walker, 1991; Tanner et al., 1991; Whigham et al., 1991; Yih et al., 1991).

18.5 Ecosystem processes

Of interest to ecologists, foresters and land managers is the impact of wind damage on ecosystem processes such as nutrient cycling, hydrology and atmospheric exchange. Of particular concern is the comparison of these impacts with the effects of human impacts, especially the salvage logging that often follows windstorms (Metropolitan District Commission, 1994). Hurricane damage results in many physical changes, including: redistribution of living and dead biomass; increase in organic inputs to the soil surface; reduction in canopy height, vertical stratification and leaf area; and exposure of mineral soil. Each of these direct impacts may alter characteristics of the soil and microenvironment that control ecosystem processes.

Information on ecosystem impacts from hurricanes is limited. Recent data have emerged primarily from the US LTER programme in which the Harvard Forest, Luquillo Experimental Forest and North Inlet projects are focussing on temperate forests in New England, tropical forests in Puerto Rico, and coastal forest and marsh ecosystems in South Carolina. Each offers an extensive, though preliminary view of ecosystem response to hurricane damage.

Severe hurricane impacts result in the immediate loss of most foliage and the collapse of vertical forest structure to boles and branches distributed within 3–4 m of the ground (Brokaw & Grear, 1991; Lodge et al., 1991). Leaf litter inputs resulting from the single event may equal 2–7 times the annual litterfall at tropical sites (Tanner et al., 1991; Walker et al., 1991). Notably, the quality (e.g. nitrogen concentration) of this windblown litter is substantially greater than normal due to the abrupt defoliation before nutrient retranslocation (Whigham et al., 1991; Tanner et al., 1991). Nitrogen contents ranging from ×1 to ×3 and phosphorus concentrations of ×1 to ×5 that of average litter contributed to a massive input of organic matter and nutrients following Hurricane Hugo in Puerto Rico (Lodge & McDowell, 1991). In tropical environments, rapid decomposition of this nutrient-rich litter may increase standing stocks of calcium, potassium, magnesium, nitrogen, phosphorus and manganese (Whigham et al., 1991).

Nutrient dynamics in hurricane-damaged forests have been incompletely documented; however, data from temperate and tropical sites suggest that nutrient retention in soils and biomass is high, nutrient losses are slight, and recovery of ecosystem processes is more rapid than following logging (Steudler et al., 1991; Lodge & McDowell, 1991; Bowden et al., 1993). In the experimental hurricane in which 36% of the mature hardwood trees were uprooted and 32% were snapped or bent, emissions of important trace gases at the soil surface were remarkably similar to control values (Bowden et al., 1993). Emissions of carbon dioxide and methane did not differ from those in the adjacent intact forest, whereas existing low emissions of nitrous oxide were lowered by 78%, but for only one season. Net nitrification was low and quite similar in control and experimental areas. A major conclusion from this experiment is that nutrient-poor temperate forests are resistant to changes in fluxes of carbon and nitrogen gases and are capable of establishing rapid control of ecosystem processes following hurricane disturbance. Despite catastrophic impact on forest structure this damage exerted little impact on nutrient cycling (Bowden et al., 1993).

Slightly more disruption of biogeochemical processes is reported for tropical sites (Tanner et al., 1991; Whigham et al., 1991; Lodge & McDowell, 1991). Following Hurricane Hugo increases were detected in ammonium pools, net nitrogen mineralisation and net nitrification, although values generally peaked within 4 months of the storm and declined rapidly thereafter. Trace-gas emissions included large but relatively short-lived (<1 year)

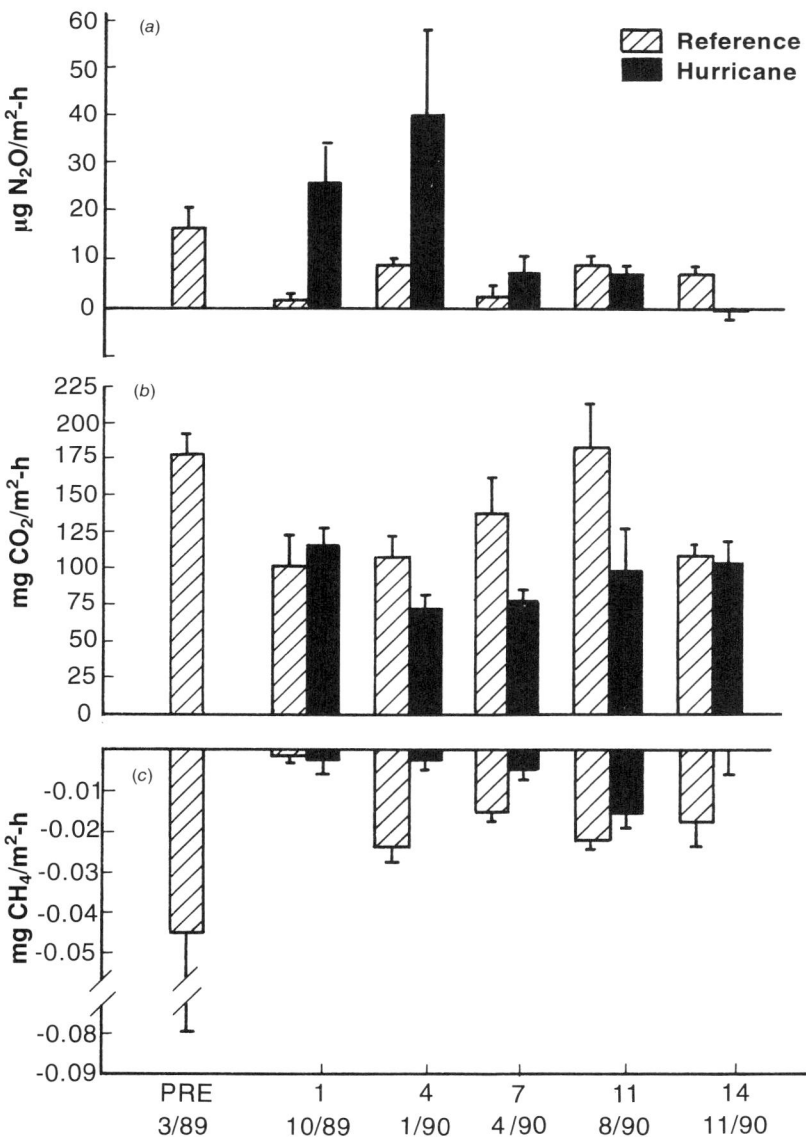

Fig. 18.14. Fluxes of (a) nitrous oxide, (b) carbon dioxide and (c) methane from reference and hurricane-damaged sites along a time sequence following Hurricane Hugo in Puerto Rico. Positive flux values indicate emissions from the soil to the atmosphere and negative values represent uptake by the soil. Flux rates are the means of four chamber measurements and bars show standard errors. From Steudler et al. (1991).

increases in nitrous oxide and decreases in methane and carbon dioxide (Fig. 18.14). The somewhat surprising decline in carbon dioxide flux probably resulted from reduced root respiration, as fine-root abundance has been noted to decrease substantially following hurricane damage (Steudler *et al.*, 1991; Tanner *et al.*, 1991). In comparison, effects upon nitrogen cycling and trace-gas fluxes were greater and more persistent after human disturbance (i.e. clearcutting of adjacent forest in Puerto Rico) than after the hurricane (Steudler *et al.*, 1991).

Numerous biotic mechanisms mitigate nutrient losses and restore pre-hurricane biogeochemical processes in forests (Vitousek *et al.*, 1979). Following windthrow most of the organic matter stays on-site and rates of vegetation survival and resprouting are remarkably high. In the New England hurricane experiment, initial survival of over 70% of the uprooted trees during the first year after damage created an extensive canopy 1–3 m above the ground. Sprouting of damaged trees, coupled with rapid expansion of fern, seedling and shrub layers, produced dense shade (Lezberg & Foster, 1994). As a consequence, soil temperatures were elevated only slightly and no change was detected in soil moisture (Bowden *et al.*, 1993). In Puerto Rico, rapid refoliation led to no detectable change in soil temperature and only a slight increase in soil moisture (Steudler *et al.*, 1991). Rapid establishment of biotic control of the soil environment, temperature and nutrient cycling appears to result from plant survival, expansion and regrowth (Bowden *et al.*, 1993; Carlton, 1993; Lezberg & Foster, 1994). In addition, microbes may immobilise and retain nutrients in the mineral soil and organic horizons (Lodge & McDowell, 1991).

As nutrient losses from ecosystems are often linked to hydrological changes, information on water balances and nutrient concentrations would be especially informative. The Luquillo Experimental Forest provides the only relevant data for hurricane impacts. Despite catastrophic damage to vegetation and great soil disturbance, nutrient exports were minimal as streams showed only minor peaks in nitrate concentrations for a 4 month period. In addition to other mechanisms discussed, low precipitation following Hurricane Hugo and nutrient retention by dense riparian vegetation may minimise stream nutrient loss (Scatena & Larsen, 1991).

The only regional analysis of hydrological impacts of hurricanes provides a striking, though perhaps misleading picture. In an ingenious use of double-mass plotting of streamflow from major watersheds following the 1938 hurricane, Patric (1974) showed that damaged watersheds (Connecticut and Merrimac Rivers) exhibited increased flows of 25 cm (4×10^6 acre feet) over a 5 year period in comparison with watersheds that were undamaged

Hurricane disturbance regimes in forests   333

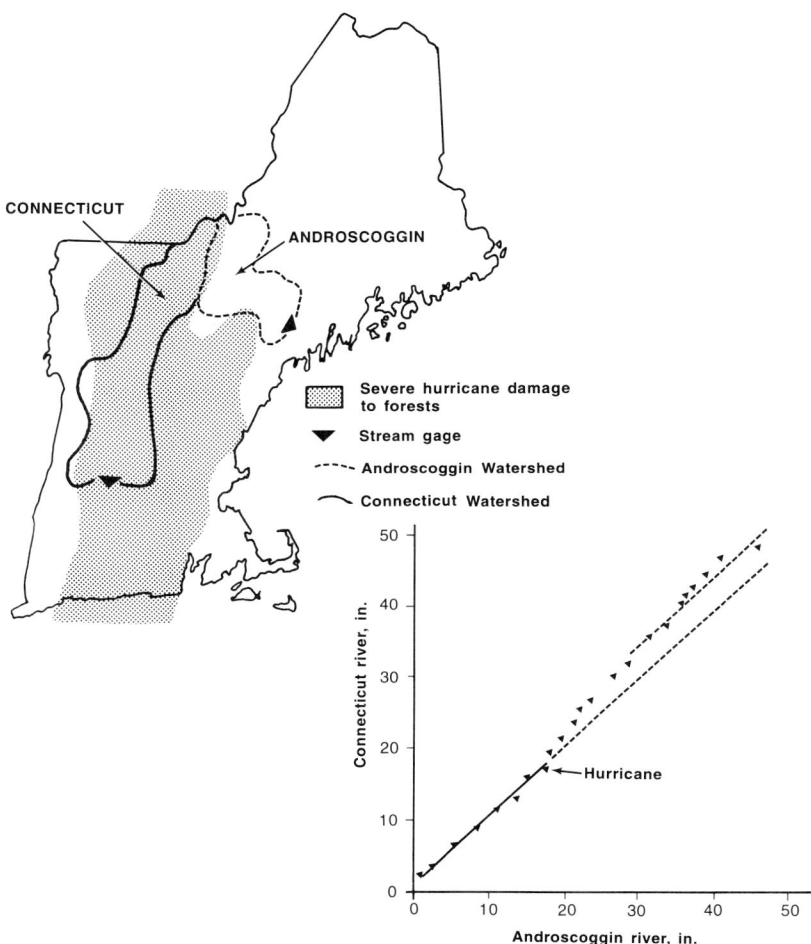

Fig. 18.15. Regional impact of the 1938 hurricane and associated salvage logging on river flow in New England. Figures show the study region with major river drainages and the extent of severe damage from the hurricane and cumulative summer flow (July, August, September) for the Connecticut river (within the damaged area) and the Androscoggin river (outside the damaged area) over a 5-year period. Modified from Patric (1974).

(Androscoggin River; Fig. 18.15). Increases peaked the first year and then declined consistently; more than 50% of the enhancement occurred in summer (Patric, 1974). Additional flow was interpreted as the result of substantial decline of evapotranspiration within the damaged area.

This interpretation appears sound; however, the additional flow may not necessarily be a direct result of the hurricane damage. Salvage operations following the hurricane harvested nearly 1 billion board feet of timber across New England (State of New Hampshire, 1940; NETSA, 1943). It is quite possible that this activity removed substantial live material, retarded the recovery of forest vegetation and leaf area, and accentuated streamflow through reduced transpiration. On a regional scale the 1938 event may have been more of a logging operation than a natural disturbance.

In summary, the available evidence suggests that despite the tremendous physical disruption generated by hurricanes, ecosystem impacts, in terms of nutrient cycling and hydrology, are generally slight and of brief duration. Survival and recovery of the vegetation results in a rapid re-establishment of leaf area that mitigates soil and microenvironment changes and promotes nutrient uptake. The result is biotic control over biogeochemical processes despite a long legacy of ensuing vegetation dynamics. Notably, hurricane effects appear to be substantially smaller than those associated with some human land-use, including heavy cutting and forest conversion.

18.6 Conclusion

There are many opportunities for increased interaction between physical and biological scientists and among ecological fields that would greatly increase the current understanding of hurricane disturbance regimes. The current level of tropical storm meteorology presents the potential to reconstruct historical distributions of hurricanes across forested regions and to relate these to vegetation dynamics and spatial patterns at landscape to regional scales. Such reconstructive approaches would enhance our understanding of the frequency, intensity and gradients of hurricane impacts and would facilitate inter-regional studies and comparisons with other disturbance processes. Initial results of modelling at landscape to regional scales suggest that predictable pathways and characteristics of storms may interact with physiography to develop long-term gradients of wind exposure.

Stand-level studies indicate that forest vegetation exhibits remarkable rates of recovery following intense canopy damage, but additional information on the response of some community-level characteristics, such as species diversity, is needed in a range of forest types. The important role of surviving plants in the re-vegetation process highlights the need for more study of plant reiteration and the physiology of damaged and sprouting plants.

The survival and vegetative reproduction of plants already established on hurricane-impacted sites explains in large part the very rapid recovery of biotic control of ecosystem processes. More study is needed of additional forest types and especially of hydrology, nutrient cycling and soil organic matter dynamics. Analysis of the role of microbial communities in nutrient immobilisation would complement the study of above-ground vegetation recovery.

Hurricanes have been recognised as an important disturbance process in a range of vegetation world wide. As a better understanding develops of the temporal and spatial variation in their impacts, and the actual range of vegetation and ecosystem response, we will be better able to identify their relative importance in controlling ecosystem structure and dynamics.

Acknowledgements

We thank A. Lezberg for providing extensive information on hurricane studies at Harvard Forest and thoughtful comments on the manuscript; R. Bowden and R. Boone for discussions on disturbance impacts to soils and nutrient cycling; M. Fluet, R. Waide, F. Scatena, A. Lugo, J. Zimmermann and P. Steudler for insights on forest dynamics in Puerto Rico; and D. Smith for technical assistance. This work was supported by NSF LTER grants to the Harvard Forest and the University of Puerto Rico.

References

Anthes, R. A. (1982). *Tropical Cyclones: Their Evolution, Structure and Effects.* Meteorological Monographs 19 (41). American Meteorological Society, Boston, MA.

Armson, R. A. & Fessenden, R. J. (1973). Forest windthrows and their influence on soil morphology. *Soil Science Society of America Proceedings*, **37**, 781–3.

Atkinson, G. D. & Holliday, C. R. (1977). Tropical cyclone minimum sea level pressure/maximum sustained wind relationship for the Western North Pacific. *Monthly Weather Review*, **105**, 421–7.

Basnet, K. (1992). Effect of topography on the pattern of trees in the tabonuco (*Dacryodes excelsa*) dominated rain forest of Puerto Rico. *Biotropica*, **24**, 31–41.

Beatty, S. W. (1984). Influence of microtopography and canopy species on spatial patterns of forest understory plants. *Ecology*, **65**, 1406–19.

Bellingham, P. J. (1991). Landforms influence patterns of hurricane damage: evidence from Jamaican montane forests. *Biotropica*, **23**, 427–33.

Boose, E. R., Foster, D. R. & Fluet, M. (1994). Hurricane impacts to tropical and temperate forest landscapes. *Ecological Monographs*, **64**, 369–400.

Bormann, F. H., Likens, G. E., Siccama, T. G., Pierce, R. S. & Eaton, J. S. (1974). The export of nutrients and recovery of stable conditions following deforestation at Hubbard Brook. *Ecological Monographs*, **44**, 255–77.

Bowden, R. D., Castro, M. C., Melillo, J. M., Steudler, P. A. & Aber, J. D. (1993). Fluxes of greenhouse gases between soils and the atmosphere in a temperate forest following a simulated hurricane blowdown. *Biogeochemistry*, **21**, 61–71.

Brake, R. W. & Post, H. A. (1941). Natural restocking of hurricane damaged 'old field' white pine areas in north central Massachusetts. MFS Thesis, Harvard University, Cambridge, M.A.

Brokaw, N. V. L. & Grear, J. S. (1991). Forest structure before and after Hurricane Hugo at three elevations in the Luquillo Mountains, Puerto Rico. *Biotropica*, **23**, 386–92.

Brokaw, N. V. L. & Walker, L. R. (1991). Summary of the effects of Caribbean hurricanes on vegetation. *Biotropica*, **23**, 442–7.

Byers, H. R. (1974). *General Meteorology*, 4th edn. McGraw-Hill, New York.

Canham, C. D. & Marks, P. L. (1985). The response of woody plants to disturbance: patterns of establishment and growth. In *The Ecology of Natural Disturbance and Patch Dynamics*, ed. S. T. A. Pickett & P. S. White, pp. 197–216. Academic Press, New York.

Carlton, G. (1993). Effects of microsite environment on tree regeneration following disturbance. PhD. Thesis, Harvard University, Cambridge, M.A.

Chazdon, R. L. & Fetcher, N. (1984). Photosynthetic light environment in a lowland tropical rain forest in Costa Rica. *Journal of Ecology*, **72**, 553–64.

Clapp, R. T. (1938). The effects of the hurricane upon New England forests. *Journal of Forestry*, **36**, 1177–81.

Clark, J. S. (1988). Effects of climate change on fire regimes in northwestern Minnesota. *Nature*, **334**, 233–5.

Clark, J. S. (1989). Ecological disturbance as a renewal process: theory and application to fire history. *Oikos*, **56**, 17–30.

Cline, A. C. & Spurr, S. H. (1942). *The Virgin Upland Forests of Central New England: A Study of Old Growth Stands in the Pisgah Mountains, Southwestern New Hampshire*. Harvard Forest Bulletin no. 21.

Committee on Global Change (1988). *Toward an Understanding of Global Change: Initial Priorities for US Contributions to the International Geosphere-Biosphere Program*. National Academy Press, Washington, DC.

Denslow, J. S. (1980). Gap partitioning among tropical rainforest trees. *Biotropica*, **12**, Supplement, 47–55.

Denslow, J. S. (1985). Disturbance-mediated coexistence of species. In *The Ecology of Natural Disturbance and Patch Dynamics*, ed. S. T. A. Pickett & P. S. White, pp. 307–23. Academic Press, New York.

Doyle, T. W. (1981). The role of disturbance in the gap dynamics of a montane rain forest: an application of a tropical forest successional model. In *Forest Succession: Concepts and Applications*, ed. D. C. West, H. H. Shugart & D. B. Botkin, pp. 56–73. Springer-Verlag, New York.

Dunn, G. E. & Miller, B. I. (1964). *Atlantic Hurricanes*, revised edn. Lousiana State University Press, Baton Rouge.

Emanuel, K. A. (1987). The dependence of hurricane intensity on climate. *Nature*, **326**, 483–5.

Foster, D. R. (1988a). Disturbance history, community organization and vegetation dynamics of the old-growth Pisgah Forest, southwestern New Hampshire, USA. *Journal of Ecology*, **76**, 105–34.

Foster, D. R. (1988b). Species and stand response to catastrophic wind in central New England, USA. *Journal of Ecology*, **76**, 135–51.

Foster, D. R. & Boose, E. R. (1992). Patterns of forest damage resulting from catastrophic wind in central New England, USA. *Journal of Ecology*, **80**, 79–98.

Frank, W. M. (1977). The structure and energetics of the tropical cyclone. I. Storm structure. *Monthly Weather Review*, **105**, 1119–35.
Fujita, T. T. (1971). *Proposed Characterization of Tornadoes and Hurricanes by Area and Intensity*. Satellite and Mesometeorology Research Project Paper 91. Department of Geophysical Sciences, University of Chicago.
Gray, W. M. (1990). Strong association between West African rainfall and US landfal! of intense hurricanes. *Science*, **249**, 1251–6.
Heinselman, M. L. (1973). Fire in the virgin forests of the Boundary Waters Canoe Area, Minnesota. *Quaternary Research*, **3**, 329–82.
Henry, J. D. & Swan, J. M. A. (1974). Reconstructing forest history from live and dead plant material: an approach to the study of forest succession in southwest New Hampshire. *Ecology*, **55**, 772–83.
Hibbs, D. E. (1983). Forty years of forest succession in central New England. *Ecology*, **64**, 1394–401.
Hobgood, J. S. & Cerveny, R. S. (1988). Ice-age hurricanes and tropical storms. *Nature*, **33**, 243–5.
Jensen, V. S. (1941). *Hurricane Damage on the Bartlett Experimental Forest*. USDA Northeastern Forest Experiment Station Technical Note 42.
Johnson, E. A. (1992). *The Ecology of Fire in the Boreal Forest*. Cambridge University Press, Cambridge.
Lamb, H. H. (1991). *Historic Storms of the North Sea, British Isles and Northwest Europe*. Cambridge University Press, Cambridge.
Lodge, D. J. & McDowell, W. H. (1991). Summary of ecosystem-level effects of Caribbean hurricanes. *Biotropica*, **23**, 373–8.
Lodge, D. J., Scatena, F. N., Asbury, C. E. & Sanchez, M. J. (1991). Fine litterfall and related nutrient inputs resulting from Hurricane Hugo in subtropical wet and lower montane rain forests in Puerto Rico. *Biotropica*, **23**, 336–42.
Lugo, A. E. (1988). Estimating reductions in the diversity of tropical forest species. In *Biodiversity*, ed. E. O. Wilson, pp. 58–70. National Academy Press, Washington, DC.
Lyford, W. H. & MacLean, D. W. (1966). *Mound and Pit Microrelief in Relation to Soil Disturbance and Tree Distribution in New Brunswick, Canada*. Harvard Forest Paper no. 15.
Marks, P. L. (1974). The role of pin cherry (*Prunus pensylvanica* L.) in the maintenance of stability in northern hardwood ecosystems. *Ecological Monographs*, **44**, 73–88.
Metropolitan District Commission (1994). *MDC Land Management Plan for the Quabbin Watershed*. State of Massachusetts.
Naka, K. (1982). Community dynamics of evergreen broadleaf forests in southwestern Japan. I. Wind damaged trees and canopy gaps in evergreen oak forest. *Botanical Magazine, Tokyo*, **95**, 385–99.
NETSA (1943). *Report of the US Forest Service Programs Resulting from the New England Hurricane of September 21, 1938*. Northeastern Timber Salvage Administration, Boston, MA.
Neumann, C. J., Jarvinen, B. R. & Pike, A. C. (1987). *Tropical Cyclones of the North Atlantic Ocean 1871–1986*, 3rd rev. NOAA–National Climatic Data Center, Asheville, NC.
O'Brien, S. T., Hayden, B. P. & Shugart, H. H. (1992). Global climatic change, hurricanes, and a tropical forest. *Climatic Change*, **22**, 175–90.
Odum, H. T. (1970). Rain forest structure and mineral-cycling homeostasis. In *A Tropical Rain Forest: A Study of Irradiation and Ecology at El Verde, Puerto*

Rico, ed. H. T. Odum & R. F. Pigeony pp. H3–H52. US Atomic Energy Commission, Springfield, VA.

Overpeck, J. T., Rind, D. & Goldberg, G. (1990). Climate-induced changes in forest disturbance and vegetation. *Nature*, **343**, 51–3.

Parton, W. J., Stewart, J. W. & Cole, C. V. (1988). Dynamics of C, N, P and S in grassland soils: a model. *Biogeochemistry*, **5**, 109–31.

Patric, J. H. (1974). River flow increases in central New England after the hurricane of 1938. *Journal of Forestry*, **72**, 21–5.

Peterson, C. J. & Pickett, S. T. A. (1991). Treefall and resprouting following catastrophic windthrow in an old-growth, hemlock–hardwoods forest. *Forest Ecology and Management*, **42**, 205–17.

Peterson, C. J., Carson, W. P., McCarthy, B. C. & Pickett, S. T. A. (1993). Microsite variation and soil dynamics within newly created treefall pits and mounds. *Oikos*, **58**, 39–46.

Powell, M. D. (1982). The transition of the Hurricane Frederic boundary-layer wind field from the open Gulf of Mexico to landfall. *Monthly Weather Review*, **110**, 1912–32.

Putz, F. E. (1983). Treefall pits and mounds, buried seeds, and the importance of soil disturbance to pioneer trees on Barro Colorado Island, Panama. *Ecology*, **64**, 1069–74.

Riggs, H. C. (1965). *Effects of Land Use on the Low Flow of Streams in Rappahannock County, Virginia*. US Geological Survey Professional Paper 525C: C169–C198.

Rowlands, W. (1941). Damage to even-aged stands in Petersham, Massachusetts by the 1938 Hurricane as influenced by stand condition. MFS Thesis, Harvard University, Cambridge, MA.

Sanford, R. L., Parton, W. J., Ojima, D. S. & Lodge, D. J. (1991). Hurricane effects on soil organic matter dynamics and forest production in the Luquillo Experimental Forest, Puerto Rico: results of simulation modeling. *Biotropica*, **23**, 364–72.

Scatena, F. N. & Larsen, M. C. (1991). Physical aspects of Hurricane Hugo in Puerto Rico. *Biotropica*, **23**, 317–23.

Shaw, W. B. (1983). Tropical cyclones: determinants of pattern and structure in New Zealand's indigenous forests. *Pacific Science*, **37**, 405–14.

Simpson, R. H. & Riehl, H. (1981). *The Hurricane and its Impact*. Louisiana State University Press, Baton Rouge.

Sipe, T. (1990). Gap partitioning among maples (*Acer*) in the forests of central New England. PhD Thesis, Harvard University, Cambridge, MA.

Smith, D. M. (1946). Storm damage in New England forests. MF. Thesis, Yale University, New Haven, CT.

State of New Hampshire (1940). *Biennial Report of the Forestry and Recreation Commission for the Two Fiscal Years Ending June 30, 1940*. Concord, NH.

Stephens, E. P. (1955). The historical-developmental method of determining forest trends. PhD Thesis, Harvard University, Cambridge, MA.

Stephens, E. P. (1956). The uprooting of trees: a forest process. *Soil Science Society of America Proceedings*, **20**, 113–16.

Steudler, P. A., Melillo, J. M., Bowden, R. D., Castro, M. S. & Lugo, A. E. (1991). The effects of natural and human disturbance on soil nitrogen dynamics and trace gas fluxes in a Puerto Rican wet forest. *Biotropica*, **23**, 356–63.

Tanner, E. V. J., Kapos, V. & Healey, J. R. (1991). Hurricane effects on forest ecosystems in the Caribbean. *Biotropica*, **23**, 513–21.

Vitousek, P. M., Gosz, J. R., Grier, C. C., Melillo, J. M., Reiners, W. A. & Todd, R. L. (1979). Nitrate losses from disturbed ecosystems. *Science*, **204**, 469–74.
Waide, R. B. & Lugo, A. E. (1992). A research perspective on disturbance and recovery of a tropical montane forest. In *Tropical Forests in Transition: Ecology of Natural and Anthropogenic Disturbance Processes*, ed. J. G. Goldharnmer, pp. 173–90. Berkhauser-Verlag, Basel.
Walker, L. R. (1991). Tree damage and recovery from Hurricane Hugo in Luquillo Experimental Forest, Puerto Rico. *Biotropica*, **23**, 379–85.
Walker, L. R., Lodge, D. J., Brokaw, N. V. L. & Waide, R. B. (1991). An introduction to hurricanes in the Caribbean. *Biotropica*, **23**, 313–16.
Walker, L. R., Voltzow, J., Ackerman, J. D., Fernandez, D. S. & Fetcher, N. (1992). Immediate impact of Hurricane Hugo on a Puerto Rican rain forest. *Ecology*, **73**, 691–4.
Webb, S. L. (1986). Windstorms and the dynamics of two northern forests. PhD Thesis, University of Minnesota, Minneapolis.
Webb, S. L. (1989). Contrasting windstorm consequences in two forests, Itasca State Park, Minnesota. *Ecology*, **70**, 1167–80.
Whigham, D. J., Olmsted, I., Cano, E. C. & Harmon, M. E. (1991). The impact of Hurricane Gilbert on trees, litterfall and woody debris in a dry tropical forest in the northeastern Yucatan Peninsula. *Biotropica*, **23**, 434–41.
Whitmore, T. C. (1978). Gaps in the forest canopy. In *Tropical Trees as Living Systems*, ed. P. B. Tomlinson & M. H. Zimmermann, pp. 639–55. Cambridge University Press, Cambridge.
Whitmore, T. C. (1982). On pattern and process in forests. In *The Plant Community as a Working Mechanism*, ed. E. I. Newman, pp. 45–59. Blackwell Scientific, Oxford.
Yih, K., Boucher, D., Vandermeer, J. & Zamora, N. (1991). Recovery of the rain forest of southeastern Nicaragua after destruction by Hurricane Joan. *Biotropica*, **23**, 106–13.
Zimmerman, J. K., Everham, E. M., Waide, R. B., Lodge, D. J., Taylor, C. M. & Brokaw, N. V. L. (1994). Responses of tree species to hurricane winds in subtropical wet forest in Puerto Rico: implications for tropical tree life histories. *Journal of Ecology*, **82**, 911–22.

19

A comparison of methods for quantifying catastrophic wind damage to forests

E. M. EVERHAM III

Abstract

Catastrophic wind events impact forests over the entire globe. Although recent examples of hurricanes in the Caribbean have led to intense examination of the impacts on, and recovery of, forests, these research efforts have largely been in isolation. Little has been done to compare the impacts of storms of varying intensity on different ecosystems. Therefore, we can not as yet fit catastrophic wind events into a general model of forest disturbance and recovery. Papers examining the impacts of 26 different wind events (cyclonic storms, tornadoes and gales) on 27 different forests are reviewed. Hurricane damage is measured as numbers or percentage: stem damage, canopy damage, biomass or stand volume loss, or mortality. The populations sampled varied from minimum stem size of 2 cm to 20 cm in diameter. Sampling methodology included small circular plots, transects, large gridded plots and remote sensing of the landscape. Plots were established 10 days to 3 years after the wind event. The implications of these different quantification systems are examined using data from a large gridded plot established in the Luquillo Mountains of Puerto Rico to study the impacts of Hurricane Hugo. Intensity of disturbance, measured as percentages of different damage types, varied depending on the minimum stem size used in the analysis. Damage to individual species also varied depending on the variable used to quantify it. Clearly, a standard measure of wind damage is needed to facilitate comparisons of the impacts of different storms on different forests. I suggest a damage measure that includes both mortality and structural loss as measured by decrease in basal area.

19.1 Introduction

Catastrophic wind events, including cyclonic storms, gales and tornadoes, affect forests around the globe in both temperate and tropical regions. Some meteorologists and some global climate change models predict an increase in both the frequency and intensity of these types of storm events (Dunn & Miller, 1960; Wentland, 1977; Emanuel, 1987; Gray, 1990). Species differences, age class or size, silvicultural treatment (even-aged stands, thinning regime, introduction of exotics), topographic features and frequency of previous wind disturbance may all influence the intensity or spatial pattern of damage. Although the influence of these factors is increasingly understood for endemic or chronic wind impacts, there is little evidence that the relationships hold for catastrophic or acute wind events that occur much less frequently. These disturbance events have been observed and quantified for over a century, but few clear generalisations have been drawn about their impacts.

Six previous papers include an attempt to compare impacts of different storm events on different forests. Liegel (1984) reported assessments of the impacts of hurricanes on 11 conifer plantations. He found windthrow the most commonly reported damage. No clear trend was found for differences between damage to conifers and broadleaf trees. Glitzenstein & Harcombe (1988) compared the impacts of a tornado, a large blowdown associated with a severe thunderstorm (Dunn et al., 1983) and a hurricane (Lugo et al., 1983) on both temperate and tropical broadleaf forests. They found the size of the area affected and the intensity of disturbance were comparable for all three catastrophic wind events. They commented that severely damaged trees often regenerate and concluded that catastrophic wind disturbance probably never approaches the intensity of disturbance of catastrophic fire.

Brokaw & Walker (1991) presented a summary of damage reported for hurricanes affecting 14 tropical and temperate conifer and broadleaf forests. They found:

1. Variation in stem breakage as opposed to uprooting varied between forests, sometimes correlated to stem size.
2. Understorey trees may be more susceptible to indirect damage from other trees falling on them and this can result in a bimodal distribution of damage versus stem diameter, large trees being more likely to be directly damaged by the wind and small trees being more likely to be indirectly damaged.
3. Previous mechanical damage to trees may increase subsequent wind damage.
4. No clear relationships between damage of conifers versus broadleaf plants, or monocots versus dicots.
5. Species differences in impacts, but no clear generalisations such as early to late successional class.

6. No clear correlation between topographic exposure and intensity of damage.
7. Generally low levels of mortality, but high structural damage.

Lugo *et al.* (1983) compared hurricanes with other natural disturbances. They proposed that hurricanes have a greater impact than either earthquakes or landslides, both because of higher frequencies and because larger areas are affected. They also predicted faster recovery from hurricane impacts. Tanner *et al.* (1991), in a summary of impacts of hurricanes on forest ecosystems in the Caribbean, stated that the effects on ecosystems depend on: (1) hurricane intensity (including rainfall), (2) storm size and movement, (3) topographic characteristics, and (4) susceptibility of the system to damage.

Lugo *et al.* (1983) stated that a barrier to the comparison of hurricane impacts is the lack of a standard for measuring storm intensity. They suggested that the duration of hurricane winds and the windspeed would be a minimum standard for comparison and that this might help separate differences in damage due to system responses from those due to variations in the intensity of the 'stressor'. However, wind intensity of these events may vary significantly over scales of kilometres (Boose *et al.*, 1994) and the lack of detailed local meteorological data has made comparisons difficult (Tanner *et al.*, 1991). Scatena & Larsen (1991) developed three indices of storm intensity using: (1) maximum sustained wind and storm duration (as suggested by Lugo *et al.*, 1983), (2) maximum sustained wind and proximity to the storm, and (3) rainfall totals (as a percentage of annual average).

Another possible explanation for a lack of generalisations of impacts of catastrophic winds is a failure to standardise our efforts to quantify these impacts. Ackerman *et al.* (1991) reported the results of a conference on ecological effects of hurricanes and mention measures of damage gradients as a missing research tool.

This chapter is a review of the literature reporting impacts of catastrophic wind events on forests, examining the ways in which damage has been quantified, investigating the implications of different methods, and attempting to draw conclusions regarding standardisation.

19.2 Literature review

I reviewed 51 papers reporting the impacts of 26 storm events on 27 forests in 15 countries. Table 19.1 is a summary of the storms, locations and references with indications of attempts to quantify the storm intensity or damage. The table is organised chronologically by storm event. In several cases, more

Table 19.1. *Papers reviewed*

Year	Storm	Location	Storm intensity	Quantify damage	Reference
1815	Hurricane	USA	Yes	Yes	Darling (1842)
1928	Hurricane	Puerto Rico	Yes	No	Bates (1930)
1938	Hurricane	MA, CT, USA	No	No	Curtis (1943)
		MA, USA	Yes	Yes	Spurr (1956)
			Yes	Yes	Foster & Boose (1992)
			Yes	Yes	Foster (1988a)
			Yes	Yes	Foster (1988b)
1945	Cyclones	Mauritius	Yes	Yes	King (1945)
1953	Gale	Scotland, UK	Yes	Yes	Anderson (1954)
1956	Agnes	Australia	Yes	No	Webb (1958)
1956	Betsy	Puerto Rico	Yes	Yes	Wadsworth & Englerth (1959)
1960	Alix, Carol	Mauritius	Yes	Yes	Sauer (1962)
1960s	Hurricanes	Western Samoa	Yes	Yes	Wood (1970)
1967	Annie	Soloman Islands	Yes	Yes	Whitmore (1974)
			Yes	Yes	Whitmore (1989)
1968	Gale	New Zealand	Yes	Yes	Irvine (1970)
1969	Camille	MS, USA	Yes	Yes	Touliatos & Roth (1971)
1975	Gale	New Zealand	Yes	Yes	Wilson (1976)
1975	Eloise	FL, USA	Yes	Yes	Wilkinson et al. (1978)
1977	Wind storm	WI, USA	Yes	Yes	Dunn et al. (1983)
1977	Wind storm	South Africa	Yes	Yes	Versfeld (1980)
1978	Cyclone	Sri Lanka	Yes	Yes	Dittus (1985)
1979	David	Dominica	Yes	Yes	Lugo et al. (1983)
1979	David, Frederic	Puerto Rico	Yes	Yes	Liegel (1984, 1982)
1980	Allen	Jamaica	Yes	Yes	Thompson (1983)
1983	Tornado	TX, USA	Yes	Yes	Glitzenstien & Harcombe (1988)
1986	Winifred	Australia	Yes	Yes	Applegate & Bragg (1992)
			Yes	Yes	Turton (1992)
1988	Gilbert	Jamaica	Yes	Yes	Bellingham et al. (1992)
			Yes	Yes	Bellingham (1991)
			Yes	Yes	Wunderle et al. (1992)
1988	Gilbert	Mexico	No	Yes	Whigham et al. (1991)
1988	Joan	Nicaragua	Yes	Yes	Boucher et al. (1990)
			Yes	Yes	Boucher (1990)
			Yes	Yes	Yih et al. (1991)
1989	Hugo	SC, USA	Yes	Yes	Putz & Sharitz (1991)
			Yes	Yes	Sheffield & Thompson (1992)
			Yes	Yes	Gresham et al. (1991)
1989	Hugo	Virgin Islands	Yes	Yes	Reilly (1991)
1989	Hugo	Puerto Rico	Yes	Yes	Walker et al. (1992)
			Yes	Yes	Basnet (1990)
			Yes	Yes	Basnet et al. (1992)
			Yes	Yes	Frangi & Lugo (1991)
			Yes	Yes	Walker (1991)
			Yes	Yes	Brokaw & Grear (1991)
			No	Yes	Fernandez & Fetcher (1991)
			No	Yes	Lodge et al. (1991)
			No	Yes	Scatena & Larsen (1991)
			Yes	Yes	Scatena & Lugo (1993)
			No	Yes	Dallmeier et al. (1991)

than one paper reports on the same storm event on the same forest. All were included whenever there were differences in plot location, sampling methodology or analysis. Forty-five papers include some description of the event intensity, usually either maximum wind gusts or average windspeed. Few, however, indicate the distance to the measuring site and/or distance from the eye of the storm. Sampling plots vary from 2.5 m^2 (litterfall studies) to 16 ha and include circular, rectilinear and gridded plots, and transects. Minimum stem diameters range from 2 to 20 cm. The sampling time after the event varies from 10 days to 3 years (not including analysis of historical storm data). The greatest variation of methodology occurs between the parameters used to quantify damage. In the 48 papers reviewed that included quantification of damage, 28 different schemes or methodologies had been used. Clearly there are no standards for assessing impacts of catastrophic wind events. One obvious explanation is that few researchers design experiments that anticipate naturally occurring catastrophic events. Instead, they capitalise on any studies established before the disturbance event or on surveys established afterwards (Lugo et al., 1983; Tanner et al., 1991).

Efforts to quantify damage can be classified into five types: (1) stem damage, (2) canopy damage, (3) volume or mass changes, (4) mortality, (5) classification categories (using one or more of the above to set arbitrary damage categories of e.g. high, low, etc.) These types are not mutually exclusive and many authors used more than one of them. Table 19.2 includes a summary of the different damage quantification methods reviewed. No paper reported damage in all categories and this increases the difficulty of developing generalisations.

19.2.1 Stem damage

Stem damage is most commonly reported as percentage of downed trees. Secondarily, the types of damage are delineated: uprooting, breakage, and bending or leaning. Of the 30 papers that reported stem damage, 16 report total downed trees, two reported only uprooted stems, and the remaining 12 distinguish between broken and uprooted stems. Most authors use actual counts from established plots, but aerial surveys and area estimates are also reported (Wilson, 1976). These numbers may be given as raw counts (Versfeld, 1980) without pre-disturbance stem numbers or densities, or as percentages. Researchers vary in their inclusion of leaning or bent stems. Wadsworth & Englerth (1959) used the measure 'lost' to include trees that were snapped, uprooted, or had stems leaning more than 15° from vertical, whereas Applegate & Bragg (1992) counted percentage 'smashed' as stems

Table 19.2. *Comparison of results. A single stem damage value indicates separate data are not reported*

Forest	Stem damage Snap	Uproot	Branch	Canopy	Dead	Loss	Reference
Temperate pine plant.	1.7	6.9	—	—	—	—	Versfeld (1980)[a]
Temperate mixed plant.	32						Wilson (1976)[b]
Temperate mixed	66						Spurr (1956)[c]
Temperate mixed	30.7					50	Foster (1988a)[b]
Temperate mixed	52				30.7	94	Foster & Boose (1992)[b]
Temperate mixed	89					94	Dunn et al. (1983)
Temperate mixed	24			65	24	52	Glitzenstein & Harcombe (1988)[b]
Temperate mixed	—	—	25			28	Sheffield & Thompson (1992)
Temperate coastal	11			27			Gresham et al. (1991)
Temperate swamp, mixed	34	20					Putz & Sharitz (1991)
Subtropical pine	—				5.5		Wilkinson et al. (1978)
Tropical pine plant.	65					36	King (1945)[b]
Tropical pine plant.	—	7.5			7.5		Liegel (1982, 1984)
Tropical pine plant.	44				58		Boucher (1990)
Tropical mixed plant.	47.1						Thompson (1983)
Tropical wet	30			15			Wadsworth & Englerth (1959)[b]
Tropical wet	36.9						Whitmore (1974, 1989)[b]
Tropical wet	37					32	Applegate & Bragg (1992)[b]
Tropical wet	12.3	43		23	7		Bellingham et al. (1992)
Tropical wet	8.5	5.8		14.1	8.1		Bellingham (1991)[b]
Tropical wet					13		Boucher (1990)
Tropical wet	75			82			Yih et al. (1991)[b]
Tropical wet	80				25		Dallmeier et al. (1991)
Tropical wet	8.2	30	30.5	78			Walker et al. (1992)
Tropical wet	17	16.5	34	99		41	Basnet et al. (1992)
Tropical wet	6	43				50	Scatena et al. (1993)
Tropical wet	—	9		56	7		Walker (1991)[b]
Tropical wet	11	2.5		29	1	10	Frangi & Lugo (1991)
Tropical flood plain	2						Reilly (1991)[b]
Tropical moist, dry		9					
Tropical wet, dry	15	27			2	43	Lugo et al. (1983)[b]
Tropical dry	—	32		33	40		Dittus (1985)
Tropical dry	11.5	4	77	100	9.7		Whigham et al. (1991)

[a] Values calculated from data.
[b] Values averaged for more than one site.
[c] Values estimated.

broken, uprooted, or leaning at greater than 40° from vertical. An alternative is to establish a separate category for leaning stems (Dallmeier et al., 1991; Basnet et al., 1992; Wunderle et al., 1992; Zimmerman et al., 1993). Glitzenstein & Harcombe (1988) used a category 'pinned' to indicate stems that are bent by other stems. Thompson (1983) also used the 15° minimum for counting leaning trees, summed total damage as leaning, snapped or uprooted stems, but separated snapped stems into those broken at the crown or lower on the stem. The former were not included in his 'effective damage' category. Putz & Sharitz (1991) separated uprooted stems into partly or completely uprooted. Walker (1991) included measures of the height at which the stem breaks and the percentage of the root system exposed in uprooted trees. Bellingham (1991) also measured the height of the stem break, but measured the angle of the uprooted stem. Zimmerman et al. (1994) measured the height of the snapped stems, and the size of the root mat exposed in uprooted stems. Wilkinson et al., (1978) utilised 12 categories of stem lean, four of stem damage, and four of root exposure, in addition to categories relating to crown damage. Gresham et al., (1991) set eight categories, from undamaged to downed stems. Scatena & Lugo (1993) used three main categories: little damage, standing damage and down damage; but the standing category had nine subcategories and the prone category had five.

19.2.2 Canopy damage

Canopy damage may be quantified directly on the basis of branch loss, canopy damage or defoliation; or be indirectly determined using techniques to measure structure or changes in light levels. Branch damage is quantified either by counting branches damaged or by assignment to arbitrarily defined categories, and it varies depending upon the size of the branches assessed. Dallmeier et al., (1991) used three categories of crown damage: snapped off (heavy), one to three branches remaining (medium), or more than four branches remaining (low). Walker et al. (1992) counted the trees which lost two-thirds of their branches. Many researchers set no clear size categories and grade branch damage subjectively. Gresham et al., (1991) counted only 'limbs' broken. Basnet (1990) distinguished between standing defoliated and branch broken categories by scoring small branch damage as opposed to 'major' branch damage, respectively. Reilly (1991) used three categories of damage based on 'small' branch damage, 'main' branch damage or stem damage. Whigham et al. (1991) used four categories of canopy damage: crown gone, only large branches remaining, most branches remaining, only small branches and twigs lost. Walker (1991) used three categories of branch damage based on clearly defined diameter classes: only 1 cm branches damaged (minor), branches less than 5 cm damaged (moderate), or damage to branches greater than 5 cm (heavy). Wunderle *et al.* (1992), Dittus (1985) and Zimmerman *et al.* (1993) all set 10 cm as the minimum size for counting damaged branches, and Scatena & Lugo (1993) used 10 cm as the boundary between light and heavy branch damage. It should be noted that when these clearly defined categories are applied to standing damaged trees, subjectivity is introduced in the assessment of the branch diameters from the ground. Different minimum branch sizes, unquantified branch sizes and categorisation of damage all make it difficult to calculate comparable branch damage values for a disturbed site.

Canopy damage is often even more subjective. Both Whitmore (1989) and Glitzenstein & Harcombe (1988) attempted to determine directly the percentage canopy change for the entire study area. Whitmore (1989) scored each plot as gap, building or mature, and was then able to determine the percentage of the area in gaps. Glitzenstein & Harcombe (1988) used aerial photographs to determine the percentage cover after a tornado. Frangi & Lugo (1991) scored each stem for percentage defoliated. More typically, researchers establish several categories for scoring canopy damage on each stem. This categorisation of canopy damage makes it difficult to determine an overall average for the disturbed site. Walker (1991) used four categories of defoliation: <25%, 25–50%, 50–75% and >75%. Putz & Sharitz used two: 25–50% or >50%. Scatena & Lugo also used two: >50% of the crown intact and <50% of the crown intact. Bellingham *et al.* (1992) used two categories without clearly defined scales: crown partly broken or crown defoliated. Wilkinson *et al.* (1978) established five categories of live crown ratio and four of crown damage. Dittus (1985) scored crown damage from 0 to 10 representing the percentage of the crown lost. Both Yih *et al.*, (1991) and Bellingham (1991) counted only stems that were completely defoliated. Walker *et al.* (1992) counted stems two-thirds or more defoliated.

Canopy change has been assessed indirectly by measuring changes in light levels. Turton (1992) used hemispheric photography to determine the percentage change in available light at the forest floor. Both Fernandez & Fletcher (1991) and Walker *et al.* (1992) measured light levels after hurricane disturbance; only the latter used a site established before the storm and were able to quantify changes. Whitmore (1989) used photographs to assess canopy damage and monitor recovery of specific trees. No effort was made to determine an average for the area of study. Applegate & Bragg (1992) also used photographs to monitor recovery of the canopy, but no attempt was made to quantify these changes. Changes in canopy structure have been measured more directly using foliar profiles, which are measures of the presence of vegetation at arbitrarily defined levels, sampled by placing a pole up through the canopy (Wunderle et al., 1992; Brokaw & Grear, 1991). Both measures of light levels and foliar profiles promise to give clearer averages of canopy changes, but both require that measurements be made before and after the storm event.

19.2.3 Volume or mass changes

Volume losses are typically reported by foresters concerned with timber yield. Biomass, or related measures such as basal area, are often measured or estimated by ecologists interested in energy or nutrient flows. Versfeld (1980) and Touliatos & Roth (1971) both expressed damage as total volume of timber lost; Foster & Boose (1992) gave both basal area and board feet per hectare lost, but in none of these three cases are pre-disturbance levels given to allow a determination of percentage change. Spurr (1956) gave estimates of percentage loss of timber volume. Sheffield & Thompson (1992) included a detailed assessment of the impacts of Hurricane Hugo on forests in South Carolina. Their assessment is also based on volume loses, but includes pre-hurricane levels, pre-hurricane mortality, hurricane mortality, non-lethal hurricane damage, and an elaborate post-hurricane risk assessment that is forest and age class specific and includes consideration of crown loss, root damage, stem damage and salt burn. Both Lugo et al. (1983) and Basnet et al. (1992) reported basal area losses for each damage class. Glitzenstein & Harcombe (1988) and Dunn et al. (1983) both gave basal area losses for each species and totals for their study sites. Applegate & Bragg (1992) reported only loss to standing basal area. Scatena et al. (1993) reported above-ground biomass loss. Frangi & Lugo (1991) gave total biomass loss and species-specific values for leaves, branches and stems. Lodge et al. (1991) and Whigham et al. (1991) both quantified hurricane impacts on the basis of the mass of litterfall associated with the storm.

19.2.4 Mortality

Mortality measurements pose a number of problems. Mortality is often assumed to be the same as blowdown, which may be reasonable for some coniferous forest but is not for many broadleaf forests – particularly in the tropics, where the percentage of resprouted stems may be as high as 64.8% for dicots (Zimmerman et al., 1994) and the proportion of snapped stems resprouting was 87% (Boucher, 1990) or 56% (Bellingham et al., 1992). Mortality can result from less severe damage (Wadsworth & Englerth, 1959;

Sauer, 1962). No clear relationship exists between damage and mortality. In addition, post-disturbance mortality appears to be elevated for months to years after a catastrophic wind event (Sauer, 1962; Bellingham, 1991). Dittus (1985) found that 19% of the canopy trees alive 1 month after a cyclone, died with in 42 months. When and how mortality is assessed influences the values found. Glitzenstein & Harcombe counted any stem that did not have green leaves or sprouts above 1.3 m as dead. Often assessment of storm-associated mortality can be obscured by standing dead trees that died before the storm. Most researchers appear to determine subjectively whether a stem was previously dead or was killed by the storm. Bellingham *et al.* (1991) used the objective criterion of the presence of bark on a dead stem to indicate that it has died recently. Thirty-three of the papers that quantified damage did not include measures of mortality. More research is required to track mortality for several years after catastrophic wind events to determine how it changes.

19.2.5 Classification categories

Five of the papers reviewed included an overall damage categorisation that could be used for comparing spatial patterns of damage. As might be expected, no two of these papers use the same system. These efforts utilise three to five categories of percentages of types of stem damage. Wadsworth & Englerth (1959) used a three-category system based on perception of stem 'loss' (broken, uprooted, or leaning more than 15°): negligible, slight and extreme. Bellingham *et al.* (1992) used a three-category system that included both stem and canopy damage: severe (up to 20% windthrow and severe defoliation), moderate (few uprooted trees but significant crown damage) and minor (moderate defoliation). Foster (1988b) established three categories based on the percentage of stems down (as viewed from aerial photographs): undamaged (few stems down), moderate (<50% of stems down) and extensive (>50% of stems down). Scatena *et al.* (1993) used a clearly defined three-category system: class 1 (crown damage only); class 2 (>50% standing stems); and class 3 (>50% downed stems). Foster & Boose (1992) utilised five categories of percentage broken or uprooted canopy stems: 0, <25, 25–50, 51–75, >75. Although this type of categorisation may facilitate analysis of spatial patterns, unless overall average values of damage are given, comparisons between sites and events are difficult.

19.3 Case study

In September 1989 Hurricane Hugo struck the Luquillo Experimental Forest in eastern Puerto Rico. Hugo was a category 4 storm with maximum sus-

Table 19.3. *Percentage damage for different damage parameters and different minimum stem sizes*

Minimum stem size (cm)	Stem broken (%)	Stem uprooted (%)	Branch damage (%)	Dead (%)	BA damaged (%)
10	8.89	6.93	15.32	9.76	18.75
15	9.97	8.63	19.92	11.61	19.52
20	10.40	11.34	30.18	12.93	20.13
25	9.65	11.49	34.46	12.99	19.59
Range	1.51	4.56	19.14	3.23	1.38
Per cent	16.99	65.80	124.93	33.09	7.36

Branch damage includes all stems with damage to at least one 10 cm or larger branch. Percentage dead includes standing and down dead stems. BA (basal area) damaged includes all stems uprooted, broken or standing dead. Range is the total change from minimum to maximum percentage values through all size classes. Per cent is the proportion of the minimum value that the range represents.

tained winds of 46 m s^{-1} and peak gusts of 194 kph (Scatena & Larsen, 1991). A 16 ha grid, the Hurricane Recovery Plot, was established to quantify the damage from the hurricane and to track recovery (Zimmerman *et al.*, 1994). All stems greater than or equal to 10 cm dbh (diameter at breast height) were identified, mapped, measured, permanently tagged, and assessed for hurricane damage. This damage assessment included categories for stem damage, branches damaged and mortality. Overall, 13 126 stems of 89 species were examined. Within 30 months after the hurricane 8.9% died and 15.8% received major stem damage (6.9% were uprooted and 8.9% snapped). An additional 15.3% received damage to at least one branch >10 cm diameter (Zimmerman *et al.*, 1994).

These totals vary depending on the stem class size used. Table 19.3 presents a comparison of the damage categories for a variety of minimum stem sizes (down to 10 cm, the minimum for this study). There is a general trend for a higher percentage damage with a larger minimum stem size. This probably indicates a greater damage level for larger stems, but the trend does not hold for broken stems or basal area lost. The largest change is in the branch damage category. This is partly caused by the branch damage minimum size of 10 cm: small trees do not have branches large enough to be counted. The smallest absolute and proportional change is in basal area lost. From a minimum stem size of 10 cm to a minimum stem size of 20 cm this damage measurement varies by less than 1.5%.

Searching for trends in individual species response (Fig. 19.1) is also confounded by different measures of damage. *Prestoea montana*, a palm, has

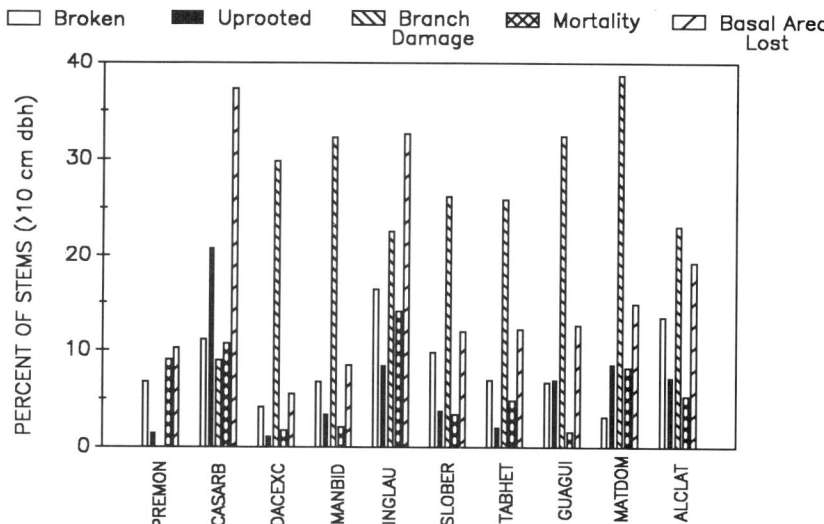

Fig. 19.1. Species differences in damage for the ten most common species. PREMON, *Prestoea montana*; CASARB, *Casearia arborea*; DACEXC, *Dacryodes excelsa*; MANBID, *Manilkara bidentata*; INGLAU, *Inga laurina*; SLOBER, *Slonea berteriana*; TABHET, *Tabebuia heterophylla*; GUAGUI, *Guarea guidonia*; MATDOM, *Matayba domingensis*; ALCLAT, *Alchornea latifolia*.

no branches, and therefore no branch damage. *Casearia arborea* has an elevated level of uprooting and a correspondingly high basal area lost. *Dacryodes excelsa* and *Manilkara bidentata*, two primary forest species, show low levels of damage except for branch damage. *Inga laurina* has a tendency to break and to die. The remaining species are similar to the primary forest species to varying degrees.

Variations in spatial patterns may be difficult to interpret when species respond differently to damage, and species distribution over the landscape also varies. Are the spatial patterns of damage driven by patterns of species distribution, or are differences in damage to species the result of their distribution in areas not equally affected by wind? Some recent efforts by Borchard *et al.* (1992) hold promise for unravelling this complex problem.

19.4 Discussion

Apparent intensity of damage from a catastrophic wind event varies depending on how the damage is measured. How it is quantified relates to the purpose of the analysis. The question of whether there are species differences in impacts from catastrophic winds has clearly been demonstrated by comparing specific types of damage. This type of analysis can be site and storm specific and still lend itself to generalisations. However, the remaining questions introduced at the beginning of this chapter – the influence of age class or size, the impact of silvicultural treatment, the control of damage by topographic features, the adaptation of forests relative to the frequency of previous wind disturbance, and the ability to predict the path of recovery of a disturbed forest – all require comparisons between forests and between storm events. These comparisons require standard measures of both storm and damage intensity, but no such standards exist.

Measurements of damage intensity are sensitive to species distributions and to stem size. Any standard for quantifying damage should minimise these influences. Species differences are most pronounced when differences in types of stem damage or in stem and branch damage are considered. Therefore a summary damage parameter that incorporates all types of stem damage would tend to minimise the species effects. One alternative for eliminating minimum stem size impacts would be for researchers to report their data at a variety of stem sizes. Comparisons could then be made using the larger stem classes, common to more studies. Another approach would be to use a damage parameter that is less sensitive to change in minimum stem size. The parameter with the smallest variation of damage intensity for all stem sizes examined in Table 19.3 is percentage basal area change. Basal area calculations minimise the effect of small stems and therefore vary little with changes in the minimum stem size.

A general measure of structural damage, such as basal area lost, may not be adequate to describe changes to a forest and in particular to predict the vector of recovery. Boucher (1990) reported the inconsistency in the impact of and recovery from Hurricane Joan in Nicaragua as: 'initial damage to pine forests was considerably less than to rainforests', but predicted 'long-term prospects for rainforest areas in Nicaragua to be considerably brighter'. Seventy-five per cent of the stems in the rainforest were broken or uprooted compared with 44% in the pine plantation, but mortality in pine was 58% compared with only 13% in the rainforest. The rainforest was affected more only if damage was quantified by percentage stems lost. In terms of mortality, the rainforest was less affected and its more rapid recovery is predictable. The impacts of a given catastrophic wind event might best be expressed as a combination of a general measure of structural damage (such as percentage basal area lost) and a measure of potential compositional change (such as percentage mortality). Fig. 19.2 illustrates this concept. The points on this graph are located using percentage mortality and percentage stem damage or basal area lost. This allows comparison of 13 disturbed forests. Mortality

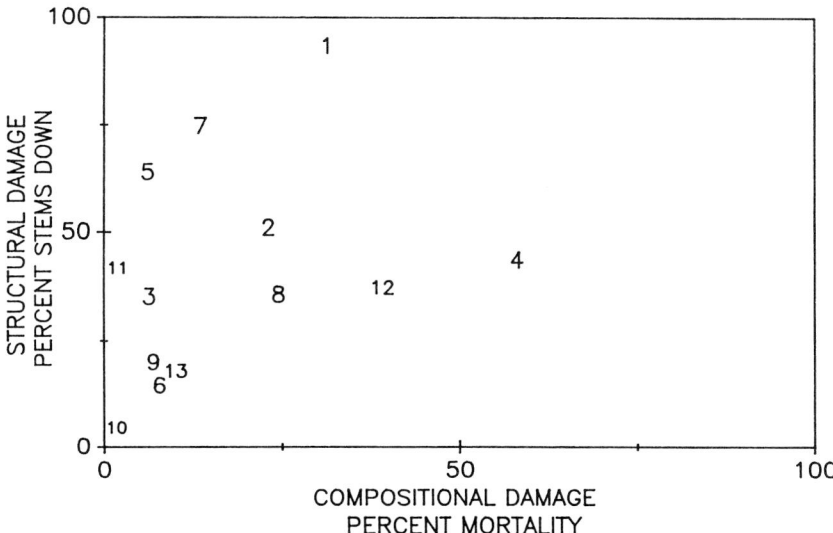

Fig. 19.2. Catastrophic wind damage measured as both structural and compositional change. 1, Foster (1988b); 2, Glitzenstein & Harcombe (1988); 3, Liegel (1984); 4, Boucher (1990); 5, Bellingham et al. (1992); 6, Bellingham (1991); 7, Boucher (1990); 8, Dallmeier et al. (1991); 9, Walker (1991); 10, Frangi & Lugo (1991); 11, Lugo et al. (1983); 12, Dittus (1985); 13, Whigham et al. 1991.

tends to be concentrated at the lower end of the gradient. The maximum reported mortality is that for a hurricane-affected pine plantation in Nicaragua (Boucher, 1990). The highest reported structural damage is 94% lost for the hurricane-affected temperate forests in New England (Foster, 1988b).

Recovery from wind disturbance might be expected to take one of three paths: regrowth, release or recruitment. Regrowth refers to the vegetative recovery of surviving stems. This type of recovery is dependent on the nature and extent of structural damage, as sprouting varies for uprooted stems versus snapped stems for some species, and on mortality. The importance of regrowth will probably decline as mortality increases. Release refers to the rapid growth of suppressed subcanopy trees. These trees may either be immature individuals of the dominant canopy species, or they may represent a significant shift in species dominance. Again this type of recovery is triggered by both structural damage and mortality to the canopy trees. Recruitment refers to recovery of the forest by the establishment of seedlings of early successional species. This type of recovery may occur in isolated parts of a wind-affected forest, but appears to be relatively rare as the principal path of recovery, possibly because of the generally low levels of mortality in this type of disturbance. The dual damage parameter presented in Fig. 19.2 would hold promise for predictions of the type of recovery that occurs at a given plot or for an affected forest overall. Unfortunately, the data concerning recovery following intensities of catastrophic wind are currently insufficient to test this concept.

To examine adaptations to wind we must be able to standardise how we examine damage. Different forests, with different trees, in different positions on the landscape, in different positions on the globe, are subjected to different frequencies of catastrophic wind. Data reported with separate categories for size classes, vegetative types (conifer versus broadleaf) and damage classifications would facilitate comparisons of the impacts of wind. A standard of all researchers reporting average values for their study sites of both structural damage and mortality should help us develop general hypotheses about the recovery of forests following catastrophic wind disturbance.

Acknowledgements

This research was supported by the US Department of Energy Global Change Fellowship Program, US National Science Foundation Long-term Ecological Research Program, Terrestrial Ecology Division of the University of Puerto Rico, and US Forest Service International Institute of Tropical Forestry. I thank Dan Ottenheimer for comments and ideas about the scope of the paper, Joe Cornell for help in the preparation of figures and Karyn Everham for help in editing, and an anonymous reviewer for clear, helpful comments.

References

Ackerman, J. D., Walker, L. R., Scatena, F. N. & Wunderle, J. (1991). Ecological effects of hurricanes. *Bulletin of the Ecological Society of America*, **72**, 178–80.

Anderson, K. F. (1954). Gales and gale damage to forests, with special reference to the effects of the storm of 31st January 1953, in the northeast of Scotland. *Forestry*, **27**, 97–121.

Applegate, G. B. & Bragg, A. L. (1992). *Recovery of Coastal Lowland Rainforest Damaged by Cyclone 'Winifred': A Photographic Record.* Queensland Forest Service, Technical Paper no. 51.

Basnet, K. (1990). Studies of ecological and geological factors controlling the pattern of tabonuco forests in the Luquilllo Experimental Forest, Puerto Rico. PhD Dissertation, Rutgers University.

Basnet, K., Likens, G. E., Scatena, F. N. & Lugo, A. E., (1992). Hurricane Hugo: damage to a tropical rain forest in Puerto Rico. *Journal of Tropical Ecology*, **8**, 47–55.

Bates, C. G. (1930). Hurricane damage to Puerto Rican forests. *Journal of Forestry*, **28**, 772–4.

Bellingham, P. J. (1991). Landforms influence patterns of hurricane damage: evidence from Jamaican montane forests. *Biotropica*, **23**, 427–33.

Bellingham, P. J., Kapos, V., Varty, N., Healey, J. R., Tanner, E. V. J., Kelly, D. L., Dalling, J. W., Burns, L. S., Lee, D. & Sidrak, G. (1992). Hurricanes need not cause high mortality: the effects of Hurricane Gilbert on forests in Jamaica. *Journal of Tropical Ecology*, **8**, 217–23.

Boose, E. R., Foster, D. R. & Fluet, M. (1994). Hurricane impacts to tropical and temperate forest landscapes. *Ecological Monographs*, **64**, 369–400.

Borchard, D., Legendre, P. & Drapeau, P. (1992). Partialling out the spatial component of ecological variation. *Ecology*, **73**, 1045–55.

Boucher, D. H. (1990). Growing back after hurricanes: catastrophes may be critical to rain forest dynamics. *Bioscience*, **40**, 163–6.

Boucher, D. H., Vandermeer, J. H., Yih, K. & Zamora, N. (1990). Contrasting hurricane damage in tropical rain forest and pine forest. *Ecology*, **7**, 2022–4.

Brokaw, N. V. L. & Grear, J. S. (1991). Forest structure before and after Hurricane Hugo at three elevations in the Luquillo Mountains, Puerto Rico. *Biotropica*, **23**, 386–92.

Brokaw, N. V. L. & Walker, L. R. (1991). Summary of the effects of Caribbean hurricanes on vegetation. *Biotropica*, **23**, 442–7.

Curtis, J. D. (1943). Some observations on wind damage. *Journal of Forestry*, **41**, 877–82.

Dallmeier, F., Taylor, C. M., Mayne, J. C., Kabel, M. & Rice, R. (1991). Effects of the Hurricane Hugo on the Bisley Biodiversity Plot, Luquillo Biosphere Reserve, Puerto Rico. UNESCO-MAB Digest II.

Darling, N. (1842). Hurricane in New England, September, 1815. *American Journal of Science and Arts*, **42**, 243–52.

Dittus, W. P. J. (1985). The influence of cyclones of the dry evergreen forest of Sri Lanka. *Biotropica*, **17**, 1–14.

Dunn, C. P., Guntenspergen, G. R. & Dorney, J. R. (1983). Catastrophic wind disturbance in an old-growth hemlock–hardwood forest, Wisconsin. *Canadian Journal of Botany*, **61**, 211–17.

Dunn, G. E. & Miller, B. I. (1960). *Atlantic Hurricanes* Louisiana State University Press, New Orleans.

Emanuel, K. A. (1987). The dependence of hurricane intensity on climate. *Nature*, **326**, 483–5.

Fernandez, D. S. & Fetcher, N. (1991). Changes in light availability following Hurricane Hugo in a subtropical montane forest in Puerto Rico. *Biotropica*, **23**, 393–9.

Foster, D. R. (1988a). Species and stand response to catastrophic wind in central New England, USA. *Journal of Ecology*, **76**, 135–51.

Foster, D. R. (1988b). Disturbance history, community organization and vegetation dynamics of the old-growth Pisgah Forest, south-western New Hampshire, USA. *Journal of Ecology*, **76**, 105–34.

Foster, D. R. & Boose, E. R. (1992). Patterns of forest damage resulting from catastrophic wind in central New England, USA. *Journal of Ecology*, **80**, 79–98.

Frangi, J. L. & Lugo, A. E. (1991). Hurricane damage to a flood plain forest in the Luquillo Mountains of Puerto Rico. *Biotropica*, **23**, 324–35.

Glitzenstein, J. S. & Harcombe, P. A. (1988). Effects of the December 1983 tornado on forest vegetation of the Big Thicket, southeast Texas, USA. *Forest Ecology and Management*, **25**, 269–90.

Gray, W. M. (1990). Strong association between West African rainfall and US landfall of intense hurricanes. *Science*, **249**, 1251–6.

Gresham, C. A., Williams, T. M. & Lipscomb, D. J. (1991). Hurricane Hugo wind damage to southeastern US coastal forest tree species. *Biotropica*, **23**, 420–6.

Irvine, R. E. (1970). The significance of windthrow for *Pinus radiata* management in the Nelson district. *New Zealand Journal of Forestry*, **15**, 57–68.

King, H. C. (1945). Notes on the three cyclones in Mauritius in 1945: their effect on exotic plantations, indigenous forest and on some timber buildings. *Empire Forestry Journal*, **24**, 192–5.

Liegel, L. H. (1982). *Growth, Development, and Hurricane Resistance of Honduras Pine in Puerto Rico.* 1982. Noveno Simposo de Recursos Naturales. Departmento de Recursos Naturales, San Juan, Puerto Rico.

Liegel, L. H. (1984). Assessment of hurricane rain/wind damage in *Pinus caribea* and *Pinus oocarpa* provenance trails in Puerto Rico. *Commonwealth Forestry Review*, **63**, 47–53.

Lodge, D. J., Scatena, F. N., Asbury, C. E. & Sanchez, M. J. (1991). Fine litterfall and related nutrient inputs resulting from Hurricane Hugo in subtropical wet and lower montane rain forests of Puerto Rico. *Biotropica*, **23**, 336–42.

Ludwig, J. A. & Reynolds, J. F. (1988). *Statistical Ecology: A Primer on Methods and Computing.* Wiley, New York.

Lugo, A. E., Applefield, M., Pool, D. J. & McDonald, R. B. (1983). The impact of Hurricane David on the forests of Dominica. *Canadian Journal of Forest Research*, **13**, 201–11.

Putz, F. E. & Sharitz, R. R. (1991). Hurricane damage to old-growth forest in Congaree Swamp National Monument, South Carolina, USA. *Canadian Journal of Forest Reserach*, **21**, 1765–70.

Reilly, A. E. (1991). The effects of Hurricane Hugo in three tropical forests in the US Virgin Islands. *Biotropica*, **23**, 414–19.

Sauer, J. D. (1962). Effects of recent tropical cyclones on the coastal vegetation of Mauritius. *Journal of Ecology*, **50**, 275–90.

Scatena, F. N. & Larsen, M. C. (1991). Physical aspects of Hurricane Hugo in Puerto Rico. *Biotropica*, **23**, 317–23.

Scatena, F. N. & Lugo, A. E. (1993). Natural disturbance and the distribution of vegetation in two subtropical wet steepland watersheds of Puerto Rico. Unpublished manuscript.

Scatena, F. N., Silver, W., Siccama, T., Johnson, A. & Sanchez, M. J. (1993). Biomass and nutrient content of the Bisley Experimental Watersheds, Luquillo Experimental Forest, Puerto Rico, before and after Hurricane Hugo, 1989. *Biotropica*, **25**, 17–27.

Sheffield, R. M. & Thompson, M. T. (1992). *Hurricane Hugo Effects on South Carolina's Forest Resource.* Research Paper SE-284, Southeastern Forest Experiment Station. USDA Forest Service.

Spurr, S. H. (1956). Natural restocking of forests following the 1938 hurricane in central New England. *Ecology*, **37**, 443–51.

Tanner, E. V. J., Kapos, V. & Healey, J. R. (1991). Hurricane effects on forest ecosystems in the Caribbean. *Biotropica*, **23**, 513–21.

Thompson, D. A. (1983). Effects of Hurricane Allen on some Jamaican forests. *Commonwealth Forestry Review*, **62**, 107–15.

Touliatos, P. & Roth, E. (1971). Hurricanes and trees: ten lessons from Camille. *Journal of Forestry*, **69**, 285–9.

Turton, S. M. (1992). Understory light environments in a north east Australian rain forest before and after a tropical cyclone. *Journal of Ecology*, **8**, 241–52.

Versfeld, D. B. (1980). An assessment of windfall damage to *Pinus radiata* in the Bosboukloof Experimental Catchment. *South African Forestry Journal*, **112**, 15–19.

Wadsworth, F. H. & Englerth, G. H. (1959). Effects of the 1956 hurricane on forests in Puerto Rico, *Caribbean Forester*, **20**, 28–51.

Walker, L. R. (1991). Tree damage and recovery from Hurricane Hugo in Luquillo Experimental Forest, Puerto Rico. *Biotropica*, **23**, 379–85.

Walker, L. R., Voltzow, J., Ackerman, J. D., Fernandez, D. S. & Fetcher, N. (1992). Immediate impact of Hurricane Hugo on a Puerto Rican rain forest. *Ecology*, **73**, 691–4.

Webb, L. J. (1958). Cyclones as an ecological factor in tropical lowland rain forest, North Queensland. *Australian Journal of Botany*, **6**, 220–30.

Wentland, W. M. (1977). Tropical storm frequencies related to sea surface temperatures. *Journal of Applied Meteorology*, **16**, 477–81.

Whigham, D. F., Olmsted, I., Cano, E. C. & Harmon, M. E. (1991). The impact of Hurricane Gilbert on trees, litterfall, and woody debris in a dry tropical forest in the northeastern Yucatan Peninsula. *Biotropica*, **23**, 434–41.

Whitmore, T. C. (1974). *Change with Time and the Role of Cyclones in Tropical Rain Forest on Kolombangara, Solomon Islands*. Commonwealth Forestry Institute Paper no. 46.

Whitmore, T. C. (1989). Changes over twenty-one years in the Kolombangara rain forests. *Journal of Ecology*, **77**, 469–83.

Wilkinson, R. C. Britt, R. W., Spence, E. A. & Seiber, S. M. (1978). Hurricane–tornado damage, mortality, and insect infestations of slash pine. *Southern Journal of Applied Forestry*, **2**, 132–4.

Wilson, H. H. (1976). The effect of the gale of August 1975 on the forests of Canterbury. *New Zealand Journal of Forestry*, **21**, 133–40.

Wood, T. W. W. (1970). Wind damage in the forest of Western Samoa. *Malayan Forester*, **33**, 92–9.

Wunderle, J. M. Jr, Lodge, D. J. & Waide, R. B. (1992). Short-term effects of Hurricane Gilbert on terrestrial bird populations of Jamaica. *The Auk*, **109**, 148–68.

Yih, K., Boucher, D. H., Vandermeer, J. H. & Zamora, N. (1991). Recovery of the rain forest of southeastern Nicaragua after destruction by Hurricane Joan. *Biotropica*, **23**, 106–13.

Zimmerman, J. K., Everham, E. M. III, Waide, R. B., Lodge, D. J., Taylor, C. M. & Brokaw, N. V. L. (1994). Responses of tree species to hurricane winds in subtropical wet forest in Puerto Rico: implications of tropical tree life histories. *Journal of Ecology* **82**, 911–922.

20
Windthrow and airflow in a subalpine forest

G. WOOLDRIDGE, R. MUSSELMAN
and W. MASSMAN

Abstract

A survey of the directions of coniferous tree windthrow in a subalpine ecosystem reveals a distinct relationship between the airflow and local terrain features. The survey covered a portion of the USDA Glacier Lakes Ecosystem Experiments Site (GLEES), located at an altitude of about 3300 m above sea level in Southeastern Wyoming, USA. The GLEES site is located in the lee of a glacial cirque basin in the upper treeline ecotone. The direction of the surface wind over the GLEES, as determined by meteorological data and the pattern of wind deformation of trees, is predominantly westerly. These winds result from channelling of larger-scale winds that blow around the south side of the mountain to the west of the GLEES. On the other hand, when the large-scale winds blow around the north side of the mountain, they may descend the ridge to the north of the GLEES at high speed from the northwest. The two modes of wind direction agree with observed bimodal windthrow directions.

20.1 Introduction

20.1.1 The GLEES Project

The Glacier Lakes Ecosystem Experiments Site (GLEES) in the Snowy Range of Southeastern Wyoming is a 600 ha alpine and subalpine watershed at an elevation of 3200–3500 m above mean sea level (msl). Research at the GLEES is coordinated by scientists from the USDA Forest Service Rocky Mountain Forest and Range Experiment Station, the University of Wyoming, and the Medicine Bow National Forest. Nearly 20 cooperating universities and governmental agencies together with scientists from other countries conduct research on ecosystem and atmospheric processes at the GLEES which relate to climate change.

The research site has a long history of grazing and recreational use, although grazing has recently been removed from this sensitive, wilderness-like ecosystem. The Forest Service has made a long-term research commitment for the GLEES, and has assembled an extensive database on ecosystem processes. The GLEES is accessible by road during the months of June to October, and by snowshoes, skis or snow machines during the remainder of the year.

20.1.2 Wind damage to trees

Stembreak and windthrow cause major economic losses in forests throughout the world. Catastrophic losses may occur during infrequent storms such as hurricanes or widespread gale-force and stronger winds associated with mid-latitude synoptic storms. Less spectacular but significant losses of forest trees and stands occur during thunderstorm downbursts or tornadoes, and in complex terrain where winds are channelled and accelerated by specific terrain features, such as along valleys or in the lee of ridges and mountains.

In the United States the susceptibility of old-growth conifer stands to smaller-scale windthrow events was studied by Veblen *et al.* (1991) in the subalpine zones along the Front Range of the Rocky Mountains. They found, through tree-ring chronology examination, that windthrows there were due mainly to frequent, low-intensity events rather than to infrequent major storms. In a survey of forest damage due to an incidence of hurricane-force winds in New England, Foster (1988) showed that windthrow direction agreed with the direction of the strongest winds, with a broad maximum in directional frequency due to easterly and northeasterly winds.

20.1.3 Terrain–airflow interactions and downslope lee side winds

Through investigations employing field experiments, numerical models and physical modelling, a large body of knowledge has developed concerning the changes in airflow as it encounters and flows over and/or around specific terrain features. The most commonly used non-dimensional parameter to characterise these flows has been the Froude number, which relates the flow structure to the height (h) of a terrain obstacle as follows:

$$F = \frac{U_\infty}{Nh} \quad (20.1)$$

where U_∞ represents the far upwind speed perpendicular to an obstacle and N the Brunt–Väisälä frequency.

The features of an airflow encountering a terrain obstacle that are of greatest interest here are the direction and speed of the downslope wind on the lee side, and the direction of the flow near the surface as it divides and flows around the obstacle. Kitabayashi *et al.* (1971) applied atmospheric wind tunnel technology to examine airflow over and downwind of an asymmetric mountain in Wyoming, USA. The mountain has a longer, more moderate approach slope than the steep lee side. Downwind, flow is often characterised by a strong downslope flow of air.

For a three-dimensional hill, Hunt & Snyder (1980) found that the highest Froude number for which lee side airflow separation was suppressed was $F = 0.8$. Strong lee side downslope winds occurred with Froude numbers of 0.4, 0.66 and 0.73 in their physical model. The values of the Froude number in airflow exhibiting an upstream block and accelerated downslope lee side flow found by Baines (1987) were in the range 0.45–0.65 for an infinitely deep, continuously stratified flow over a mountain. He concluded that for such flows the specific shape of an obstacle was of less importance than the structure of the approach flow.

The speed of the downslope flow behind a mountain ridge may accelerate by a factor of 2 or more (Smith, 1989). A high steep ridge is not required for this phenomenon to prevail; such hydraulic flow may exist behind small ridges, with attendant strong downslope winds (Bacmeister & Pierrehumbert, 1988). Aircraft data from the lee of a 300 m escarpment near Perth, Australia, showed the formation there of strong downslope winds, with a maximum speed in the lowest 300 m of the high-speed flow (Pitts & Lyons, 1989).

20.2 Airflows and topographic effects at the GLEES

At the beginning of the GLEES studies, a number of fixed sites were instrumented to provide long-term climatologies of wind direction and speed, precipitation, and other meteorological variables. To investigate transient phenomena, field investigations involving the use of a tethered balloon system and balloon soundings at a number of locations within and in the vicinity of the GLEES were pursued during the summer and autumn seasons of 1988–91. Data from these sources have proved useful for this study and are briefly described below as they relate to the topography and the airflow over the GLEES.

20.2.1 Local topography

The mountain ridge which lies upwind of the GLEES may be characterised as irregular and gently curved, with several peaks along its length (Fig. 20.1).

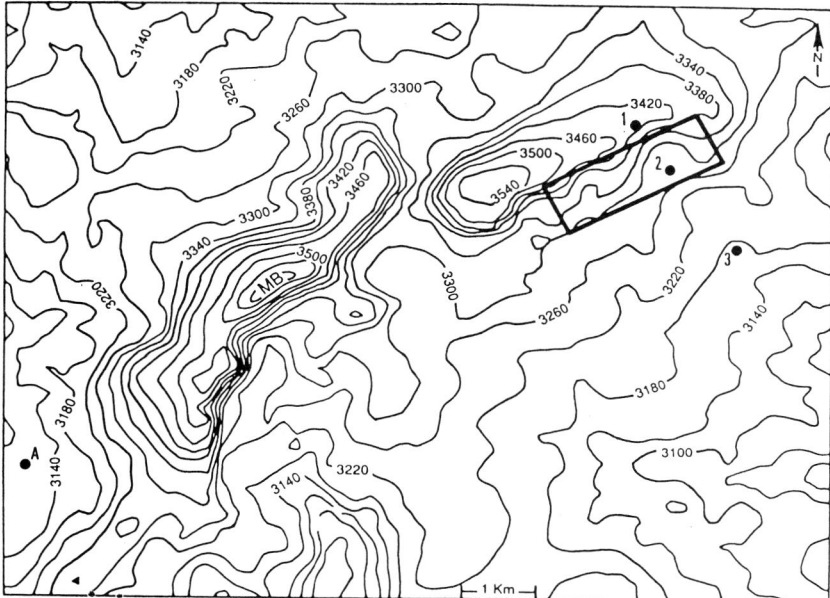

Fig. 20.1. Topography of the GLEES and the surrounding region, with elevations in metres above sea level. Anemometer sites are 1, 2 and 3; A is an upwind radiosonde launch site.

The upwind terrain first slopes upward gradually from west to east for more than 50 km before rising sharply from an elevation of about 3200 m at the base of the ridge to 3662 m at Medicine Bow Peak. Downwind to the east of the ridge lie several steeply sloped glacial cirques, and then the slopes become less steep. To the north, the ridge has two low peaks. On the south-southeast aspect of this second, lower portion of the ridge are located two permanent snow fields.

20.2.2 Winds at the GLEES

A 3 m tower was located at the summit of the ridge above the snow fields (site 1 in Fig. 20.1). There are no trees near this site, and bushes common to the top of the ridge are located farther to the north. Winter winds here blow predominantly from the west-northwest and northwest (Fig. 20.2). A second tower (hereafter called the GLEES tower and located at site 2 in Fig. 20.1) rises from a slight promontory near the centre of the GLEES. The wind instruments here are mounted at a height of 17 m above the ground, several metres above a sparse tree canopy. The winter wind rose (Fig. 20.3) indicates

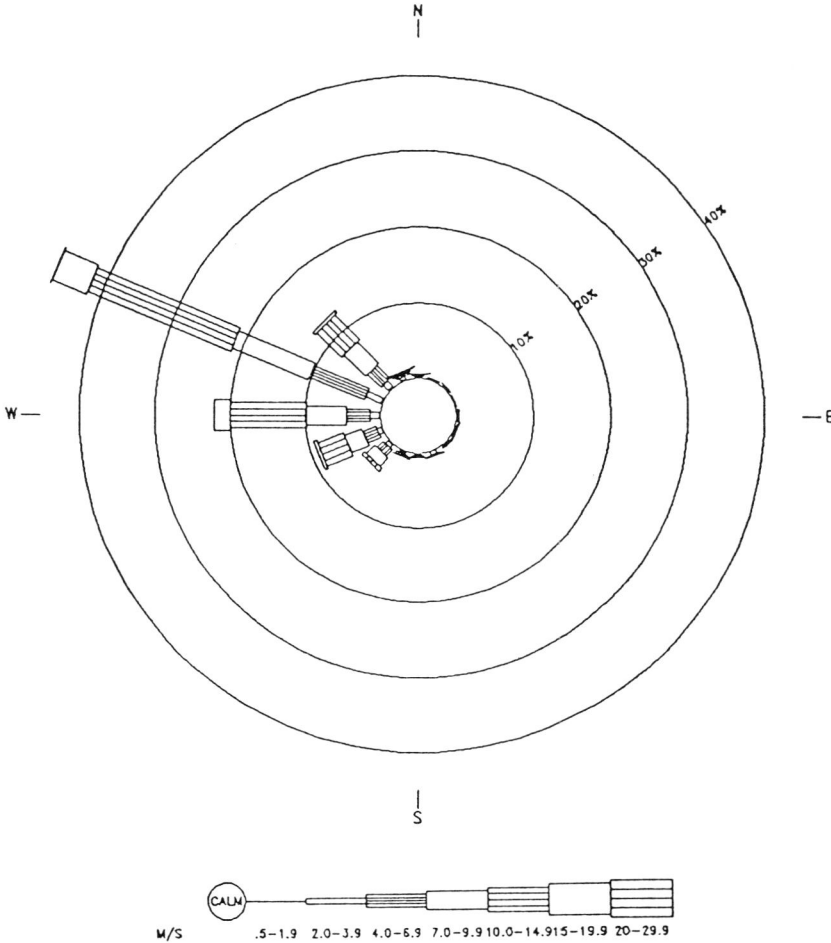

Fig. 20.2. Winter wind rose for the summit north-northwest of the GLEES (site 1 in Fig. 20.1).

a clear dominance of westerlies at this location. The third, or Brooklyn tower, at the south-southeast edge of the GLEES, is in a small clearing in a mature Engelmann spruce forest (site 3 in Fig. 20.1). The wind instruments are fixed at 30 m above the surface, about 12 m above the surrounding forest canopy. The most frequent wind direction is west-northwest, with west the second most frequent (Fig. 20.4).

A comparison between the exposures of the three wind sensor locations is difficult. The terrain at each site differs significantly, and the vegetation,

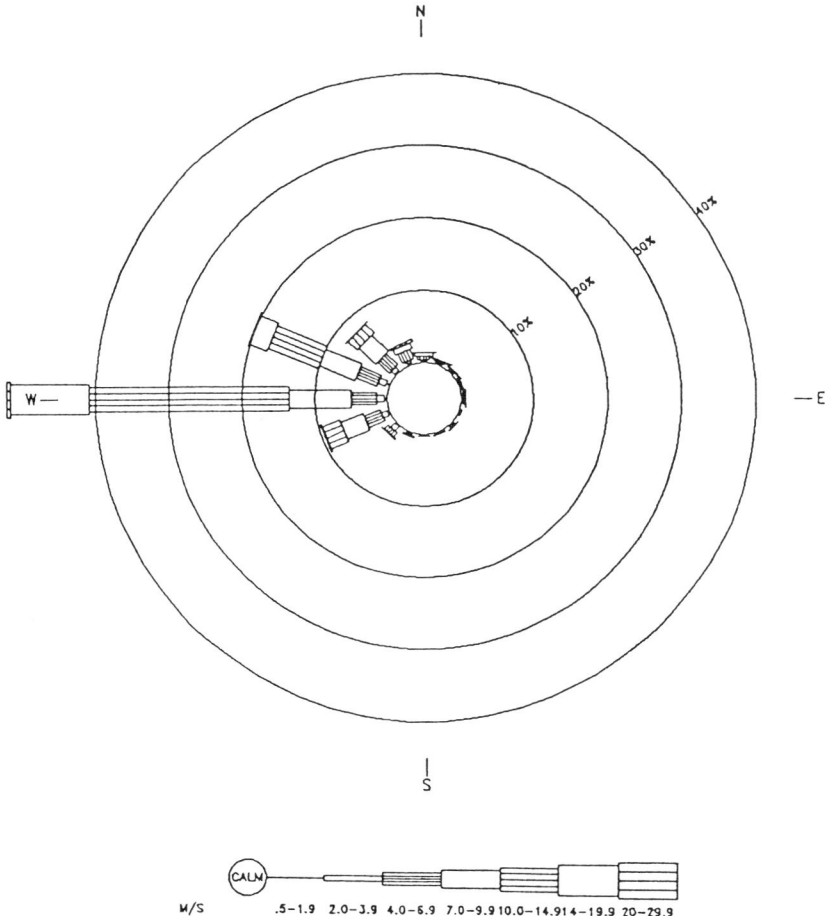

Fig. 20.3. Winter wind rose for the GLEES tower (site 2 in Fig. 20.1).

or trees, are of different heights and densities. However, at all three sites the sensors are above the immediate vegetation or tree canopies.

Analyses of wind directional frequencies for winds with gusts equal to or greater than 20 m s^{-1} for the three tower sites were completed, since research has indicated that gusts may be considered a significant aspect of windspeed producing wind damage to trees. The wind gusts reveal sharper directional maxima than the mean wind roses, but blow from much the same directions. The wind data from the short tower on the ridge to the north of the GLEES has been taken for reference since it is less likely to be affected by terrain

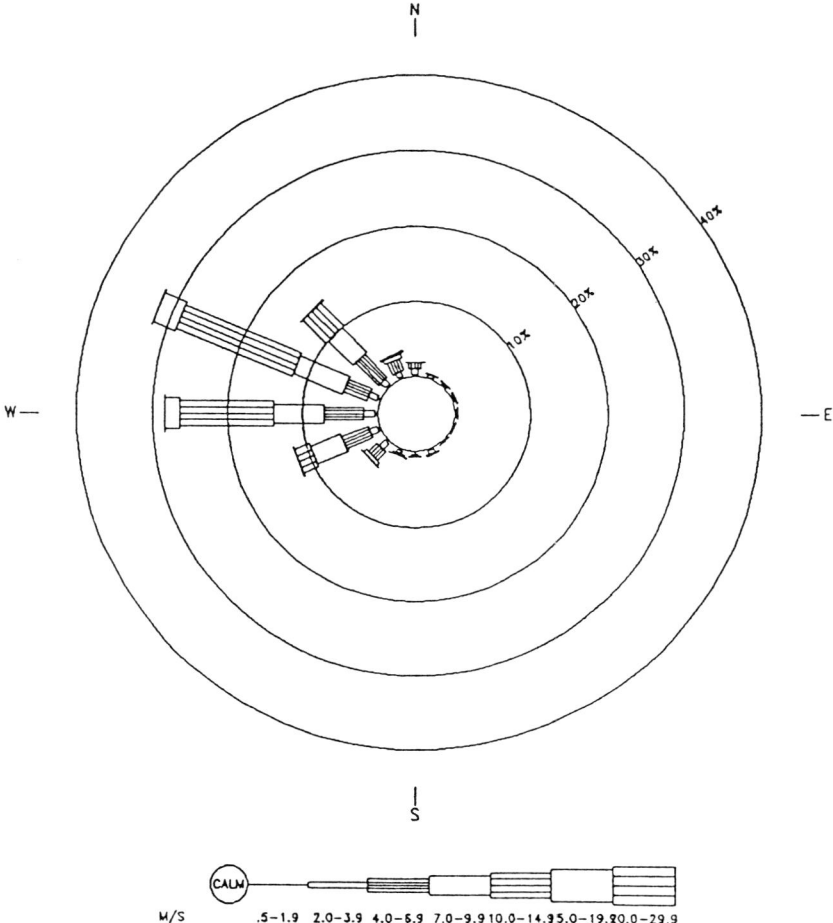

Fig. 20.4. Winter wind rose for the Brooklyn tower (site 3 in Fig. 20.1).

than the other tower wind data. The 20 m s^{-1} threshold value has been selected as a gust speed capable of damaging trees. No wind gusts over 20 m s^{-1} were recorded at any of the tower sites during the 3 year period from an easterly direction or during summer or early autumn months.

Fig. 20.5 indicates the channelling of strong winds which takes place over site 2 at the GLEES under the ridge to the north. For sites 1 and 3 the frequency peak is broader and shifted northward. A comparison of synoptic differences in wind directions between the GLEES tower and the Summit tower reveals the turning of the approach airflow as a function of wind direc-

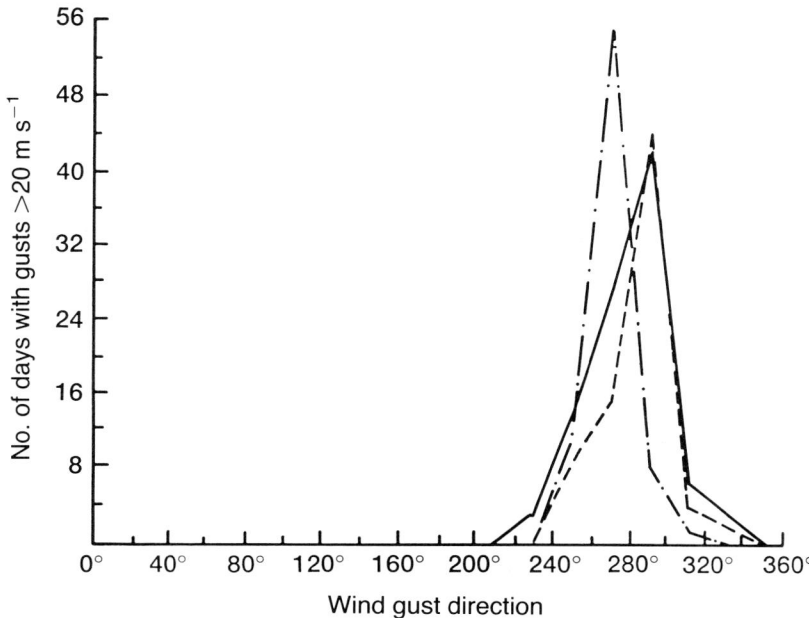

Fig. 20.5. Strong wind gust frequencies as a function of direction over: (1) the summit (continuous line); (2) Brooklyn Lakes (dashed line); and (3) the GLEES (dot-dashed line).

tion. When the approach direction is less than 265° the flow is deflected to the right as it meets the escarpment above the GLEES. As the approach flow veers more northerly the wind over the GLEES backs relative to the direction at the summit, with the greatest channelling effect at an approach direction of 310°. The effect quickly diminishes for flows from a more northerly approach.

Winds gusting at or over 20 m s^{-1} occur more often during nighttime than during daytime, with a broad maximum frequency during pre-dawn hours. A minimum frequency of occurrence prevails during afternoon hours (Fig. 20.6). Together with the directional frequency of strong gusty winds, this indicates that these winds occur frequently during periods when the atmosphere may be stable, rather than during afternoon thunderstorms.

A study of trees deformed by the prevailing winds over the GLEES (Wooldridge et al., 1992) has provided climatic patterns of airflow and windspeeds there (Figs. 20.7, 20.8). Trees were examined, photographed, and categorised according to deformation on a 100 m × 100 m grid. Two species,

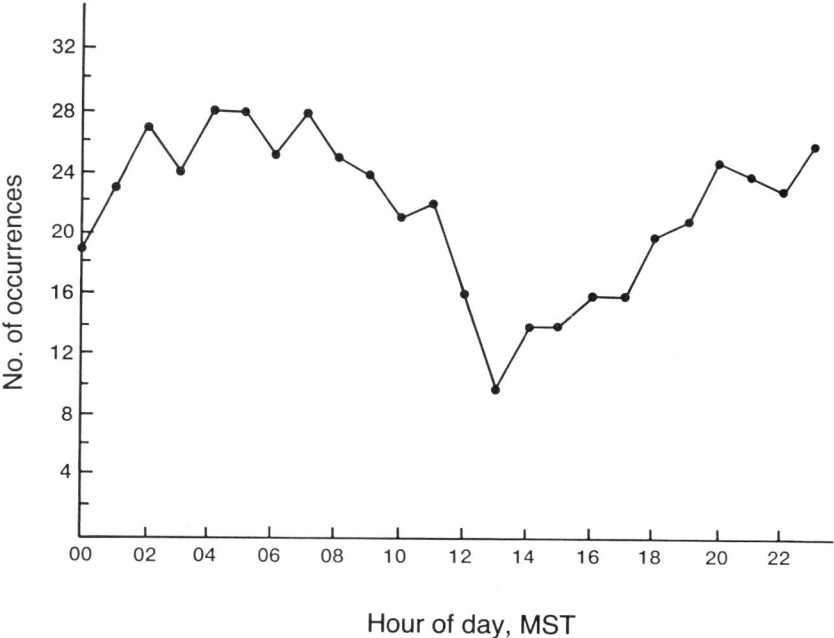

Fig. 20.6. Strong wind frequencies at the summit as a function of the hour of the day.

Engelmann spruce (*Picea engelmannii*) and subalpine fir (*Abies lasiocarpa*), are present over the lower portion of the GLEES in appropriate density to permit adequate sampling for this purpose. Krummholz and bare ground over the higher elevations prevent evaluation of mean windspeeds greater than about 12 m s^{-1}. Although small-scale topographic features provide some sheltering in certain locations, as seen in the isotach patterns in Fig. 20.8, the airflows reveal no leeside recirculation features.

An analysis of snow drifting discernible in aerial photographs reveals two wind patterns over and at the surface of the GLEES (Fig. 20.9). Larger drifts form in the deeper snow found in the GLEES wind channel, with more limited drifting from the northwest along the ridge and the southeastern border. These drifting patterns are present in aerial photographs taken over several years, indicating consistency in the prevailing winds and subsequent drifting.

20.2.3 Airflow above the surface layer at the GLEES

Three series of two-station radiosonde sequences were completed approximately along streamlines as air flowed eastward from a site upwind (A in Fig. 20.1) of the mountain ridge west of the GLEES. These paired soundings

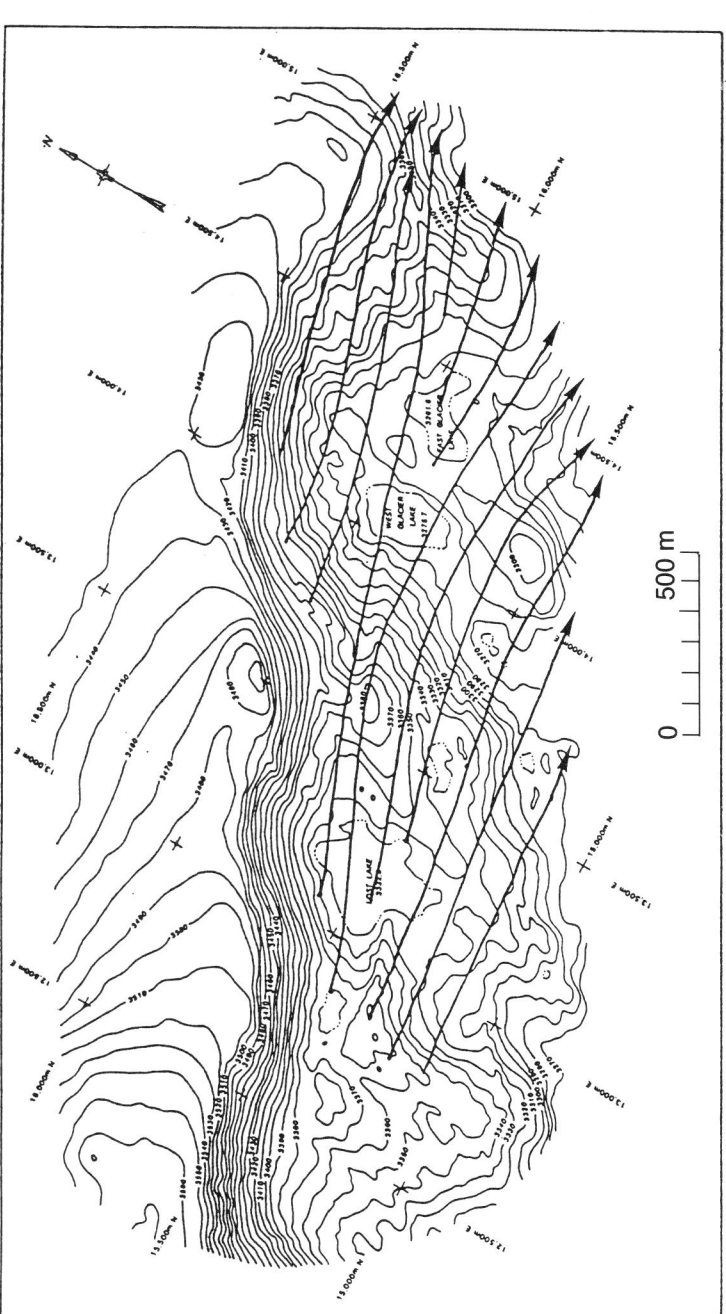

Fig. 20.7. Surface airflow over the GLEES as determined from tree deformation. After Wooldridge *et al.* (1992).

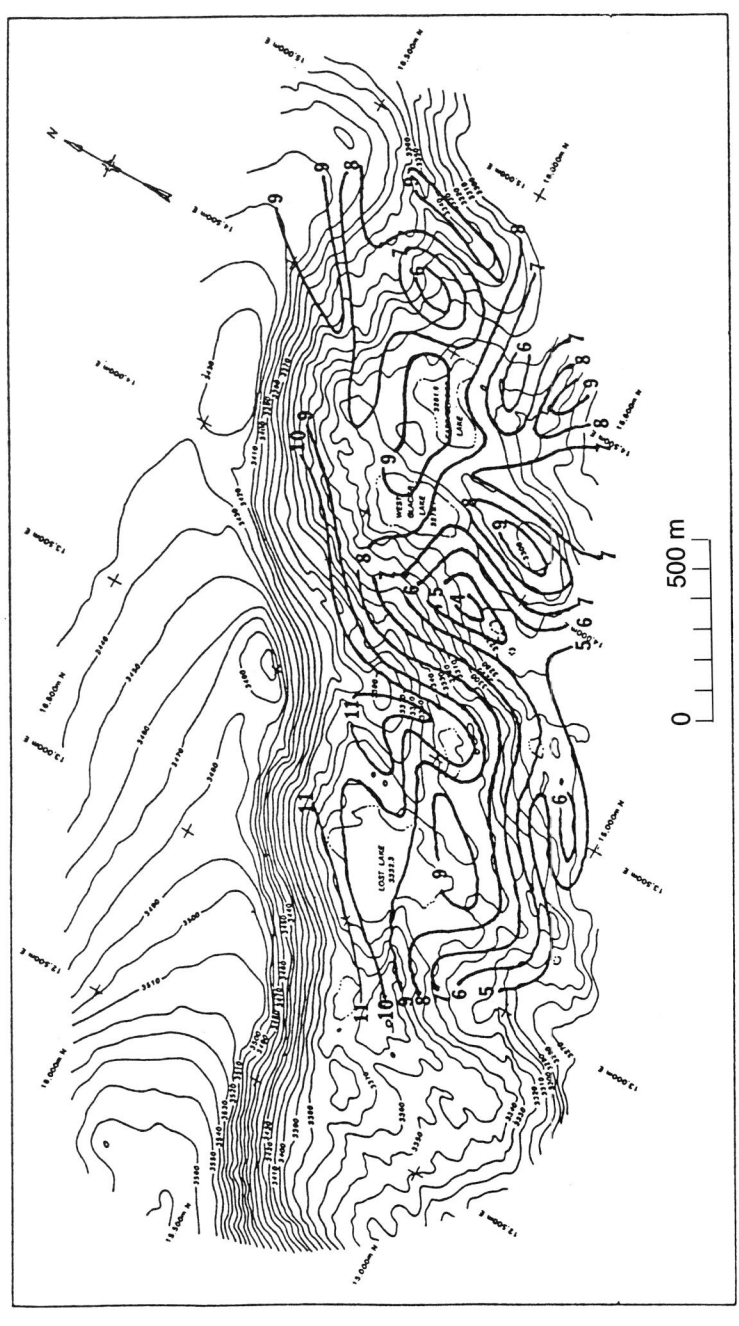

Fig. 20.8. Surface isotach pattern over the GLEES as determined from tree deformation (wind speed in m s^{-1}). After Wooldridge *et al.* (1992).

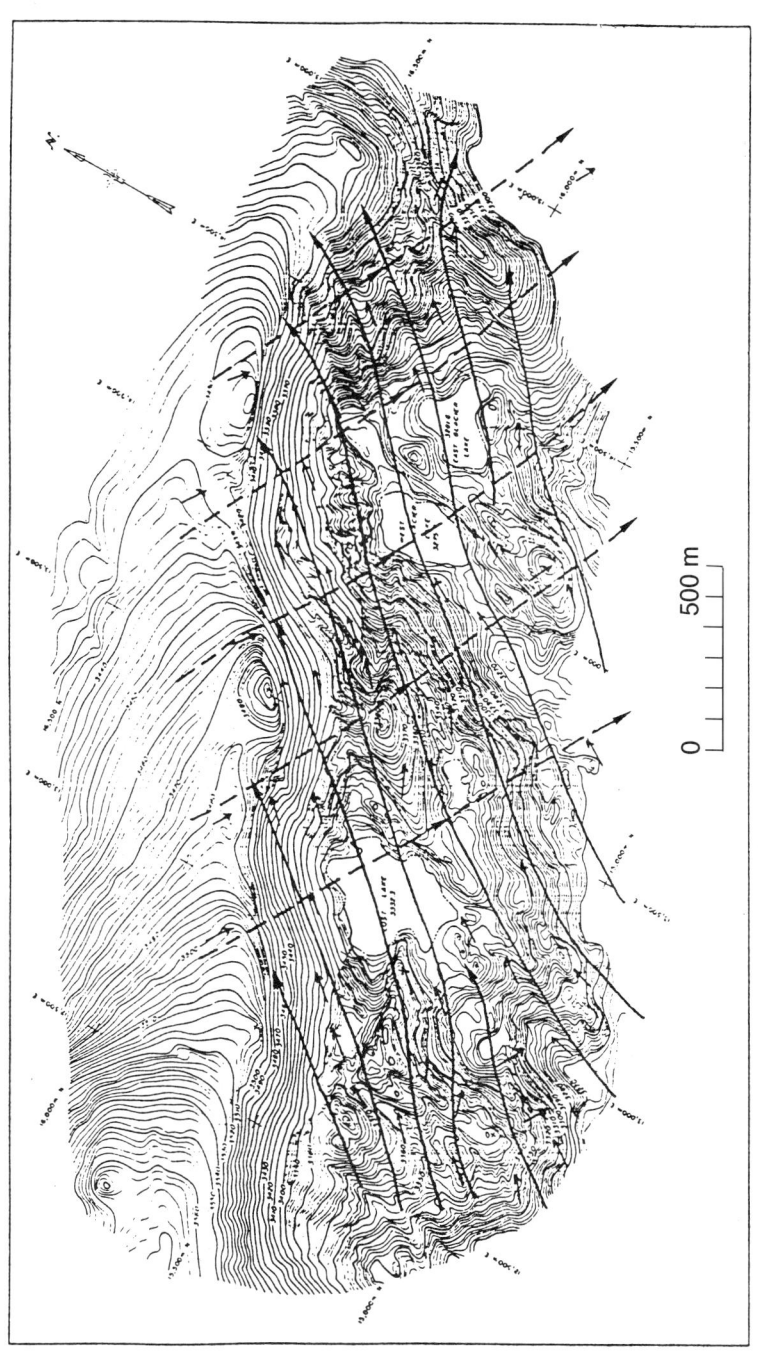

Fig. 20.9. Bi-level airflow patterns over the GLEES as shown in snowdrift orientations in aerial photographs. The flow in the 'channel' is given in continuous lines; the flow overhead is given in dashed lines. Smaller arrows indicate the orientation of individual drifts.

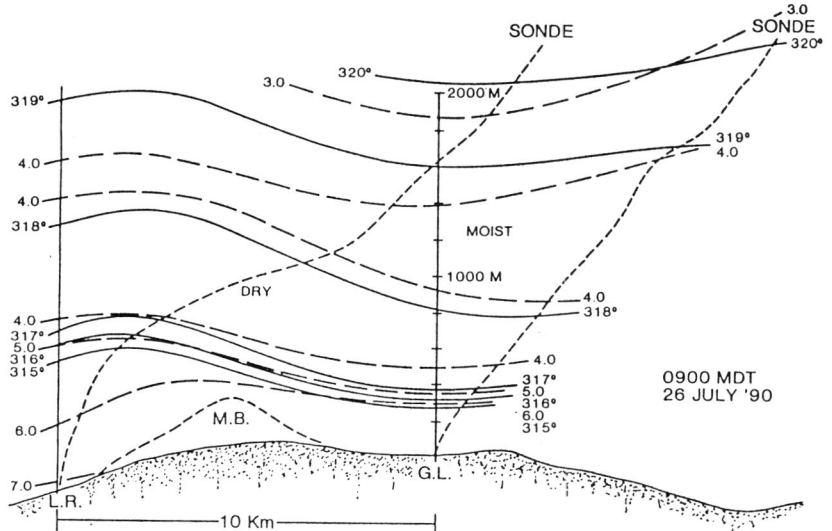

Fig. 20.10. Vertical section of the airflow from an upwind site (L. R. is A in Fig. 20.1) to the GLEES (G.L.). Potential temperature surfaces are labelled in degrees Kelvin; specific humidities in g kg^{-1}. Sonde trajectories are given in short dashed lines.

enabled representations of vertical sections of airflow over the terrain to the west of the GLEES as the Froude number was changing during morning hours (Wooldridge et al., 1992).

A vertical section of the flow structure is shown in Fig. 20.10 for 0803 hours Mountain Standard Time (MST) on 26 July 1990. The downwind sonde encountered windspeeds of 23 m s^{-1} at an elevation of 220 m above ground level (agl) over the GLEES. Windspeeds below and upwind of Medicine Bow Peak were very light; at elevations above 1000 m agl over both launch sites windspeeds were about 12 m s^{-1}.

At 0713 MST an upwind sounding showed slightly stable air up to an elevation of about 300 m agl with light winds; the Froude number was calculated at 0.45. At 0803 MST the Froude number had increased to 0.6, and by 0902 MST had reached 1.2. By 0902 MST the vertical section (not included here) showed that the lee side wave trough no longer dipped below the elevation of Medicine Bow Peak. The windspeed maximum was less distinct over the GLEES, weaker, and located at a elevation nearly equal to that of the Peak.

20.3 Windthrow at the GLEES

A survey and analysis of the orientation of windthrown trees in the northeast portion of the GLEES was conducted to relate the directions to the prevailing

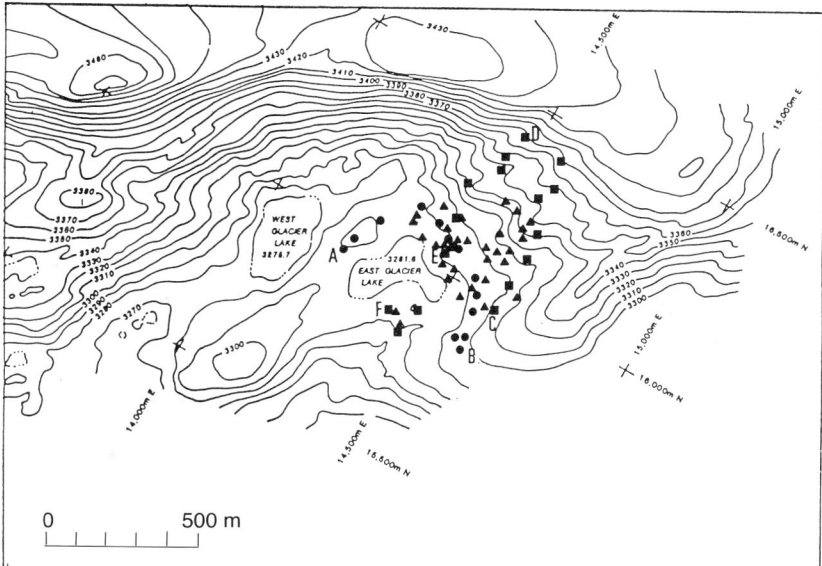

Fig. 20.11. Windthrown tree survey locations at the GLEES. Ground survey sites are at circles and squares; aerial photograph survey sites are at triangles.

wind directions. Aerial and ground survey data showed comparable results, and were combined.

20.3.1 Ground survey

The first part of the ground survey began between West and East Glacier Lakes (A in Fig. 20.11). Trees are sparse on this exposed part of the terrain, and have broken tops. This section of the survey line continued in an arc (filled circles) to point B. Windthrow was less frequent than stembreak near point A, but windthrow was almost exclusively the cause of tree loss along the rest of the survey line. Downwind, along this arc, the terrain slopes upward. A second ground survey line started at point C (squares) on higher terrain, then turned at D, descended toward East Glacier Lake at point E, and ended at point F. The individual sites for the ground-level survey were not chosen by any special random sampling scheme, but according to the presence of one or more windthrown trees.

Windthrown trees were photographed at each site, and the orientation of the trunk determined with a hand-held compass, corrected for magnetic declination. The frequency of windthrow direction is shown as the long-dashed line in Fig. 20.12. Two distinct peaks indicate the bidirectionality of windthrow. One peak agrees with the channelled wind direction across the GLEES; the

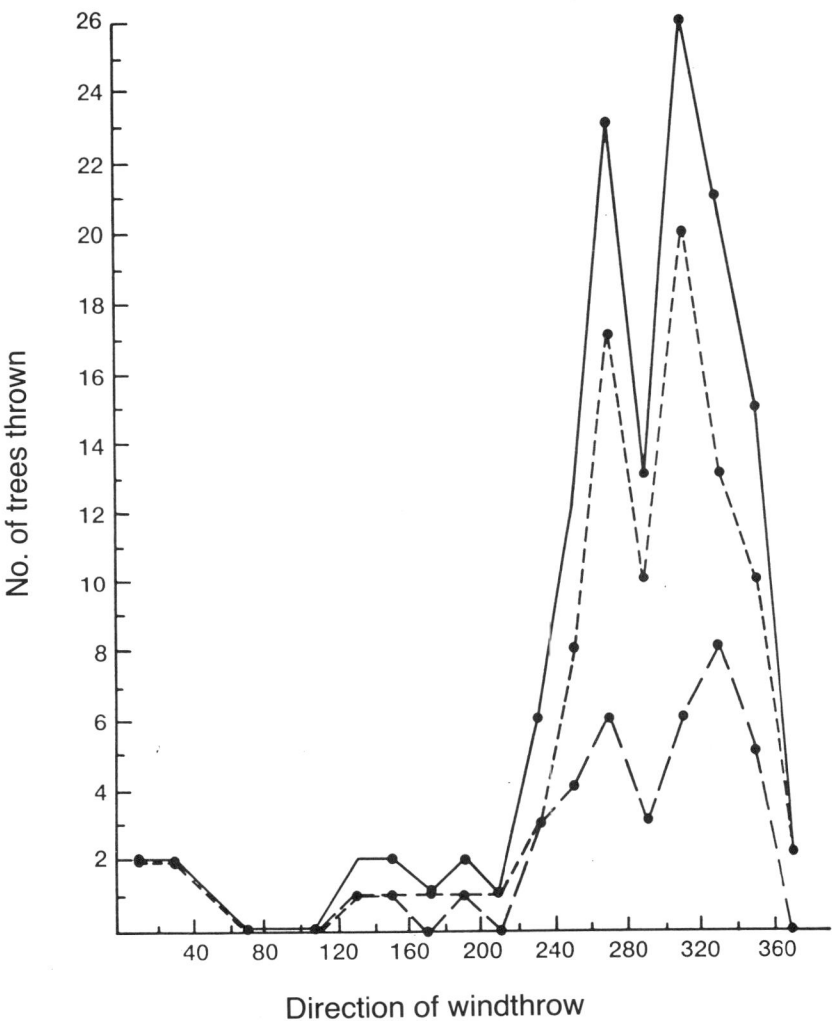

Fig. 20.12. Frequencies of windthrown tree orientations at the GLEES. Ground survey results are shown as a long-dashed line; results from analyses of aerial photographs as a short-dashed line; and combined results in the continuous line.

second peak occurs slightly more northerly than the most frequent wind direction recorded at the summit and on the tower at Brooklyn Lakes.

20.3.2 Aerial photography survey

The snow survey research group for the GLEES has been recording the depositional and melt patterns of the snow cover for spring seasons through application of aerial stereophotography. Thirty-five locations on the photographs (filled triangles), clustered in the same region of the GLEES as the ground survey, were identified and transformed into digital images for study. Windthrow directions of 92 trees at these locations were determined, and are presented as the short-dashed line in Fig. 20.12. The peak in the directional frequency along the mean wind is the same as for windthrown trees examined in the ground survey, but the second, higher peak frequency at 310° is slightly less northerly than that for the ground survey. It was determined that four trees were duplicated in the surface and the aerial photographs; these were removed from the aerial survey population. The average difference in direction recorded for these four windthrown trees was 4°. The combined population of 131 windthrown trees has the directional frequency given by the continuous line in Fig. 20.12.

20.4 Discussion

The windthrow pattern observed at the GLEES results from the effects of terrain on airflow at several scales and the morphological reaction of the trees to the winds. As persistent westerly winds blow over the Medicine Bow Mountains they diverge around and/or flow over Medicine Bow Peak, to the west of the GLEES. Some convergence takes place in the lee of the Peak, along the ridge that lies in a southwest–northeast orientation to the northwest of the GLEES. The southwesterly branch of the flow is channelled along the lee of the ridge at the northwest edge of the GLEES. Air which flows over the Peak during periods when the Froude number ranges between about 0.4 and 0.8 'shoots' down the lee side to flow eastward over the GLEES. When the macroscale airflow in this same Froude number range veers towards the northwest, it may descend the lee side of the ridge and accelerate, depending on the Froude number of the flow. These terrain-affected flows greatly affect the forest in the eastern section of the GLEES.

The bidirectional pattern of windthrow differs slightly from the airflow pattern. Although the most frequent strong winds recorded at the GLEES tower blow from the west, the most frequent windthrow direction in the survey site is from the northwest. This indicates that the majority of the trees are thrown by strong winds descending the ridge along the northwestern border of the GLEES.

Two factors may account for the small difference between the dominant strong west-northwesterly winds at the Summit tower on the ridge and the northwesterly direction of the tree windthrow. One may be an asymmetry of the tree root structure due to stronger root development in response to the predominant westerly winds near the surface of the GLEES. This could result in a change in the moments exerted on the roots and cause the tree to fall at an angle to the winds coming over the ridge. A second factor involves the turning of the wind as it accelerates down the lee side of the ridge. If a west-northwesterly 20 m s^{-1} wind approaches the ridge, and its component perpendicular to the ridge doubles as it descends on the lee slope, as suggested by Smith (1989), it will veer approximately 20°.

It appears that in the complex terrain in which many trees grow, definitive airflow studies specific to the area in question will be required to determine the potential for wind damage. The studies should consider all meteorological and terrain scales and the appropriate terrain–airflow interactions; the detail of the study will depend on the complexity of the terrain and the characteristics of the approach flow.

Acknowledgements

The authors wish to acknowledge the support of the United States Department of Agriculture Forest Service for the Glacier Lakes Ecosystem Experiment Project 4452. They also wish to acknowledge the support of, and interactions with, the other scientists and technicians who form the study team of this project. In particular, we are indebted to Dr Richard Sommerfeld for his analysis of aerial photographs of the GLEES and to Ms Bernadette Connell for making wind data and summaries available. The suggestions offered by reviewers Professors Robert Meroney and David Miller and by Ms Connell are appreciated and have been useful in the preparation of this manuscript.

References

Bacmeister, J. T. & Pierrehumbert, R. T. (1988). On high-drag states on nonlinear stratified flow over an obstacle. *Journal of the Atmospheric Sciences*, **45**, 63–80.

Baines, P. G. (1987). Upstream blocking and airflow over mountains. *Annual Review of Fluid Mechanics*, **19**, 75–97.

Foster, D. R. (1988). Species and stand response to catastrophic wind in Central New England, USA. *Journal of Ecology*, **76**, 135–51.

Hunt, J. C. R. & Snyder, W. H. (1980). Experiments on stably and neutrally stratified flow over a model three-dimensional hill. *Journal of Fluid Mechanics*, **96**, 671–704.

Kitabayashi, K., Orgill, M. M. & Cermak, J. E. (1971). *Laboratory Simulation and Atmospheric Transport-Dispersion over Elk Mountain, Wyoming*. Fluid Dynamics and Diffusion. Laboratory Report 16, CER 70–71 KK-MMO-JEC-65, Colorado State University, Fort Collins, CO.

Pitts, R. O. & Lyons, T. J. (1989). Airflow over a two-dimensional escarpment. I. Observations. *Quarterly Journal of the Royal Meteorological Society*, **115**, 965–81.
Smith, R. B. (1989). Hydrostatic airflow over mountains. *Advances in Geophysics*, **31**, 1–38.
Veblen, T. T., Hadley, K. S. & Reid, M. S. (1991). Disturbance and stand development of a Colorado subalpine forest. *Journal of Biogeography*, **18**, 707–16.
Wooldridge, G. L., Musselman, R. C., Connell, B. B. & Fox, D. G. (1992). Airflow patterns in a small subalpine basine. *Theoretical and Applied Climatology*, **45**, 37–41.

Part V

Risk assessment and management response

21

Assessing the risk of wind damage to forests: practice and pitfalls

C. P. QUINE

Abstract

The nature of wind damage as a risk to managed forests is reviewed, based primarily upon experience in Britain. Risk assessment is required to guide site selection, choice of techniques to counter the damage, and plans to respond to damage. Assessment may simply rank sites but prediction of timing of damage is also needed for many decisions. Subjective assessments can be unreliable where there is little history of forest management in that locality or when the damage is determined by many factors. Objective assessments are preferred but are not easy to derive. The existing British windthrow hazard classification is readily applied to sites and can be mapped. However, both its treatment of the wind climate and its estimates of timing of damage are deterministic. The potential to improve the system is discussed using a conceptual model of the assessment of risk which highlights a number of important research issues. Wind damage occurs as a result of an interaction between a variable wind climate and changing tree vulnerability. Vulnerability includes persistent, progressive and episodic components. There can be substantial variability in the frequency and magnitude of damaging storms which is not adequately represented by the conventional separation into storms causing endemic and catastrophic damage. The change in vulnerability with time can have a very marked effect on the frequency of damage. Recent validation has indicated that the estimates of damage based on the hazard classification of damage tend to be pessimistic, and that there is more variability in rate of windthrow than the classification allows. The variability requires a stochastic model and necessitates flexible forest planning; the latter conflicts with many pressures on forest managers. Small changes in tree vulnerability caused by the actions of forest managers or a small shift in wind climate can have a major influence on the likelihood of damage. An improved understanding of the change in vulnerability of trees with time is needed as this is crucial to many management responses.

21.1 Why assess the risk of wind damage?

Damage to trees by strong winds threatens the productivity of managed forests because it normally occurs in mature or semi-mature stands; it is therefore one of the most important abiotic hazards. Losses can be substantial and include reduction in harvestable timber due to breakage and degradation, increased harvesting costs, loss of potential production, and temporary glutting of the market; the windblown timber may also become a breeding source for damaging insect pests. The profitability of forestry is very sensitive to the rate of windthrow. Manley & Wakelin (1989) showed that a 1% annual windthrow rate reduces forest productivity to 90% of that of the undamaged forest, while a 4% rate results in only 63% of the optimum productivity. In Britain premature felling in anticipation of windthrow was implicated as a major factor in the £5.6 million revenue foregone per year due to harvesting at non-economic optimum age; 37% of stands were felled more than 5 years before the age of maximum economic return (NAO, 1993). Approximately 8% of all serious harvesting accidents in the Forestry Commission in the period 1987–92 occurred to operators working windblown trees (E. Ramsay, personal communication).

The likelihood of damage due to strong winds can be regarded as a risk, and the nature of this risk can be quantified in a risk classification. Quantification of risk permits managers to consider the hazard and their response. One strategy is hazard avoidance (Thompson, 1976) by choosing to invest only in the most stable sites. The risk can also be spread within an organisation by obtaining insurance, by selecting sites with a national, regional or local spread, or by managing the age-class distribution as a normal forest. The risk can be confronted by adopting treatments to site or crop which seek to mitigate the damage as opposed to those which may exacerbate damage; choice of ground preparation, spacing, plant type, species, thinning regime, stand edge treatments and length of rotation are all influential. In some instances the most appropriate response may simply involve formulation of contingency plans in the event of disaster (Holtam, 1971; Christensen, 1987; Grayson, 1989; Quine & Gardiner, 1991), which may include definition of survey techniques, clearance priorities, marketing methods and replanting policy.

Many of the preventive measures are costly due to loss of production or additional cost of operations. An estimate of timing of damage is necessary if the value of such operations is to be assessed – simple ranking of treatments is insufficient as the cost of the most stabilising treatment may far outweigh the benefits (Wardle, 1970). Therefore a common question is how much more stable a given treatment will make the stand. Identification of the location and type of stands most at risk is also needed to target preventive measures to best effect.

21.2 How should risk be assessed?

Risk may be judged informally through personal perception or quantified in a formal classification. The former will vary markedly between individuals and will be strongly influenced by direct experience of the damaging agent. Perceived risk may differ significantly from that assessed objectively (Smith, 1992). Formal risk assessment is common in many other disciplines, most notably in engineering for design of major buildings against potential flood and storm damage. In these circumstances it is possible to calculate the magnitude of the event that will damage the structure and then the probability of this event occurring from climatic statistics. Many public buildings are designed to withstand an event expected to occur once in 50 years (Cook, 1985). Climatic risk studies are well advanced in agriculture (Muchow & Bellamy, 1991), where annual cropping allows rapid development of datasets that identify the links between climate and crop growth. This understanding permits the impact of differing management options to be identified.

The probability of damage may also be derived from magnitude/frequency data on the occurrence of the phenomena under study. However, a substantial time series is required and such assessments cannot forecast the effectiveness of new techniques or operations. Time series of areas lost to fire have been used to derive the probability of fire losses (Dempster & Stevens, 1987) but depend upon careful recording over many years and few changes in the structure or extent of the forest. However, this technique is inappropriate in those countries such as Britain where the forest area has changed substantially.

21.3 Wind risk assessment in practice

The dynamic behaviour of trees in the wind (stem and branch swaying, streamlining) has frustrated simple assessments of the windspeed required to blow over trees of different form and anchorage. Where windthrow is progressive, the enhanced risk due to the presence of existing windthrow is also a problem largely avoided in other risks (cf. fire, where the occurrence reduces the fuel build-up, and where damage to buildings is typically repaired). Furthermore, a detailed wind climatology of forest areas is rare and this makes the calculation of the probability of occurrence of damaging winds difficult.

The occurrence of wind damage has been studied by forest ecologists and the results can be regarded as similar to risk assessment; turnover rates (the time taken for an area equal to that of the forest to be disturbed) varying from tens to hundreds of years have been identified for forests in the United States (Canham & Loucks, 1984; Runkle, 1990).

In managed forests and plantations a number of classification systems have been devised to address the risk of wind damage. Many of these seek to rank sites or to rank treatments but do not specify the overall likelihood of damage. For example, Mayer (1988) provided a mapping method for assessing storm risk in Bavaria, as did Mackenzie (1974) in Northern Ireland. Hutte (1968) emphasised the role of topographic position and soils. Prien (1974) and Konopka (1973) reviewed historical data to estimate regional hazards. Prien (1974) emphasised the importance of soil type, proportion of Norway spruce and past damage, and recommended mapping, forest level management strategy and detailed silvicultural techniques as different levels of response. Moore (1977) listed factors contributing to the stability of riparian strips in British Columbia but did not attempt to rank or quantify the factors other than as high or low risk; Elling & Verry (1978) also considered the likelihood of leave strips blowing down in relation to their size. Many other authors have considered the impact of operations such as thinning (Savill, 1983; Hendrick, 1988; Valinger & Lundqvist, 1992) or have attempted to rank species or treatments following the evidence of individual storms (Cutler *et al.*, 1990).

One of the most ambitious systems is the British windthrow hazard classification (Booth, 1977; Miller, 1985; Quine & White, 1993), which combines an assessment of where damage is likely with an estimate of when in the life of the crop it will occur. It is applicable from the stand to the national level. The classification is both pragmatic and deterministic and has been shaped by the particular circumstances of British forestry in the twentieth century.

21.4 The windthrow hazard classification

21.4.1 Pertinent aspects of climate and forest history

Britain's wind climate is simple and severe and is dominated by extra-tropical cyclones which form in the Atlantic and pass west to east across or close to Britain. Approximately 160 such depressions/weather systems affect Britain per year, with the strongest winds usually experienced during winter months. The forest area of Britain has changed substantially during the twentieth century; the area of high forest declined from 0.9 million ha in 1913 to 0.6 million ha in 1924 and then rose to 2.1 million in 1992. Forest expansion has largely been in upland areas marginal for agriculture and for which there was no recent forest history. The wet soils and exposed conditions have meant that the risk of windthrow has become a major concern on such sites to the extent that it dominates management thinking.

21.4.2 Development of the classification

Research was prompted after damage to the earliest of the new plantations (Petrie, 1951; Maxwell Macdonald, 1952) and a number of major storms which caused extensive damage to all types of woodland (Andersen, 1954; Holtam, 1971; Quine 1988, 1991). It was believed that forest management could reduce the damage from typical storms (Petrie, 1951). A classification was developed based on site and damage surveys (Pyatt, 1968a, b), tree pulling (Fraser & Gardiner, 1967) and wind exposure measurements (Lines & Howell, 1963; Miller *et al.*, 1987). The classification was first applied to individual forests and was used to predict the likelihood of damage in terms of the area at risk above a certain critical threshold height (Pyatt, 1968b) and to identify thin/no thin zones (Foot, 1975). This was subsequently developed into the windthrow hazard classification, a national system that permits comparisons between sites across the country. Few data on actual rate of windthrow were available to develop the initial model because of the small area of mature forest.

The national classification originally combined three scores reflecting the windiness of the site and one score reflecting the rooting depth of the soil and hence anchorage of the trees (Booth, 1977; Miller, 1985). These scores were used to define six classes and there was an indicative height (critical height) per hazard class at which windthrow was believed to commence (i.e. commencement of risk). However, the demand for investment appraisal of afforestation projects and the need to forecast the production of existing stands for 20 years ahead required an estimate of the length of the rotation in stands potentially subject to windthrow. This estimate (terminal height per hazard class) was defined as being the stage when 40% (initially 33%) of the stand was damaged, at which point it was assumed that it would be most economic to clearfell the remaining trees (Busby & Grayson, 1981). The terminal heights were derived from a study forest, monitored between 1959 and 1976 (see discussion on Kershope Forest below). Terminal heights were presented as guides requiring local validation (Miller, 1986a) but, because of the lack of error bars attached to the figures, the simplifying assumption of endemic damage (see below) and the overriding influence of investment appraisal models, they were taken as precise estimates. The ready acceptance of the accuracy of the figures may have been influenced by the fact that they were published during a period in which several major storms occurred. The level of constraint indicated for the highest hazard class (hazard class 6) was substantial; felling an unthinned Sitka spruce stand of yield class 16 at a terminal height of 16 m gave a tree size only 36% of that achieved at the economic optimum, total volume 55% of the optimum, and price per cubic metre 50% of the optimum.

A number of damage surveys were carried out (Miller, 1986b) but these tended to concentrate on damaged stands with a consequent risk of bias. With much felling deliberately being anticipatory there was also no information on damage levels in stands approaching or exceeding terminal height. One forest that was monitored showed substantial damage but the monitoring was hampered by harvesting (Quine & Reynard, 1990).

21.4.3 The key assumptions of the classification

One of the key assumptions in the classification is the distinction between catastrophic and endemic damage. Exceptional storms caused catastrophic damage and the pattern of this overrode site differences. The storms did not have a simple meteorological definition but their identification depended upon the coincidence of strong winds (gusts of 40 m s^{-1} or more) with large areas of forest at risk. In some respects they represented the scale of storm that caused market disruption, and were largely events that blew down 1 million cubic metres of timber or more.

Endemic damage was characterised as the result of 'normal' winter storms (gusts around 30 m s^{-1}) and displayed a pattern that reflected site differences. The windthrow hazard classification was held to apply to endemic damage. Endemic damage was interpreted as the result of the intersection of increasing tree vulnerability (declining threshold windspeed) and a wind climate consisting only of 'normal' winter storms. Stands on stable sites were assumed to take longer to reach this intersection than those on unstable sites (Fraser, 1965), and thus initial damage was assumed to occur at a greater (critical) height (Fig. 21.1). A further assumption was that following initial windthrow there would be a regular progression to terminal height as stands were repeatedly damaged in successive winters. The annual rates of windthrow implied by the estimated critical and terminal heights ranged from 3% to 8% depending upon yield class.

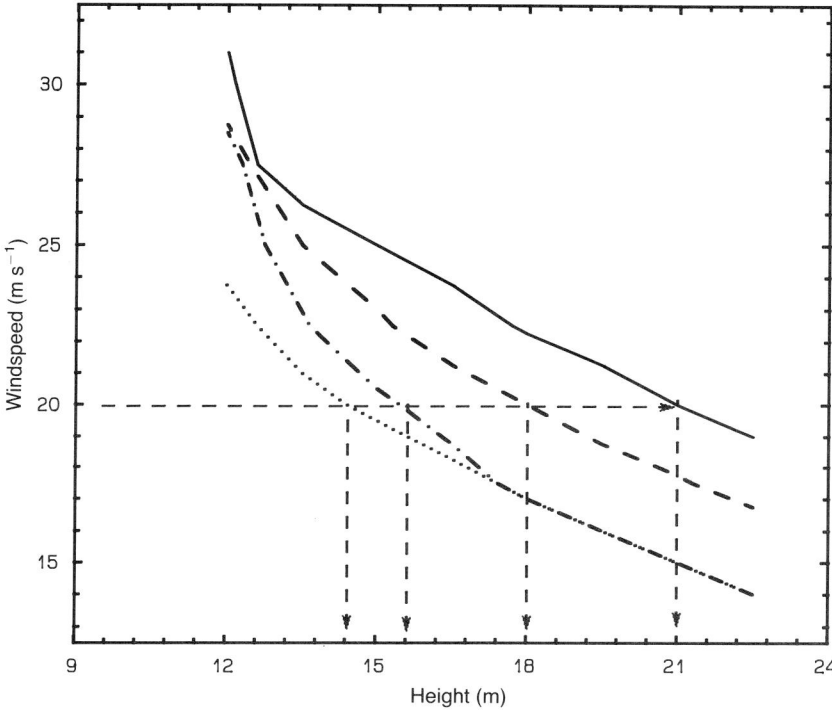

Fig. 21.1. The intersection of tree vulnerability with unvarying (endemic) wind climate to produce endemic damage. Curves show the relationship between critical velocity and tree size, for trees on different soil types: continuous line, brown earth; dashed line, deep peat; dot-dashed line, surface water gley; dotted line, peaty gley. Adapted from Fraser & Gardiner (1967).

21.5 Examining the hazard classification using a conceptual model

21.5.1 The conceptual model

Before probing the key assumptions of the hazard classification it is helpful to consider how the numerous factors affecting the stability of a tree contribute to the risk of windthrow. The siting, growth and management of a tree contribute to its vulnerability, which can be expressed as a critical threshold windspeed required to blow the tree over. The risk of a tree being blown over can then be defined as the probability of the threshold windspeed being exceeded on that particular site. The probability will increase (i.e. damage will become more likely) as the vulnerability increases and the threshold windspeed declines.

Three components can be identified to simplify our understanding of the change of tree vulnerability across space and through time (Fig. 21.2). The

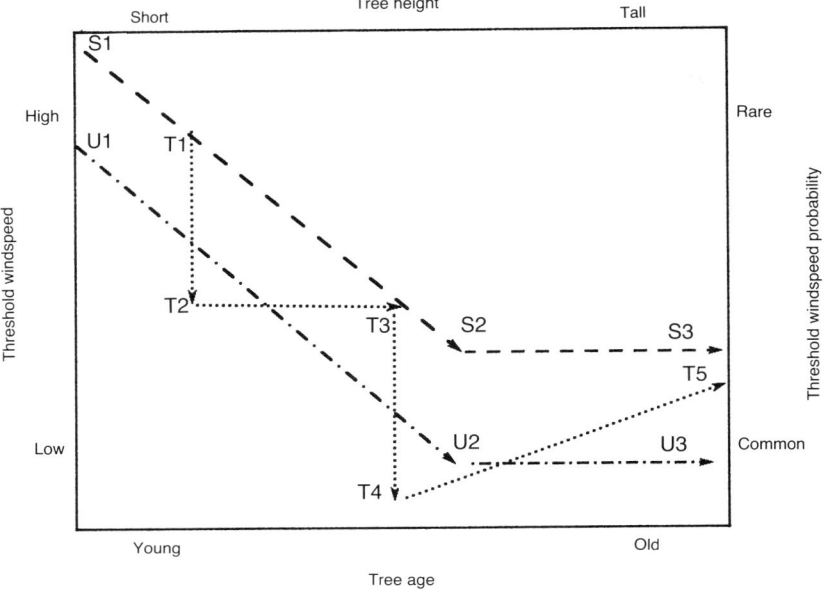

Fig. 21.2. The change in tree vulnerability with time expressed as three components: persistent, progressive and episodic. S1–U1 represents the difference in persistent vulnerability between a stable and unstable soil. S1–S2 and U1–U2 represent the change in progressive vulnerability due to the growth of the tree, and S2–S3 and U2–U3 a period when progressive change levels off. T1–T2 represents an episodic change in vulnerability induced by thinning, and T2–T3 the recovery period after thinning; T3–T4 represent a further thinning.

site contribution (due to the soil and its initial modification by site preparation) although spatially variable, changes little over the life of a tree, and is therefore a *persistent* component of vulnerability. Soils are classed as stable and unstable, reflecting the physical limitations on rooting and the strength of the soil. The persistent component establishes the baseline from which other changes occur. As a tree grows, the particular form it develops (e.g. height, diameter, crown form, root system) results in changes in vulnerability. These changes represent a *progressive* component of vulnerability. Finally, there are events that contribute to vulnerability for a period ranging from hours to years (e.g. removal of neighbours in thinning, loading of crowns by wet snow, soil saturation, loss of neighbours in storms, leafiness in deciduous trees). These form an *episodic* component of vulnerability.

Vulnerability therefore varies over time among trees in a stand, due to the particular combination of persistent, progressive and episodic components;

the importance of each component is dependent upon the additional contribution of the others. Thus in Fig. 21.2, in the absence of thinning, a hypothetical stand on the stable site (S1–S3) remains less vulnerable than an identical stand on the unstable site (U1–U3), despite having the same rate of change in progressive vulnerability. However, when thinned the stand on the stable site (which follows the vulnerability path S1–T1–T2–T3) is more vulnerable for a period than the unthinned stand on the unstable site, before recovering to the pathway of the unthinned at T3. A subsequent thinning (T3–T4) will cause the vulnerability to increase once again, and result in a different position at which vulnerability returns to that of the unthinned stand.

The strength of wind experienced by a tree, and hence the probability of the critical threshold windspeed being exceeded, will depend upon its location (e.g. in relation to the passage of weather systems), site factors such as elevation, topography, presence of upwind stands, and also the position of a given tree in the stand. The pattern of damage left after a storm will reflect the spread of vulnerability within a stand and characteristics of the gust structure of the storm.

21.5.2 Model specification for simulation

We can tailor this conceptual model to British conditions to explore the assumptions of the windthrow hazard classification. For the purposes of the examples the progressive component of vulnerability is treated at a linear decline of threshold windspeed with age (*Rapid* 1.0, *Moderate* 0.5, and *Slow* 0.25 m s^{-1} per year); the starting point of the progressive component is determined by the persistent component (*Stable* 58 m s^{-1}, *Unstable* 48 m s^{-1}); and the endpoint of the progressive component is unspecified or treated as the point at which the decline in threshold windspeed levels off (*Low* 30, *Mean* 33, *High* 36 m s^{-1}). Episodic vulnerability is not included in the simulations that follow.

The timing of first damage occurs when the threshold windspeed (persistent and progressive vulnerability) is exceeded or equalled by the windspeed that occurs that year. Only the single maximum gust per year is assumed to be important in causing damage. The windspeed is either the mean of the annual maximum gusts (the *endemic windspeed*) or a time series of the annual maximum gusts (the *time series windspeed*). The windspeeds have been obtained from records of annual maximum gusts 1914–91 for Eskdalemuir contained in Meteorological Office Monthly Weather Reports. Eskdalemuir is a Meteorological Office observatory situated in a mid-slope position in the Southern Uplands of Scotland (elevation 259 m, latitude

55° 19′ N, longitude 03° 12′ W). A return period analysis was calculated for the period 1958–91 using the Lieblein method outlined by Cook (1985).

Terminal damage occurs when 40% damage of the stand is first recorded, i.e. equalled or exceeded. The amount of damage is generated by the magnitude of the exceedance of the threshold windspeed by the endemic or time series windspeed. The threshold windspeed is taken to refer to the most vulnerable tree in the stand and the distribution of vulnerability within the stand is described by a logistic function; this is chosen so that 50% damage (i.e. damage to 50% of the trees in the stand) occurs when the threshold windspeed of the most vulnerable tree is exceeded by 15 m s^{-1}. The distribution of vulnerability in the stand is assumed to remain unchanged after damage; this can be interpreted as enhanced vulnerability of remaining trees in the stand following damage to the most vulnerable. If this assumption does not hold it implies that a windspeed of greater magnitude than that preceding is required to extend damage, i.e. the sequence of winds becomes important, or that the delay between winds is sufficient to cause progressive vulnerability to lower the threshold windspeed.

21.5.3 An endemic windthrow version of the conceptual model

Although the application of the conceptual model is not how the hazard classification was explicitly constructed, it helps to highlight the assumptions and inadequacies. The classification treats persistent vulnerability through selection of a soil score, progressive vulnerability is linked to height, and the windspeed likely to be encountered at a site is defined as a wind score that does not vary. We can therefore set up the model in an endemic form, such that damage occurs when the threshold windspeed intersects the endemic windspeed (Fig. 21.3). Note that episodic vulnerability is essentially ignored by the classification and stands which are thinned are treated as though they never recover to the unthinned progressive vulnerability.

Taking the *moderate* rate of change in progressive vulnerability, an age of first damage of 31 years for an *unstable* site and 51 years for a *stable* site at Eskdalemuir is obtained. The differential in timing between *unstable* and *stable* would be maintained to terminal height 12 years after first damage, giving the same annual windthrow rates for the two sites. Changes in the rate of threshold windspeed decline alter the age of first damage, the progression time to terminal height, and therefore the annual windthrow rate (Table 21.1). There is no scope for variation in the timing of first or terminal damage for stands planted in different years and the model can therefore be regarded as deterministic.

Table 21.1. *Age of first windthrow and terminal damage (years) plus mean annual windthrow rate (%) for a stand at Eskdalemuir on an* unstable *and a* stable *site with* endemic windspeed

Rate of change:	Slow		Moderate		Rapid	
Endpoint:	None		None		None	
Site:	Unstable	Stable	Unstable	Stable	Unstable	Stable
Mean age at first damage:	61	101	31	51	16	26
SD:	n.a.	n.a.	n.a.	n.a.	n.a.	n.a.
Mean age at terminal damage:	78	118	43	61	24	34
SD:	n.a.	n.a.	n.a.	n.a.	n.a.	n.a.
Mean annual windthrow rate (after first damage):	2.3	2.3	3.3	3.3	5.0	5.0

Comparisons are made between different rates of change of progressive vulnerability. n.a., not applicable.

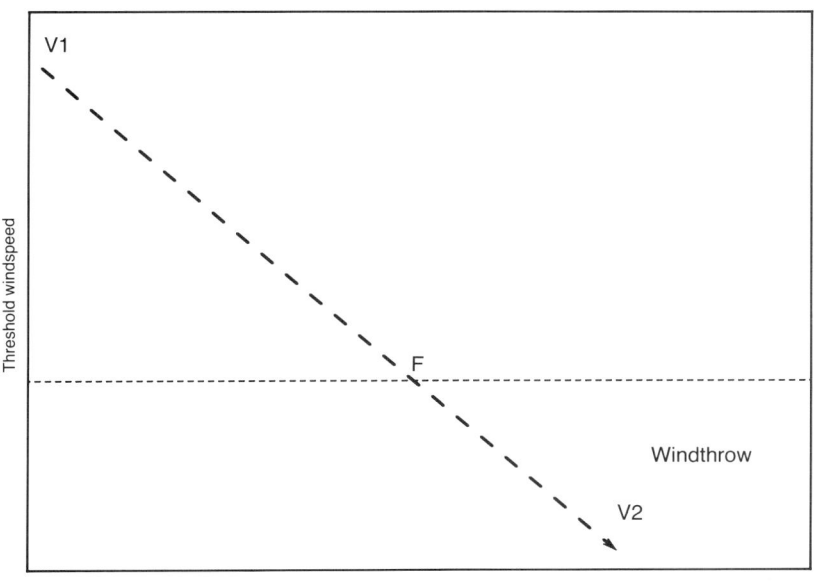

Fig. 21.3. The conceptual model in endemic form. V1–V2 indicates a linear decline in threshold windspeed (persistent and progressive) and the dotted line the endemic windspeed. F identifies the point of first damage.

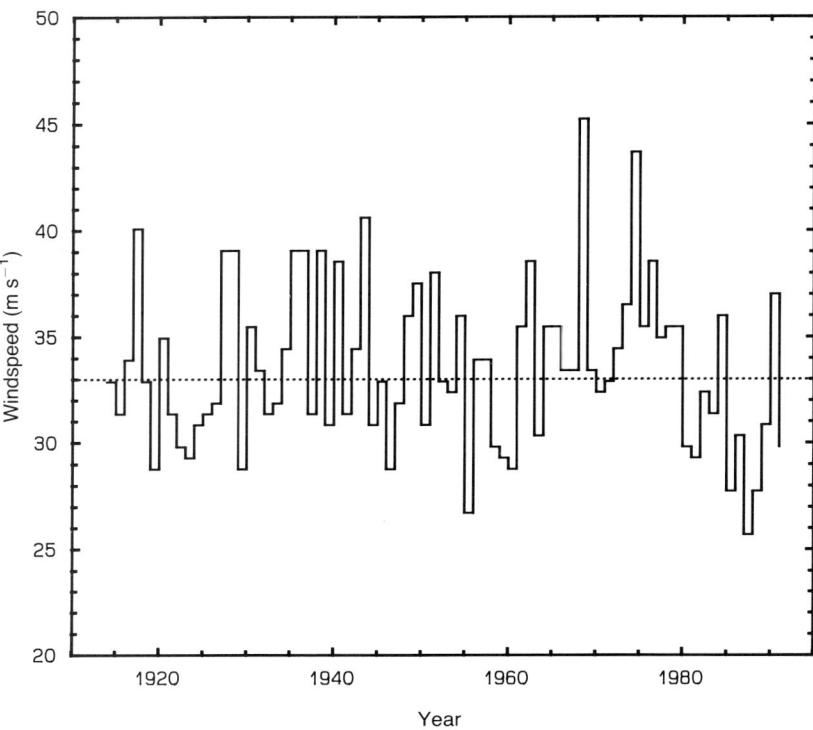

Fig. 21.4. Annual maximum gust (m s^{-1}) recorded at Eskdalemuir Meteorological Office observatory for the period 1914–91. The dotted line shows the mean for the period.

21.5.4 Incorporating a time series

A more realistic model can be obtained by replacing the uniform endemic windspeed with a time series windspeed. When the wind climate in Britain is examined carefully it is hard to sustain the notion that there are two distinct categories of storms (endemic and catastrophic). The time series of windspeeds from Eskdalemuir is shown in Fig. 21.4. There are two peaks which might be interpreted as catastrophic events, but there is also considerable year-to-year variation in annual maximum gust speed; there is little sign of the normal winter storm. The variation means that the intersection of threshold windspeed with windspeed occurrence can vary, which will affect the predictability of timing of damage.

The previous model can be adapted to incorporate a time series to demonstrate the importance of annual variability in windspeed (Fig. 21.5). The calculations assume an *unstable* site and stands with *slow*, *moderate* and *rapid*

Assessing the risk of wind damage to forests 391

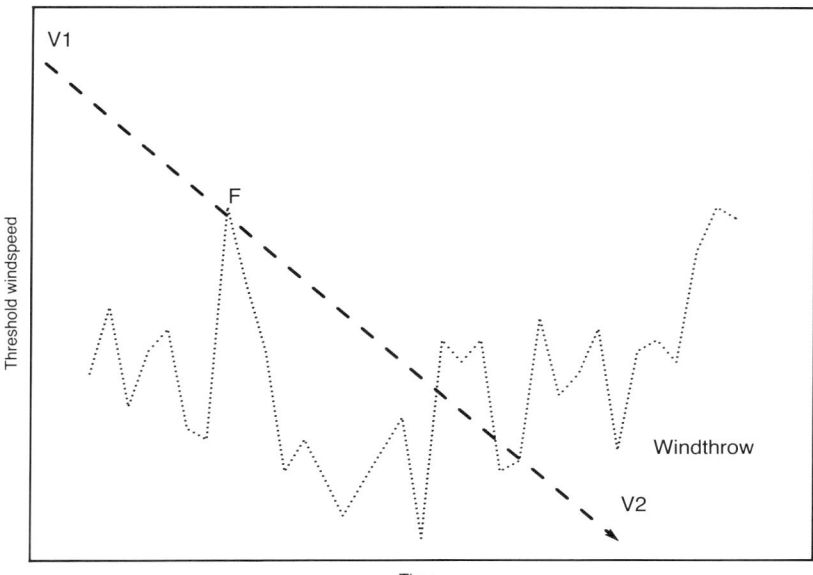

Fig. 21.5. The conceptual model adapted to include a time series of windspeeds. V1–V2 indicates a linear decline in threshold windspeed (persistent and progressive) and the dotted line the time series of windspeed. F indicates where first damage occurs, and subsequent damage occurs wherever the time series line exceeds V1–V2.

change in vulnerability. However, the *endemic windspeed* is replaced with the *time series windspeed*, and a year of planting is specified; stands planted in successive years from 1900 to 1991 are investigated.

The variability in storm magnitude results in a considerable range of age of first damage and terminal damage compared with the endemic model (Table 21.2). The variation in age of terminal damage is less than the variation in age of first damage because of the moderating effect of the series of storms required for damage to progress, but it is nevertheless substantial.

The rate of change in vulnerability is influential. If rate of change is slow then major storms will dominate the occurrence and progression of damage and result in major differences in timing of damage. Only with a very rapid increase in vulnerability relative to magnitude of storm variability will the behaviour of stands approach that of the endemic (e.g. suffer first damage in low windspeed years).

The average age of first damage and terminal damage occurs earlier with the time series than with the endemic model because the occasional high windspeeds affect many stands.

Table 21.2. *Age of first windthrow and terminal damage (years) plus mean annual windthrow rate (%) for a stand at Eskdalemuir on an unstable* site with *time series windspeeds*

Rate of change:	Slow	Moderate	Rapid
Endpoint:	None	None	None
Mean age at first damage	36	21	12
SD	9.8	5.4	3.0
Mean age at terminal damage	66	39	22
SD	4.2	3.1	1.9
Mean annual windthrow rate (after first damage)	1.3	2.3	4.0

Comparisons are made between different rates of change of progressive vulnerability. This table can be compared directly with the *endemic windspeed* results for the *unstable* site in Table 21.1.

Variability of timing of terminal damage of the order suggested would necessitate flexibility in planning the harvesting programmes for stands threatened by wind. Plans would need to accommodate early damage, but also stands remaining undamaged; general estimates could be used for medium-term planning but these would require updating as the deadline approached. However, these assumptions still imply a certain inevitability about damage which also can be questioned.

21.5.5 Limiting the change in progressive vulnerability

The examples above have assumed a linear decline in the threshold windspeed due to the progressive component of vulnerability. However, the validity of this assumption is questionable. Observations of old stands, the decline in the height/diameter ratio of trees with time, and the potential for adaptive growth all indicate that there could be a point beyond which there is little decline in threshold windspeed until, for example, fungal diseases affect root or stem strength. If the vulnerability were to level off in this manner (V2–V3 in Fig. 21.6) this would have important implications, particularly when combined with the non-linearity of the relationship between a given windspeed and its recurrence.

The effects of a levelling off in an endemic model are easy to identify. If the level is above the endemic mean, then no damage will occur. If the level is below the endemic mean, the timing of first damage will be identical to that obtained when there is no levelling off, but a slower rate of subsequent progression will occur.

Assessing the risk of wind damage to forests 393

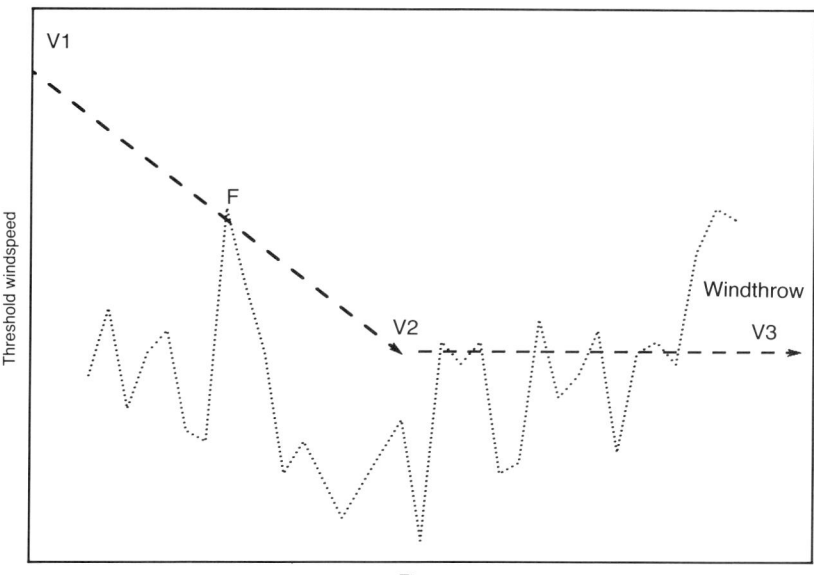

Fig. 21.6. The conceptual model adapted to include a time series of windspeeds and a levelling off in progressive vulnerability. V1–V2 indicates a linear decline in threshold windspeed (persistent and progressive), and V2–V3 the levelling off or endpoint. The dotted line represents the time series of windspeed. F identifies the first damage, and subsequent damage occurs whenever the time series line exceeds V1–V3.

The time series model can be adapted to explore the implications. Calculations will be presented for an *unstable* site, a *moderate* rate of change of vulnerability and three windspeeds at which the level occurs (*low, mean, high* endpoints). The *time series* and stands planted successively from 1900 to 1991 will once again be used.

The age of first damage is unaffected by the endpoint options because the major storms cause first damage before any of the thresholds is reached (Table 21.3). However, there are major differences in terminal damage, because the *high* level always requires high windspeeds to cause and progress damage, whereas threshold windspeed of stands with a *low* level can fall below the annual mean and therefore be damaged frequently.

It becomes conceivable that not all stands will be damaged during a normal rotation despite occurring on an unstable site. In the simulation (Fig. 21.7), stands with a *high* endpoint have not reached terminal damage levels if they were planted after 1918, but with a *low* endpoint all stands planted up to 1947 have reached terminal levels. The endpoint of the vulnerability decline is thus of considerable importance to the inevitability of damage, and highlights the relationship between windspeed and recurrence.

Table 21.3. *Age of first windthrow and terminal damage (years) plus mean annual windthrow rate (%) for a stand at Eskdalemuir on an* unstable *site with* time series windspeeds

Rate of change:	*Moderate*	*Moderate*	*Moderate*	*Moderate*
Endpoint:	*None*	*Low*	*Mean*	*High*
Mean age at first damage:	21	22	22	22
SD	5.4	5.5	5.2	5.3
Mean age at terminal damage	39	40	44	65
SD	3.1	4.3	5.8	4.5
Mean annual windthrow rate (after first damage)	2.3	2.3	1.8	0.9

Comparisons are made between differing points at which the rate of change of moderate progressive vulnerability levels off. The *none* endpoint is repeated from Table 21.2 to ease comparison.

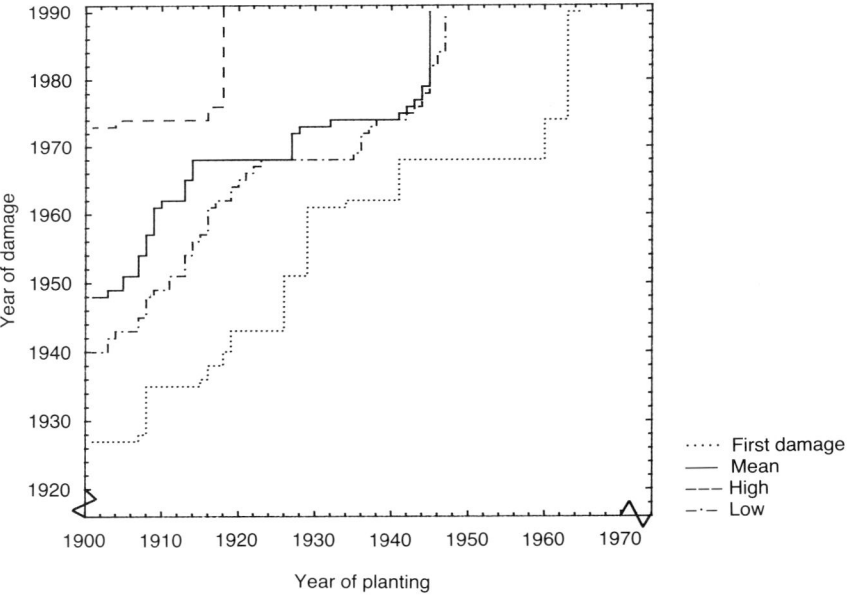

Fig. 21.7. Simulated age of first and terminal damage using Eskdalemuir time series (Fig. 21.4) and simple assumptions of change in tree vulnerability (*unstable* soil; *moderate* rate of progressive vulnerability; and endpoints of *high*, *mean* and *low* threshold windspeeds). Note that a high threshold windspeed means the risk of damage is less. Stands planted each year from 1900 to 1991, and winds occurring in years 1914–91, are included.

Three other points can be taken from the time series simulation (Fig. 21.7). Major storms (e.g. 1968, 1974) override a wide range of vulnerabilities and therefore widespread damage irrespective of site or stand differences is possible. For example, the storm of 1968 results in terminal damage to stands planted between 1914 and 1926 with a *moderate* endpoint, stands planted between 1923 and 1934 with a *low* endpoint, and first damage to stands planted between 1941 and 1959 regardless of endpoint. The variability of the time series means that there is not necessarily a good relationship between age of first damage and age of subsequent damage; thus, in the simulation a stand with a *low* endpoint planted in 1941 becomes first damaged in 1968, and terminally damaged 6 years later in 1974, whereas a stand planted in 1959 is also first damaged in 1968 but has yet to reach terminal levels 23 years later in 1991. The period representing the formative years (late 1960s to mid 1970s) of the windthrow hazard classification show the smallest intervals between first and terminal damage, and the youngest ages of terminal damage for the simulated period 1914–91.

21.5.6 Recurrence

The recurrence of storms therefore becomes important. This will be emphasised by comparing Eskdalemuir with two other Meteorological Office stations: one windier (Tiree, an island off the west coast of Scotland; elevation 21 m, latitude 56° 30′ N, longitude 06° 53′ W), and one less windy (Larkhill, a lowland inland site in southern England; elevation 155 m, latitude 51° 12′ N, longitude 01° 48′ W). The relationship between windspeed and recurrence of these stations is non-linear (Fig. 21.8).

The threshold windspeed of a stand can influence the chance of damage very dramatically. For instance, the shift in threshold windspeed level from 30 to 36 m s^{-1} (*low* to *high* options above) represents an increase in return period for Eskdalemuir from approximately 1.5 years to 4.5 years. A similar shift at Tiree would only change the period from 1 to 1.5 years, but at Larkhill the change would be dramatic (from 2.5 to 12.5 years). A greater shift to 40 m s^{-1} would result in return periods of 3, 14 and 52 years being required at Tiree, Eskdalemuir and Larkhill, respectively. Management practices that promoted tree stability could have a major impact on damage occurrence at sites such as Eskdalemuir and Larkhill. Progression of damage would not be inevitable as under the endemic assumption (e.g. terminal damage to be reached at age 23, 33 and 37 for an *unstable* site at Tiree, Eskdalemuir and Larkhill; or 41, 51 and 56 for a *stable* site), but would be dependent upon the occurrence of infrequent storms. Some stands could be vulnerable to

396　　　　　　　　　　　　　C. P. Quine

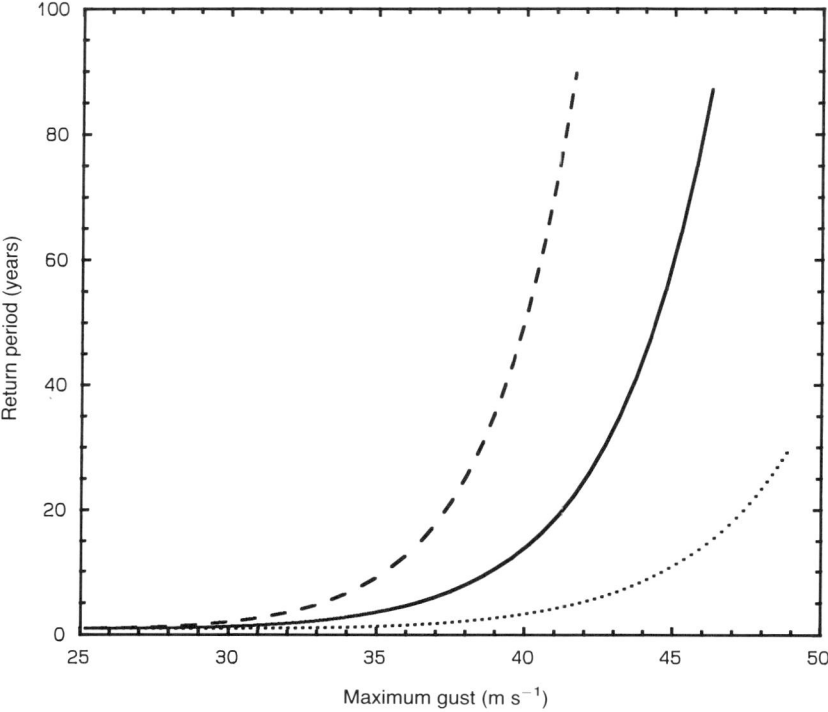

Fig. 21.8. Relationship between windspeed (maximum gust) and recurrence (expressed as return period, i.e. average interval in years between exceedances) for three Meteorological Office sites: Eskdalemuir (continuous line), Tiree (dotted line) and Larkhill (dashed line). The return periods were calculated using annual maximum gusts from 1958 to 1991 and the Lieblein method outlined by Cook (1985).

storms that recur less than once in a typical rotation. Site treatment or management practice can therefore be more influential in determining the occurrence of damage than the endemic assumptions would indicate.

The apparent sensitivity of the amount of damage to tree vulnerability raises important questions over the degree of control that the forester can exert, and the nature of adaptive growth. Is the contribution of adaptive growth in influencing the threshold windspeed limited by the ability of the tree to respond or by the nature of the wind it experiences? Certain forms of site preparation may prevent the formation of a uniform buttressed root system. Different wind climates (e.g. the relationship of extreme to mean winds) could cause different growth responses. The rate of climate change could have an important effect.

21.5.7 Recent evidence in the light of the conceptual model

The Eskdalemuir time series and conceptual model demonstrate the imprecision of the catastrophic and endemic definitions and the inadequacies of a deterministic model. It is not possible to separate endemic and catastrophic damage because these are two points on a continuum which are interlinked in a way dependent upon the sequence of occurrence.

Kershope Forest, the area in which the terminal heights were derived (see above), is close to Eskdalemuir. In the period of monitoring that resulted in the terminal height estimates (1959–76) there were at least two storms with high windspeeds not exceeded in recent years. Examining the return periods shows storms expected once every 40 and 65 years occurred in the 16 year monitoring period. The storms of 1968 and 1974 were not identified as catastrophic at Kershope because they did not cause sufficient damage in the vicinity, although the 1968 storm caused catastrophic damage further to the north and west (Holtam, 1971). The annual maximum gust figures also hide within-year variation. For instance, in 1974 there were two storms with return periods in excess of 30 years; neither of these was interpreted as catastrophic. The estimated terminal heights thus reflect rare events as well as more usual winter storms. The impact of this is shown in Fig. 21.9; overprediction of damage is apparent in Kershope after the period in which the terminal heights were defined due to a return to more normal winter storms/absence of rare events.

The time series simulations also accord with the results of a number of monitoring areas established to study the onset and progression of windthrow (Quine & Reynard, 1990). These have shown that substantial overprediction of damage has occurred across a number of sites, and that the rate of damage is extremely variable from year to year. For example, annual rates of windthrow over 5 years are less than 1% in upland monitoring areas, but a rate of 16% was recorded in a single year in a lowland forest affected by the 1987 severe storm. Some sites with appreciable initial damage have nevertheless shown little progression of damage for periods in excess of 8 years. Interpreting the discrepancies between these observations and predictions based on the classification is difficult because of the sources of variation. A major source of variation appears to be the variability in frequency of occurrence of winds of damaging speed that has been emphasised above. However, other factors could include a shift to non-thin treatment of vulnerable crops (guided by the hazard classification), changes in the stand structure due to wider spacing, better site preparation methods, or more rapid rates of growth. All these will affect the stand vulnerability and hence the chance of

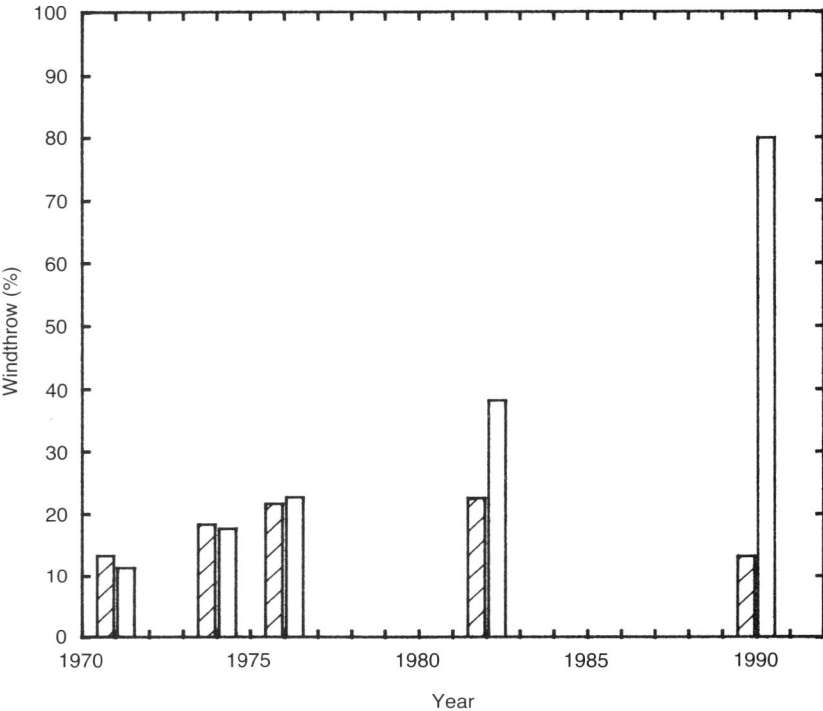

Fig. 21.9. Actual windthrow (hatched columns; recorded from aerial photographs) and predicted windthrow (open columns; from hazard class and stand top height) for Kershope Forest, north England. The information up to and including 1976 was used in the definition of the terminal heights used to derive the predicted windthrow. The subsequent divergence represents the return to a period of more typical storms following a period containing a number of rare windspeeds. Note that only 36% of the forest area sampled in 1982 remained in 1990 due to harvesting of stands.

damage occurring. The impossibility of unravelling these numerous influences in the data has emphasised the need for a new approach.

21.5.8 Future development of the hazard classification

The hazard classification is in need of improvement. The endemic assumption is flawed and key estimates provided by the classification were derived during a period of rare windspeeds. A deterministic model appears inappropriate where stochastic effects are clearly so influential. However, the existence of the classification has been a positive influence in providing a structured approach to the problem and in objectively ranking sites. It is the treatment of timing of damage that is most inadequate. Until a probability-based model is developed, emphasis is being placed on the need to treat the estimates of timing of damage as guides and not as certainties, and to review the evidence in the forest when translating medium-term plans into operational programmes.

A risk-based rather than a hazard-based classification is now required. Such a model should incorporate the considerable progress made in understanding the mechanics of windthrow to derive threshold windspeeds. Progress in our understanding of the nature of airflow in relation to topography will permit the probability of the threshold windspeed to be calculated for particular sites; this will incorporate all windspeeds and will not rely on artificial distinctions between endemic and catastrophic storms. To provide a useful classification all this must be achieved using variables easily acquired by the forest manager and preferably coupled to appropriate appraisal models. Geographic information systems (GIS) will be an important tool in applying the model to forest situations.

21.6 General conclusions

It is clear that the non-linear relationship between windspeed and recurrence means that small changes in one of many factors can lead to major changes in damage occurrence (e.g. a small shift in climate could render apparently stable stands unstable). Wigley (1985) showed that a change in mean of 1 standard deviation would lead to a 1 : 20 year storm becoming 5 times more likely, a 1 : 100 becoming 9 times more likely, and multiple recurrences being even more probable. In this respect wind climate is similar to many other hazards; Suffling (1992) indicated that a 2 °C rise in temperature could lead to a fivefold increase in fire area in Sweden.

Change in tree vulnerability with time is poorly understood and yet is crucial to management decisions such as thinning and retentions. Changes in practice (e.g. thinning), breeding strategies and fungal infection could have an important and unanticipated impact on damage occurrence, particularly where the climate is windy and sites are poor. Conversely, major improvements in stand stability are also possible. Preventive measures may be best aimed at those stands most at risk or where a large shift in damage likelihood can be achieved. Episodic changes such as thinning will be most significant when the duration of their effect (i.e. time for crowns to grow back) is longer than the recurrence interval of the appropriate threshold windspeed, or where they result in an important shift in probability.

Timing of damage will be most predictable for stands which have rapidly increasing vulnerability and where threshold windspeed approaches or falls below that of the annual storms. Conversely, timing of damage will be extremely variable for stands with a slow increase in vulnerability as the damage will depend upon the occurrence of extreme storms. Ironically, the more successful the preventive measures the more uncertainty will be associated with subsequent predictions of timing of damage. In a probabilistic environment it is vital that planning is flexible to permit response to damage or lack of it (Dempster & Stevens, 1987). However, such frequent replanning is rarely considered in current forest design plans where the landscape sequence is anticipated over decades.

Where wind damage is rare, forest managers must guard against complacency because it makes large-scale damage more likely. This is typified by maintaining large proportions of the forest in a vulnerable stage, keeping valuable stands clustered in space (e.g. the concentration of arboreta in South-East England), and failing to make any contingency planning. Windthrow of almost any tree is possible given a sufficient windspeed, or a severe episodic increase in vulnerability through soil saturation or snow loading. In contrast, where wind damage is common forest managers must guard against fatalism; change in management practice *can* significantly alter the frequency of damage. In both cases it may be instructive to consider worst case scenarios.

The complex factors affecting the timing of damage and the scale of occurrence highlight the dangers of inferring trends from short-run records and of attributing these to single causes such as recent silvicultural practice. Observed windthrow may be the result of innumerable combinations of vulnerability and windspeed. This emphasises the importance of a formal classification, as simple observations and subjective evaluation of risk may be misleading. The calibration of the exceedance probabilities of storms is desirable, and long-term records of the occurrence of damage are clearly needed.

Although management practices such as shortened rotations (Brumelle *et al.*, 1990) may be sound if based on deterministic models, the variability of storm occurrence makes a probabilistic model highly desirable. The situation is summarised well in a recent report (SAF, 1993): 'Models need to be more probabilistic, less deterministic. Forest planning models, by and large, take a deterministic view of the world in which natural forces are treated as if they were under complete control, i.e. occurring with predictable frequency and intensity. Reflecting patterns for natural disturbance will require a fundamental restructuring of forest planning methodology to recognise the probabilistic nature of these events and to measure risk of departure from these natural patterns.'

Acknowledgements

Jim Wright helped compute the examples, Sarah MacArthur assisted with the recurrence calculations, and John Mayhew carried out the Kershope comparisons. The text improved after comments from Bill Mason and Barry Gardiner but remaining inadequacies are the responsibility of the author.

References

Andersen, K. F. (1954). Gales and gale damage to forests, with special reference to the effects of the storm of 31st January 1953, in the north east of Scotland. *Forestry*, **27**, 97–121.

Booth, T. C. (1977). *Windthrow Hazard Classification*. Forestry Commission Research Information Note 22/77/SILN.

Brumelle, S., Stanbury, W. T., Thompson, W. A., Vertinsky, I. & Wehrung, D. (1990). Framework for the analysis of risks in forest management and silvicultural investments. *Forest Ecology and Management*, **35**, 279–99.

Busby, R. J. N. & Grayson, A. J. (1981). *Investment Appraisal in Forestry with Particular Reference to Conifers in Britain*. Forestry Commisison Booklet 47. HMSO, London.

Canham, C. D. & Loucks, O. L. (1984). Catastrophic windthrow in the presettlement forests of Wisconsin. *Ecology*, **65**, 803–9.

Christensen, J. B. (1987). Plans for specific aspects of forestry and their integration with the general management plan for the enterprise for the impact of catastrophic events such as storm damage (in Dutch). *Nederlands Bosbouwtijdschrift*, **59**, 58–62.

Cook, N. J. (1985). *The Designer's Guide to Wind Loading of Building Structures: Background, Damage Survey, Wind Data and Structural Classification*. Building Research Establishment Report. Butterworth, Sevenoaks.

Cutler, D. F., Gasson, P. E. & Farmer, M. C. (1990). The wind blown tree survey: analysis of results. *Arboricultural Journal*, **14**, 265–86.

Dempster, W. R. & Stevens, N. A. (1987). *Risk Management in Forest Planning*. Forestry Canada, Alberta, Paper FO 41-91/17-1987E.

Elling, A. E. & Verry, E. S. (1978). Predicting wind-caused mortality in strip-cut stands of peatland black spruce. *Forestry Chronicle*, **54**, 249–52.

Foot, D. L. (1975). Forest management in Craik: a review of current thinking with special reference to windblow. *Scottish Forestry*, **29**, 129–34.

Fraser, A. I. (1965). The uncertainties of wind damage in forest management. *Irish Forestry*, **22**, 23–30.

Fraser, A. I. & Gardiner, J. B. H. (1967). *Rooting and Tree Stability in Sitka Spruce*. Forestry Commission Bulletin 40. HMSO, London.

Grayson, A. J. (1989). *The 1987 Storm: Impact and Responses*. Forestry Commission Bulletin 87. HMSO, London.

Hendrick, E. (1988), Windthrow risk classification for thinning. Unpublished report. Irish Forest Service, Bray.

Holtam, B. W. (ed.) (1971). *Windblow of Scottish Forests in January 1968.* Forestry Commission Bulletin 45. HMSO, London.

Hutte, P. (1968). Experiments on windflow and wind damage in Germany: site and susceptibility of spruce forests to storm damage. In *Wind Effects on the Forest. Forestry* (Supplement), **41**, 20–6.

Konopka, J. (1973). Basic demarcation of regions of wind and snow damage in Slovakia (in Slovak). *Lesnicky Casopis*, **19**, 147–61.

Lines, R. & Howell, R. S. (1963). *The Use of Flags to Estimate Relative Exposure of Trial Plantations.* Forestry Commission Forest Record 51. HMSO, London.

Mackenzie, R. F. (1974). Some factors influencing the stability of Sitka spruce in Northern Ireland. *Irish Forestry*, **31**, 110–29.

Manley, B. & Wakelin, S. (1989). Modelling the effect of windthrow at the estate level. In *Workshop on Wind Damage in New Zealand Exotic Forests*, ed. A. Somerville, S. Wakelin & L. Whitehouse. FRI Bulletin 146. FRI, Rotorua, New Zealand.

Maxwell Macdonald, J. (1952). Wind damage in middle aged crops of Sitka spruce. *Scottish Forestry*, **6**, 82–5.

Mayer, H. (1988). Mapping of potential storm risk of forest sites in Bavaria (in German). *Forstwissenschaftliches Centralblatt*, **107**, 239–51.

Miller, K. F. (1985). *Windthrow Hazard Classification.* Forestry Commission Leaflet 85. HMSO, London.

Miller, K. F. (1986*a*). Wind. In *Forestry Practice*, ed. B. Hibberd, chap. 8. Forestry Commission Bulletin 14. HMSO, London.

Miller, K. F. (1986*b*). Windthrow hazard in conifer plantations. *Irish Forestry*, **43**, 66–78.

Miller, K. F., Quine, C. P. & Hunt, J. (1987). The assessment of wind exposure for forestry in upland Britain. *Forestry*, **60**, 179–92.

Moore, M. K. (1977). *Factors Contributing to Blowdown in Streamside Leave Strips on Vancouver Island.* Land Management Report no. 3. British Colombia Forest Service, Victoria, Canada.

Muchow, R. C. & Bellamy, J. A. (eds.) (1991). *Climatic Risk in Crop Production: Models and Management for the Semi-arid Tropics and Subtropics.* CAB International, Wallingford.

NAO (1993). *Forestry Commission: Timber Harvesting and Marketing.* Report by the controller and auditor general, National Audit Office. HMSO, London.

Petrie, S. M. (1951). Gale warning! Windblow in western spruce plantations. *Journal of the Forestry Commission*, **22**, 81–90.

Prien, S. (1974). Protective measures against wind damage on a basis of regional wind damage classes (in German). *Beiträge für die Forstwirtschaft*, **8**, 65–8.

Pyatt, D. G. (1968*a*). Forest management surveys in forests affected by winds. In *Wind Effects on the Forest. Forestry* (Supplement), **41**, 67–76.

Pyatt, D. G. (1968*b*). The soil and windthrow surveys of Newcastleton forest, Roxburghshire. *Scottish Forestry*, **20**, 175–83.

Quine, C. P. (1988). Damage to trees and woodlands in the storm of 15–16 October 1987. *Weather*, **43**, 114–18.

Quine, C. P. (1991). Recent storm damage to trees and woodlands in southern Britain. In *Research for Practical Arboriculture*, ed. S. J. Hodge, pp. 83–9. Forestry Commission Bulletin 97. HMSO, London.

Quine, C. P. & Gardiner, B. A. (1991). Storm damage to forests: an abiotic threat with global occurrence. In *Proceedings of the Tenth World Forestry Conference*, Paris, September 1991, vol. 2, pp. 339–45.

Quine, C. P. & Reynard, B. R. (1990). *A New Series of Windthrow Monitoring Areas in Upland Britain*. Forestry Commission Occasional Paper 25. Forestry Commission, Edinburgh.

Quine, C. P. & White, I. M. S. (1993). *Revised Windiness Scores for the Windthrow Hazard Classifiction: The Revised Scoring Method*. Forestry Commission Research Information Note 230. Forestry Commission, Edinburgh.

Runkle, J. R. (1990). Gap dynamics in an Ohio *Acer–Fagus* forest and speculations on the geography of disturbance. *Canadian Journal of Forest Research*, **20**, 632–41.

SAF (1993). *Task Force Report on Sustaining Long-term Forest Health and Productivity*. Society of American Foresters, Bethesda, MD.

Savill, P. S. (1983). Silviculture in windy climates. *Forestry Abstracts*, **44**, 473–88.

Smith, K. (1992). *Environmental Hazards: Assessing Risk and Reducing Disaster*. Routledge, London.

Suffling, R. (1992). Climate change and boreal forest fires in Fennoscandia and central Canada. In *Greenhouse Impact on Cold Climate Ecosystems and Landscapes*, ed M. Boer & E. Koster. *Catena* (Supplement), **22**, 111–32.

Thompson, A. P. (1976). 500 year evidence of gales. *Forest Industries Review*, **7**, 11–16.

Valinger, E. & Lundqvist, L. (1992). The influence of thinning and nitrogen fertilisation on the frequency of snow and wind-induced damage in forests. *Scottish Forestry*, **46**, 311–20.

Wardle, P. A. (1970). Weather and risk in forestry. In *Weather Economics*, ed. J. A. Taylor, pp. 67–82. Pergamon Press, Oxford.

Wigley, T. M. L. (1985). The impact of extreme events. *Nature*, **316**, 106–7.

22

Forest wind damage risk assessment for environmental impact studies

G. C. WOLLENWEBER and F. G. WOLLENWEBER

Abstract

In most European countries, forests are a natural resource of timber. Neglecting information about tree growth, forestry planning and landscape modifications for construction work (e.g. for industrial areas) may result in large economic losses from damage to trees. This chapter presents a method for assessing areas of forest with an increased risk of storm damage due to a proposed landscape modification for a new motorway. The planned new road will cut through a forested mountain ridge, changing both the topography and land-use. A hierarchy of models is applied to simulate winds over hills and changing terrain properties. The first coarse-grid model is on the need to determine wind modification according to the overall topography of the area using synoptic input data. These results are then included as input data for further calculations with fine-scale models. The models are applied to a high-resolution grid which covers the area of the motorway to be built and includes the modifications of landscape and land-use. Finally, wind fields and turbulent kinetic energy are determined at a high spatial resolution for different weather regimes, topographies and land-uses. Areas with an increase or decrease in windspeed will be readily defined as a function of synoptic windspeed and wind direction. The computing facilities required are those of a 486 PC or a workstation.

22.1 Introduction

Meteorological variables such as high windspeed and precipitation are known to have an important influence on the risk of damage to trees, together with topography, silvicultural methods, species and stand structure, and the health of the trees (Mayer, 1985). Wind-related damage to trees includes the effects of high mean windspeeds as well as the impact of the intensity and frequency spectrum of gusts.

Soil conditions can increase the danger of storm damage, especially where there is a thin humus layer and particularly in wet regions. In areas with high water-tables the development of the tree root system is reduced. The growth reduction increases with decreasing vertical distance between the upper water level and the soil surface. Spruces, for example, tend to have a flat root system on wet soils. High windspeeds induce vibration, which results in pumping effects of the root system in wet soils; if these reach a certain amplitude trees are uprooted. In dry soils, the resistance of the tree to a storm is determined partly by the leeward horizontal roots near the stem.

A dependence of storm damage on tree height and soil type was found by Wangler (1974). Critical windspeeds for Sitka spruces of different heights, growing on various types of soil, were established by pulling trees over (Fraser, 1965). In contrast to storms with varying winds and turbulent wind fluctuations, Fraser's definition relates to constant or slowly changing mean windspeed.

Storms are most likely during winter and spring. In Potsdam (Germany), for example, the probability of a storm is highest in January and February with a maximum around noon (Geiger, 1950). The probability of a coincidence between storms and wet soil varies according to season. A maximum probability of 4% can be found in January and February for Potsdam.

Besides storms and soil conditions, the risk of damage to forests is also influenced by topography. Storm damage can be found at different places within the same area depending on the position of trees relative to the terrain.

Many plans for new industrial areas and motorways include the modification of landscape and land-use with a subsequent influence on trees. To estimate possible impacts of a planned motorway, model calculations are valuable firstly for the assessment of the current situation and secondly to estimate the impact of proposed changes. Rather than investigating a large number of parameters, all of which determine the probability of damage to the forest, this study will concentrate only on the effect of changes in mean wind flow and turbulent kinetic energy.

In the present study a sequence of three models is applied. A mesoscale model supplies the fine-scale models with meteorological input data, which already include the modification of the surface wind by the overall terrain features. The calculations begin on a coarse grid with a horizontal resolution of 1 km, which covers a region of 50 km × 50 km. A high-resolution model on a 40 m × 40 m grid scale, which is based on a variational approach, is the fundamental method used to investigate the air flow in the area affected by the new motorway. The results will establish estimated areas with higher windspeeds which are expected to be a risk, especially for spruce forests. A third model is used to determine turbulent kinetic energy for a cross-section of the new motorway. This is a high-resolution two-dimensional model based on a turbulent closure of higher order.

22.2 Model descriptions

No wind measurements were available from the area of interest. Consequently the first model computations are performed for a larger area to provide wind values, already influenced by the overall terrain of the area, for the subsequent small-scale models.

22.2.1 The mesoscale model

The framework for these calculations is a three-dimensional atmospheric boundary layer model (Heimann, 1985) which consists of three layers. The lowest layer extends from the top of the topography and includes the surface layer of the atmosphere. An Ekman layer with a variable upper boundary ranges above the surface layer. A third layer with a fixed upper boundary is above the Ekman layer. The depth of the surface layer, relative to the height of topography and the model height, is kept constant.

The inclusion of entrainment processes at the top of the Ekman layer can lead to an increase in its vertical extension up to the upper model height, mixing with the air of this layer. During the night a stable atmospheric boundary layer can develop. Its thickness can increase, decoupling air in higher layers from the air below. To allow for this effect, detrainment is included by decreasing the height of the Ekman layer during the night, returning air from the Ekman layer to the third layer. The increase in height of the Ekman layer with time is described according to the formulation of Deardorff (1974) for unstable conditions and according to Smeda (1979) for stable conditions.

For each layer, the prognostic equations of momentum, continuity and heat have to be solved. Turbulent fluxes are described as a function of drag coefficients, including mixing length and stability functions (Businger, 1973). Prescribed changes in surface temperature and horizontal pressure gradient permit the simulation of wind and temperature with time. To start the model calculations, topography, land-use classes and meteorological input parameters have to be provided. Surface characteristics are related to surface roughness, according to Stull (1991) (Table 22.1).

The main interest in this chapter concerns high windspeeds. Simulations are conducted for a neutrally stratified atmosphere with constant surface temperatures. The synoptic horizontal pressure gradient is included by using the geostrophic wind. For each single run this wind is kept constant with time

Table 22.1. *Aerodynamic roughness length for different terrain types*

Terrain type	Roughness length, z_0 (m)
Villages	0.6
Forests	0.6
Bushes	0.06
Agricultural areas, meadows	0.03
Roads, rivers	0.02

and independent of height. The prognostic equations are integrated forward in time until a stationary solution is reached.

The results of the model simulations provide input data for the high-resolution model. Before these data can be included in the fine-scale models, the meteorological parameters are interpolated onto a terrain grid with high horizontal resolution (40 m) based on an inverse distance squared weight.

22.2.2 The fine-scale model to determine wind flow

Starting from the meteorological input data provided by the first model, a direct variational relaxation analysis is performed to determine the wind field (Ball & Johnson, 1978). This high-resolution wind analysis method is based on a surface layer model with constant height above an area of changing elevation. The thickness of the layer covers parts of the atmospheric Prandtl layer containing winds close to the surface. The calculations are based on the physical equations of the conservation of momentum and mass. Gauss's Principle of Least Constraints is applied to the momentum equation, with the incompressibility condition as a constraint. Thermal effects are included by a Businger-type approach but this is not described here. The calculations require information about profiles of temperature and wind perpendicular to the surface. It is assumed that the empirical relations of Businger (1973) can be applied in the direction normal to the surface, even in rough terrain. All calculations are done in a terrain-following coordinate system. Wind and temperature fields are adjusted according to terrain geometry and conservation laws to minimise the dynamic constraints following the variational relaxation algorithm of the Gaussian Principle.

22.2.3 A fine-scale model to determine turbulent kinetic energy

Wind damage to trees is caused not only by high mean windspeeds but also by the effect of wind fluctuations, turbulent kinetic energy and gusts.

Information about the effect of changed terrain and roughness conditions on turbulent kinetic energy is found from energy calculations carried out for cross-sections of the motorway. These estimates are based on a multilayer, two-dimensional, atmospheric boundary layer model, which extends up to heights well above the Ekman layer (Wollenweber, 1990). The calculations are based on the equation of momentum, heat, conservation of mass and the equation of turbulent energy. Turbulent fluxes are parameterised as a function of turbulent energy with a closure of order 1.5. A coordinate transformation is performed which accounts for changes in topography, surface roughness and zero plane displacement height. The vertical resolution is high close to the surface and decreases with height.

To start the calculations, surface elevation, roughness and zero plane displacement have to be known as a function of position. The meteorological input parameter is a vertical profile of wind, which includes geostrophic wind as an upper boundary value and winds determined with the variational model already described. During the integration along the transect, winds, which were calculated with the variation method, are included at each grid point. Since the main interest is in high windspeed, neutral atmospheric stability is assumed with constant potential temperature.

22.3 Topography and land-use

The region covered by the coarse grid is the Hunsrück/Mosel/Eifel area, situated in the south-west part of Germany close to the border with Luxembourg. The grid of the mesoscale model covers a region of 50 km × 50 km. It extends from Trier on the south side to close to Cochem on its north side (Fig. 22.1). The region covers parts of the Hunsrück area, which extends from south-west to north-east on the east side, and parts of the Eifel area at the west side. The two mountains are separated by the Mosel valley and its tributaries. The surface elevations range from between 100 m in the Mosel valley to 790 m above sea level in the Hunsrück. The area includes forests, farms, cities and villages, rivers and lakes, bushes and vineyards in the Mosel valley. The economy of this area depends on agriculture, forests and tourism. A map of land-use characteristics is shown in Fig. 22.2, with its lower left point located on the lower south-west corner of the model grid (Fig. 22.1).

The main area of interest is situated south of Wittlich extending west of the Asberg–Mundwald ridge towards Altrich with UTM coordinates at its lower south-west corner of 2561000/5533000. It covers a region of 5 km × 5 km around the projected line of the motorway. A small part of this region is shown in Fig. 22.3 with a horizontal resolution of 40 m. The

Wind in environmental impact studies 409

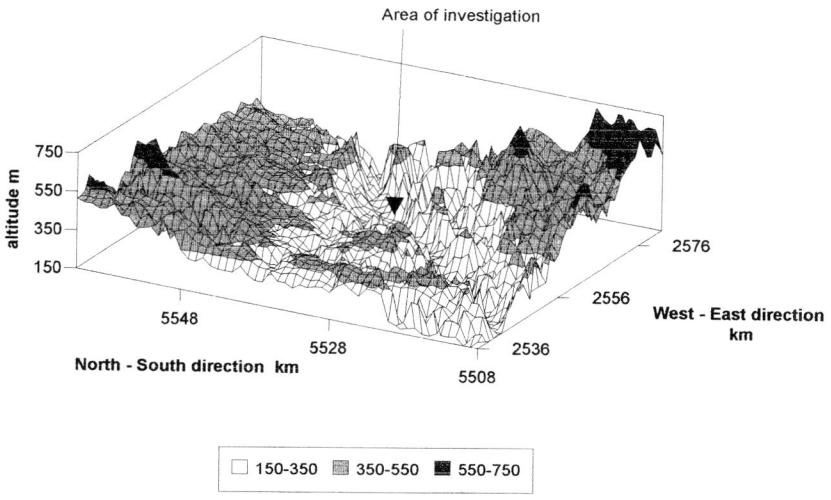

Fig. 22.1. Topography of the overall terrain (1 km grid).

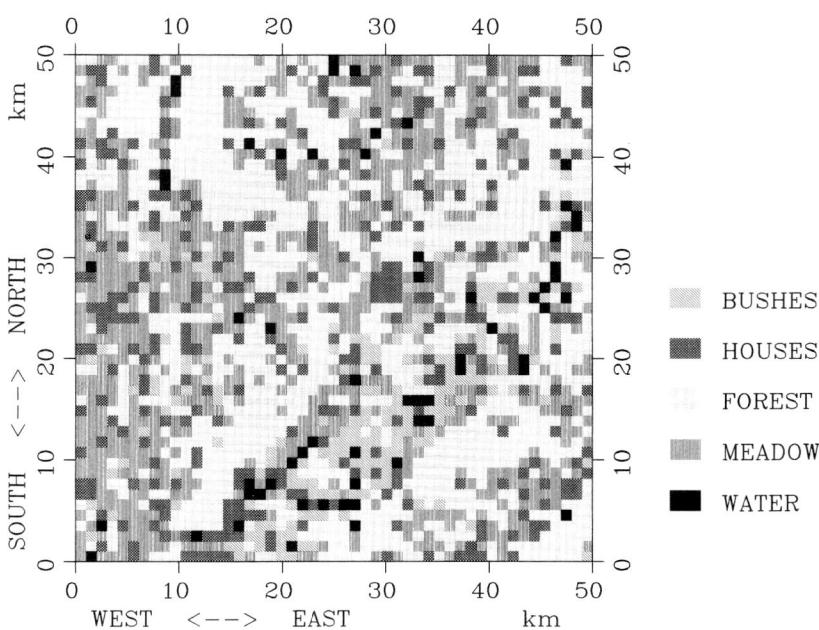

Fig. 22.2. Land-use of the overall terrain (1 km grid).

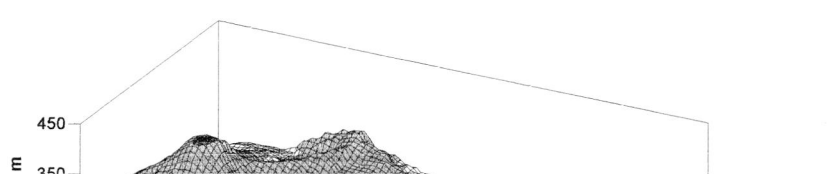
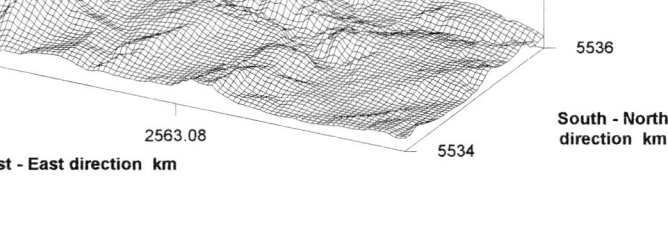

Fig. 22.3. Topography of the area surrounding the planned motorway (40 m grid).

orientation of the Asberg–Mundwald ridge is south-west/north-east and the surface elevation varies between 355 m and 160 m above sea level. The area is mainly characterised by forests, meadows, agricultural areas, farms and villages. This is shown in Fig. 22.4, which has its origin fixed at the south-west corner of the whole area (Fig. 22.3) of interest. A major existing road is the A1/A48 motorway passing north to south. The new motorway will cut through the Asberg–Mundwald ridge, separating the Asberg from the Mundwald ridge. The deepest part of the cut will be about 30 m. East of the Mundwald ridge the motorway will head south-east towards the A1/A48, requiring smaller cuts through the terrain and also some bridges. For the computations the intersection with the A1/A48 has been simplified and the extension towards Altrich has not been included. The forest, which will be divided into two by the road, consists of oaks, beeches, ash, small patches of willow, hickory, larch, pine, silver fir and Douglas fir.

22.4 Results

Before the effects of terrain modification were studied the probability of storm events was investigated from an annual wind distribution measured at a meteorological station in the Hunsrück region.

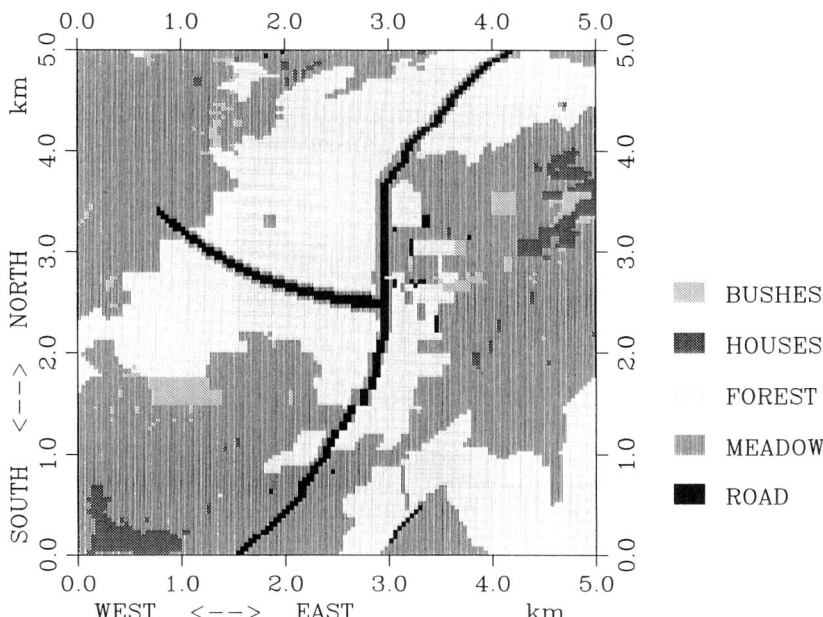

Fig. 22.4. Land-use of the area in the vicinity of the new motoway (40 m grid).

22.4.1 Wind distribution

Storms, which are defined by windspeeds above $17\,\mathrm{m\,s^{-1}}$, do not occur often in this part of Germany. They are more likely in higher mountains and less likely inland. Since no wind measurements are available close to the projected motorway, the probability of high windspeeds was studied by analysing the wind distribution from a site in Deuselbach, 20 km away. This station is situated in the Hunsrück region and is operated by the German Weather Service. The basis for the computation of wind distribution is a set of 10 min averages measured at 3-hourly intervals between 0600 hours and 1800 hours GMT over 15 years (1977–91). The annual distribution of wind direction (Fig. 22.5) shows a bias towards winds from the south-west (19.3%). Windspeeds below $2\,\mathrm{m\,s^{-1}}$ can be found in 21% of the measurements, with a northerly main wind direction. Winds above $16\,\mathrm{m\,s^{-1}}$ are rare, and are found only during autumn and winter. High-velocity wind directions are mainly west, east, south-east and south-west. It can be concluded that windspeeds in excess of $10\,\mathrm{m\,s^{-1}}$ will occur rarely (<3% of the cases), and that when they do they will be from westerly or easterly directions.

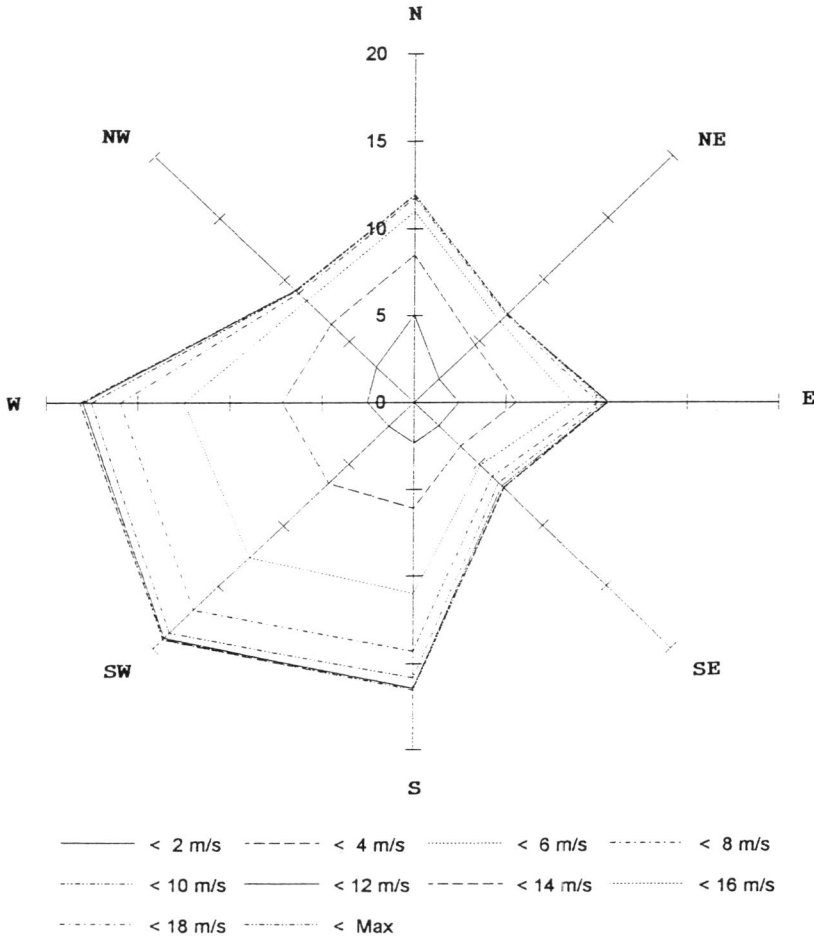

Fig. 22.5. Annual wind distribution for Deuselbach (1977–91).

22.4.2 Determination of wind in the overall area

To include the effect of the overall topography and land-use in the estimates of local wind modifications, model calculations are performed with the meso-scale model. Fig. 22.6 shows the calculated wind vectors, assuming a geostrophic wind of 12 m s^{-1} from the south-east as input. The airflow is down the Hunsrück slopes (south-east and east) converging with the flow along the Eifel foothills (east). Changes in windspeed further reflect the local topography.

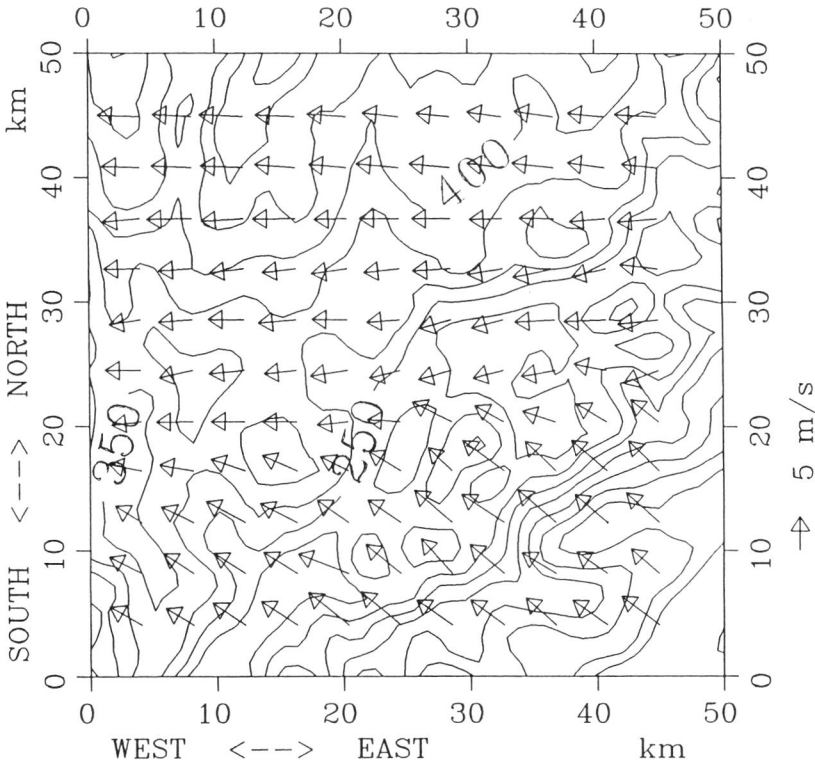

Fig. 22.6. Model results of the mesoscale model for a geostrophic wind of 12 m s^{-1} from the south-east.

22.4.3 Impact on local wind

The effect of the construction of the motorway on wind flow is studied for three cases. Case A corresponds to a geostrophic south-east wind, case B to a south-westerly and case C to geostrophic wind from the north-west. In all three cases a geostrophic windspeed of 24 m s^{-1} is assumed. The computations are performed for the original terrain and land-use and then for the modifications of terrain and land-use in the area of the new motorway. Changes in land-use are considered by changing the values for surface roughness and displacement height. The effect of the modification will be shown in the form of wind differences, [wind (with original terrain) − wind (with the modified terrain)], for differences above 0.5 m s^{-1}. In all cases the input data for the calculations are the output data from the mesoscale model interpolated onto the high-resolution grid points.

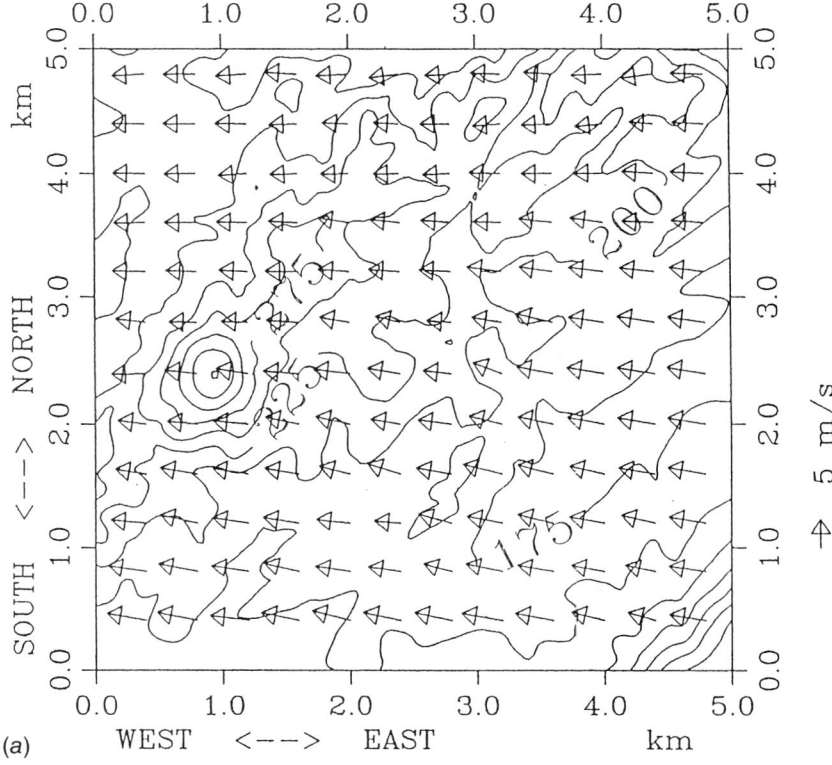

Fig. 22.7(a). For caption see opposite.

For the case of a geostrophic wind from the south-east (case A) the principal surface wind is easterly, with a reduced speed in front of the Mundwald ridge (Fig. 22.7). The effects of the modification of terrain and land-use are concentrated within a narrow band along the new motorway. This band is shifted towards the south in comparison with case B. At the intersection with the A1/A48 motorway the 'difference winds' show an increase in speed. East of the Mundwald ridge a reduction in windspeed can be found with a change in wind direction towards the south. At the south side of the area, where the motorway cuts through the Asberg–Mundwald ridge, the windspeed increases. At the end and at the south side of the cut the wind difference turns into a more northerly wind together with a reduction in windspeed.

With a geostrophic wind coming from south-west (case B) the air flows over the changed terrain from a southerly direction (Fig. 22.8). The windspeed decreases as the air approaches the Mundwald, turning slightly

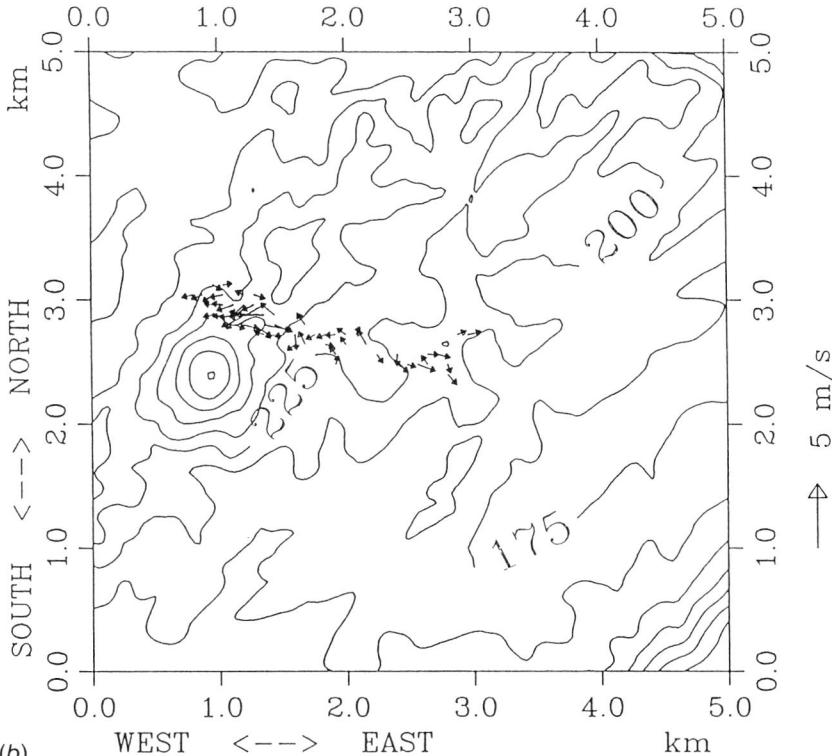

Fig. 22.7. (a) Surface wind for a geostrophic wind of 24 m s^{-1} from the south-east without terrain modification. (b) Wind difference for the geostrophic wind of (a): [v(original terrain) – v(modified terrain)].

into a westerly wind at the north side of the hill. The terrain modification leads to a more westerly wind and a reduction in speed as the air rises along the upslope side of the Mundwald ridge. On the east side of the ridge a veering of the modified wind towards the west occurs. Near the junction with the A1/A48, changes in surface characteristics induce a reduction in windspeed and an easterly shift in the modified wind direction.

The calculations with the variational approach show westerly surface winds (case C) for a geostrophic wind from the north-west (Fig. 22.9). As the air approaches the Asberg–Mundwald ridge the windspeed is reduced and a slight turning of the flow can be seen. The wind differences, which show the effect of the modification of the terrain and land-use, are concentrated along the new motorway. The cut through the Asberg–Mundwald ridge,

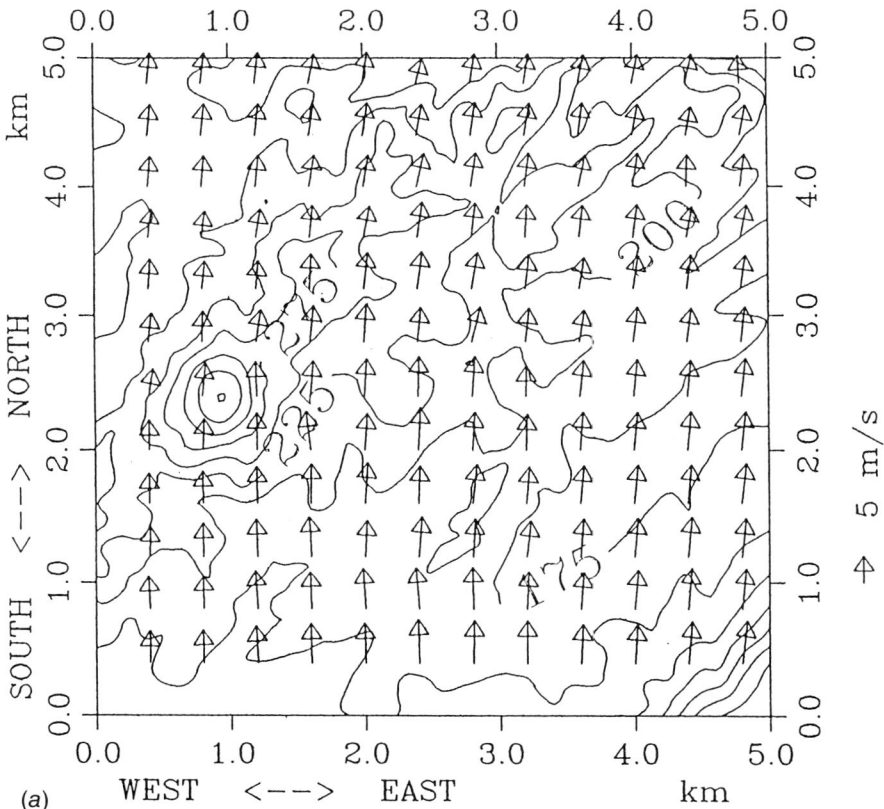

Fig. 22.8(a). For caption see opposite.

which separates the Asberg hill from the Mundwald ridge, coincides with a southerly change in modified wind direction and a reduction in speed at the north side of the new road.

Within the modified area and towards the south side of the cleft, surface winds increase in speed, showing difference winds opposite to the original wind flow and with a component towards the south. Close to the intersection with the A1/A48 the modifications of the surface lead to an increase in windspeed in many areas on the south side and a decrease on the north side.

For planning purposes the size of areas with increased or reduced windspeeds is of interest, as well as the mean increase or decrease of windspeed. To estimate the relative size of areas with changes in windspeed, all regions with an increase as well as a decrease in wind above $1\ \mathrm{m\ s^{-1}}$ are counted

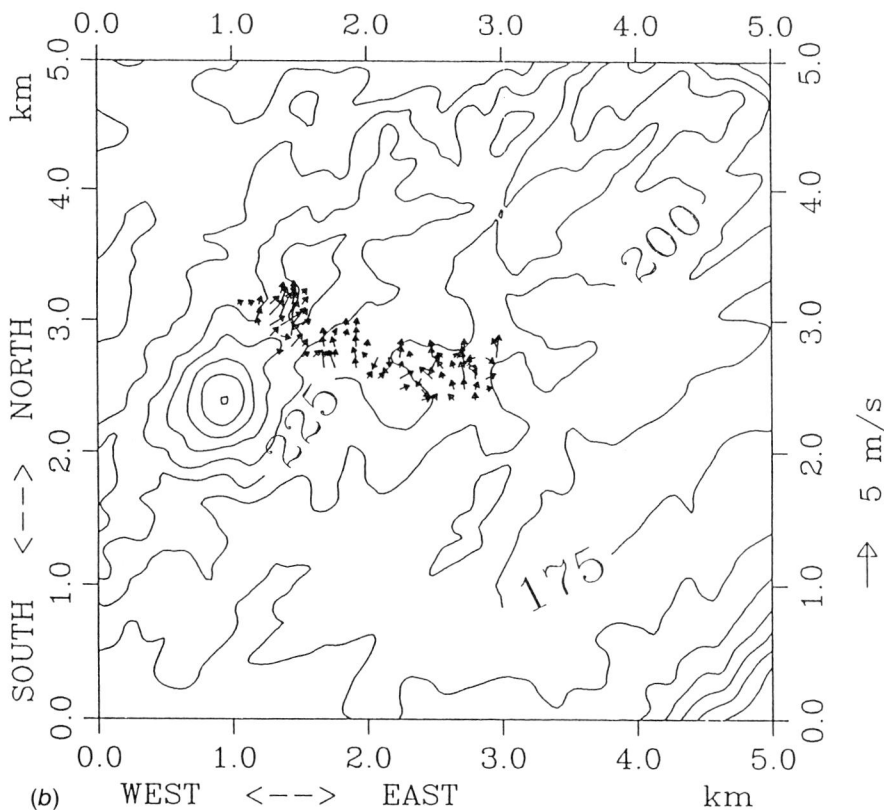

Fig. 22.8. (a) As for Fig. 22.7a but for a geostrophic wind from the south-west. (b) Wind difference for the geostrophic wind of (a): [v(original terrain) − v(modified terrain)].

separately. These computations are pursued for a smaller area 3 km (east–west) × 1.8 km (north–south) which covers the areas with wind changes. In relation to the south-west corner of the fine grid, this region is shifted 600 m eastwards and 1.8 km northwards to cover the area of the planned motorway. Table 22.2 gives the percentages of areas within this window where windspeed changes greater than this lower limit can be found. As an example, for a geostrophic wind of 36 m s^{-1} from the north-west, 3.9% of the area shows an increase in windspeed, with an average increase of 2.3 m s^{-1} (Table 22.4). A geostrophic wind from the west (36 m s^{-1}) shows a decrease in velocity within 11.3% of the area (Table 22.3). The size of the area affected

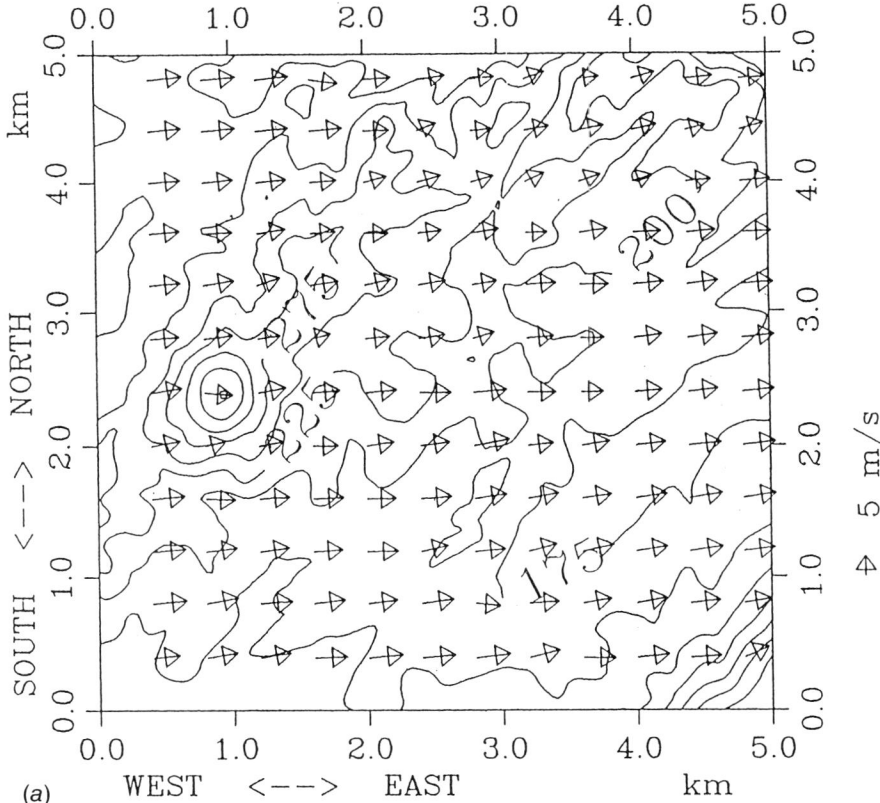

Fig. 22.9(a). For caption see opposite.

varies with windspeed and wind direction, generally affecting larger areas for higher windspeeds. For geostrophic winds from the north, north-west, west and, to a lesser degree, from the south and east, larger areas are affected by topography and land-use changes. A comparison with the annual wind distribution shows that windspeeds above 16 m s^{-1} are most likely to be from west, south-west, south-east and east. A closer examination of results from the model reveals that the total area with reduced windspeeds is considerably larger than the total area with increased windspeeds. An appropriate scheme to assess the resulting impact on forest damage has yet to be established.

22.4.4 Impact on turbulent kinetic energy

To estimate the changes in turbulent kinetic energy resulting from the land-

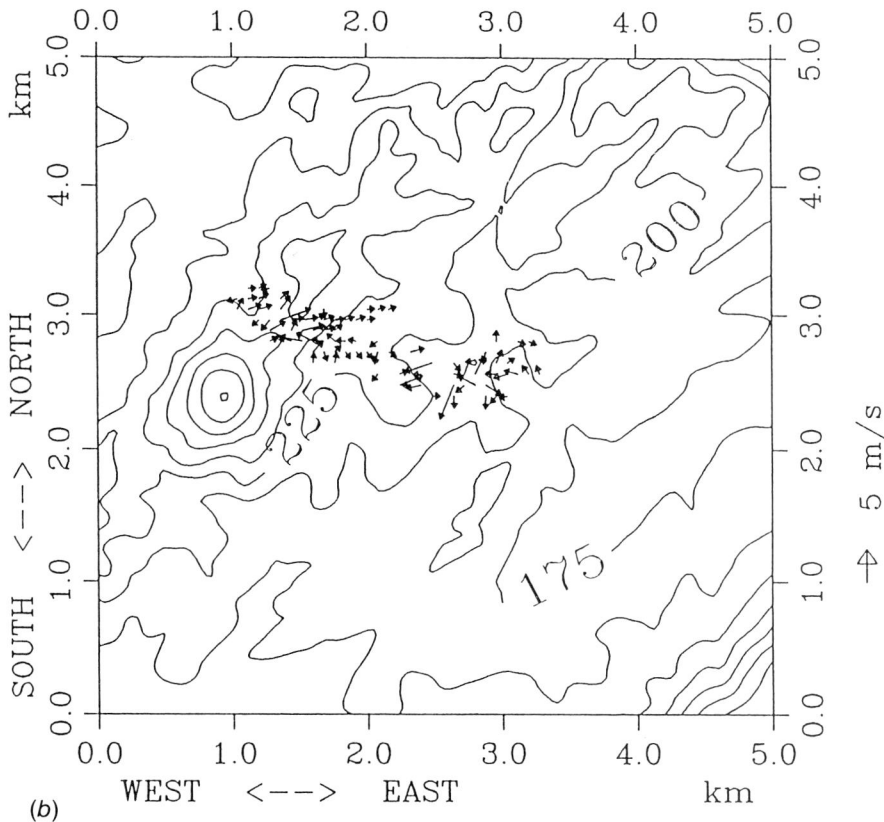

(b)

Fig. 22.9. (a) As for Fig. 22.7a but for a geostrophic wind from the north-west. (b) Wind difference for the geostrophic wind of (a): [v(original terrain) – v(modified terrain)].

Table 22.2. *Percentage of area with an increase in windspeed above 1 m s^{-1} for three windspeed classes and eight wind directions*

Windspeed (m s^{-1})	Wind direction							
	N	NE	E	SE	S	SW	W	NW
12	0	0.0	0.1	0.3	0.2	0.0	0.9	0.8
24	0.4	0.2	3.5	0.6	1.2	0.1	0.7	2.3
36	3.0	0.5	0.3	1.0	1.8	0.4	2.1	3.9

Table 22.3. *Percentage of area with a decrease in windspeed exceeding 1 m s^{-1} for three windspeed classes and eight wind directions*

Windspeed (m s^{-1})	Wind direction							
	N	NE	E	SE	S	SW	W	NW
12	0.1	0.2	0.8	1.3	1.9	0.9	2.4	1.0
24	0.7	1.7	2.0	1.0	30.8	3.2	36.3	4.3
36	1.7	3.3	7.5	2.6	17.3	3.1	11.3	4.4

Table 22.4. *Mean increase in windspeed for the areas with an increase in wind of more than 1 m s^{-1} as a function of windspeed and wind direction*

Windspeed (m s^{-1})	Wind direction							
	N	NE	E	SE	S	SW	W	NW
12	0.0	0.0	1.3	1.8	1.1	0.0	1.8	1.6
24	1.4	1.3	1.2	1.5	1.8	1.6	2.4	2.0
36	1.6	1.3	1.4	1.8	1.5	1.3	2.3	2.3

scape modifications, model computations are done with a two-dimensional multilayer boundary-layer model for sections across the planned motorway at right angles to the road's direction. The section is situated close to the Asberg hill, oriented south–north. A southerly geostrophic wind of 24 m s^{-1} is assumed. The calculations are done without (Fig. 22.10a) and with (Fig. 22.10b) terrain modifications, which include a change in terrain height, zero plane displacement and roughness length. For the case of unmodified terrain, a maximum of turbulent energy near the surface can be found on the upslope side of the Asberg–Mundwald ridge covering the upper parts of the descending and ascending side of the hill and its peak. With the terrain modifications included, the position of the turbulent kinetic energy maximum shifts to the ascending side of the cut through the valley. This will lead to an increase in turbulence in an area of new forest growth and near the new border of the old forest. Studies on storm damage to trees (Hütte, 1967; Rottmann, 1986) show a risk of damage along the upper part of the upwind slope and in the area of the peak if the wind flow is perpendicular to the ridge. In valleys, problems with tree damage arise in the middle of the upslope side, especially if the wind is orientated perpendicular to the valley. This supports the results of the model.

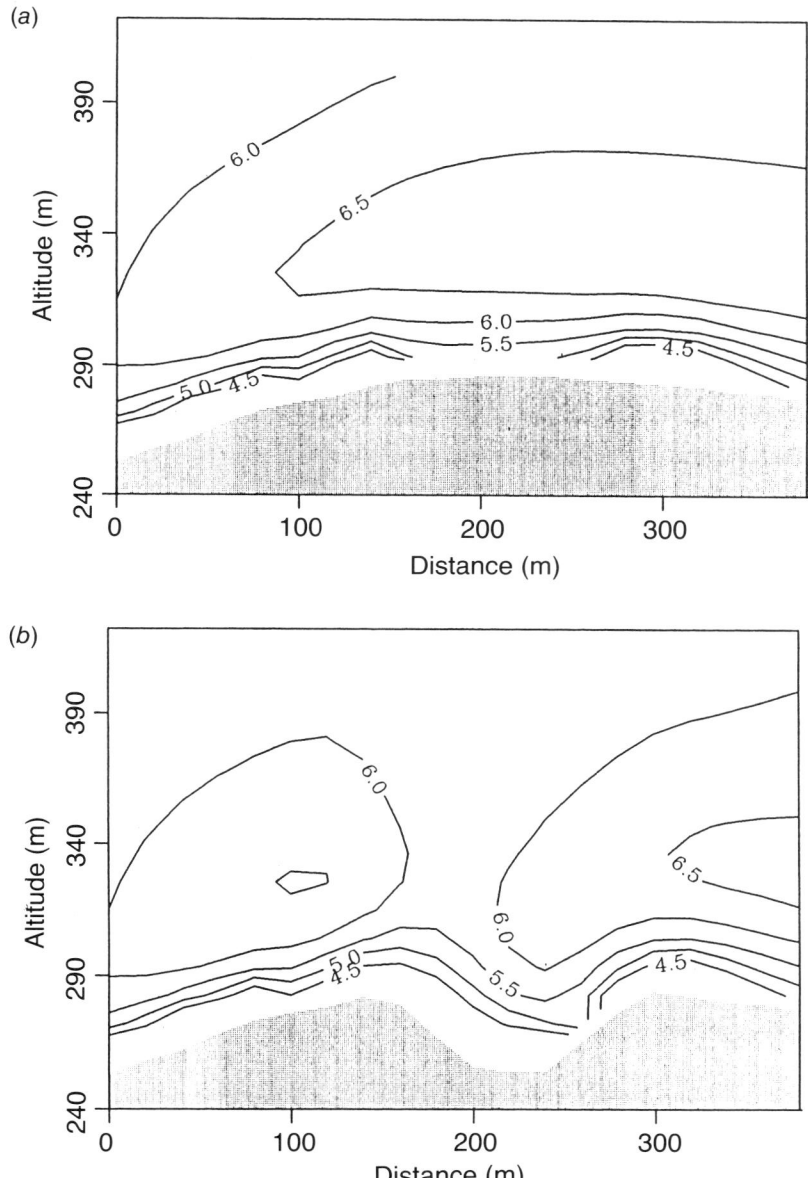

Fig. 22.10. Contour lines of turbulent kinetic energy ($m^2\ s^{-2}$) along a cross-section without (a) and with (b) terrain modifications.

22.5 Summary

To investigate the effects of the construction of a new motorway a hierarchy of models is adapted to describe the changes in mean wind flow and to estimate the additional risk of damage to trees from high windspeeds. The calculations start with a meso-scale model on a coarse grid. Surface winds are determined as a function of meteorological input parameters, topography and land-use. On a high-resolution grid, covering the area of interest, the difference in wind and turbulent energy is estimated by including the expected modifications of landscape and land-use. Wind vortices or gusts are not considered. Effects such as thermal circulation systems, which are induced by thermal inhomogeneity and non-neutral atmospheric stability, and which can only be found together with low geostrophic winds, are not included in this analysis. The method used is a fast and efficient sequence of models, giving cost-effective assessments of environmental problems. Specifically this allows for defining areas which will be affected by an increase or decrease in wind, or for minimising the mean increase in windspeed due to the expected changes in terrain and land-use.

Acknowledgements

This study has been performed under a contract with the Strassenprojektgruppe Wittlich and the Strassenneubauamt Trier, who provided planning information. The terrain data were collected and made available by the Landesvermessungsamt Rheinland-Pfalz in Koblenz. The German Weather Service provided wind data from its measurement site in Deuselbach.

References

Ball, J. A. & Johnson, S. A. (1978). *Physically Based High Resolution Surface Wind and Temperature Analysis for EPAMS*. Technical Report US Army Atmospheric Science Laboratory, White Sands Missile Range, NN 88002, ASL-CR-78-0043-1.

Businger, A. (1973). Turbulent transfer in the atmospheric surface layer. In *Workshop on Micrometeorology*, ed. D. A. Haugen. American Meteorological Society, Boston, MA.

Deardorff, J. W. (1974). Three dimensional numerical study of the height and mean structure of the heated boundary layer. *Boundary-Layer Meteorology*, **7**, 81–106.

Fraser, A. (1965). Wind stability. Tree pulling investigations; survey of crop stability. *Report on Forest Research*, Forestry Commission. 1963/1964.

Geiger, R. (1950). Die meteorologischen Voraussetzungen der Sturmgefährdung. *Forstwissenschaftliches Centralblatt*, **69**, 71–80.

Heimann, D. (1985). Verkürzte Fassung der REWIMET Modell-Beschreibung. DFVLR/PA, Oberpfaffenhofen, 8031 Wessling.

Hütte, P. (1967). Die standörtlichen Voraussetzungen für Sturmschäden. *Forstwissenschaftliches Centralblatt*, **86**, 276–95.

Mayer, H. (1985). *Baumschwingungen und Sturmgefährdungdes Waldes.* Wissenschaftliche Mitteilungen Meteorologisches Institut Universität München no. 51.

Rottmann, M. (1986). *Wind- und Sturmschäden im Wald.* Beiträge zur Beurteilung der Bruchgefährdung, zur Schadensvorbeugung und zur Behandlung sturmgeschädigter Nadelholzbestände. J. D. Sauerländer's Verlag, Frankfurt am Main.

Smeda, M. S. (1979). Incorporation of planetary boundary layer processes into numerical forecasting models. *Boundary-Layer Meteorology*, **16**, 115–29.

Stull, B. B. (1991). *An Introduction to Boundary Layer Meteorology.* Kluwer, Dordrecht.

Wangler, F. (1974). Die Sturmgefährung der Wälder in Südwestdeutschland: eine waldbauliche Auswertung der Sturmkatastrophe 1967. PhD Thesis, University of Freiburg.

Wollenweber, G. C. (1990). Modellierung der Struktur der internen Grenzschicht. PhD Thesis, University of Bonn.

23

Recommendations for stabilisation of Norway spruce stands based on ecological surveys

C. C. N. NIELSEN

Abstract

Wind stability is considered the major problem in Norway spruce silviculture. Traditional thinning models are characterised by a constancy of thinning intensity throughout the rotation. These models will generally result in low wind stability in spruce stands because they violate the important processes involved in wind stability: (1) The regulation of stem number is usually delayed until pulp wood or timber products can be extracted. This causes a high density of trees and high root competition in young stands, where the branching pattern and the long-term increment capacity of the structural root system are fixed. The production of finer roots during the second half of the rotation is thus strongly inhibited by high stem density in the pre-commercial phase. (2) Stem biomass extraction during the middle and last part of the rotation causes severe loss of 'anchorage biomass', as stumps do not contribute to neutralising the 'storm energy' transferred to the stand canopy. (3) The removal of trees destroys canopy closure: thinning reduces wind protection by neighbouring trees, because wind gusts penetrate deeper into the canopy, and physical crown contact is reduced. (4) A tree adapts to the very specific wind climate defined by the surrounding trees; therefore a change in wind flow in the canopy caused by removal of neighbouring trees makes it susceptible to damage. The negative effects of these last three factors increase with increasing age and with increasing thinning intensity. Regarding wind stability, a 'D- to A-degree' stem number reduction model is recommended. This model prescribes low stem numbers in young stands (through wide spacing or pre-commercial thinning operations), decreasing thinning intensity during the middle part of the rotation and no thinning at all in the last third

Dedicated to Dr Mike Coutts on his sixtieth birthday and retirement from the Forestry Commission.

of the rotation. This model avoids all the negative effects caused by traditional thinning models.

23.1 Introduction

Norway spruce (*Picea abies* Karst.), because of its high wood quality and fast growth, is the most widespread commercial tree species in Northern and Central Europe. The main problem in the silviculture of this species is the low wind stability of medium-aged and old stands. Wind damage in spruce forests causes great economic losses, but lack of flexibility in establishing the succeeding generation might be of even greater importance. Spruce stands are usually not capable of sustaining a stable cover for regeneration, hence clearcutting is in most cases the only possible regime. The choice of tree species in the succeeding generation is thus often limited.

23.2 Introduction to stem number regulation models

The change in the number of trees per hectare over time has a natural upper limit, which will differ depending mainly on site and early plant density. The limit is encountered in control plots (also called A-degree or non-thinning plots) in thinning experiments. Different thinning models aim at controlling the development of stem number with respect to production goals. This chapter deals with a model aiming at high wind stability.

Although stem number can be regulated by *commercial thinning* operations in middle-aged and older stands, the economy and ecology of thinning models depend strongly on the stand density in youth. Since early stand density is very dependent on *planting density* and *pre-commercial thinning (respacing)*, these should be included in any discussion of thinning models. As a consequence, discussion here has been expanded from 'thinning' to 'stem number regulation', a concept which includes the stem number regulation pattern in the pre-commercial phase. This broader concept is of great importance when dealing with wind stability.

23.2.1 Stem number regulation models with constant versus varying thinning intensity

It is important to distinguish between stem regulation models with constant or varying thinning intensity over time. Most thinning schedules in older thinning experiments prescribe a *constant stem regulation intensity*, as implemented in several research plots in Britain, Germany, Sweden and

Denmark (Bornebusch, 1933; Carbonnier, 1957; Henriksen, 1961; Assmann, 1964; Bryndum, 1969, 1978; Hamilton, 1976; Kramer, 1978a; Schober, 1979, 1980). Such thinning schedules may be strictly defined as a certain percentage of the maximum basal area (as by Bryndum, 1969, 1978) or may be expressed in more subjective terms of stand closure (as in most German thinning plots; Kramer, 1988). The schedules are named A-, B-, C- or D-degree (in Scandinavia), or schwache, mässige and starke Niederdurchforstung (in Germany).

Varying stem regulation intensity over time has become more dominant in practical silviculture within the last three decades. Schedules with increasing thinning intensity (B_f, or B- to C-degree) were represented in the 'Gludsted IS' Norway spruce experiment and were not recommended (Bryndum, 1969). Models with decreasing thinning intensities were implemented in some experiments (D- to B-degree (Bryndum, 1978) and gestaffelte Niederdurchforstung (Wiedemann 1942, 1951; Schober, 1979, 1980)), and these models of stand treatment found wide acceptance in Denmark and Germany.

Since 1970 the importance of *wider spacing and/or pre-commercial regulation of stem number* for stand stability and crop economy has been stressed by several authors (Pollanschütz, 1971, 1974; Johann & Pollanschütz, 1974; Forstliche Versuchs, 1975; Persson, 1975; Kramer, 1975, 1978b, 1980, 1988; Kramer & Spellmann, 1980; Johann, 1980; Burschel, 1981; Bryndum, 1982; Nielsen, 1990b,c, 1991). These newer models of treatment tend to reduce the stem number to a maximum of 2.500 trees per hectare at a stand height of 6–10 m.

23.3 Introduction to the fundamental physics of wind stability of trees

The wind stability of a tree depends on the ratio between anchorage and wind load. The forces, lever arms and functions which constitute these two components of stability are discussed in detail by Nielsen (1990a) and Coutts (1986). All below-ground forces influencing anchorage are concentrated in the moment of anchorage (Coutts, 1986; Nielsen, 1991) and all above-ground forces (including the wind force) constitute the destructive storm moment (Fig. 23.1). Since both definitions are turning moments acting on a common turning axis, the actual value of wind stability can be expressed as the ratio 'moment of anchorage/destructive storm moment'. As long as this ratio is greater than 1, the anchorage system is undamaged. This principle is elaborated on by Nielsen (1990a,c).

The moment of anchorage of a tree depends mainly on soil factors and the biomass of all root classes (probably including fine roots). Where there

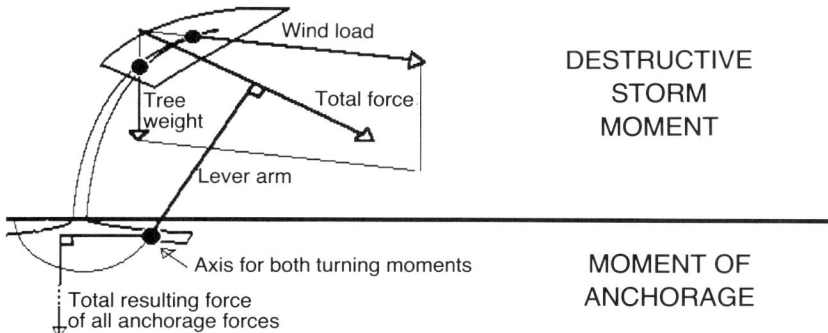

Fig. 23.1. Illustration of the concepts 'moment of anchorage' and 'destructive storm moment'.

is inadequate root symmetry or vertical rooting, root architecture may be considered an important factor. The destructive storm moment of a tree depends (besides storm velocity) mainly on topography, height and crown biomass of the tree and stand closure. All the factors influencing these two moments are discussed in detail by Nielsen (1990*a*).

23.4 The short-term influence of a thinning operation on wind stability

After stem reduction the stand stability is reduced. The degree of destabilisation depends on the tree species, soil type, water content of the soil, age of the stand, type and intensity of the thinning and earlier stand treatments. The destabilisation can be explained in terms of the interaction of trees and wind (see papers by Wood and by Gardiner, this volume) and can be summarised for single trees or populations as follows:

23.4.1 Single tree level

The removal of neighbouring trees results in an opening of the canopy, and the wind gusts penetrate deeper. The wind protection from neighbouring trees is reduced. Mechanical crown contact is also reduced. Both mechanisms contribute to increased tree sway amplitudes (and the destructive storm moment: Fig. 23.2). The anchorage of a single tree is usually not influenced by a thinning operation in the short term. Only on wet soils the loss of foliage may increase waterlogging, which might kill the deeper roots.

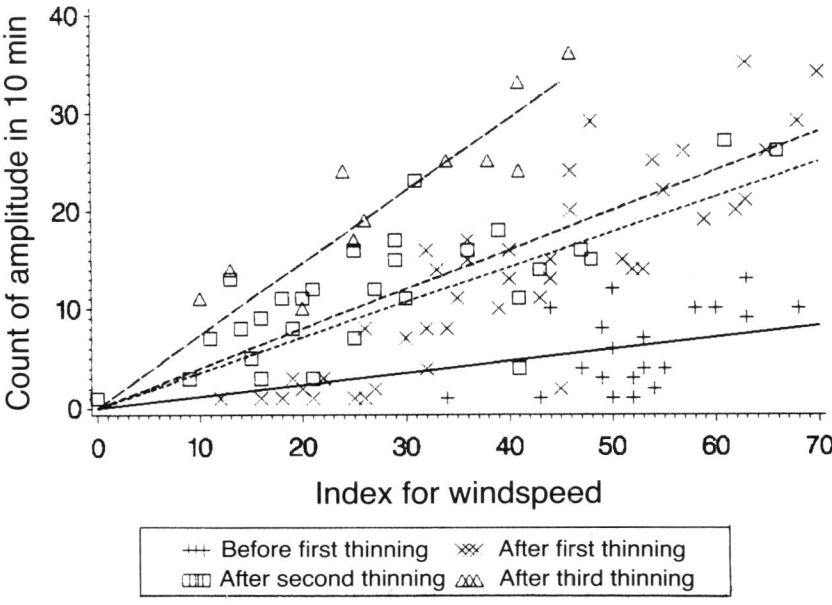

Fig. 23.2. The change in frequency of tree sway amplitudes greater than 2 mm in 2 m stem height, following three thinning operations in one winter (Nielsen, unpublished data).

23.4.2 Population level

Owing to a thinning operation the canopy surface becomes rougher. This will enhance the momentum transfer from wind to stand (Miller, 1986). This *increased* amount of transferred energy is distributed to a *reduced* number of trees per hectare, which increases the destructive storm moment of the remaining trees (Nielsen, 1990a). Because energy neutralisation due to mechanical crown contact is reduced, a higher proportion of the absorbed energy is transferred to the anchorage systems. All above-ground factors are negatively influenced by stem reduction; hence any thinning operation will increase the destructive storm moment.

The roots of the removed trees are interwoven with roots from the remaining trees (and will as such contribute to their moment of anchorage), but the primary anchorage function of the new stumps is lost. Because of this loss of active anchorage biomass the sum of anchorage moments per hectare is reduced. *The combination of lost forces of anchorage and increased wind energy uptake on stand level causes a reduction in the moment of anchorage/ destructive storm moment ratio. Hence the short-term effect of any stem reduction is a destabilisation of the stand.*

23.5 Long-term strategies for wind stability based on stem number regulation

23.5.1 Complete canopy closure in older stands

The effects discussed of a thinning operation on the destructive storm moment increase dramatically with age: the capacity for crown expansion decreases with increasing age. Total closure of the canopy after a thinning operation will typically take from 10 to 15 years in a stand with a mean height of 20 m, but complete closure will not be achieved again after thinning in old stands (where height growth has almost ceased). With respect to the destructive storm moment it seems appropriate to decrease the thinning intensity with increasing age and to avoid any extractions during the last third of the rotation (independent of rotation length, actual height growth will be restricted during the last third of the rotation, hence this recommendation is applicable to most sites). This concept will provide a smooth canopy surface, low momentum transfer to the canopy, high wind protection between neighbouring trees and higher reductions of the tree sway amplitudes due to rigid physical crown contact between trees.

23.5.2 Accumulation of anchorage biomass throughout the rotation

Because every extraction of stem biomass leads to a comparable loss of root biomass, the sum of moments of anchorage of all trees on a hectare is reduced by stem reductions. At the single tree level this may be formulated as follows: A wind gust of a specific volume will be neutralised by the root systems of two trees before the thinning. After removal of one of the trees, the root system of the remaining tree will have to neutralise the energy of the same (or even a bigger) gust of wind.

This statement explains why the total moments of anchorage in a stand increase with the amount of living root biomass per hectare (It seems probable that all root classes, including fine roots, contribute to anchorage: Nielsen 1990a,c, 1991). *From the point of view of anchorage, it is important to avoid extraction of (stem) biomass during the middle and last part of the rotation as much as possible.* Such a strategy enables the complete increment of root biomass throughout the last decades of the rotation to stay 'alive' on active anchorage systems instead of being lost as stumps. Such strategies were implemented in research plots in the Hauersteig spacing experiment (2×2 m plot) and in the Paderborn thinning experiment (the Schnellwuchs plot), as illustrated in Fig. 23.3. This figure also illustrates the ranges from several traditional yield tables (Gehrhardt, 1930; Möller, 1933; Wiedemann, 1936/42 *a, b*, 1951; Assmann & Franz, 1965; Forstliche Versuchs, 1975).

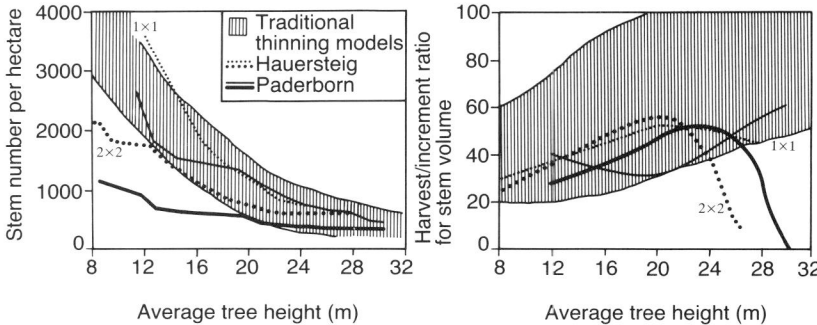

Fig. 23.3. The development of stem number and harvest/increment ratio (harvest = removal of stem biomass per hectare and year; increment = stem increment per hectare and year) from traditional yield tables and two wind-stable research plots. From Nielsen (1990c).

23.5.3 Increasing the long-term increment capacity of the 'thin root' compartment

Biomass investigations in old spacing plots on different sites have shown that intraspecific root competition *in youth* has a very marked effect on the biomass of thin roots (1–20 mm diameter) and on the thin root/stem ratio at the end of the rotation, where the number of trees per hectare was completely equalised between research plots (Fig. 23.4; Nielsen, 1990b,c). Root competition is especially high in young stands with high stem numbers, and branching intensity of root systems was shown to increase with decreasing root competition (Nielsen, 1990b). It is likely that the thin root increment in the long term is predetermined by the branching intensity of the structural root system, which apparently becomes fixed during the first third of the rotation. Since the root compartments between 1 and 20 mm in diameter are important for the size of the root ball and for the windward anchoring of thick roots, it may be important from an anchorage point of view to stimulate root branching through low stem numbers in young stands.

23.5.4 Undisturbed adaption processes to the wind structure in the canopy

Several processes of adaption to wind load in Norway spruce were investigated by Nielsen (1990a–d). First, with increasing wind load an increasing proportion of total wood increment is allocated to the thick roots, which enhances the thick root/stem ratio. Secondly, radial root growth is to a large

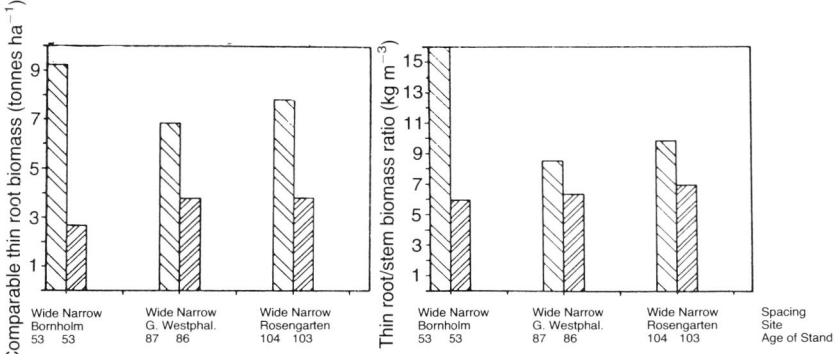

Fig. 23.4. Influence of original spacing on the amount of thin root biomass (1–20 mm diameter, adjusted for different mean stand heights) and on the thin root/stem ratio of Norway spruce in old stands on three sites. Wide and narrow spacing: less than 1700 and more than 6000 stems per hectare respectively (Nielsen, 1990b).

extent controlled by mechanical stress. Thus roots allocate radial increment in accordance with the forces transferred from the crown, to optimise the strength of the anchorage system against the predominating wind climate. Finally, tree crowns expand (often asymmetrically) to close the canopy. As a result all trees have crown contact with one another.

The architecture of a tree adapts to the wind environment over a long period. Thinnings alter the airflow in the canopy, and thereby expose trees to an environment to which they are not adapted. The adaption processes must start again after every thinning operation, and year-ring studies in roots show that such adaptions require decades without disturbance to maximise the strength of the anchorage system (Nielsen, 1990c). Because the adaption processes take longer with increasing age and with increasing thinning intensity, it is recommended that thinning operations are reduced in number and intensity with increasing age.

23.6 The recommended discontinued 'D- to A-degree' stem number regulation model

The strategies discussed above for enhancing wind stability stress the importance of decreasing the thinning intensity with increasing age and of the total absence of thinnings during the last decades of the rotation. The importance

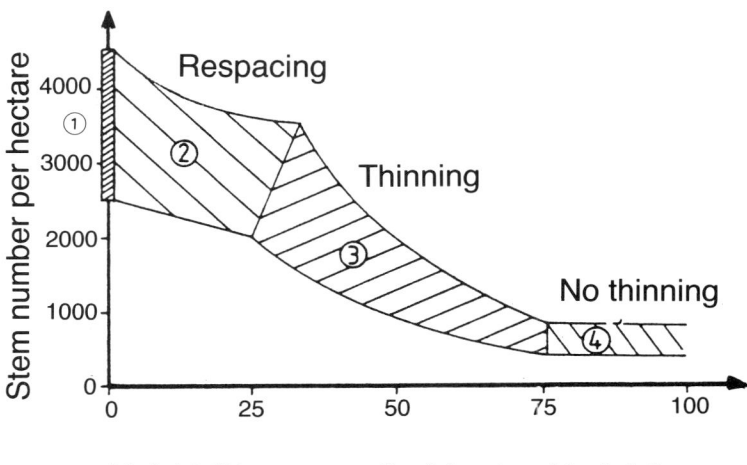

Fig. 23.5. The recommended range for the development of stem number over time according to the 'D- to A-degree' stem number regulation model. See text for explanation of 1–4. After Nielsen (1990c).

is stressed of regulating stem number in young stands at an age where storm risk is limited. Since the long-term productivity of the thin root compartment depends strongly on root competition, it is recommended to aim at a stem number of 2000–3000 trees per hectare as a maximum at an average stand height of 6–10 m. This can be carried out through wide spacing or pre-commercial thinning (points 1 and 2 in Fig. 23.5). The highest timber quality is attained through dense planting and intensive pre-commercial thinning. Hence the model of 'D- to A-degree' stem number regulation prescribes 'D-degree respacing and thinning' (high intensity) in youth, decreasing thinning intensity in middle-aged stands (point 3 in Fig. 23.5) and no thinning at all ('A-degree thinning') during the last decades before clearcutting (point 4 in Fig. 23.5).

The model described above must be adjusted to the actual site (yield class and storm exposure) and production goal (volume production, economy, stability and wood quality). These topics are discussed in detail by Nielsen (1990c). The range for stem number development for a more or less wind stable stand is illustrated in Fig. 23.5. If wind stability has the highest priority, stem number should be as low in youth and as high as possible at the end of the rotation. The most stable stand is wide spaced initially and is not thinned at all. A stem number development along the upper range in Fig. 23.5 secures a high timber quality and should only be implemented on protected sites with high yield class, where long rotations are possible.

23.7 Models with constant thinning intensities and wind stability

Stem regulation models with constant thinning intensity over time violate the stabilising growth processes discussed in Section 23.5. Most traditional thinning models thus decrease wind stability. Since the violation of the stabilising processes becomes more dramatic with increasing thinning intensity, wind stability tends to decrease with increasing thinning intensity. This conclusion receives empirical support from observations on damage in several thinning research plots, but the conclusion presumes complete stands without edges or holes in the stand (Nielsen, 1990c). If the stand is damaged or ripped up by previous storms, the resistance against total destruction may increase with thinning intensity, because the 'plasticity' of the root systems increases with thinning intensity. A detailed discussion of the ecological basis for wind stability related to continued thinning models is given by Nielsen (1990c).

References

Assmann, E. (1964). Der Fichten-Durchforstungsversuch Bowmont. *Allegemeine Forst- und Jagdzeitung*, **135**.

Assmann, E. & Franz, F. (1965). Vorläufige Fichten-Ertragstafel für Bayern. *Forstwissenschaftliches Centralblatt*, **84**, 13–43.

Bornebusch, C. H. (1933). Ein Durchforstungsversuch in Fichte. *Forstlige Forsogsvaesen i Danmark*, **13**.

Bryndum, H. (1978). Durchforstungsversuche in jungen Fichtenbeständen auf feinlehmigen, jungdiluvialen Moränenböden. *Forstlige Forsogsvaesen i Danmark*, **36**.

Bryndum, H. (1982). Vorläufige Beobachtungen in den dänischen Durchforstungsversuchen nach dem orkanartigen Sturm 24/25 Nov. 1981. Deutsch. Verb. forst. Forschungsanstalten, Sektion Ertragskunde, 159, Weibersbrunn.

Burschel, P. (1981). Neue Erziehungskonzepte für Fichtenbestände, *Allgemeine Forst- und Jagdzeitung*, **92**, 1386–95.

Carbonnier, C. (1957). Ett gallringsförsök i planterad granskog. *Statens skogsforskningsinstitut-Uppsatser*, **55**, 463–76.

Coutts, M. (1986). Components of tree stability in Sitka spruce on peaty gley soil, *Forestry*, **59**, 173–97.

Forstliche Versuchs Forschungsantalt, Baden-Württemberg (1975). *Entscheidungshilfen für die Durchforstung von Fichtenbeständen, Mitteilungen*, Merkblatt 13.

Gehrhardt, G. (1930). *Ertragstafeln für reine und gleichartige Hochwaldbestände von Eiche, Buche, Tanne, Fichte, Kiefer, grüner Douglasie und Lärche.* Julius Springer, Berlin.

Hamilton, G. J. (1976). The Bowmont Norway spruce thinning experiment 1930–1974. *Forestry*, **49**, 49.

Henriksen, H. A. (1961). A thinning experiment with Sitka spruce in Nystrup Dune Forest. *Forstlige Forsogsvaesen i Danmark*, **27**.

Johann, K. (1980). Bestandesbehandlung und Schneebruchgefährdung, *Allgemeine Forstzeitung*, **91**, 269–86.

Johann, K. & Pollanschütz, J. (1974). Durchforstungsmodelle als Entscheidungshilfen. *Allgemeine Forstzeitung*, **85**, 307–13.

Kramer, H. (1975). Erhöhung der Produktionssicherheit zur Förderung einer nachhaltigen Fichtenwirtschaft. *Forstarchiv*, **46**, 9–13.

Kramer, H. (1978a). Der Bowmont-Durchforstungsversuch. *Forstwissenschaftliches Centralblatt*, **97**, 131–41.

Kramer, H. (1978b). Bestandesbegründung unter dem Aspekt der künftigen Durchforstung. *Forst und Holzwirt*, **30**, 324–30.

Kramer, H. (1980). *Einfluss verschiedenartiger Durchforstungen auf Bestandessicherheit und Zuwachs in einem weitständig begründeten Fichtenbestand*. Schriften aus der Forstlichen Fakultät der Universität Göttingen und der Niedersächsischen Forstlichen Versuchsanstalt no. 67.

Kramer, H. (1988). *Waldwachstumslehre*. Verlag Paul Parey, Hamburg.

Kramer, H. & Spellmann, H. (1980). *Beiträge zur Bestandesbegründung der Fichte*. Schriften aus der Forstlichen Fakultät der Universität Göttingen und der Niedersächsischen Forstlichen Versuchsanstalt no. 64.

Miller, K. F. (1986). Recent aeromechanical research in forest plantations. In *Proceedings of the Workshop: Minimizing Wind Damage to Coniferous Stands*. Directorate-General Science, Research and Development, Commission of the European Community.

Möller, C. M. (1933). Bonitetsvise Tilvækstoversigter for Bøg, Eg og Rødgran i Danmark. *Dansk Skovforenings Tidsskrift*, **18**.

Nielsen, C. C. N. (1990a). Eine theoretische Grundlage zur Sturmfestigkeitsforschung auf Einzelbaumebene. In *Einflüsse von Pflanzenabstand und Stammzahlhaltung auf Wurzelform, Wurzelbiomasse, Verankerung sowie auf die Biomassenverteilung im Hinblick auf die Sturmfestigkeit der Fichte*, ed. C. C. N. Nielsen. Schriften aus der Forstlichen Fakultät der Universität Göttingen und der Niedersächsischen Forstlichen Versuchsanstalt no. 100.

Nielsen, C. C. N. (1990b). Methodische und ökologische Untersuchungen zur Sturmfestigkeit der Fichte. In *Einflüsse fon Pflanzenabstand und Stammzahlhaltung auf Wurzelform, Wurzelbiomasse Verankerung sowie auf die Biomassenverteilung im Hinblick auf die Sturmfestigkeit der Fichte*, ed. C. C. N. Nielsen. Schriften aus der Forstlichen Fakultät der Universiäat Göttingen und der Niedersächsischen Forstlichen Versuchsanstalt no. 100.

Nielsen, C. C. N. (1990c). Methodische, ökologische und waldbauliche Beiträge zur Sturmfestigkeit der Fichte. In *Einflüsse von Pflanzenabstand und Stammzahlhaltung auf Wurzelform, Wurzelbiomasse, Verankerung sowie auf die Biomassenverteilung im Hinblick auf die Sturmfestigkeit der Fichte*, ed. C. C. N. Nielsen. Schriften aus der Forstlichen Fakultät der Universität Göttingen und der Niedersächsischen Forstlichen Versuchsanstalt no. 100.

Nielsen, C. C. N. (1990d). Bevoksningsstruktur og stormstabilitet. *Dansk Skovbrugs Tidsskrift*, **75**, 72–80.

Nielsen, C. C. N. (1991). Zur Verankerungsökologie der Fichte. *Forst und Holz*, **46**, 178–82.

Persson, P. (1975). *Windthrow in Forest: Its Causes and the Effect on Forestry Measures*. Department of Forest Yield, Royal College of Forestry, Stockholm.

Pollanschütz, J. (1971). Durchforstung von Stangen- und Baumhölzern. *Forstarchiv*, **42**, 257–9.

Pollanschütz, J. (1974). Erste ertragskundliche und wirtschaftliche Ergebnisse des Fichten-Pflanzweiteversuchs Hauersteg. 100 Jahre Forstlichen Bundesversuchsanstalt Wien.
Schober, R. (1979). Massen-, Sorten-, und Wertertrag der Fichte bei verschiedener Durchforstung: I. *Allgemeine Forst- und Jagdzeitung*, **150**, 129–53.
Schober, R. (1980). Massen-, Sorten- und Wertertrag der Fichte bie verschiedener Durchforstung: II. *Allgemeine Forst- und Jagdzeitung*, **151**, 1–21.
Wiedemann, E. (1936/42*a*). Fichten-Ertragstafeln. Cited in Schober (1975). *Ertragstafeln wichtiger Baumarten*. Sauerländer's Verlag, Frankfurt am Main.
Wiedemann, E. (1936/42*b*). Die Fichte 1936. *Mitteilungen aus Forstwirtschaft und Forstwissenschaft* [1937]. (Correction in *Mitteilungen aus Forstwirtschaft und Forstwissenschaft* [1942], 287.)
Wiedemann, E. (1951). *Ertragskundliche und waldbauliche Grundlagen der Forstwirtschaft*. Sauerländer's Verlag, Frankfurt am Main.

24
Thinning regime in stands of Norway spruce subjected to snow and wind damage

M. SLODIČÁK

Abstract

Between 1981 and 1990 more than 50% of the total timber yield in the Czech Republic had to be cut down due to some kind of injury, mostly windthrow or snowbreak. Most damage occurred in stands of spruce (*Picea abies* Karst.), a species occupying 55% of the forest area. Snow is the main damage factor in younger stands, in the period of height growth culmination. The best protective measure, proven by many experiments, is stimulation of diameter growth by wide spacing or heavy thinning in the period of canopy closure. On the other hand, wind damage starts when the top height of the stands exceeds 10–15 m, and the most reliable protection is a closed canopy and mutual shelter of individuals, with resistance to snow damage created by the previous treatment at a younger stage. A thinning regime is proposed for spruce stands suffering from both snow and wind. Heavy thinning in young stands is recommended, changing to light thinning and growth with full canopy closure in the second half of the rotation. This thinning regime is supported by the results of three long-term experiments in north-eastern Bohemia.

24.1 Introduction

Damage to forests by abiotic factors (snow, ice and especially wind) is a very frequent phenomenon in countries where forestry has been based on artificial regeneration with one prevailing species and where the relevant weather conditions are severe. This situation can be found in all Middle European countries (Germany, Switzerland, Austria, Slovakia, the Czech Republic and Poland) as well as in Great Britain and elsewhere (Australia, New Zealand and Japan).

In the Czech Republic, with a total forest area of more than 2.6×10^6 ha (33% of the country), huge salvage cuttings have been needed (Fig. 24.1).

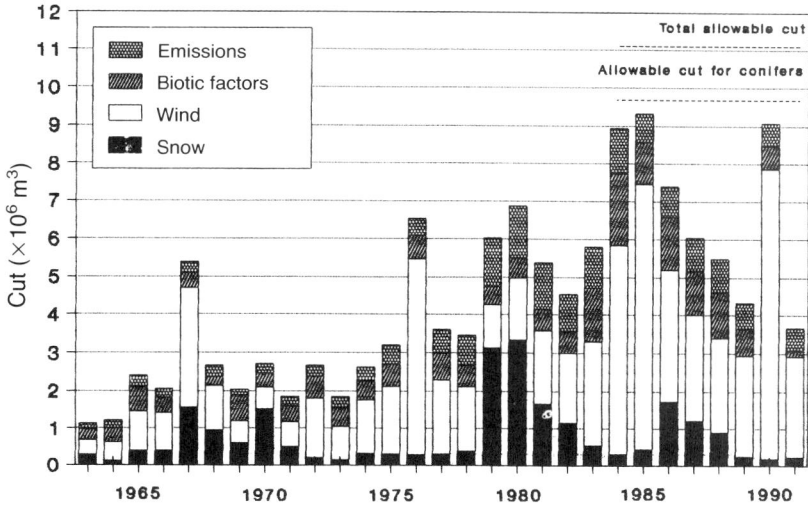
Fig. 24.1. Salvage cuttings in forests of the Czech Republic from 1963 to 1991.

Between 1981 and 1990 more than 50% of total yield had to be cut down because of some kind of injury (mostly windthrow or snowbreak). The greatest damage occurred in pure, even-aged spruce stands (*Picea abies* Karst.). In some years (e.g. 1984, 1985 and 1990) the salvage cutting represented nearly 90% of the allowable cut for conifers. The situation has reached the point where normal forest management is being gradually changed to coping with the consequences of catastrophes. Much attention is therefore being paid to preventive measures against snow and wind damage.

This chapter deals with thinning as a silvicultural method for stabilising spruce stands subjected to snow and wind.

24.2 History

Observation and experiment have shown that snow and wind differ in their influence. Snow affects all stands at elevations of 500–800 m, i.e. the sites with a high incidence of heavy wet snowfall (Pařez, 1972). Young, even-aged spruce stands at the period of height growth culmination are the most vulnerable. As the influence of snow on a tree is more static with prevailing vertical pressure (Petty & Worrell, 1981; Marsch, 1989), stembreak is the most frequent type of damage. Trees from lower tree classes with a high height/diameter (h/d) ratio are, as a rule, broken first (Slodičák, 1983) and damage increases with the amount of wet snow; sometimes damage may achieve catastrophic dimensions.

On the other hand, wind damage increases with stand age, especially when the top height exceeds 15–20 m (Vicena, 1964; Mitscherlich, 1974; Chroust, 1980). The stands on sites with wet and waterlogged soils are much more susceptible than those on well-drained soils (Málek, 1976). The influence of wind on a tree is a dynamic, prevailingly horizontal pressure. The type of damage depends mainly on site conditions. On wet and waterlogged soils, especially in stands with well-developed tree crowns, uprooting prevails, whereas in better-rooted stands on well-drained soils stembreak may dominate.

From the brief analysis of snow and wind damage given above, it is apparent that there is no simple recipe for preventive silvicultural measures. Two main strategies of forest stabilisation were developed in the past. The first deals with a forest as a unit. The main principles are mutual shelter of adjacent stands and neighbouring trees within a stand. The stabilising elements are windfirm edges, increased proportion of windfirm species (beech, larch), regeneration from the leeward part of the stand and drainage of forest soils. This strategy was widely used in the Czech Republic but it has its weaknesses. First, all systems of mutual shelter are developed with respect to a prevailing wind direction (mainly west or north-west). However, the wind may come from other directions (Heger, 1953) and once damage has started all elements become susceptible (Heger, 1953; Busby, 1965). A second preventive silvicultural strategy based on the stability of individuals was therefore worked out (Heger, 1953; Mitscherlich, 1974). The main principle of this is the growth of spruce stands in a loose canopy so that the trees can form large crowns, stronger root systems and stable stems with a low h/d ratio. For this purpose, wide initial spacing or early respacing and heavy thinning is recommended. This second strategy proved successful, especially against snow damage (Pařez 1972; Mráček, 1979; Chroust, 1980; Slodičák, 1987), but *later* heavy thinning as a part of this strategy was frequently followed by increased wind damage (Vicena, 1964; Persson, 1969; Rottmann, 1985; Lohmander & Helles, 1987; Slodičák, 1987).

Therefore a third strategy, using the advantages of both previous ones, was created. It is the so-called gradual thinning (gestaffelte Durchforstung) proposed by Wiedemann (1955) for improvement of quality and amount of production, though the idea of different treatment for younger and older spruce stands is much older. It was first mentioned by Bohdanecký (1890), who recommended growing spruce in a loose canopy in the first half of the rotation (to create quantity) and in a full canopy in the second half of rotation (to achieve quality).

According to this third strategy, the stability of individuals can be built only in young stands, by keeping them in a loose canopy by wide spacing or respacing and early heavy thinning. The aim of this is tall, well-developed crowns, large root systems and stable stems with a low h/d ratio. Stands treated in this way are, as a rule, very resistant to snow, but a tall crown represents a larger area for wind to act on a tree. Although there is some evidence that the stable stem can outweigh increased drag forces (Blackburn & Petty, 1988), the best measure seems to be reducing crown size either by pruning (Brünig, 1973; MacCurrach, 1991) or naturally by changing the strategy of a loose canopy to that of a closed canopy (Wiedemann, 1955; Mitscherlich, 1974; Johann, 1981; Chroust, 1980, 1989; Slodičák, 1987). Choosing the method of natural reduction of crown size by lower thinning intensity, a system of mutual shelter is gradually built, but in this case it is a mutual shelter of individually stable trees (Marsch, 1989) because the effect of low stocking during the early years is permanent after the trees reach a height of 15 m (Johann, 1981; Cremer *et al.*, 1982).

Particular points of this third strategy of stabilisation of spruce forest stands to snow and wind damage may be supported by experiments carried out by the Forest Research Station, Opočno. These are described below.

24.3 Experiment I

The experiment started in 1952 in the north-west part of the Orlické hory Mountains. The site is at an elevation of 700 m with western slope of 5%. Mean annual temperature is 5.5 °C, and annual precipitation 1090 mm. Spruce thicket originated by sowing on medium rich, slightly podzolic forest soil was experimentally thinned at the age of 12 years (1952). The aim of the experiment was to study the effect of various thinning regimes on growth and development of young spruce stands in mountainous conditions with the emphasis on stability (Chroust, 1968). The initial density of the stand, *c.* 30 000 trees ha^{-1}, was therefore reduced by heavy low thinning to 4300 individuals ha^{-1} on plot 2, and by very heavy low thinning to 1750 trees ha^{-1} on plot 3. The control plot 1 was left without thinning (Fig. 24.2). At the age of 22 years the number of trees decreased on plot 1 to 10 340 by self-thinning and on plot 2 to 2400 stems ha^{-1} by low thinning. The number of trees on plot 3 did not change (1750). In 1966, at the age of 26 years, the experimental series was heavily damaged by snow. The most densely grown control (stand 1) was completely destroyed (only 19% basal area (ba) left) and heavy damage occurred on plot 2 (54% ba left). On the other hand stand 3, developing in a loose canopy, was damaged only slightly (80% ba left). More detailed information about this damage can be found in Chroust (1968, 1969).

The snow destroyed adjacent stands also, and so the experimental series created a sharp forest edge oriented to the North. Five years later, in 1971,

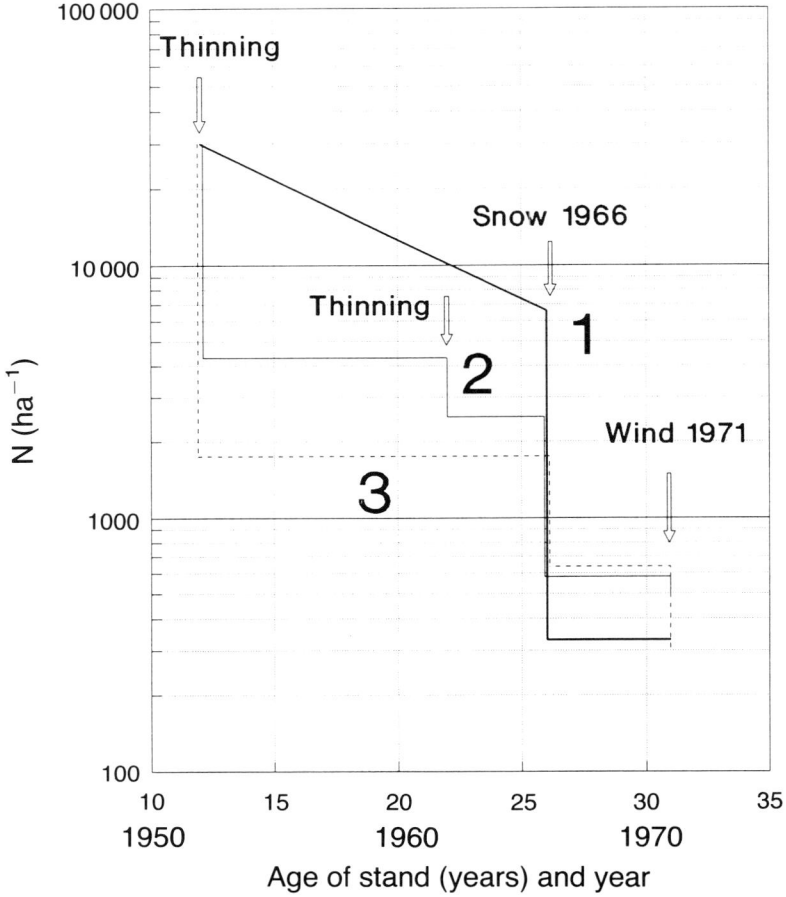

Fig. 24.2. Development of stand density in plots 1–3 of experimental series I.

wind of speed c. 80 km h^{-1} virtually destroyed stand 3 and partially destroyed stand 2. The trees left on plot 1 after snow damage in 1966 (330 individuals ha^{-1}) were not damaged at this time.

After this catastrophe, the experiment ended. The detailed study was published by Chroust (1980). The main conclusions from this experiment are: (1) very heavy early thinning (or respacing) is a reliable silvicultural measure for stabilisation of spruce stands to snow damage; (2) the large, well-developed crowns and stable stems of trees kept in a loose canopy are not a reliable protection against wind damage.

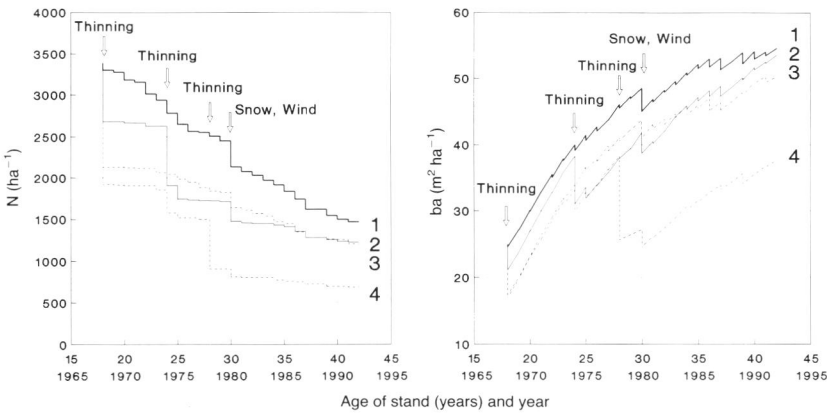

Fig. 24.3. Development of stand density (N) and basal area (ba) in plots 1–4 of experimental series II.

24.4 Experiment II

The second experiment was established in 1969 in the south-east part of the Orlické hory Mountains in order to study silvicultural treatments in spruce stands on former agricultural land (Chroust, 1979). The site is at an elevation of 570 m with a slope of 2%. Mean temperature is 6.1 °C, mean precipitation 1220 mm. The spruce stand had been established by sowing on grazing land. When the trees had achieved a top height of c. 1 m, the stand was respaced so that before the first experimental thinning there were 3300 trees ha^{-1} in irregular spacing. The experimental series consists of four plots: a control (plot 1) and three plots where various thinning regimes are being investigated. In plot 2 there is light thinning from below at 5 year intervals; in plot 3 there is heavy thinning initially; and on plot 4 there is very heavy thinning from below.

At the age of 19 years (1969) the first thinnings were done on plots 2, 3 and 4 where 19%, 36% and 44% of trees were removed respectively. The second thinnings were done 5 years later (1974) on plots 2 and 4, with 33% and 16% of trees removed respectively. By the third treatment on plot 4 at the age of 29 years, 37% of trees were cut down. The development of the experiment up to 1990 is given in Fig. 24.3 and more detailed results have been published (Slodičák, 1983, 1987).

In winter 1980/81, i.e. at age 30 years, the stands of the experimental series were damaged, firstly by wind (November 1980) and then by snow (January 1981). Trees damaged by wind were of medium size and medium h/d ratio (Fig. 24.4) with well-developed crowns; the heavily thinned stand on plot 4

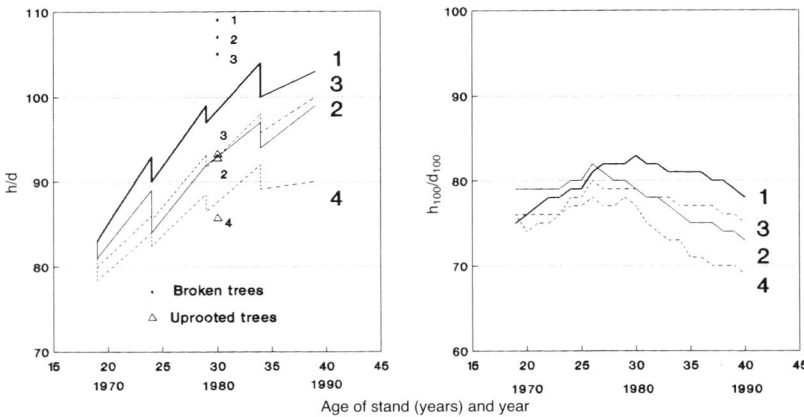

Fig. 24.4. Ratio h/d of the plots and for 100 trees with the largest dbh (h_{100}/d_{100}) in plots 1–4 experimental series II.

was most affected (8% uprooted). The trees were uprooted in groups and large gaps formed in the canopy. Plots 2 and 3 were damaged less (1% and 3.5% of trees respectively), whereas the most densely grown plot (control plot 1, without thinning) was not damaged at all.

Trees damaged by snow were mainly subdominant or overtopped individuals with an above-average h/d ratio (Fig. 24.4). The densely grown control (plot 1), and plot 2 with light thinning every 5 years, were most affected (nearly 5% of each, i.e. 231 and 188 trees ha^{-1} respectively). The snow did not damage the heavily thinned plot 4.

As can be seen in Fig. 24.4, thinning influenced slenderness of the trees in the series of experiments. The h/d ratio tended to increase on all the experimental plots. On dominant trees (d_{100}) it reached a peak at the age of 25–30 years; on average it is expected to take 40 years to peak. Thinning of various intensities did not change the tendency of h/d to increase, but it reduced the rate of increase. Therefore, the sooner the first heavy thinning is carried out, the lower the starting position of h/d and the lower the final h/d.

Two conclusions can be drawn from this experiment: (1) heavy early thinning was confirmed as a suitable measure for stabilising spruce stands to snow damage; (2) the best protection against wind was the close canopy of the unthinned centre plot, though the h/d ratio of both the average and the dominant trees was highest in this stand.

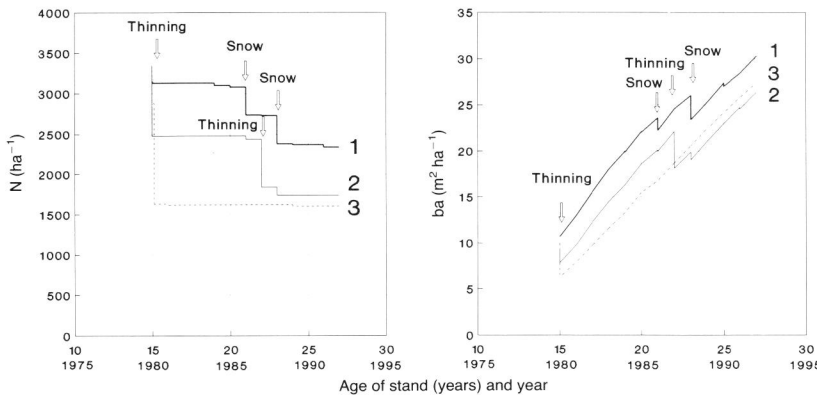

Fig. 24.5. Development of stand density (N) and basal area (ba) in plots 1–3 of experimental series III.

24.5 Experiment III

The third experiment reflects the conditions in which spruce stands have been established since the 1970s, i.e. planting of well-developed plants (bare root or containerised) at lower densities (3000–4000 trees ha^{-1}).

The experiment started in 1980 in the north-western part of the Orlické hory Mountains at an elevation of 800 m. The aim was to find out the influence of various thinning methods on the stability of young stands of Norway spruce. The experimental series was founded in a 15-year-old plantation with an original density of 3500–4000 trees ha^{-1} in irregular spacing. The experiment consists of three plots with different thinning regimes.

Plot 1 is a control without intentional thinnings. The programme with heavy thinning from below at 5 year intervals is tested on plot 2 and the programme based on one very heavy thinning at a young age is applied on plot 3 (Fig. 24.5). The first experimental thinnings on plots 2 and 3 were done at the age of 15 years (in 1980), and the second treatment on plot 2 at 22 years (in 1987).

The initial number of trees varied from 2950 ha^{-1} on plot 3 to 3150 ha^{-1} on plot 1. Density of the stand on plot 2 was decreased by thinning with negative selection from below, from 3110 to 2490 trees ha^{-1}. The second thinning in 1987 (negative selection from below) was postponed by 2 years and joined the third one planned for 1990. The number of trees decreased in this treatment to 1840 ha^{-1}. The programme on plot 3 started by opening the canopy and loosening the crowns of trees by very heavy thinning. The

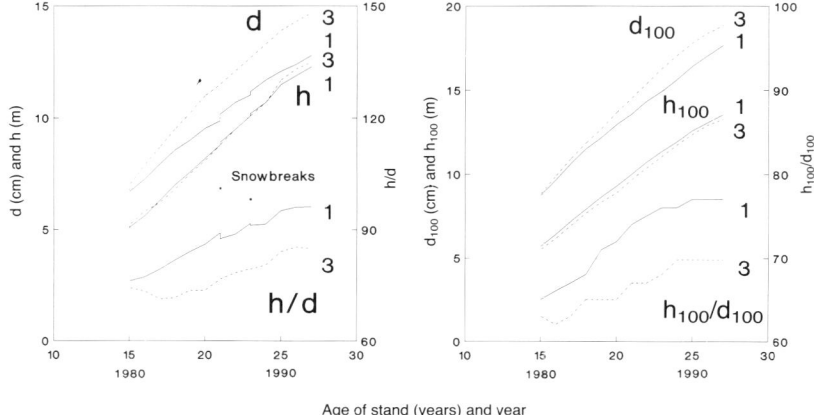

Fig. 24.6. Ratio h/d and diameter and height growth of plots 1 and 3, and of 100 trees with the largest dbh, in experimental series III.

number of trees was reduced from an initial 2950 to 1630 ha^{-1}, i.e. by 45%.

Before the first experimental treatments in 1980, the differences between investigated characteristics (stand density, basal area, diameter) were not significant, but after the first thinnings they had become clearly differentiated. In the period studied (age 16–27 years), basal area of the control plot 1 increased nearly 3-fold (from 10.7 m^2 to 30.5 m^2) and basal area of plot 3, with very heavy thinning, by more than 4-fold (from 6.3 m^2 to 27.3 m^2). At this time the stand height only doubled and was not influenced by thinning (Figs. 24.5, 24.6).

The basal area in absolute units was highest in the unthinned plot (plot 1) up to the age of 27 years, but the differences in basal area between plots 1 and 3 were gradually getting smaller and 12 years after the first thinnings they became statistically insignificant. At the same time, mean diameter of the experimental stands differed very significantly: in plots 2 and 3 it was larger by approximately 1 and 2 cm respectively than in the control (plot 1).

The development of diameter and height growth influenced the h/d ratio, which had a tendency to increase up to the age of 25 years when it appeared to reach a maximum (Fig. 24.6). The h/d of trees in the heavily thinned stand was significantly lower at all times.

Since the beginning of observations, the experimental series has been damaged by snow twice (Fig. 24.5). The unthinned control plot was most affected (700 trees ha^{-1} broken); in plot 2 snow damaged only 150 trees ha^{-1} and the heavily thinned stand (plot 3) was not damaged at all.

From the results of this experiment it can be concluded: (1) spruce stands originating from planting 3000–4000 trees ha^{-1} in regions endangered by snow must be stabilised by respacing or heavy thinning in the period of canopy closure; (2) respacing to c. 2500 trees ha^{-1} was not sufficient to stabilise the stand against snow; (3) respacing to 1630 trees ha^{-1} prevented all damage up to the age of 27 years.

24.6 Discussion

In all three experiments described above, heavy early thinning at the period of canopy closure increased the resistance of Norway spruce stands to snow damage. These results can be supported by many other studies recommending early thinning (e.g. Persson, 1969; Somerville, 1980; Cremer et al., 1982; Rollinson, 1988; Konôpka et al., 1989; MacCurrach, 1991) or wide spacing (Pollanschütz, 1974; Mráček, 1979; Blackburn & Petty, 1988) for stabilisation of forest stands to stembreak. On the other hand, later thinning has no or little effect on stability (Pařez, 1972; Huss, 1990) and the risk of wind damage may even be increased (Cremer et al., 1982; Lohmander & Helles, 1987; Slodičák, 1987). Consequently, the need for changing the silvicultural strategy from a loose canopy in young stands to a full canopy in older stands is evident. In addition, the value of the gradual thinning is supported by the fact that the stability of the stems (low h/d ratio) achieved in youth is maintained in older stands (Johann, 1981). The results of some experiments even show a decreasing tendency of h/d after the latter has reached its maximum (Blackburn & Petty, 1988; Handler, 1990).

The results of our experiments confirm that the stability of individuals does not completely protect the stand against wind. In the system using gradual thinning, the stability of individuals created by early heavy thinning is combined and strengthened by mutual shelter as stands become older. In such old stands thinning intensity ought to be decreased or thinning stopped altogether.

Acknowledgements

This work was supported by the Ministry of Agriculture of the Czech Republic. I thank Luděk Chroust for fruitful discussions, and Christian Nielsen and Mike Coutts for reviewing the manuscript.

References

Blackburn, P. & Petty, J. A. (1988). Theoretical calculations of the influence of spacing on stand stability. *Forestry*, **61**, 235–44.

Bohdanecký, J. (1890). Statistisch-topographische und forstliche Beschreibung der Karl Fürst zu Schwarzenberg'schen Herrschaft Worlik (Böhmen). *Vereinsschrift für Forst-, Jagd- und Naturkunde*, **1**, 3–89.

Brünig, E. F. (1973). Sturmschäden als Risikofaktor bei der Holzproduktion in den wichtigsten Holzerzeugungsgebieten der Erde. *Forstarchiv*, **44**, 137–40.

Busby, J. A. (1965). Studies on the stability of conifer stands. *Journal of the Royal Scottish Forestry Society*, **19**, 86–102.

Chroust, L. (1968). Význam silných prořezávek pro zvýšení odolnosti smrkových porostů proti škodám sněhem. [Importance of heavy thinnings for increasing the resistance of spruce stands to snow damage.] *Lesnícky časopis (Bratislava)*, **14**, 943–60.

Chroust, L. (1969). Der Einfluss starker Eingriffe in einem jungen Fichtenbestand. *Forstwissenschaftliches Centralblatt*, **88**, 309–19.

Chroust, L. (1979). *Plánovací jednotky a výchovné programy pro mladé smrkové porosty v 5. vegetačním stupni*. [Planning units and thinning programmes for young spruce stands in 5th forest vegetation degree.] Research report, VÚLHM, Jíloviště-Strnady.

Chroust, L. (1980). Tvar kmene a velikost korun při výchově smrkových porostů ve vztahu ke škodám působeným sněhem a větrem. [Stand form and crown size in thinning of spruce stands in connection with damage caused by snow and wind.] *Práce VÚLHM (Jíloviště-Strnady)*, **56**, 31–52.

Cremer, K. W., Borough, C. J., McKinnell, F. H. & Carter, P. R. (1982). Effects of stocking and thinning on wind damage in plantations. *New Zealand Journal of Forestry Science*, **12**, 244–68.

Handler, M. M. (1990). *Et planteafstandsforsog med gran, Picea abies (L.) Karst. Forsog 626/627, Spikkestad*. [Spacing experiment with Norway spruce, *Picea abies* (L.) Karst. Experiment 626/627, Spikkestad.] Norsk institutt for skogforskning, Ås.

Heger, A. (1953). *Die Sicherung des Fichtenwaldes gegen Sturmschäden*, 2nd edn. Neumann Verlag, Berlin.

Huss, J. (1990). Zur Durchforstung engbegründeter Fichtenjungbestände. *Forstwissenschaftliches Centralblatt*, **109**, 101–18.

Johann, K. (1981). Nicht Schnee, sondern falsche Bestandesbehandlung verursacht Katastrophen. *Allgemeine Forstzeitung*, **92**, 163–71.

Konôpka, J., Lehotský, L. & Toma, R. (1989). Vplyv výchovy na statickú stabilitu smrekových porastov. [Influence of thinning on stability of spruce stands.] *Lesnícky časopsis (Bratislava)*, **35**, 87–99.

Lohmander, P. & Helles, F. (1987). Windthrow probability as a function of stand characteristics and shelter. *Scandinavian Journal of Forest Research*, **2**, 227–38.

MacCurrach, R. S. (1991). Spacing: an option for reducing storm damage. *Scottish Forestry*, **45**, 285–97.

Málek, J. (1976). Intenzita větrného polomu podle souborů lesních typů. [Intensity of wind damage in connection with units of forest types.] *Informace ÚHÚL Brandýs nad Labem*, **19**, 21–4.

Marsch, M. (1989). Stabilisierung von Fichtenbeständen gegenüber Schnee und Sturm durch Dichteregulierung in der Jugend. In *Treatment of Young Forest Stands: Proceedings*, pp. 96–119. IUFRO, Dresden.

Mitscherlich, G. (1974). Sturmgefahr und Sturmsicherung. *Schweizerische Zeitschrift für Forstwesen*, **125**, 199–216.

Mráček, Z. (1979). *Hustota kultur a její vliv na vývoj a produkci porostů borovice lesní a smrku*. [Density of plantations and its influence on development and wood production of Scots pine and Norway spruce forest stands.] Research report, VÚLHM, Jíloviště-Strnady.

Pařez, J. (1972). Vliv podúrovňové probírky na výši škod sněhem v porostech pokusných probírkových ploch v období 1959–1968. [Influence of thinning from below and from above on snow damage in experimental stands in the period 1959–1968.] *Lesnictví (Praha)*, **18**, 143–54.

Persson, P. (1969). The influence of various thinning methods on the risk of windfalls, snowbreaks and insect attacks. In *Thinning and Mechanization: Proceedings*, pp. 169–74. IUFRO, Stockholm.

Petty, J. A. & Worrell, R. (1981). Stability of coniferous tree stems in relation to damage by snow. *Forestry*, **54**, 115–28.

Pollanschütz, J. (1974). Ergebnisse eines 84 jährigen Fichten-Pflanzweiteversuches im Wienerwald. *Allgemeine Forstzeitung*, **29**, 816–18.

Rollinson, T. J. D. (1988). Respacing Sitka spruce. *Forestry*, **61**, 1–21.

Rottmann, M. (1985). Waldbauliche Konsequenzen aus Schneebruchkatastrophen. *Schweizerische Zeitschrift für Forstwesen*, **136**, 167–84.

Slodičák, M. (1983). Výskyt poškození sněhem a větrem v rozdílně vychovávaných smrkových porostech. [Incidence of snow and wind damage in variously thinned spruce stands.] *Práce VÚLHM (Jíloviště-Strnady)*, **62**, 151–78.

Slodičák, M. (1987). Resistance of young spruce stands to snow and wind damage in dependence on thinning. *Communicationes Institute Forestalis Čechosloveniae (Jíloviště-Strnady)*, **15**, 75–86.

Somerville, A. (1980). Wind stability: forest layout and silviculture. *New Zealand Journal of Forestry Science*, **10**, 476–501.

Vicena, I. (1964). *Ochrana proti polomům.* [Protection from windfalls.] Státní zemědělské nakladatelství, Prague.

Wiedemann, E. (1955). *Ertragskundliche und waldbauliche Grundlagen der Forstwirtschaft*, 2nd edn. J. D. Sauerländer's Verlag, Frankfurt am Main.

25
A synopsis of windthrow in British Columbia: occurrence, implications, assessment and management

S. J. MITCHELL

Abstract

In 1992 the British Columbia Ministry of Forests conducted a province-wide survey of windthrow. Windthrow damage is estimated to amount to 4% of the provincial annual allowable cut. Seventy five per cent of this windthrow is scheduled for salvage, the remainder is not considered to be commercially viable. Loss of streamside buffers and forested corridors disrupts planning for wildlife and aesthetic concerns. As alternatives to clearcutting are investigated in each region, many foresters are becoming concerned that partial cutting in previously unmanaged stands will leave these stands susceptible to windthrow. In response to these concerns, pre-harvest windthrow risk assessments are now required in some regions of British Columbia, and a windthrow risk assessment procedure for field foresters is in preparation. In keeping with the present approach to ecosystem classification and interpretation in previously unmanaged ecosystems, the presence and nature of pit-moulding, existing windthrown material, and wind-induced crown and bole modifications will be considered in assessing windthrow risk. Historically, management of windthrow has focused on cutblock configuration and the use of swamps, lakes, creeks and immature types for cutblock boundaries. Wind-proofing of streamside buffers through topping and pruning is being tested in coastal old-growth stands.

25.1 Nature of British Columbia's forests

25.1.1 Biogeoclimatic zones

British Columbia (BC) is a large and geographically diverse province with a total land area of 948 600 km^2. The province is divided into five physiographic regions (Valentine *et al.*, 1978): the Coast Mountains and Islands,

Fig. 25.1. The six Forest Regions administered by the British Columbia Ministry of Forests: 1, Cariboo; 2, Kamloops; 3, Nelson; 4, Prince Rupert; 5, Prince George; 6, Vancouver.

the Interior Plateau, the Columbia Mountains and Southern Rockies, the Northern and Central Plateaus and Mountains, and the Great Plains. Since the mid-1970s BC's forests have been managed within an ecosystem classification and interpretation framework. The classification system is based, with some modifications, on the work of Krajina (1965, 1973) and incorporates climate, soil and vegetation data. Twelve forested biogeoclimatic zones, each characterised by a distinct climax forest community, are recognised.

25.1.2 Land ownership and tenure

The provincial government owns 95% of BC's 470 000 km^2 of productive forest land. The BC Ministry of Forests (MOF) administers six Forest Regions (Fig. 25.1) each of which is further subdivided into forest Districts. Cutting rights to Crown timber are held by forest products companies under various forms of tenure. These include volume-based Forest Licences and area-based Tree Farm Licences. Licensees are responsible for harvesting and reforestation under MOF supervision.

25.1.3 Utilisation

BC's timber inventory is dominated by previously unmanaged mature and over-mature age classes. The annual allowable cut (AAC) from Crown Land is 74.4×10^6 m^3. While the majority of the area harvested annually in BC is clearcut (Ministry of Forests, 1992), partial cutting is well established in the dry Douglas fir and western larch forest types of the southern interior (De Long, 1991), and is currently being extended on an experimental basis to other forest types. The proportion of the area partially cut is expected to increase as the industry responds to public demands for increased emphasis on biodiversity and aesthetics (Canadian Pulp and Paper Association, 1992). To date commercial thinning is localised and provides a very small amount of timber. This is expected to change as the economics of harvesting and processing small logs improves. In 1990, the Research Branch of the Ministry of Forests established the Silviculture Systems Program to investigate alternatives to conventional clearcutting. This program funds research, development, demonstration and extension of partial cutting systems in each Forest Region.

25.2 Occurrence of windthrow

Prior to 1992, windthrow recovery could be estimated from stumpage receipts for salvaged timber, but this represented only a portion of total windthrow. In August 1992 a province-wide survey of windthrow was conducted by the MOF. Each Forest District was asked to estimate by licence category the volume of commercially viable and non-viable windthrow. The data were largely obtained through aerial and ground reconnaissance and represent the best estimates of local licensee and ministry staff.

The estimates of viable and non-viable windthrow for each forest Region are shown as a percentage of provincial AAC in Fig. 25.2. In many cases a portion of standing timber is harvested along with the windthrown timber. This is done in order to improve the windfirmness of residual boundaries, to ensure that small volumes of standing wood are not isolated, and in some cases to improve the economics of accessing small patches. This standing wood component is reported below as 'green' wood.

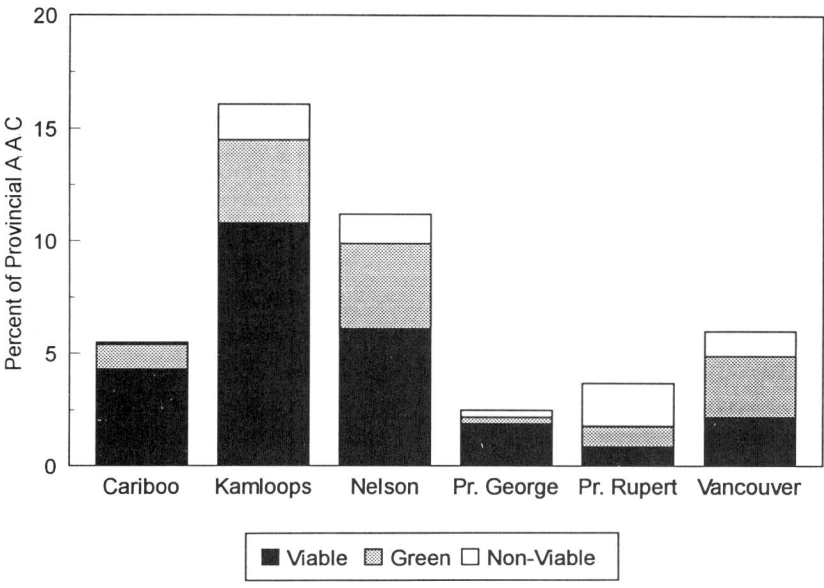

Fig. 25.2. Results of the BC Ministry of Forests 1992 windthrow survey summarised by Region. Windthrown timber that is not salvaged for economic or environmental reasons is designated as non-viable. In the course of salvaging viable windthrow, standing green timber may be harvested for practical or economic reasons. AAC, annual allowable cut.

25.3 Implications of windthrow

25.3.1 Economic losses

Wind damage affects a volume of timber equivalent to 4% of the provincial AAC. This level of damage is similar to that caused by wildfire or insect infestations (Fig. 25.3). In the face of declining AAC levels in most districts, windthrow is salvaged where viable. In coastal BC, approximately two-thirds of identified windthrow is salvaged. This proportion increases to 95% in the central interior where access is easier and the threat of bark beetle epidemics promotes vigilance. Salvaging windthrow is more expensive than harvesting standing green timber and resulting log values are generally lower due to breakage and deterioration. In the BC interior these additional costs are recognised through an appraisal cost allowance. In coastal BC the majority of windthrow is salvaged with no direct allowance for additional costs.

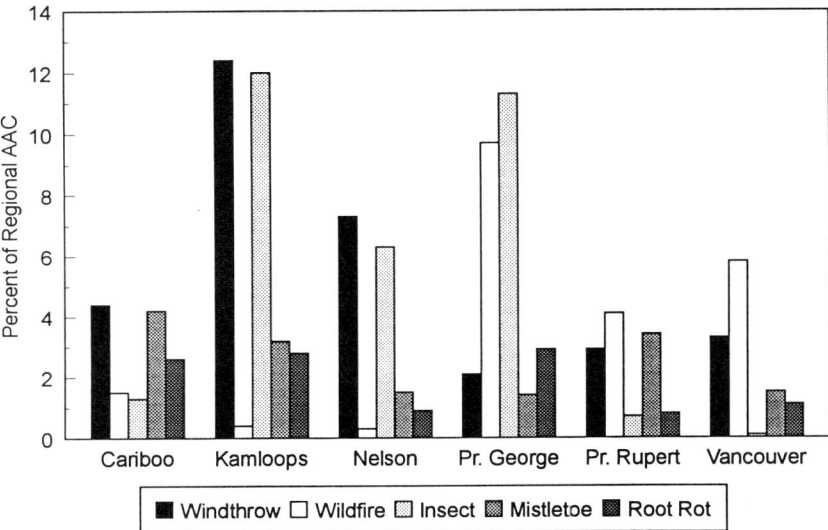

Fig. 25.3. A comparison by Region of the volume of timber damaged by windthrow and other causes. These are pre-harvest estimates. Much of this volume will subsequently be salvaged. Windthrow damage estimates from Ministry of Forest 1992 windthrow survey; other damage estimates from Annual Report for 1990–1991 (Ministry of Forests, 1992).

25.3.2 Planning of harvesting

The harvesting of public timber is carried out under permit in accordance with Pre-Harvest Silviculture Prescriptions (PHSPs) signed by Professional Foresters. These PHSPs are prepared within the context of a development planning process. During the preparation of 5-Year Development Plans, interagency and public consultation is conducted to enable integration of timber, wildlife, fisheries, hydrology and aesthetic concerns. Subsequent loss of designated forested streamside buffers, wildlife corridors and visual quality to windthrow seriously disrupts the intent and implementation of integrated resource planning. The short period during which salvage of windthrown timber is worth-while adds further pressure to planning staff.

There is some concern amongst foresters that the risk of windthrow and the ensuing disruption of resource plans may increase where shelterwood or selection cutting are introduced to replace clearcutting. Persistent wind damage to selection and seed block stands throughout the 1940s and 1950s, culminating in a catastrophic wind storm in 1964, was an important factor in the abandonment of partial cutting and the introduction of clearcutting in north-central BC during the 1970s (Benskin, 1975).

25.3.3 Site degradation

Site degradation through road construction and skidding operations is a growing concern. Soil disturbance guidelines are now in place throughout BC with penalties for non-compliance. Ground-based withdthrow salvage operations often result in higher than normal levels of soil disturbance because of the inability of managers to pre-plan felling and skidding patterns. The construction of roads to isolated patches may be precluded due to excessive site degradation in areas of steep or unstable terrain.

25.3.4 Insect outbreaks

In the late 1970s, major outbreaks of spruce beetle (*Dendroctonus rufipennis* Monk.) developed from non-salvaged windthrow in the spruce forests of the central interior (Ministry of Forests, 1980). In the southern interior, Douglas fir bark beetle (*Dendroctonus pseudotsugae* Hopkins) rapidly invades windthrown Douglas fir trees and frequently spreads to standing green trees, necessitating prompt salvage and slash management. In coastal BC, windthrown conifers play host to ambrosia beetles (*Trypodendron* spp. and *Gnathotrichus* spp.). While ambrosia beetles do not attack standing timber, the damage to windthrown, felled and bucked, and stored timber causes estimated losses of $63 million annually (McLean, 1985).

25.4 Constraints to salvaging windthrow

In the MOF windthrow survey, respondents were asked to comment as to why windthrow was classified as non-viable. Their comments indicated three principal constraints: accessibility, commercial viability, and compatibility with non-timber management objectives.

The time and cost of building road access to isolated patches of windthrow precludes conventional harvesting. Helicopter yarding has been used with some success for salvage in coastal BC but requires high-value logs and nearby open water drop zones. Commercial viability is a function of logging costs and timber value. Logging costs are higher where volumes are small and yarding distances are large. The value of windthrown logs is lower due to breakage and deterioration, particularly by ambrosia beetles. In most Regions, isolated patches of windthrow less than 2 ha in size are considered non-viable.

Protection of non-timber resources constrains the salvage of windthrow in a number of ways. The salvage of edge windthrow along critical streamside buffer strips and wildlife corridors may be disallowed. Within scenic areas, harvesting may be restricted in order to maintain visual quality. When the inter-agency and public consultation process is lengthy, windthrown timber may deteriorate beyond the point of commercial viability before harvesting plans are approved.

25.5 Windthrow monitoring

An early result of the 1992 windthrow survey has been the recognition of the need for a formal process of windthrow monitoring and salvage. Discussion currently centres on a four-step process which provides for early identification of windthrow following major wind events through aerial reconnaissance; prompt decision as to the viability of salvage; assignation of responsibility for salvage and penalties for non-compliance; and development of a specific approval process for salvage harvesting.

25.6 Windthrow assessment

25.6.1 Pre-harvest assessment

In addition to the proposals to improve windthrow monitoring and salvage, there is increasing recognition of the need for measures to reduce windthrow losses. Pre-harvest assessments of windthrow risk are now required for partial cuts in some Regions (Ministry of Forests, 1991). Assessments of windthrow risk for sensitive clearcut boundaries are periodically requested during the Development Planning referral process.

At present, however, there is no formal assessment procedure, and few data on which to base assessment or prescription decisions. Several empirical models for predicting edge windthrow have been developed for coastal forests (Rollerson, 1979; Holmes, 1985). But these models cover a limited range of terrain and forest types. General application of an empirically derived quantitative approach to windthrow risk assessment in BC is hindered by the diversity of climate, terrain and forest types and limited wind and windthrow data. The current approach, therefore, is to develop a relative risk assessment procedure based on a functional understanding of windthrow.

25.6.2 Relative risk assessment procedures

Procedures for assessing relative windthrow risk are currently under development. Alexander's (1987) suggestions for assessing windthrow risk in the sub-alpine forests of the Pacific Northwest are being extended and generalised for broad application to forests throughout BC. Three site components – topographic exposure, soil conditions and stand features – are to be assessed and ranked as low, moderate or high risk. The individual risks for each of these three components will then be combined to yield an overall risk for the site.

Much of the soils and stand information required to assess component risks is already being collected during routine PHSP field work. The presence and nature of pit-mounding, historic windthrown material, and wind-induced crown and bole modifications can be used to reconstruct past windthrow history and improve the estimates of the risk categories for the topographic and soil components. This approach is in keeping with the present use of site indicators for ecosystem classification and interpretation in previously unmanaged ecosystems (e.g. Lewis *et al.*, 1986). The assessment of stand risk focuses on initial stand structure and density. The information requirements are similar to those for selection of silvicultural system (e.g. Weetman *et al.*, 1990). For clearcuts the key considerations in the prescriptions are opening size, opening orientation, boundary placement and boundary modification. For partial cuts they are basal area removal, tree selection, edge buffering and access placement.

25.7 Windthrow management

25.7.1 Windthrow in partial cuts

Recurrent windthrow in seed-block and seed-tree cuts in the north-central interior was in part responsible for the introduction of clearcutting in the spruce forests there (Benskin, 1975). As partial cutting is reintroduced and extended to previously untested forest types, emphasis is being placed on removal levels and better selection of leave trees. For example, Moore (1991) noted that windthrow losses in coastal hemlock–spruce partial cuts were lower when taller trees and a greater range of diameters were retained and patch sizes were reduced.

25.7.2 Windthrow of edges

Historically, management of windthrow has focused on cutblock configuration and the use of swamps, lakes, creeks and immature types for cutblock boundaries. Where such measures failed, clearcuts were extended progressively. There is concern that the current reduction in average clearcut size, intended to accommodate non-timber resources, may intensify the problem of edge windthrow through increasing the total area of exposed edges and roads. Feathering of edges is applied in areas with windthrow histories. Windproofing of streamside buffers through topping and pruning is being tested in coastal old-growth stands (Rollerson *et al.*, 1992). Non-salvage of windthrow in sensitive stream management zones and corridors is becoming increasingly accepted where bark beetle outbreaks are not expected and edges appear to have stabilised.

25.8 Windthrow research

25.8.1 Wind in British Columbia

Both Environment Canada and the BC Ministry of Environment maintain permanent weather stations throughout BC. However, the majority of the weather stations are at coastal lighthouses or in major valleys near airports. The duration and quality of wind records varies between stations. Many Environment Canada weather stations do not record wind continuously. In these cases the reported 'hourly mean wind' is a 1–2 min reading taken on the hour and the 'peak gust' is the peak instantaneous reading which occurred during the 15 min prior to the hour. There are no published data for extreme winds in BC; however, Coatta (1992) has conducted an unofficial analysis based on wind data collected prior to 1980. Climate normals are periodically published. The most recent edition provides wind data up to 1990 (Environment Canada, 1992).

25.8.2 Studies of windthrow occurrence

25.8.2.1 Windthrow of edges

With the exception of the 1992 MOF windthrow survey there have been no comprehensive studies of windthrow in BC. There have, however, been a number of local studies. Moore (1977) summarised the factors contributing to windthrow in streamside leave strips on Vancouver Island. Rollerson (1979) studied windthrow of edges in the Skidegate Plateau of the Queen Charlotte Islands and developed an empirical model to rank sites as having more or less than a 50% risk of windthrow. Holmes (1985) studied windthrow along clearcut boundaries in the Tsitika watershed on Vancouver Island. In each case the key features were exposure relative to prevailing storm winds, rooting depth and stand height.

25.8.2.2 Windthrow in partial cuts

Windthrow in 14 coastal partial cuts was monitored by Moore (1991). Butt & Bancroft (1990) documented mortality due to windthrow in a retrospective study of 40 partial cuts in the southern interior. Windthrow incidence was low on most of these sites, with mixed-species stands showing greater damage than single-species stands. Mitchell (1992) summarised the occurrence of windthrow in newly established Silviculture System partial cutting trials throughout BC. Zielke & Deverney (1993) evaluated windthrow in partial cuts in a southern interior Forest District.

25.8.3 Studies of windthrow process

Wind-proofing of streamside buffers through topping and pruning is being tested in coastal old-growth stands (Rollerson *et al.*, 1992). A wind tunnel and field study of the wind regime and windthrow in high elevation clearcuts is under way (Black, 1993). The future stability of unmanaged thrifty stands is also of concern and research has commenced on this front.

25.9 Conclusion

The 1992 MOF windthrow survey indicates the magnitude and extent of windthrow in BC and the problems created by windthrow and its salvage. A unified programme of windthrow reporting and administration is under discussion. A guide to windthrow risk assessment for field foresters is in preparation. Windthrow research in BC is in its infancy but it is possible to provide immediate advice to practitioners responsible for the management of windthrow.

References

Alexander, R. (1987). *Ecology, Silviculture and Management of Engelmann Spruce and Subalpine Fir Type in Central and Southern Rocky Mountains.* United States Department of Agriculture Forest Service, Agriculture Handbook no. 659.

Benskin, H. J. (1975). *Preliminary Study of Windthrow Losses in the White Spruce–Subalpine Fir Forests of the Prince George Forest District.* British Columbia Forest Service E.P. 755. Victoria, British Columbia.

Black, T. A. (1993). Personal communication. University of British Columbia Faculty of Agriculture Department of Soil Science.

Butt, G. & Bancroft, B. (1990). Natural regeneration under partially cut Douglas-fir and mixed conifer stands in the Nelson Forest Region. Unpublished Draft Report, Forest Resource Development Agreement 3.66. Victoria, British Columbia.

Canadian Pulp and Paper Association (1992). *Canadian Pulp and Paper Association Woodland Section Executive Committee Meeting Minutes, March 1992.* Ottawa, Ontario.

Coatta, E. (1992). Personal communication. Environment Canada Atmospheric Environment Service, Vancouver, B.C.

De Long, D. L. (1991). *Partial Cutting Prescriptions in Southern British Columbia.* Forest Engineering Research Institute of Canada Interim Report. Vancouver, B.C.

Environment Canda (1992). *British Columbia Climate Normals 1961–1990.* Atmospheric Environment Service, Downsview, Ontario.

Holmes, S. R. (1985). An analysis of windthrow along clearcut boundaries in the Tsitika watershed. Unpublished BSF Thesis, Faculty of Forestry, University of British Columbia.

Krajina, V. J. (1965). Biogeoclimatic zones and biogeocoenoses of British Columbia. *Ecology of Western North America,* **1**, 1–17.

Krajina, V. J. (1973). *Biogeoclimatic Zones of British Columbia.* British Columbia Department of Lands, Forests and Water Resources, Ecological Reserves Committee, Victoria, British Columbia.

Lewis, T., Pojar, J., Holmes, D., Trowbridge, R. and Coates, K. D. (1986). *A Field Guide for the Identification and Interpretation of the Sub-Boreal Spruce Zone in the Prince Rupert Forest Region.* MOF Land Management Handbook no. 10. Victoria, British Columbia.

McLean, J. A. (1985). Ambrosia beetles: a multimillion dollar degrade problem of sawlogs in British Columbia. *Forestry Chronicle,* **61**, 296–8.

Ministry of Forests (1980). *Spruce Beetle Management Seminar and Workshop: Proceedings in Abstract.* Ministry of Forests Pest Management Report no. 1. Victoria, British Columbia.

Ministry of Forests (1991). Silviculture manual insert 'L'. Ministry of Forests Prince Rupert Forest Region, Smithers, British Columbia.

Ministry of Forests (1992). *Annual Report of the Ministry of Forests 1990–1991.* Ministry of Forests, Victoria, British Columbia.

Mitchell, S. J. (1992). *A Synoptic Study of Windthrow in British Columbia Partial Cuts.* Ministry of Forests Silviculture Branch, Silviculture Systems Project, Contract Report SS066.

Moore, K. (1991). *Partial Cutting and Helicopter Yarding on Environmentally Sensitive Floodplains in Old-growth Hemlock–Spruce Forests.* Economic and Regional Development Agreement Contract Report. Victoria, British Columbia.

Moore, M. K. (1977). *Factors Contributing to Blowdown in Streamside Leave Strips on Vancouver Island.* Ministry of Forests Land Management Report no. 3. Victoria, British Columbia.

Rollerson, T. P. (1979). *Queen Charlotte Woodlands Division Windthrow Study.* MacMillan-Bloedel Limited Woodland Services, Nanaimo, British Columbia.

Rollerson, T. P. (1982). Relationships between windthrow and various environmental variables. In *Soil Interpretations for Forestry.* Proceedings of the British Columbia Soil Survey Workshop on Soil Interpretations for Forestry, Victoria, British Columbia. Ministry of Forests Land Management Report no. 10.

Rollerson, T. P., Butt, G. & Chatwin, S. (1992). *Stand Edge Modification to Control Windthrow.* Working Plan, Ministry of Forests. Silviculture Branch, Silviculture Systems Project, Victoria, British Columbia.

Valentine, K. W., Sprout, P. N., Baker T. E. & Lavkulich, L. M. (1978). *The Soil Landscapes of British Columbia.* Ministry of Environment, Resource Analysis Branch, Victoria, British Columbia.

Weetman, G. F., Panozzo, E., Jull, M. & Marek, K. (1990). An assessment of opportunities for alternative silvicultural systems in the SBS, ICH and ESSF biogeoclimatic zones of the Prince Rupert Region. Ministry of Forests Prince Rupert Forest Region. Unpublished Contract Report.

Zielke, K. & Deverney, S. (1993). Evaluation of silvicultural systems in the Vernon Forest District. Ministry of Forests, Vernon Forest District. Unpublished Contract Report.

26

Wind damage to New Zealand State plantation forests

A. SOMERVILLE

Abstract

New Zealand's conifer plantations have suffered at least 50 000 ha of catastrophic wind damage. This is wind damage to stands over 5 years of age where most stems are either windthrown or broken. Damaged areas have ranged in size from around 1 ha to up to many thousand hectares, and where these have occurred in mature or semi-mature plantation they have usually been harvested. Also of major importance has been continuing attritional wind damage resulting from lesser winds. This is scattered endemic damage which is rarely recoverable. Quantification of the risk of wind damage is important when considering profitability, yield and cash flow projections. For 17 previously State-owned forests, where past wind damage events have been documented, risk of catastrophic wind damage was calculated as the average percentage of net stocked area lost per year. This work considered 50 years of records documenting 30 860 ha of catastrophic wind damage within the 259 950 ha of softwood plantations. Estimates of attritional losses to stands over 14 years of age have been obtained from the extensive permanent growth sample plots located throughout these forests. The overall forest catastrophic and attritional levels of damage were found to be 0.38% and 0.25% of net stocked area lost per annum respectively. These data correspond to an average 12% of forest area lost over a 28 year rotation. While the least-affected forests would lose only 5–6%, the worst-affected would lose most of its stocked area. Forests subjected to high windspeeds associated with upwind mountainous terrain have had far greater levels of wind damage.

26.1 Introduction

In New Zealand (NZ), large-scale plantings of introduced softwoods were well under way by the mid 1920s. By 1951 there was a total plantation estate of 356 000 ha of which 52% was owned by the State. At that time, 40% of the State's plantations were radiata pine with most of the balance being in other introduced softwood species. In 1990 the State began to divest itself of the main portion of its forestry interests in an asset sales programme. By then the State owned around 554 000 ha of the country's total 1.2×10^6 ha of plantation forests of which 80% was radiata pine (*Pinus radiata* D. Don.).

Records of wind damage date back to the 1940s, with many of the main events documented in published accounts (Wendleken, 1955, 1966; Prior, 1959; Chandler, 1968; Irvine, 1970; Wilson, 1976; Somerville, 1989) supplemented with numerous unpublished records, inventories and aerial photographs. Most records of damage corresponded to single storm events where mature or semi-mature plantations were totally or partially destroyed. Records indicate there has been something of the order of 50 000 ha of catastrophic wind damage (defined here as any continuous area of wind damage over 1 ha in size) to NZ's plantation estate. Damage to young plantations, very often appearing as severely leaning stems, has largely remained undocumented. Also remaining undocumented have been continuing attritional losses (isolated and scattered tree losses) in stands from lesser storms. However, in permanent growth sample plots (PSPs), stand condition in NZ forests has been monitored from the onset of plantation forestry and as trees lost to windthrow are described, these records do offer some insight into attritional wind damage losses.

Wind damage in many of the worst catastrophic wind storm events has been aggravated by orographic lee waves and wind channelling associated with mountainous topography. The lee waves result in zones of increased windspeed on lower-lying forests downwind from mountainous areas. This type of damage is typified by events on the Canterbury Plains and winds off the Southern Alps and also more recently in forest areas in the central North Island with winds off mountainous terrain to the East (Littlejohn, 1984). Much of the past catastrophic damage in NZ has also been associated with management activities, particularly exposure downwind from recent clearfelling (for example the 1964 Eyrewell wind damage), exposure resulting from late thinning and 'edge effect' (i.e. the high wind-risk zone at or immediately in the lee of a stand windward edge: Somerville, 1989). In some forests the length of exposed edge has been increasing as stand size has decreased (New, 1989).

The general attitude amongst NZ forest managers to the risk of wind damage has been one of acceptance and there has not been an adequate awareness of actions that can be taken to alleviate wind risk, although this is not universally true (e.g. see Studholme, this volume). There is, however, an appreciation of the importance of the quality of tree stocks and planting technique and the consequences of delayed and severe thinning, but there is often little regard to other factors: the location of forests, species, soils and site factors in general, age class structure, exposure from upwind clearfelling, and creation of edge effect. The attitude of managers to the risk may well reflect both the general lack of research within NZ into this phenomenon along with the absence of a quantification of wind risk which would allow forest managers to judge the importance of wind in terms of financial and yield considerations.

The work reported here examines and quantifies the historical catastrophic and attritional type wind risk in the State-owned forests most seriously affected by wind from 1940 to 1990 when the bulk of these forests were sold. Risk is quantified as the average percentage of net stocked area (NSA) lost per annum to both catastrophic and attritional damage for the period over which records exist.

26.2 Methods

Seventeen forests consisting of 259 950 ha (in 1990) of plantation were examined.

26.2.1 Catastrophic damage

The approximate NSA of softwood plantation over 5 years of age was determined for each year for each forest. Plantations of 5 years and under were excluded as it was considered that damage in these would be in the form of leaning stems and not only would it be difficult to decide on the significance of any such damage but most events would not have been recorded. Areas were estimated from annual published statistics (Government Printer, 1987) and are approximations only.

For each forest, all available records of catastrophic wind damage to stands over 5 years of age were examined. The accumulated area of damage for each forest was divided by the summed annual total NSA (for stands over 5 years of age) and converted to a percentage. This gave, by forest, the average annual percentage of NSA area lost to catastrophic wind events to 1990.

26.2.2 Attritional damage

Information on attritional damage was obtained from the PSP system. This was interrogated to extract for each plot, by year, the total number of undamaged and wind-damaged trees within each of the study forests. Only stands over 14 years of age were considered as the effects in younger stands would probably be negated by final thinning occurring at around 12–14 years. Data for the two main species, radiata pine and Douglas fir, were considered. Re-measurement data for any plot showing 25% or more wind damage were discarded as this was likely to be catastrophic damage already accounted for. Risk of attritional wind damage was determined as the average annual percentage of trees lost (for the purpose of this analysis considered here to equate to % NSA lost) by forest and region.

26.3 Results

Table 26.1 provides, by forest, the year when significant plantings began, the NSA in 1990, the sum of annual NSA for stands over 5 years of age (described as 'hectare-years'), the accumulated area of catastrophic damage in stands over 5 years of age, the sum of the periods in years for which PSPs were re-measured (described here as 'plot-years') and the average annual percentage of NSA lost as both catastrophic and attritional types of wind damage to 1990. Regional totals and averages are also presented along with weighted average losses based on NSA of forest in 1990. Catastrophic losses incorporate records back to 1940 only, while data in brackets include all records.

Results show very considerable between-forest variation in catastrophic risk. Some data sets for younger forests are sparse and should not be considered in isolation (Glen Dhu, Herbert and Rankleburn) while other forests and the regional averages are based on substantial data (for instance, greater than 100 000 hectare-years). With the exception of Canterbury, the regional average losses as catastrophic damage are quite similar at 0.20–0.26% of NSA lost per annum. Canterbury, however, has been subjected to very high levels of catastrophic damage.

The PSP records for Kaingaroa, Ashley, Balmoral, Hanmer, Berwick and the Tapanui district forests along with the regional totals have over 400 plot-year observations each and are likely to indicate the general level of attritional wind damage. This is around 0.14–0.24% of NSA lost per annum with a slightly lower figure for Berwick and higher figures for Balmoral and Golden Downs Forests. Although attritional damage is generally higher in the forests with high catastrophic damage, the comparative increase for the Canterbury Plains Forests (Balmoral and Eyrewell) is small and the Canterbury total is only slightly ahead of Central North Island and Otago. Golden Downs Forest has been subjected to a high rate of attritional damage.

Table 26.1. *Catastrophic and attritional wind damage in selected New Zealand forests*

Region	Forest	Planting began	NSA in 1990 (ha)	Area* years (Σ over 5 yrs.) (ha-years)	Area damaged (ha)	Catastrophic risk (% loss yr^{-1})	Attritional risk plot-years	Attritional risk % loss yr^{-1}
Central North Is.	Kaingaroa	1901	122 000	4 101 500	6 750	0.16 (0.14)	11 523	0.14
	Waimihia	1934	18 800	676 900	5 320	0.79	216	0.73
	Lake Taupo	1970	22 900	197 950	790	0.40	125	0.37
	Total		**163 700**	**4 976 350**	**12 860**	**0.26**	**11 864**	**0.15**
Nelson	Golden Downs	1927	30 600	726 600	1 680	0.23 (0.22)	2 615	0.35
Canterbury	Ashley	1939	5 200	114 450	745	0.65	608	0.16
	Balmoral	1916	8 050	294 350	5 000	1.70 (1.34)	512	0.34
	Eyrewell	1928	6 180	239 950	8 750	3.65 (3.23)	140	0.47
	Hanmer	1901	4 070	150 950	375	0.25 (0.20)	430	0.14
	Total		**23 500**	**799 700**	**14 870**	**1.86**	**1690**	**0.24**
Otago	Beaumont	1927	2 300	112 150	285	0.25 (0.23)		
	Berwick	1946	11 300	152 500	210	0.14	824	0.08
	Conical	1903	1 740	79 250	50	0.06 (0.05)		
	Dusky	1898	1 740	76 950	225	0.29 (0.23)		
	Glen Dhu	1980	4 670	21 950	20	0.09		
	Herbert	1948	4 030	36 850	165	0.45	312	0.18
	Naseby	1900	2 140	59 500	30	0.05 (0.04)	24	0.14
	Otago Coast	1960	9 370	113 900	160	0.14	166	0.21
	Rankleburn	1960	3 930	63 850	10	0.02		
	Tapanui	1901	930	45 750	295	0.64 (0.63)	1 840[a]	0.22[a]
	Total		**42 150**	**762 650**	**1 450**	**0.20**	**3 166**	**0.18**
Overall forests			**259 950**	**7 265 330**	**30 860**	**0.42**	**19 335**	**0.19**
Weighted average						**0.38**		**0.25**

[a] Includes Beaumont, Conical, Dusky, Rankleburn and Tapanui.

Table 26.2. *Forests with and without orographic effects*

	Area in 1990 (ha)	Catastrophic loss (% yr^{-1})	Attritional Loss (% yr^{-1})	Total Loss (% yr^{-1})
Orographic	101 800	0.85	0.20	1.05
Others	158 150	0.14	0.19	0.33

Kaingaroa is a very large forest and the average loss of 0.16% catastrophic damage masks a high within-forest variation, with the south of the forest (the southernmost third) having around 3 times the risk of the north. The high-risk forests – South Kaingaroa, Waimihia, Lake Taupo, and the Canterbury forests of Balmoral, Eyrewell (and to a lesser extent Ashley) – are also the forests most clearly subjected to enhanced windspeeds associated with upwind mountainous terrain. (The small Tapanui forest has had a high rate of catastrophic damage which is not apparently associated with any orographic phenomenon.) Table 26.2 shows data summarised into sets of forests with and without orographic effects.

26.4 Discussion

Average results presented in Table 26.1 are based on a substantial database and provide an indication of the level of historical catastrophic and attritional damage in some of NZ's worst wind-affected forests. However, the application of these data to existing and future situations must be done with caution. Risk of wind damage to a stand is influenced by factors such as the species, stand age, exposure from recent felling, thinning severity, time since thinning, stand arrangement and induced edge effect, and both large-and small-scale variations in topography, soils, soil moisture and site preparation. Each of these factors will have influenced the extent or occurrence of any particular wind damage event and each will have a unique effect within any future event.

Typically the main portion of catastrophic wind damage in merchantable stands is recovered, usually at higher harvesting costs and with some losses due to breakage. An assessment of recoverable yields of windthrow resulting from the 1982 windstorm at Kaingaroa Forest determined a 7% loss in recoverable volume because of denied access and a further 6% loss resulting from stem breakage. Damage in unmerchantable stands is a total loss.

In NZ to date, catastrophic wind losses have rarely been considered in economic analyses and in growth and yield forecasting. In the analysis used here it is assumed that attritional losses before the age of 15 years are largely discounted by thinning. Losses after this point may be partly balanced by a growth response from immediate neighbours. Attritional damage is incorporated into those growth models developed from data sets that include this sort of wind damage. Yield tables based on inventory information obtained from late in the life of a stand also account for attritional wind losses prior to the inventory.

There are other losses from wind damage which are rarely considered or attributed to wind. One of these is butt log malformation, largely resulting from severe lean commonly inflicted on planted radiata at age 2 years. In stands which are later butt log pruned, the resulting sinuosity and sweep can cause a substantial enlargement of the effective knotty zone which in turn means reduced high-value clear-grade yields of sawn timber and veneer. Severe persistent wind reduces stand growth performance. Wind also causes leader damage and the resulting malformation means lower yields at harvest. Resin pockets are a significant degrading factor in radiata pine over parts of NZ. Wind is likely to be a contributing factor in their formation (Clifton, 1969).

Wind has and will continue to have a major impact on plantation forestry in NZ. Quantifying historical wind damage may provide an indication of potential risk from wind. Ideally, however, an assessment of wind risk should be based on an understanding of the interaction of the many stand and site variables and the likelihood of high-speed wind events.

26.5 Summary

Wind damage to the year 1990 in the 17 previously State-owned forests investigated was found to vary considerably between forests. Catastrophic-type damage ranged from 0.02% to 3.65% of forest NSA lost per year and an overall weighted average (based on area in 1990) of 0.38% was determined. Corresponding levels of attritional damage losses were a range of 0.08–0.34% (for forests with sufficient observations) and a weighted average of 0.25% NSA lost per annum.

On the basis of the average of the historical levels of wind damage in the forests studied, a plantation with a 28 year rotation could be expected to lose 12.2% of NSA over its life. Corresponding estimates for forests least affected by wind (e.g. Northern Kaingaroa, Hanmer and most of the Otago Forests) indicate losses of around 5–6%, while the worst-affected Canterbury Plains forest would lose an estimated 90% of its NSA.

The group of forests examined that have been subjected to high-speed winds associated with upwind mountainous terrain have had similar levels of attritional losses to other forests but more than 4 times the amount of catastrophic damage.

Acknowledgments

The author wishes to thank Tasman Forestry Ltd who commissioned the early stages of this work. Thanks are also due to NZ FRI Ltd staff members Piers Maclaren, Judy Dunlop and Simon Papps who assisted in the examination of permanent sample plot data, and to the British Council for supplying travel funds as part of their Higher Education Link programme.

References

Chandler, K. C. (1968). Climate damage to forests of the Tapanui district. *New Zealand Journal of Forestry*, **13**, 98–110.

Clifton, N. C. (1969). Resin pockets in Canterbury radiata pine. *New Zealand Journal of Forestry*, **14**, 38–49.

Government Printer (1987). *Statics of the Forests and Forest Industries of New Zealand 1953–1987*. New Zealand Forest Service and Ministry of Forestry, Wellington.

Irvine, A. I. (1970). The significance of wind throw for *Pinus radiata* management in the Nelson district. *New Zealand Journal of Forestry*, **15**, 57–68.

Littlejohn, R. N. (1984). Extreme winds and forest devastation resulting from Cyclone 'Bernie'. *Weather and Climate*, **4**, 47–52.

New, D. (1989). Accounting for New Zealand plantation's risk to wind damage: facing the facts. In Proceedings of the workshop on wind damage in New Zealand exotic forests. *FRI Bulletin*, **146**, 62–5.

Prior, K. W. (1959). Wind damage in exotic forests in Canterbury. *New Zealand Journal of Forestry*, **8**, 57–68.

Somerville, A. (1989). Tree wind stability and forest management practices. In proceedings of a workshop on wind damage in New Zealand exotic forests. *FRI Bulletin*, **146**, 38–58.

Wendleken, W. J. (1955). Root development and wind-firmness of the shallow gravel soils of the Canterbury Plains. *New Zealand Journal of Forestry*, **7**, 71–6.

Wendleken, W. J. (1966). Eyrewell Forest: a search for stable management. *New Zealand Journal of Forestry*, **10**, 43–65.

Wilson, H. H. (1976). The effect of the gale of August 1975 on forestry in Canterbury. *New Zealand Journal of Forestry*, **21**, 133–40.

27

The experience of and management strategy adopted by the Selwyn Plantation Board, New Zealand

W. P. STUDHOLME

Abstract

Wind damage is frequent in conifer plantations in the Canterbury Plains of New Zealand; 90% of all wood harvested has followed windthrow. Methods have been evolved to cope with this difficult situation. Forestry can now be carried out profitably so long as proper attention is paid to the location of stands and plantation layout, the selection of species, cultivation, tree spacing, planting pattern, thinning, pruning and harvesting. The methods used are described in relation to the prevailing climatic, edaphic and economic constraints.

27.1 Introduction

The Selwyn Plantation Board has managed forests on the plains of Canterbury since 1911. At present some 10 000 ha of plantation forests are managed for both shelter and wood production. This chapter details the management experiences built up since 1911, and the strategies adopted in the light of that experience. This is reviewed against the background of the national and regional environment.

27.2 New Zealand

New Zealand lies in the south-west section of the Pacific, 1600 km to the east of Australia, 10 000 km from Panama and San Francisco, and a similar distance from Tokyo and Singapore. The area of the country measures 26.9×10^6 ha. It is similar in size to the British Isles and to Japan.

New Zealand is a long (1600 km) narrow country characterised by a ridge of mountains running down its north-northwest to south-south-east axis. In the North Island the mountains tend to be lower. However, high mountains still manage to occupy approximately 10% of the North Island surface. With the exception of four peaks, the mountains of the North Island do not exceed an altitude of 1800 m. The South Island is much more mountainous. The Southern Alps run almost the entire length of the island. This mountain chain has 19 peaks exceeding 3000 m.

Terrain and location have a significant effect on New Zealand's weather. The country is located between latitudes 34° and 47° south, just to the south of the subtropical mean high-pressure belt. In the south New Zealand penetrates into the hemispheric westerly air stream (the 'roaring forties'). Hot air masses from the interior of Australia in summer, and freezing air masses from the Antarctic, have a further significant effect on the weather.

The chain of high mountains extending from the south-west to the north-east through the length of the country rises as a formidable barrier to the path of the prevailing westerly winds. The mountains produce a sharper east–west climatic contrast than the contrast from north to south. In some South Island areas this can produce a distinctly continental climate, despite the fact that no part of New Zealand is more than 130 km from the sea.

Westerlies are the predominant winds. Mountain systems and the heating or cooling contrast between land and sea cause important modifications to the wind patterns. The north-western foehn wind in eastern areas gives rise to a characteristic weather type. The blocking effect of the mountains tends to decrease windspeeds on the upwind side and increase precipitation. Windspeeds to the east of the mountains tend to be generally higher and the humidity of the winds drier.

27.3 Forests

A thousand years ago, under a climate which was slightly milder than today, New Zealand's forest development was at its short-term maximum. Approximately 78% of the country was forest covered. The forest cover in the recent past extended down to the 650 mm per year rainfall isohyet. The original Polynesian settlers arrived in the mid-tenth century AD. For the next 800 years a combination of climatic change and volcanic activity eroded some of the natural forest base. However, by the time of the main European settlement in the 1840s it was estimated that wildfire, particularly in the drier, windier regions, originating from Maori land clearance fires had reduced forest cover to 53% of the total land mass. Between the middle of the last century and the present day, European settlers have further reduced the natural forest as land was developed for farming and the forest was harvested for wood. Only 23% of the natural forest cover remains today.

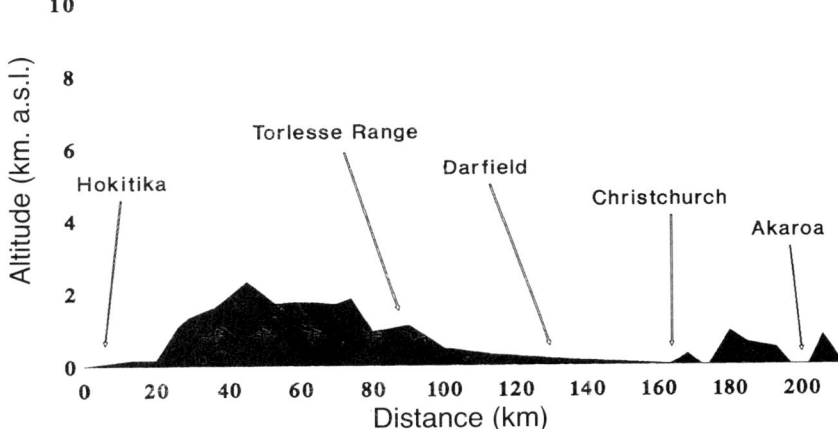

Fig. 27.1. Profile of the Canterbury plains, New Zealand.

A series of government initiatives in the late nineteenth and early twentieth century slowed down the removal of natural forests. Planted production forests, principally of the fast-growing radiata pine species (*Pinus radiata* D. Don.), gradually replaced exploitation of and the dependence on natural forests. The 23% of the land area of natural forests that remain today is supplemented by 5% of the total land area of New Zealand under planted production forests. These plantations produced, in the year ended 31 March 1992, 14.2×10^6 m^3 of roundwood logs, contrasting with the rapidly diminishing production from natural forests which in that year yielded only 200 000 m^3 of roundwood logs.

27.4 Canterbury and the plains plantations

The development, characteristics and silvicultural treatment of the Selwyn Plantation Board Limited's (SPBL) plantations will be viewed against this background. It is a story of afforestation in an area of high windthrow risk.

The plains of Canterbury lie in the rainshadow of the Southern Alps and are swept by frequent strong, turbulent and dry north-west winds. Fig. 27.1, a typical profile of the plains, shows the relative positions of plains and mountains. The profile is drawn through the volcanic hills of Banks Peninsula which mark the location of the major South Island city of Christchurch in the east, to the west coast town of Hokitika.

With early European settlement came an appreciation of the inherent capacity of the plains soils and climate to produce crops of wheat, oats and

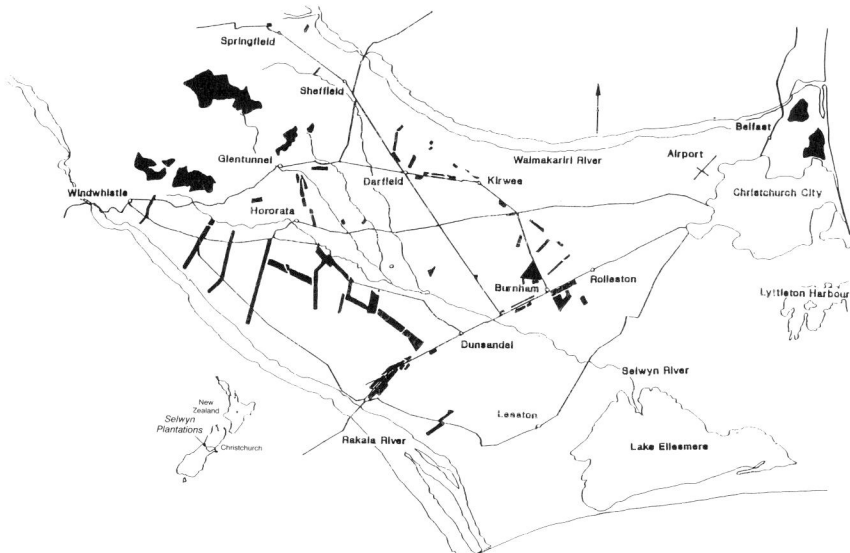

Fig. 27.2. Selwyn Plantation Board Ltd (SPBL) plantations on the Canterbury plains.

barley. In the 25 years until 1883 there was a rapid escalation of the amount of crop grown. By then some 160 000 ha of these crops were grown each year. Much of this was in effect mining the soil and as successive crops reduced soil structure and the lack of shelter exposed these now fragile soils to wind, much of the topsoil was blown. From the early days of settlement the planting of exotic trees for shade and shelter was actively promoted. It was not until the turn of the century, however, that any large-scale plantings were actually pursued.

The soils of the Canterbury plains tend to be shallow and gravelly. The risk of windthrow is alway high. Plantation growth is not easy to achieve on what are essentially non-tree growing sites subject not only to high wind risk but also other elements such as heavy frosts, droughts, low humidity, and high winds with a consequent risk of fire damage.

The plantation pattern developed by the SPBL may be divided into three broad classes as seen in Fig. 27.2. The large forests to the west are planted in the foothill country of the Southern Alps. Plantations on the plains form an extensive shelter pattern and are responsible for the overall silvicultural pattern which has been developed. To the east are two forests planted in sand country. Located near the coast these forests were originally developed to control the spread of the sand inland over rich market gardening soils to the north of Christchurch city.

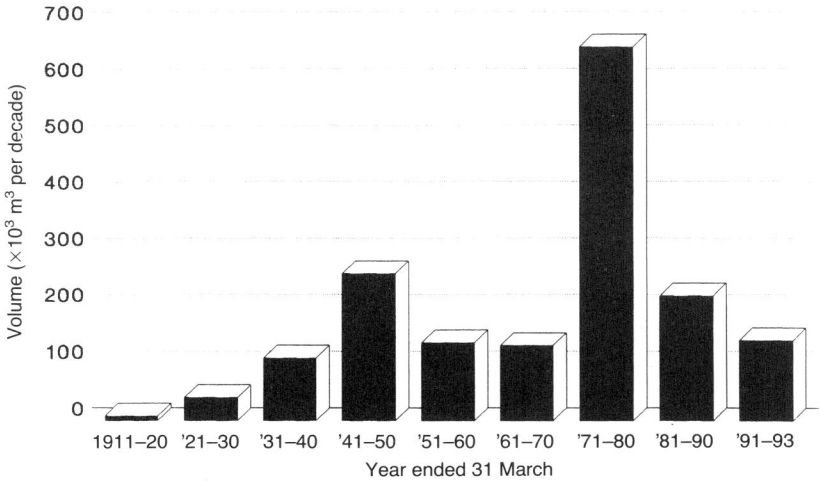

Fig. 27.3. Volume production from SPBL plantations since 1911.

27.5 Wind effects

Wind is the major silvicultural factor taken into account in the management of the SPBL's stands. Shallow plains soils and the topography of Canterbury are factors that contribute to the risk of wind damage. To counter these factors are a number of characteristics that make forestry on the Canterbury plains inherently simple. Access is rarely a problem. Logging and extraction are, because of the flat topography, relatively uncomplicated. There is a well-developed infrastructure of high-quality roads in the province. The plains plantations are well served by the road system and well located in relation to markets and deep-water ports.

Major gales have, over the years, resulted in significant damage to SPBL stands. This damage arrives with an almost predetermined regularity. Northwest gale damage has been recorded in 1914, 1930, 1945, 1956 and 1975, and in 1968 a south-westerly storm also resulted in significant plantation damage. Since the turn of the century 90% of all wood harvested from the SPBL's stands has been following windthrow.

Fig. 27.3 shows the production of wood since 1911. Present silvicultural practice takes a number of steps to reduce the impact of windthrow and damage. In the decades 1941–50 and 1971–80 significant increases in volumes harvested can be seen. This is the result of the major wind storm in 1945 when speeds of 40 m s^{-1} were recorded and in 1975 when the measured windspeeds peaked at 47 m s^{-1}. The effect such storms can have on sustained forest management may be seen in the age class profile at the time of the

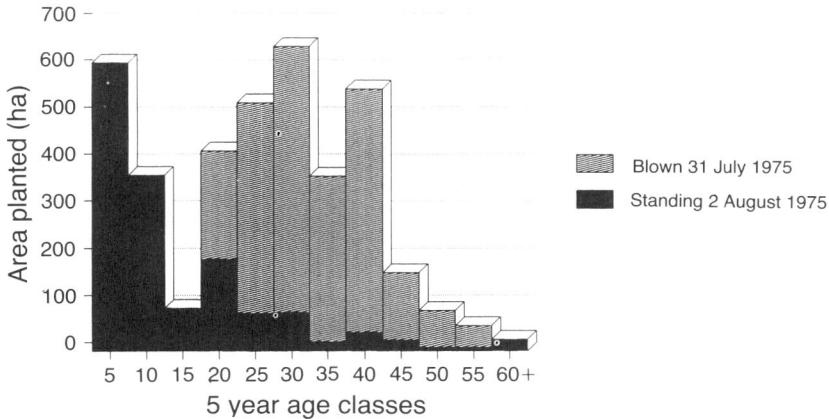

Fig. 27.4. Effect of the August 1975 windstorm: age class profile.

1975 windstorm (Fig. 27.4). As a result of the 1975 storm 60% of all SPBL stands were blown down in a 12 h period. Over the whole province some 11 000 ha were blown. The SPBL stands contributed to 22% of this total blown area.

27.6 Present silvicultural practice

Over the years silvicultural regimes and management practices have been evolved to minimise the damage from catastrophic wind storms and to reduce the impact that such storms have and will have on sustained forest production.

The Selwyn Plantation Board Empowering Act 1992 details SPBL's overall management philosophy in that it charges the Company to: 'operate as a successful business consistent with the principles of conservation and the provision of shelter on the plains of Canterbury'. During the last 10 years New Zealand has seen a number of central and local government trading enterprises formed. All of these have had the single objective of being a successful business. The statement in the SPB Empowering Act is a significant departure from the norm inasmuch as it recognises that in addition to providing direct financial returns the SPBL's forests need to continue to provide both shelter and wood in what is a relatively unattractive tree-growing environment.

The overall objective of SPBL's forest management is to grow forests that: yield the maximum volume of wood per hectare, of the highest quality possible, in the shortest possible time, for the minimum financial input. The silvicultural practice keeps these objectives very much to the fore as it accommodates the general environmental constraints of the Canterbury plains and minimises the risk of further catastrophic wind damage to SPBL stands. These overall objectives are accommodated in a variety of ways that are discussed below.

27.6.1 Species selection

Radiata pine is the predominant species grown in New Zealand. Its wood quality, growth rates and marketability make it the preferred species on most sites. It does, however, suffer from damage from wind on the Canterbury plains. SPBL have determined that 90% of its total forest area will be planted with radiata pine, but that 10% will be established in Douglas fir. Douglas fir is grown on the sites where the rainfall is higher than 800 mm per year, i.e. back into the hills. Douglas fir exhibits better wind stability characteristics than does radiata. The species has a softer crown, is easier shaped by the wind and therefore presents a reduced sail area. The inherently strong stem of Douglas fir accommodates the greater wind stability generated from the extensive root grafting and root systems of the species without exhibiting the windbreak tendencies of radiata.

27.6.2 Rotation

Rotation of radiata pine is kept to 25 years to minimise the risk of wind damage. Trees over age 15 years tend to be at heightened risk from wind. A shorter rotation therefore reduces the risk of windthrow.

27.6.3 Financial return

Because of the high risk of windthrow on the Canterbury plains the fashionable New Zealand goal of defect-free wood production has been restricted and no attempt is made to grow clearwood on these sites. Resin pockets are endemic because of the edaphic and climatic environment of the plains – a further reason to resist heavy investment in high pruning and low stockings. Keeping investment low ensures a reasonable return from wind-damaged stands. To gain any real benefit from heavy investment in silviculture requires a longer rotation to maximise the return from that investment. This therefore increases the risk quite significantly.

SPBL grows Douglas fir on a nominal rotation of 100 years. This yields large, high-quality Douglas fir stems which can be sold to supplement cashflow in times of damage to the major radiata pine crop.

27.6.4 Plantation layout

All planting lines are orientated into the north-west wind. This, coupled with a 4×2 m spacing (4 m between, 2 m in rows), gives mutual support while the stand is growing and enables orderly harvesting after windthrow. The straight lines permit some cultivation and site amelioration at planting time, and facilitate access for mid-rotational silvicultural or harvesting operations.

SPBL stands (unlike many in Canterbury) are not ripped to induce deep rooting into the gravelly C horizons of the plains soils. To induce deep rooting is an invitation to increase the probability of stem breakage. There will inevitably be gale force winds over the Canterbury plains. It is SPBL's intention that stands should blow down before they break.

Under Canterbury climatic conditions there are no more than about 6 months to recover radiata pine logs following break. Wood has been successfully recovered up to 5 years after a blow when some of the roots are left in the ground. Whilst there is an increasing amount of reaction wood produced relative to the time of blow, the amount of fibre which can be recovered 5 years after a blow may be approximately 60–70% of the total assessed volume.

27.6.5 Plantation location

A balance of plantations between hills, plains and coastal sites is maintained. This covers a spectrum of wind conditions and acknowledges that there are locational variations in the pattern of wind with any windstorm. Risk is spread by spreading plantation cover over a range of soils and climatic zones. There are added advantages to dispersal of stands in that it also spreads risk from both fire and pathogen attack.

27.6.6 Silvicultural treatment

The silvicultural treatment of radiata pine aims to complete thinning by age 12 years on a 25 year rotation. It has been demonstrated that after height 12 m radiata tends to be at greater risk. Sporadic erosive windthrow is reduced by the elimination of late thinning, as is the incidence of catastrophic windthrow.

High stocking is maintained. Despite the advances in inherent genetic quality of radiata pine seed available over the last few years, SPBL maintains a planting regime of 1200 stems ha^{-1}. These are thinned to 600 stems ha^{-1} at age 7–8 years. A low pruning, to height 3 m, is performed at the same time.

Pruning is to improve access into the stand and therefore ultimate marketability, to reduce the probability of a crown fire, and to increase the value of the butt log. A light thinning to 550 stems ha^{-1} may be conducted at age 11 years if market conditions allow. Stands are then left to grow until age 25 years.

It is the intention of the silvicultural regime that stembreak, as a result of wind, is minimised. Shallow cultivation on plains soils enables vigorous early growth and provides a maximum shallow root stability. Thinning and pruning provide sufficient space for the stems to grow, by reducing intense intra-stand competition, but still allow them to be mutually supportive. SPBL prefers throw rather than break and it is the intention of the silvicultural regime that this occurs.

27.7 Conclusions

Over the last 80 years silviculture regimes and management techniques have been evolved which permit successful forest growth on the plains of Canterbury. There is a need for shelter and wood to be provided at minimum cost to the recipients. The plantations themselves generate a cashflow which ensures that they have a viable future.

In developing the regime much has been achieved. The longer rotations have been modified to short rotations. The species mix has been altered to provide cashflow at times when the forests are recovering from catastrophic wind storm. The trees are planted in a way that minimises break and in a variety of locations to spread the risk of wind damage.

SPBL has been successful in achieving forest growth in this very wind-prone environment. The financial returns are not as spectacular as they are from the rest of New Zealand but given the environmental constraints of the Canterbury region, are more than acceptable.

Index

Abies concolor, spiral compression wood, 252
Abies fraseri
  foliage studies, 244
  radial growth effects, 249
Abies lasiocarpa, GLEES project, 366
adaptive growth hypothesis, 134
aerodynamic roughness length, estimates, 96, 170, 177, 407
aeroelastic models
  modelling at small scale, 143
  modelling of bending at small scale, 143, 145
ambrosia beetles, 453
amyloplasts, 253
anemometer
  hot-film constant pressure, wind tunnel studies, 75–6
  overspeeding error, 92
  sonic, 43–4
  three-axis, 42–3
  Vector A100R, 92
  wind loading tests, 209
ANOVA, 279–81
Arabidopsis, calmodulin, 256
Atlantic, North, hurricanes, 307–11
atmospheric stress, 244–5
auxins, compression wood formation, 255
Average Tree *see* Gardiner Standard Tree
averaging operator, overbar, 5–6
axial splitting of hazard beams, 195–201

beetle pests, 453
bending
  bending strength, finite element method (FEM), 199
  dynamic response and damping
    damping in steady-state forced vibrations, 159–60
    effects of strong wind, 160–2
    transient response, 160
  engineering theories
    bending in circular arc, 135–4
    bending with large deflections, 136–8
    bending with small deflections, 136
    Second Moment Area theorem, 136
  field theories
    distribution of bending stiffness, 138–9
    Gardiner Standard Tree, 141–2, 149
    mass distribution, 136
    sway frequency and damping, 139–40
  modelling at small scale
    aeroelastic models, 143, 145
    overturning moments, 143–6
    wind flow, 143
  modelling mechanical stresses in Sitka spruce, 165–81, 165–94
    static and dynamic stem bending, 171–3
  peak stresses on stems, 174–6
  pull and sway computations
    analysis of sway vibrations, Stodola's method, 155–7
    pull-deflection results, 154–5
    stem bending, steady point force, 152–4
    summary and conclusions, 157–8
  stress in stems, maximum values, 179
  thinned vs unthinned stands, modelling mechanical stresses, Sitka spruce, 165–94
  wind forces on single tree, 147–52
    canopy drag and Reynolds stress, 150–1
    drag coefficients, 147–50
    gust excitation and sway response in frequency domain, 152
    Honami gusts, 151–2
  Young's modulus, 136–8
  summary and conclusions, 162–3, 177–80
  see also sway, wind-induced

477

478 Index

Betula, seedling establishment, hurricane effects, 325–6
Bordeaux, Laboratory of Wood Rheology, 182–3
British Columbia, windthrow studies, 448–59
   biogeoclimatic zones, 448–9
   harvest planning, 452
   insect outbreaks, 453
   land ownership and tenure, 449–50
   ocurrence and implications of windthrow, 450–1
   salvage, 453–4
   utilisation, 450
Brunt–Väisälä frequency N, 95, 359
Building Research Establishment wind direction analysis, 90
buttress development, root growth, 293–301
buttresses in rainforest trees, 293–301

calcium transport
   Nicotiana, 256
   stretch-activation, 256
calmodulin, expression, 256
cambium
   activity
      and mechanical stress, 266
      thigmomorphogenesis, 238, 247–51, 266
   surface strain transduction, 257
Canada *see* British Columbia
canopy(ies)
   Sitka spruce, 167–70
   various
      leaf area index, 7
      physical and aerodynamic properties, 7
      velocity moments, 8–9
canopy closure, and thinning, 429
canopy damage, hurricanes, literature review, 346–8
canopy eddies
   'inactive motion', 21
   integral length scales, 19
   primary and secondary instability processes, 19, 20
   scale, 10–14
   two-point space–time correlation, 19
canopy turbulence
   2D ridge
      flow characteristics, 31–7
      wind tunnel model, 29–31
   canopy drag, 26–7
      and Reynolds stress, 150–1
   drag, Reynolds shear stress, 150–1
   interactions of wind and trees, 41–59
      instrumentation, 42–4
      Rivox forest, Scotland, 41–59
   mean canopy flow, response to pressure gradient, 25–9

   momentum absorption capacity, 28
   roughness layer, edge effects on diffusivity, 60–70
   roughness length, 96, 170, 177, 407
   turbulent airflow, mixing layer as model, 17–21
   wake production, 16–17
carbon dioxide concentration changes, 245
   fluxes, hurricane damage, 331
Caribbean, hurricanes, 305–39
Castledaly State Forest, Ireland, field studies of effects of wet mineral soils on dynamic loading, Sitka spruce, 204–19
catastrophic damage *see* hurricanes; windthrow
cats-eyes, Stuart vortices, 19
city trees, visual tree assessment, 233
clearcutting, wind tunnel studies, 71–87
   flow in opening, mean flow field, 80–1
   vertical diffusivity, 81–5
compression elements (tracheids), 255–6
compression wood
   angiosperms, 253–4
   defined, 253
   gravitropic response, 254
   lateral branches, 254
Coriolis parameter, 169
Counihan spires, 72–3
Czech republic, thinning regimes, snow and wind damage, 436–47

damping measurements, 208, 216
deformation of trees, surface airflow and, 367–70
dehydration stress, transpiration and wind theory, 238
Dendroctonus pseudotsugae, fir bark beetle, 453
Dendroctonus rufipennis, spruce beetle, 453
desiccation *see* transpiration and wind
developmental responses to wind, 241–63
diffusivity, edge effects, roughness layer, 60–70
disturbance processes, discussion, 306–7
drag
   canopy drag, 26–7
   canopy drag and Reynolds stress, 150–1
   coefficients, 147–50
   Mayhead formula, 148–50
dynamic flexure loading *see* sway
dynamic loading *see* stem, dynamic loading

ecosystem processes, hurricanes, 329–34
eddies *see* canopy eddies
edge effects, diffusivity *see* roughness layer
Ekman layer, 406

elastic modulus (E), 249–50
elastic potential energy $W_e(t)$, 185
elastic strain, defined, 238
energy spectrum $E(k)$, 14
England see Eskdalemuir; Kershope Forest; Larkhill
environmental impact studies, risk assessment, 404–23
Eskdalemuir
 annual maximum gusts, 390
 storm recurrence, 395–6
 windthrow
  age at first, and terminal damage, 389
  plus mean windthrow rate, 392
ethylene
 block of auxin transport, 255
 precursor in compression wood cambium, 255
Euler Bernouilli theory of bending, 135
Eulerian integral length scales, defined, 10
Eulerian time scale, and Lagrangian time scale, 82

failure of hollow trees by cross-sectional flattening, 195, 196–7
failure by windthrow, 201–3
fallen tree see windthrow
finite element method (FEM), bending strength, 199
fir bark beetle, *Dendroctonus pseudotsugae*, 453
flexure responses, summary and conclusions, 255–6
flexure stiffness (EI), 249
 see also elastic
Flowstar, linear airflow model for complex terrain, 88–112
foliage
 mechanical abrasion strain, 244
 random variations, 5–6
 stress, strain and injury, 241–5
 algorithm, 242
forest clearcuts see clearcutting
FORICO, stand development model, 329
Fractometer, 230–3
fracture moment and angle, species lists, 230
Froude numbers, 94, 360
frozen turbulence hypothesis, 10

Gardiner Standard (Average) Tree, 141–2, 149
 computed pull–deflection results, 154–5
 first mode sway vibration, 158
 mass distribution vs height, 141
 second mode sway vibration, 159
 stress distribution vs height, 154
 surface stress distributions, computed sway vibration, 157

Germany, Hunsrück/Mosel/Eifel area, environmental impact studies, 404–23
Gill propeller arrays, 62–3
Glacier Lakes Ecosystem Experiments Site (GLEES project), 358–75
glucans, compression wood, 253
Gnathotrichus (ambrosia beetle), 453
gravistimulus, perception, 253
gravitropic response
 formation of reaction wood, 246, 251
 reaction wood, 254
 statoliths, 253
 summary and conclusions, 255–6
gusts
 annual maximum, Eskdalemuir, 390
 excitation and sway response in frequency domain, 152
 frequency spectrum in wind tunnel, 146
 Honami gusts, 151–2
 Sk(u) values, 10
 variable time-averaging technique (VITA), 49–50

hazard beams, axial splitting, 195–201
hills and ridges
 2D ridge canopy flow, 31–7
 consecutive profiles of $U(z)$, 32
 isolated hills, turbulent airflow, 21–5
  buoyancy forces, 24
  development of velocity profiles, 23
  linear analytical model, 24
  perturbation in pressure, Poisson equation, 24
  streamlines, 32–7
 see also topography
Holland see Netherlands
hollow trees, failure by cross-sectional flattening, 195, 196–7
Honami gusts, 151–2
Hooke's law, strain gauge results, 297–300
hot-film constant pressure anemometer, 75–6
hurricanes, 305–39
 classification, 349
 community-level effects, 320–9
 ecosystem processes, 329–34
  foliage effects, 330
  hydrological effects, 332–4
  nutrient dynamics, 330–32
 experimental models, 326–9
 HURRECON model, 314–19
 literature review, 342–51
  canopy damage, 346–8
  list, 343
  mortality, 348–9, 352–4
  stem damage, 344–6
  volume changes, 348
 meteorology, 307–11
  formation, 308

hurricanes (cont.)
  North Atlantic, 307–11
  Puerto Rico, 349–51
  quantification of damage, 340–57
  regional and landscape effects, 311–19
  salvage logging effects, 333
  summary of damage, 341–2
  tropical vs temperate forest effects, 311
  see also salvage logging; windthrow

indole-3-acetic acid, compression wood, 255
isotach patterns, deformation of trees, 368

Juglans nigra, walnut
  foliage studies, 244
  windspeed studies, 267

K-theory, 61
Kansas spectra, 64, 66–7, 69
Kelvin–Helmholtz waves, Stuart vortices, 18–19
Kershope Forest, England, actual windthrow, 398
kinetic energy $W_c(t)$, 184
Kintyre, Scotland
  contour plot of terrain, 91
  Flowstar linear airflow model, 88–112
    instrumentation, 92–94
    solution of equations, 94–107
    inner region results, 105–7
    outer layer flow, 97–105
    synoptic situation, 97
Kolmogoroff scaling laws, 14–17
K_c, turbulent diffusivity, 61

Lagrangian time scale, and Eulerian time scale, 82
laricinan (glucan), compression wood, 253
Larix decidua, wind tunnel studies, 264–75
Larkhill, storm recurrence, 395–6
lateral branches, compression wood, 254
leaf area index, 7
  hurricane damage effects, 326
leaves see foliage
length scales, 10–14
  Eulerian integral, defined, 10
  Kolmogoroff scaling laws, 14–17
linear airflow model, complex terrain, 88–112
Liquidamber styraciflua, woody tissue changes, 249–50
load cell
  calibration, 223
  measurement of wind-induced root stresses, 220–6
  schematic diagram, 222

Malaysia, rainforest trees, buttress development, 293–301
management response, and risk assessment, 379–474
Massgebender factor, 237
Mattheck's model of buttress development, 295
mechanical abrasion strain, 244
mechanical perturbations (MP), effects, various species, 239
Medicine Bow National Forest, Wyoming, SE, 358–75
Meteorological Office UK (UKMO) stations, 114–16
meteorology of hurricanes, 307–11
methane, fluxes, hurricane damage, 331
Metriguard Stress Wave Timer, 227, 228–32
mixing layer
  development to turbulent state, 19
  linear stability analysis, 17–21
  mean velocity development, 18
  model of turbulent airflow, 17–21
  plane, velocity moment, 18
models
  aeroelastic models, 143, 145
  conceptual model of windthrow hazard classification, 385–99
  Flowstar linear airflow model, Kintyre, Scotland, 88–112
  forest stand, 74–5
    development, 329
    FORICO stand development model, 329
    mesoscale, 406
    scale model, principles, 72–3
  hurricanes, 326–9
    HURRECON model, 314–19
  linear analytical model, isolated hills, turbulent airflow, 24
  mechanical flexure, *Phaseolus vulgaris*, 250–1
  modelling at small scale, 143–6, 407–8
  modelling mechanical stresses, Sitka spruce, 165–94
  modelling of tree sway under dynamic bending, 183–8
  scale model behaviour, wind-induced sway, 188–9
  specification, simulation of wind stress, 387–8
stem number regulation, 425–6, 431–3
thinning, Norway spruce, 425
thinning intensity, 425–8, 433

turbulent airflow, mixing layer as model, 17–21
see also wind tunnels
moments
  destructive storm moment, 427, 428
  fracture moment and angle, 230
  moment of anchorage, 427
  overturning moments, 143–6
  Second Moment Area theorem, 136
  velocity moments, 8–9, 18
momentum absorption capacity, canopy turbulence, 28
momentum flux, tree movement, 48–9, 52
momentum transfer model, 169, 177
Monin–Obhukov similarity theory, 60, 64
mortality, 348–9, 352–4
motorway construction, environmental impact studies, 404–23

Netherlands, Sleen, roughness layer, edge effects on diffusivity, 60–70
Neutral Kansas spectra, 64, 66–9
New Zealand
  50 years of damage to State plantation forests, 460–7
    attritional damage, 462–4
    catastrophic damage, 462, 464
    Canterbury plains, 470–1
    experience and management strategy, 468–76
    forest characteristics, 469–70
    species selection, 474
    geographical characteristics, 468–9
    salvage logging, 475
    wind-induced root stresses, 220–6
Nicaragua, pine mortality, 352–4
Nicotiana, calcium transport, 256
nitrous oxide, fluxes, hurricane damage, 331
North Atlantic, hurricanes, 307–11
notation, meteorological, 5

oak, *Quercus rubra*, field experiments, dynamic spectrum analysis, 192–3
orographic lee waves, 461, 465
overbar, averaging operator, 5–6

pathogens, and stem structural failure, 240–1
Phaseolus vulgaris (bean), mechanical flexure model, 250–1
PHSPs (Pre-Harvest Silviculture Prescriptions), 452
physiological responses to wind, 237–301
  buttresses in rainforest trees, 293–301
  photosynthesis, 241–5, 325
  root growth in young trees, 264–75
  root system biomass, 276–92
  wind-induced, 237–63

Picea abies, Norway spruce
  adaptation processes to wind load, 430–1
  planting density, 426
  recommendations for stabilisation, 424–35
  root system, biomass, 429
  thinning intensity, 425–8, 433
  thinning regimes, 436–47
Picea engelmannii, Engelmann spruce
  GLEES project, 366
Picea sitchensis, Sitka spruce
  characteristics, 173
  critical windspeeds, 405
  drag coefficient, 149
  field studies, effects of wet mineral soils on dynamic loading, 204–19
  Gardiner Average Tree measurement, 141–2
  modelling mechanical stresses, 165–81
    dynamic stem bending, 171–3
    static stem bending, 171
    windspeeds, 167–70
  needle area density distributions, 172
  planting density, 166
  Queen Charlotte Islands 'plus-tree', 278
  root growth, wind tunnel studies, 264–75
  roots, wind tunnel studies, 264–75
  selection, 276–7
  wind stability factors, distribution of biomass in root systems, 276–92
  SSST, comparison with *Pinus pinaster*, maritime pine, 189
Pinus contorta, contorted pine
  ethylene precursor in compression wood cambium, 255
Pinus pinaster, maritime pine,
  comparison with Standard Sitka spruce, 189
  wind-induced sway, 183
Pinus radiata, radiate pine
  measurement of wind-induced root stresses, 220–6
  in New Zealand, 474
  radial growth effects, 247–51
  root anchorage, 266
Pinus sylvestris, Scots pine
  sway, and root effects, 250
Pinus taeda
  radial growth effects, 248–9
  spiral compression wood, 252
plastic strain, defined, 238
Poisson equation, perturbation in pressure over isolated hills, 24
pollarding, effect on wind-induced sway, 189–91

pore water pressure
  data, 210–15
  pressure transducers, 207–17
Power Law tree, 142, 154
Pre-Harvest Silviculture Prescriptions, 452
prediction of windspeeds over complex terrain, 113–29
principal components analysis (PCA), 114–15
pruning, effect on wind-induced sway, 189–91
Pseudotsuga menziesii, Douglas fir, in New Zealand, 474
Puerto Rico
  case study, hurricane Hugo, 349–51
  hurricanes, 312–39

Quercus, seedling establishment, hurricane effects, 325–6
Quercus rubra, red oak
  field experiments, dynamic spectrum analysis, 192–3

radial growth
  sway effects, 247–9
  see also cambium
radial stress wave velocity, various species, 229
rainforest trees, buttress development, 293–301
reaction wood, 246, 251–2
  defined, 253
  gravitropic response, 254
regeneration, and clearcutting, 71
regression analysis
  prediction of windspeeds in complex terrain, 121–6
    multiple regression techniques, 122–4
    simple regression techniques, 121–2
    test on independent data, 124–6
Reynolds shear stress, canopy measurements, 10–11
Reynolds stress profile, wind tunnel studies, 73
rheology, Laboratory of Wood Rheology, Bordeaux, 182–3
risk assessment
  for environmental impact studies, 404–23
    fine-scale models, 407–8
    mesoscale model, 406–7
    model descriptions, 406–8
  hazard classification, conceptual model, 385–99
  how to assess, 381
  and management response, 379–474
  in practice, 381–2

and profitability, 380–1
windthrow hazard classification, 382–4
Rivox forest, Scotland, interactions of wind and trees, 41–59
rocking device, measurement of dynamic loading, 206–7
root anchorage–destructive storm moment ratio, 428
root growth
  buttress development, rainforest trees, 293–301
  cross-sectional area calculations
    distribution between roots, 287
    selected spruce, 280–1
    spuce and larch, 270–2
  root : shoot ratio, 281–3
  root, comparison with beam, 134
  wind tunnel studies, 264–75
    numbers of small laterals, 269
  in young trees
    Larix decidua, 264–75
    Picea sitchensis, Sitka spruce, 264–75
root plate, radius, vs stem radius, in failure by windthrow, 202
root stresses, New Zealand, 220–6
root system
  anchorage biomass, accumulation through rotation, 429
  anchorage components resisting overturning, 265
  biomass distribution, 276–92, 429
  clonal variation, 288
  and failure by windthrow, 201–3
  stresses, wind-induced, New Zealand studies, 220–6
  'thin root' compartment, increasing long-term increment capacity, 430
  wind stability factors, distribution of biomass, 276–92
  see also root growth
roughness layer
  edge effects on diffusivity, 60–70
    data analysis, 63–5
    results, 65–8
    site and instrumentation, 62–3
    summary and conclusions, 68–9
roughness length, estimates, 96, 170, 177, 407

salt spray, 244
salvage logging
  British Columbia, 449, 453–4
  constraints, 453–4
  effects, 333
  snow and wind damage, 436–47
  see also windthrow
Scorer equation, 94

Index

Scotland
  Flowstar linear airflow model, Kintyre, 88–112
  Moffat Forest, 167–94
  Rivox Forest, 41–59
  sites selected for wind regime determinations, 115–19
  Teindland Forest, 277–8
  Tiree, storm recurrence, 395–6
Second Moment Area theorem, 136
secondary wind-induced displacement, static loading, 251–5
seedling establishment, hurricane damage effects, 325–6
Selwyn Plantation Board, New Zealand, management strategy, 468–76
SHEAR project, 62
silviculture practices, simulation, 189–92
Sitka spruce *see* Gardiner Standard (Average) Tree; *Picea sitchensis*
skewness, Sk(u) values, gustiness, 10
snow and wind damage, thinning regime, 436–47
soils
  pore water, pressure transducers, 207–17
  shear zones, 201
  wet mineral soils, effects of dynamic loading, 204–19
species diversity, FORICO stand development model, 329
species lists
  fracture moment and angle, 230
  mechanical perturbations (MP) effects, 239
  radial stress wave velocity, 229
spectral analysis, 52–6
  power spectral densities (PSDs), 52–6
splitter plate, 17–18
spruce beetle (*Dendroctonus rufipennis*), 453
stand development model, FORICO, 329
Standard Tree, Gardiner *see* Gardiner Standard Tree
static loading
  agencies, 252
  secondary wind-induced displacement, 251–5
statolith theory, gravity, 253
stem
  dynamic loading, 246–7
    effects in wet mineral soils, 204–19
    rocking device, 206–7
    see also sway
  radial growth effects, 247–51
  static loading stress, 246–7
stem damage, hurricanes, literature review, 344–6
stem displacement
  measurement by transducers, 207, 208
  stiffness value and Young's modulus, 208–17
stem number
  regulation models
    constant vs varied thinning, 425–6
    D to A-degree model, 431–3
stem tissues, stress, strain and injury, algorithm, 242
stembreak, snow and wind damage, 436–47
storms
  monitoring, 217
  recurrence, 395–6
  risk assessment, motorway construction studies, 404–23
  see also hurricanes; wind stress; windthrow
strain
  damping measurements, 208, 216
  elastic, defined, 238
  extension strain behaviour, 211, 213
  plastic, defined, 238
  rootplate, vertical movement, 212
  tree pulling arrangement, 209
strain gauges, applied to rainforest trees, Malaysia, 297
stratification types
  Flowstar linear airflow model, 88–112
  upstream options, 95
stress *see* wind stress; windspeed
stress wave velocity, Metriguard Stress Wave Timer, 227, 228–32
Stuart vortices, cats-eyes, 19
subalpine forest, Glacier Lakes Ecosystem Experiments Site (GLEES project), 358–75
surface airflow, deformation of trees, 367–70
sway, Stodola's method, analysis of sway vibrations, 155–7
sway, wind-induced
  analysis of sway vibrations, Stodola's method, 155–7
  effects on growth and development, 247–51
  experimental analysis, 182–94
    displacement function $U(z)$, 185
    elastic potential energy $W_e(t)$, 185
    equations of motion and approximate solution method, 184–6
    kinetic energy $W_c(t)$, 184
    mechanical measurement and modelling of tree sway under dynamic bending, 183–8
    scale model behaviour, 188–9
  frequency and damping, 139–40
    free frequencies and damping measurements, 186–8

484 Index

sway, wind-induced (cont.)
  pruning, effect on wind-induced sway, 189–91
  pull and sway computations, 152–8
  see also bending

temperature contouring technique, 91, 93–4
  isopleths, 91
tension wood, 253–4
terrain see topography
thigmomorphogenesis, 247–51, 266–73
  defined, 238
thinning intensity
  and canopy closure, 429
  constant vs varying, 425–6
  and onset of wind damage, 166
  short-term influence on wind stability, 427–8
  snow and wind damage, 436–47
    strategies, 438–9
  vs unthinned stands, modelling mechanical stresses, Sitka spruce, 165–94
Tiree, storm recurrence, 395–6
topographic exposure (topex)
  defined, 116–17
  inter-variable correlations, 120
  variables, 117–19
topography
  complex terrain
    linear airflow model testing, 88–112
    prediction of wind speeds, 113–29
  and land-use, 408–410
  orographic lee waves, 461, 465
  terrain–airflow interactions, 359–60
    GLEES project, 360–75
  windspeeds prediction, 114
  see also hills and ridges
tracheids, compression elements, 255–6
transpiration and wind, 238, 241–5
  atmospheric stress, 244–5
tree movement
  displacement, 43–7
  gust passage, 48–52
    variable time-averaging technique (VITA), 49–50
  instrumentation, 42–4
  momentum flux, 49
    coherence, 56
    instantaneous momentum flux, 48, 52
  secondary static loading, 251–5
  spectral analysis, 52–6
  wind loading, 44–8
  summary and conclusions, 56–8
tree roots see roots
tree structure, for mechanical measurement and modelling of tree sway under dynamic bending, 183–8

tree vulnerability
  and wind climate, 385
  persistent, progressive, and episodic, 386
tropical forest, see also Puerto Rico case study
trunk see stem
Trypodendron (ambrosia beetle), 453
turbulent airflow
  forest canopies
    2D ridge, 29–37
    mixing layer as model, 17–21
    see also canopy turbulence
  horizontally homogeneous forests, 6–17
    energy spectrum $E(k)$, 14–17
    length scales, 10–14
    single point velocity, 6–10
  isolated hills, 21–5
  models
    mixing layer, 17–21
    wind tunnel of canopy on 2D ridge, 29–31
  turbulent kinetic energy (TKE), 14–17, 68, 77, 82–4
  see also canopy turbulence; wind tunnel studies

United Kingdom Meteorological Office (UKMO) stations, 114–16
United States, Wyoming, SE, Glacier Lakes Ecosystem Experiments Site (GLEES project), 358–75
United States, north-east, hurricanes, 313–39
USDA, Glacier Lakes Ecosystem Experiments Site, 358–75

variance, ANOVA, 279–81
velocity, components, 5
visual tree assessment, 227–34
VITA see gusts
von Karman's constant, 63, 79, 169
vortices
  hairpin, 19, 20
  Stuart, 18–19

wake production, wind tunnel study, 16
wet mineral soils, effects of dynamic loading, 204–19
wheat field, wind tunnel model, 29–31
wildlife corridors, 453
wind
  fetch, stability and direction, 65–8
  see also windspeeds; windthrow
wind flow
  effect of motorway construction, 404–23
  modelling at small scale, 143

Index

wind hazard *see* windthrow, hazard classification
wind loading, test instrumentation, 209
wind stability
  distribution of biomass in root systems, 276–92
  physics of, 426–7
wind stress
  acute, 240–1
  chronic vs acute, 240–1
  foliar responses, 241–4
  simulation
    model specification, 387–8
    strain gauge results, 297–300
  stems, maximum values, 179
  stress, strain and injury algorithm, 242
  and tree vulnerability, 385
  see also strain
wind tunnel studies
  aeroelastic models, 143, 145
  bluff bodies, 73–4
  canopy on 2D ridge, 29–31
  flow control devices, 146
  forest clearcuts, 71–87
  instrumentation, 75
  model forest, 74–5
  results and discussion
    flow above model forest, 76–80
    flow in the opening, 80–5
  Reynolds stress profile, 73
  root growth, 264–75
  wake production, 16
windspeeds
  effects on trees, 133–64
    see also bending; roots
  lee waves and, 461
  prediction in complex terrain, 113–29
    analysis of topographic variables, 119–21
    regression analysis, 121–6
    sites in Scotland, 115–19, 173–4
    summary and conclusions, 126–7
    topography, 114–15
  UK profiles, 168
  vertical, 93
windthrow
  age at first, and terminal damage, 389
  assessment, 454–5
  British Columbia studies, 448–59

chronic winds, 252
constraints to salvage logging, 453–4
failure by, 201–3
  root plate radius vs stem radius, 202
hazard classification, 382–4
  conceptual model, 385–99; endemic windthrow version, 388–90; incorporating a time series, 390–2; model specification, 387–8; recent evidence, 397–9
  development, 383–4
  future development, 398–9
  key assumptions, 384
limiting change in progressive vulnerablity, 392–5
management, 455–6
  partial cuts, 457
microtopography, 323
mixed forests, 320–3
percentage, and profitability, 380
relationship with age, height, density, 320–4
research, 456–7
snow and wind damage, Czech Republic, 436–47
streamside leave (buffer) strips, 453, 456
subalpine forest, GLEES project, 358, 370–4
treefall orientation, 318
see also salvage logging
woody tissue
  compression and tension wood, 253–4
  compression wood, spiral, 252
  quality, assessment in standing trees, 227–32
  reaction wood, 246, 251–2
  response to stress, 245–51
  tracheids, compression elements, 255–6
Wyoming, SE
  Glacier Lakes Ecosystem Experiments Site (GLEES project), 358–75
  see also radial growth

xylem *see* woody tissue

Young's modulus
  measurement, 136–8
  stiffness value, stem displacement, 208, 216